Pesticide Chemistry

Edited by

Hideo Ohkawa,

Hisashi Miyagawa,

and Philip W. Lee

1807–2007 Knowledge for Generations

Each generation has its unique needs and aspirations. When Charles Wiley first opened his small printing shop in lower Manhattan in 1807, it was a generation of boundless potential searching for an identity. And we were there, helping to define a new American literary tradition. Over half a century later, in the midst of the Second Industrial Revolution, it was a generation focused on building the future. Once again, we were there, supplying the critical scientific, technical, and engineering knowledge that helped frame the world. Throughout the 20th Century, and into the new millennium, nations began to reach out beyond their own borders and a new international community was born. Wiley was there, expanding its operations around the world to enable a global exchange of ideas, opinions, and know-how.

For 200 years, Wiley has been an integral part of each generation's journey, enabling the flow of information and understanding necessary to meet their needs and fulfill their aspirations. Today, bold new technologies are changing the way we live and learn. Wiley will be there, providing you the must-have knowledge you need to imagine new worlds, new possibilities, and new opportunities.

Generations come and go, but you can always count on Wiley to provide you the knowledge you need, when and where you need it!

William J. Pesce
President and Chief Executive Officer

Peter Booth Wiley
Chairman of the Board

Pesticide Chemistry

Crop Protection, Public Health, Environmental Safety

Edited by
Hideo Ohkawa, Hisashi Miyagawa, and Philip W. Lee

BICENTENNIAL 1807 WILEY 2007 BICENTENNIAL

WILEY-VCH Verlag GmbH & Co. KGaA

The Editors

Prof. Dr. Hideo Ohkawa
Fukuyama University
Green Science Res. Center
Sanzo, Gakuen-cho 1
Hiroshima, 729-0292
Japan

Dr. Hisashi Miyagawa
Kyoto University
Graduate School of Agriculture
Kyoto, 606-8502
Japan

Dr. Philip W. Lee
DuPont Ctrl. R & D
Biochem. Sc. and Engineering
P.O. Box 6300
Newark, DE 19714-6300
USA

Library of Congress Card No.:
applied for

British Library Cataloguing-in-Publication Data
A catalogue record for this book is available from the British Library.

Bibliographic information published by the Deutsche Nationalbibliothek
The Deutsche Nationalbibliothek lists this publication in the Deutsche Nationalbibliografie; detailed bibliographic data are available in the Internet at <http://dnb.d-nb.de>.

Composition Manuela Treindl, Laaber
Printing betz-druck GmbH, Darmstadt
Bookbinding Litges & Dopf GmbH, Heppenheim
Cover Adam Design, Weinheim
WILEY Bicentennial Logo Richard J. Pacifico

Printed in the Federal Republic of Germany
Printed on acid-free paper

ISBN 978-3-527-31663-2

11th IUPAC International Congress of Pesticide Chemistry
August 6–11, 2006, Kobe, Japan

Committees and Sponsoring Organizations

Executive Committee
K. Mori (Japan, Chairperson)
Representatives from
Japan Society for Bioscience, Biotechnology and Agrochemistry
Japan Society for Environmental Chemistry
Japan Society of Applied Entomology and Zoology
Japan Society of Environmental Entomology and Zoology
Society of Environmental Science, Japan
The Food Hygienics Society of Japan
The Japanese Society for Chemical Regulation of Plants
The Japan Society of Medical Entomology and Zoology
The Phytopathological Society of Japan
The Weed Science Society of Japan

Advisory Board
T. Ueno (Japan)
K. D. Racke (IUPAC, USA)
J. Harr (Switzerland)
I. Yamaguchi (Japan)

Organizing Committee
H. Ohkawa (Japan, Chairperson)
I. Ueyama (Japan)
H. Abe (Japan)
N. K. Umetsu (Japan)
H. Miyagawa (Japan)
A. Katayama (Japan)
H. Matsumoto (Japan)
K. Tanaka (Japan)
M. Sasaki (Japan, Chief Secretariat)

Pesticide Chemistry. Crop Protection, Public Health, Environmental Safety
Edited by Hideo Ohkawa, Hisashi Miyagawa, and Philip W. Lee

ISBN: 978-3-527-31663-2

Scientific Program Committee

H. Miyagawa (Japan, Chairperson)
T. Ando (Japan)
T. Asami (Japan)
E. Carazo (Costa Rica)
J. M. Clark (USA)
G. Donn (Germany)
A. Felsot (USA)
R. Feyereisen (France)
Y. Hashidoko (Japan)
S.-R. Jiang (China)
A. Katayama (Japan)
Y.-H. Kim (Korea)
S. Kuwahara (Japan)
K. Kuwano (Japan)
P. W. Lee (USA)
H. Matsumoto (Japan)
H. Miyoshi (Japan)
S. Powles (Australia)
U. Schirmer (Germany)
M. Skidmore (UK)
K. Tanaka (Japan)
T. Teraoka (Japan)
H. Yamamoto (Japan)
J. A. Zabkiewicz (New Zealand)

Preface

The 11th IUPAC International Congress of Pesticide Chemistry was held from August 6–11, 2006, in Kobe, Japan. Since the 5th Congress held in Kyoto in 1982, this was the second time that the Congress took place in Japan. During this 24-year time period, we witnessed dramatic changes in science and technology around pesticides. The Congress' subtitle, "Evolution for Crop Protection, Public Health and Environmental Safety", focused on the current situation surrounding pesticides, which are now more commonly referred to as agrochemicals.

Pesticides or agrochemicals have played not only a critical role in the production of food and feed to support the growing world population's demands, but also in control of infectious diseases transmitted by insect vectors and microorganisms. Advances in technologies such as computational chemistry, automated high-throughput biological screens, crop genetics, biotechnology, formulations, and precision agriculture have offered novel tools in the discovery and development of new agrochemicals. For example, new chemical classes such as the sulfonylurea and imidazolinone herbicides, the neonicotinoid insecticides, and the strobilurin fungicides were significant discoveries during this 24-year period.

Public concerns pay more attention to risk assessment and risk management of pesticide use regarding human health and environmental safety. The discovery of "safer" pesticides with new modes of action becomes the challenge of the new generation of pesticide scientists. One of the highlights of this Kobe Congress was the presentation of a novel class of insecticides (Rynaxypyr™ and flubendiamide) which target specifically the insect ryanodine receptor. For example, Rynaxypyr™ showed laboratory and field activities on all major lepidoptera pests at a low rate of 0.01 ppm (equivalent to approximately 10 g/ha). Furthermore, a 350^{+}-fold difference in safety/selectivity was observed between mammalian and insect cell targets. Both compounds show a great safety margin to non-target organisms. These attributes set new standards for new safe and active insecticides.

The high cost of conducting pesticide research and discovery, along with the ever-increasing pesticide registration requirements and regulations, also have a negative impact on the advances of pesticide science. With an estimated cost of approximately 200 million USD and 11 years to bring a new pesticide from discovery to the market place, this economic pressure resulted in a significant consolidation of the agrochemical business during the past 10^{+} years. Today, fewer than 10 major agrochemical companies remain. Important products

Pesticide Chemistry. Crop Protection, Public Health, Environmental Safety
Edited by Hideo Ohkawa, Hisashi Miyagawa, and Philip W. Lee

ISBN: 978-3-527-31663-2

such as the organophosphate and carbamate insecticides, chloroacetanilide and triazine herbicides are withdrawn/canceled due to the high cost of re-registration, unfavorable acute human toxicity, and/or environmental persistence issues. The cancellation of these products has a significant impact on farmers, especially in developing countries. It is a continuous loss of useful tools in crop protection and public health protection. The topics of global harmonization of pesticide residues (MRL) and vector-borne communicable diseases were addressed in this Congress.

On a more positive note, presentations on new technologies and approaches from this Congress offer a bright future for discovery of a new generation of low use rate, highly selective, and environmentally friendly products.

The Congress had 1142 participants from 52 countries. The Congress' scientific program included the keynote address, 4 plenary lectures, and approximately 110 session lectures. In addition to the platform lectures, more than 550 posters were submitted. The poster award committee selected three excellent papers from each of three categorized presentations. The first paper in each category was also awarded by IUPAC. This publication includes the keynote address, 4 plenary lectures, and one or two papers in each of 20 sessions, and the posters awarded by IUPAC.

We take this opportunity to acknowledge the efforts from members of the organizing and programming committees, along with the help from countless volunteers. We also need to point out several novel programs implemented at this Congress, including the Research Director Forum, the programming of the special luncheon and evening seminars, and the publication of the Congress News Letter (Kobe Gazette).

The publication of the Congress Proceeding is a time-honored tradition for the IUPAC International Congress of Pesticide Chemistry. This Proceeding not only accurately documents the scientific contents of this Conference, but also highlights significant advances and important issues facing pesticide science and technology. The Editorial Team expresses our appreciation to Wiley-VCH (Germany) and Dr. Frank O. Weinreich for the publication of this Proceeding. In particular, we gratefully acknowledge the dedicated editorial support provided by Ms. Carol Ashman.

Finally, this Proceeding is dedicated to all past and present pesticide scientists; it is their vision and creativity that continue to push back the frontier of pesticide sciences. We look forward to seeing you at the 12th IUPAC International Congress of Pesticide Chemistry, in Melbourne, Australia, 2010.

Editorial Team
Hideo Ohkawa (Kobe, Japan)
Hisashi Miyagawa (Kyoto, Japan)
Philip W. Lee (Newark, DE, USA)

Contents

11th IUPAC International Congress of Pesticide Chemistry *V*

Preface *VII*

List of Contributors *XXVII*

I Keynote and Plenary Lectures *1*

1 Challenges and Opportunities in Crop Production Over the Next Decade *3*
James C. Collins, Jr.
1.1 Meeting Society's Agricultural Needs *3*
1.2 Global Trends and Uncertainties *3*
1.3 Grain Stocks *5*
1.4 Exchange Rates *6*
1.5 Biofuels *8*
1.6 Counterfeit Products *9*
1.7 Product Commercialization *10*
1.8 Convergence of Factors *11*

2 Searching Environmentally Benign Methods for Pest Control: Reflections of a Synthetic Chemist *13*
Kenji Mori
2.1 Introduction *13*
2.2 Pesticides and Our Daily Life *13*
2.3 Contributions in Pesticides Discovery by Japanese Scientists *14*
2.4 Natural Products Synthesis and Pesticide Science *16*
2.5 Enantioselective Pheromone Synthesis and Pesticide Science *18*
2.6 Conclusion *21*
2.7 References *21*

Pesticide Chemistry. Crop Protection, Public Health, Environmental Safety
Edited by Hideo Ohkawa, Hisashi Miyagawa, and Philip W. Lee

ISBN: 978-3-527-31663-2

3 The Current Status of Pesticide Management in China *23*
Yong Zhen Yang
3.1 Introduction *23*
3.2 The Current Direction of Pesticide R&D in China *23*
3.2.1 The Status of Pesticide Production and Usage in China *24*
3.2.2 Pesticide Regulation and Management Systems in China *24*
3.3 China's Policies in Pesticide Regulation and Management *26*
3.4 The Regulatory Infrastructure within China in the Regulation of Pesticide *27*
3.5 Key Administrative Actions on Pesticide Management *27*
3.6 Future Direction of Pesticide Regulation in China *28*
3.7 Conclusion *28*

4 Pesticide Residues in Food and International Trade: Regulation and Safety Considerations *29*
Kenneth D. Racke
4.1 Introduction *29*
4.2 Globalization of the Food Chain *30*
4.3 Regulation of Pesticide Residues in Food *31*
4.3.1 The World Food Code and Codex MRLs *31*
4.3.2 U.S. Tolerances *33*
4.3.3 Japan MRLs *33*
4.3.4 EU MRLs *34*
4.4 Disharmonized MRLs, Monitoring, and Consumer Safety *35*
4.5 Recent Trends *37*
4.5.1 Improvements in the Codex Sytem *37*
4.5.2 Regionalization of MRL Policies *37*
4.5.3 Growth of Private Standards *38*
4.5.4 Communication of MRL Information *38*
4.5.5 Adoption of Practices to Preempt or Mitigate Residue Issues *39*
4.6 Conclusion *40*
4.7 Acknowledgments *40*
4.8 References *41*

5 Hunger and Malnutrition Amidst Plenty: What Must be Done? *43*
Shivaji Pandey, Prabhu Pingali
5.1 Introduction *43*
5.2 The Current Situation *44*
5.3 What Must be Done? *46*
5.3.1 National Commitment and Good Governance *46*
5.3.2 Investment in Rural Infrastructure *47*
5.3.3 Improving Irrigation Infrastructure *47*
5.3.4 Improving Soil Fertility *47*
5.3.5 Improved Agricultural Technologies *48*
5.3.6 Energy Supply Needs to be Improved *48*

5.3.7 Development Assistance is Needed *49*
5.3.8 Trade Helps Rural Poor *49*
5.3.9 Implementing Policies that Promote Protection of Natural Resource Base *50*
5.3.10 Preparing for the Future *50*
5.4 References *50*

II New Chemistry *53*

6 Modern Tools for Drug Discovery in Agricultural Research *55*
Alexander Klausener, Klaus Raming, Klaus Stenzel
6.1 Introduction *55*
6.2 Tools and Their Integration in the Drug Discovery Process *56*
6.3 Mode of Action Elucidation – An Example for the Integration of New Technologies *58*
6.4 Conclusion *62*
6.5 Acknowledgments *63*
6.6 References *63*

7 Target-Based Research: A Critical Review of Its Impact on Agrochemical Invention, Focusing on Examples Drawn from Fungicides *65*
Stuart J. Dunbar, Andrew J. Corran
7.1 Introduction *65*
7.2 Selection of Targets for *In Vitro* Screening *66*
7.3 Assay Design and Implementation *68*
7.4 *In Vitro* to *In Vivo* Translation *70*
7.5 Structure Based Design *71*
7.6 Conclusion and a Forward Look *72*
7.7 References *73*

8 Virtual Screening in Crop Protection Research *77*
Klaus-Jürgen Schleifer
8.1 Introduction *77*
8.2 General Lead Identification Strategies *77*
8.3 Virtual Screening Based on 1-D and 2-D Descriptors *78*
8.4 Virtual Screening Based on 3-D Descriptors *81*
8.4.1 Ligand-Based Screening Strategies *81*
8.4.2 Structure-Based Screening Strategies *83*
8.5 Conclusion *87*
8.6 References *87*

9 Synthesis of Triazolo[1,5-c]pyrimidine Sulfonamides Leading to the Discovery of Penoxsulam, a New Rice Herbicide *89*
Timothy C. Johnson, Timothy P. Martin, Rick K. Mann
9.1 Introduction *89*
9.2 Chemistry *89*
9.3 Biology *93*
9.4 Selection of Penoxsulam for Development *98*
9.5 Conclusion *99*
9.6 References *99*

10 Discovery and SAR of Pinoxaden: A New Broad Spectrum, Postemergence Cereal Herbicide *101*
Michel Muehlebach, Hans-Georg Brunner, Fredrik Cederbaum, Thomas Maetzke, René Mutti, Anita Schnyder, André Stoller, Sebastian Wendeborn, Jean Wenger, Peter Boutsalis, Derek Cornes, Adrian A. Friedmann, Jutta Glock, Urs Hofer, Stephen Hole, Thierry Niderman, Marco Quadranti
10.1 Introduction *101*
10.2 Optimization Phase and Discovery of Pinoxaden *102*
10.3 Chemistry *103*
10.4 Mode of Action *105*
10.5 Structure-Activity Relationships *105*
10.6 Biological Performance *106*
10.6.1 Grass Weed Spectrum *106*
10.6.2 Crop Tolerance in Wheat and Barley *106*
10.6.3 Adjuvant Effect – Adigor® *107*
10.6.4 Introducing Axial® *108*
10.7 Conclusion *109*
10.8 Acknowledgments *109*
10.9 References *109*

11 Rynaxypyr™: A New Anthranilic Diamide Insecticide Acting at the Ryanodine Receptor *111*
George P. Lahm, Thomas M. Stevenson, Thomas P. Selby, John H. Freudenberger, Christine M. Dubas, Ben K. Smith, Daniel Cordova, Lindsey Flexner, Christopher E. Clark, Cheryl A. Bellin, J. Gary Hollingshaus
11.1 Introduction *111*
11.2 Discovery of the Anthranilic Diamide Insecticides *112*
11.3 Discovery of Rynaxypyr™ *115*
11.4 Biological Attributes *117*
11.5 Toxicology *117*
11.6 Mechanism of Action *118*
11.7 Conclusion *119*
11.8 Dedication *120*
11.9 References *120*

12 **Elucidation of the Mode of Action of Rynaxypyr™, a Selective Ryanodine Receptor Activator** *121*
Daniel Cordova, Eric A. Benner, Matthew D. Sacher, James J. Rauh, Jeffrey S. Sopa, George P. Lahm, Thomas P. Selby, Thomas M. Stevenson, Lindsey Flexner, Timothy Caspar, James J. Ragghianti, Steve Gutteridge, Daniel F. Rhoades, Lihong Wu, Rejane M. Smith, Yong Tao
12.1 Introduction *121*
12.2 Symptomology Associated with Anthranilic Diamides *121*
12.3 Rynaxypyr™ Stimulates Release of RyR-Mediated Internal Ca^{2+} Stores *122*
12.4 Rynaxypyr™ Binds to a Unique Site on the RyR *123*
12.5 Cloning and Expression of Pest Insect RyRs *124*
12.6 Rynaxypyr™ is Highly Selective for Insect RyRs *124*
12.7 Conclusion *125*
12.8 References *125*

13 **Flubendiamide, a New Insecticide Characterized by Its Novel Chemistry and Biology** *127*
Akira Seo, Masanori Tohnishi, Hayami Nakao, Takashi Furuya, Hiroki Kodama, Kenji Tsubata, Shinsuke Fujioka, Hiroshi Kodama, Tetsuyoshi Nishimatsu, Takashi Hirooka
13.1 Introduction *127*
13.2 Structure-Activity Relationship *128*
13.2.1 Lead Generation *128*
13.2.2 Lead Optimization *128*
13.3 Chemistry *130*
13.4 Mode of Action *132*
13.5 Biological Profile *132*
13.6 Toxicological Properties *134*
13.7 Conclusion *135*
13.8 Acknowledgments *135*
13.9 References *135*

14 **Flubendiamide Stimulates Ca^{2+} Pump Activity Coupled to RyR-Mediated Calcium Release in Lepidopterous Insects** *137*
Takao Masaki, Noriaki Yasokawa, Ulrich Ebbinghaus-Kintscher, Peter Luemmen
14.1 Introduction *137*
14.2 Calcium Release Induced by Flubendiamide *138*
14.3 Specific Stimulation of Ca^{2+} Pump *138*
14.4 Luminal Ca^{2+} Mediated Ca^{2+} Pump Stimulation *139*
14.5 Conclusion *140*
14.6 References *140*

15 **Novel Arylpyrazole and Arylpyrimidine Anthranilic Diamide Insecticides** *141*
Thomas P. Selby, Kenneth A. Hughes, George P. Lahm
15.1 Introduction *141*
15.2 Synthesis of Anthranilic Diamides *142*
15.3 Insecticidal Activity *144*
15.4 Conclusion *145*
15.5 Acknowledgment *148*
15.6 References *148*

16 **Metofluthrin: Novel Pyrethroid Insecticide and Innovative Mosquito Control Agent** *149*
Yoshinori Shono, Kazuya Ujihara, Tomonori Iwasaki, Masayo Sugano, Tatsuya Mori, Tadahiro Matsunaga, Noritada Matsuo
16.1 Introduction *149*
16.2 Discovery *149*
16.3 Efficacy *152*
16.3.1 Intrinsic Insecticidal Activity *152*
16.3.2 Activity in Devices *153*
16.3.2.1 Heated Formulations *153*
16.3.2.2 Non-heated Formulations *155*
16.4 Conclusion *158*
16.5 Acknowledgment *158*
16.6 References *158*

17 **Design and Structure-Activity Relationship of Novel Neonicotinoids** *159*
Xuhong Qian, Yanli Wang, Zhongzhen Tian, Xusheng Shao, Zhong Li, Jinliang Shen, Qingchun Huang
17.1 Introduction *159*
17.2 Selectivity Mechanism and Binding Model of Neonicotinoids *159*
17.2.1 Bioinformatic Analysis *159*
17.2.2 *Ab initio* Quantum Chemical Calculation *161*
17.3 Chemical Modification for *cis* Nitro Configuration *165*
17.3.1 Synthesis *165*
17.3.2 Biological Activity *166*
17.3.3 QSAR Analysis *167*
17.4 Conclusion *168*
17.5 References *168*

18 **Synthesis and Inhibitory Action of Novel Acetogenin Mimics Δlac-Acetogenins: A New Class of Inhibitors of Mitochondrial NADH-Ubiquinone Oxidoreductase (Complex-I)** *171*
Hideto Miyoshi, Naoya Ichimaru, Masatoshi Murai
18.1 Introduction *171*
18.2 Mode of Action of Δlac-Acetogenins *172*

18.3 SAR of Δlac-Acetogenins *173*
18.4 Conclusion *173*
18.5 References *174*

III Biology, Natural Products and Biotechnology *175*

19 Plant Chemical Biology: Development of Small Active Molecules and Their Application to Plant Physiology, Genetics, and Pesticide Science *177*
Tadao Asami, Nobutaka Kitahata, Takeshi Nakano
19.1 Introduction *177*
19.2 Development of BR Biosynthesis Inhibitors *178*
19.2.1 Assay Methods for BR Biosynthesis Inhibitors *179*
19.2.2 Structure-Activity Relationship Study *180*
19.2.3 Target Site(s) of BR Biosynthesis Inhibitor *180*
19.2.4 Searching for Novel BR Biosynthesis Inhibitors *181*
19.3 Functions of BRs in Plant Development Unveiled by BR Biosynthesis Inhibitors *182*
19.4 BR Biosynthesis Inhibitors as a Useful Screening Tool for BR Signaling Mutants *183*
19.5 Usefulness of Biosynthesis Inhibitors of Biologically Active Molecules in Plant Biology *184*
19.6 Abscisic Acid Biosynthesis Inhibitors Targeting 9-cis-Epoxycarotenoid Dioxygenase (NCED) *184*
19.7 Conclusion *186*
19.8 References *186*

20 An Overview of Biopesticides and Transgenic Crops *189*
Takashi Yamamoto, Jack Kiser
20.1 Introduction *189*
20.2 *Bacillus thuringiensis* *190*
20.3 Spray-On Bt Insecticide Formulations *190*
20.4 Discovery of Multiple Toxins in One Bt Strain *191*
20.5 Mode of Action of Bt Insecticidal Proteins *192*
20.6 Transgenic Bt-Crops *193*
20.7 Selection of Bt Genes for Transgenic Cotton *194*
20.8 Corn Insect Pests and Bt Genes *195*
20.9 Potential Issues of Bt-Crops *195*
20.10 Insect Resistance to Bt *196*
20.11 Resistance Mechanism *196*
20.12 Resistance Management Program for Bt Transgenic Crops *197*
20.13 Conclusion *197*
20.14 References *198*

21 **Essential Oil-Based Pesticides: New Insights from Old Chemistry** *201*
Murray B. Isman, Cristina M. Machial, Saber Miresmailli, Luke D. Bainard
21.1 Introduction *201*
21.2 Essential Oil Composition *201*
21.3 Biological Activities of Essential Oils *202*
21.3.1 Insecticidal/Deterrent Effects *203*
21.3.2 Herbicidal Activity *205*
21.3.3 Antimicrobial Activity *207*
21.4 Challenges and Future Opportunities *208*
21.5 Conclusion *208*
21.6 References *209*

22 **Eco-Chemical Control of the Potato Cyst Nematode by a Hatching Stimulator from Solanaceae Plants** *211*
Akio Fukuzawa
22.1 Introduction *211*
22.2 Classification of Cyst Nematodes and Research History to Elucidate the Naturally Occurring Hatching Stimulants *211*
22.3 Involvement of Multiple Factors in Hatching Stimulation for Cyst Nematodes *212*
22.4 Isolation of Hatching Stimulators and Stimulation Synergists in TRD *214*
22.5 Application of TRD to Decrease PCN Density in Soil *214*
22.6 References *215*

23 **Vector Competence of Japanese Mosquitoes for Dengue and West Nile Viruses** *217*
Yuki Eshita, Tomohiko Takasaki, Ikuo Takashima, Narumon Komalamisra, Hiroshi Ushijima, Ichiro Kurane
23.1 Introduction *217*
23.2 Possible Vector of Japanese Mosquitoes Against Dengue Virus *217*
23.2.1 Susceptibility of Orally Infected Japanese Mosquitoes *218*
23.2.2 Transmission of Mouse-Adapted or Non-mouse Passaged Dengue Viruses by the Japanese Mosquito Species *219*
23.2.3 Further Analysis for Vector Mosquitoes to Dengue Viruses in Japan *220*
23.3 Possible Vector of Japanese Mosquitoes Against West Nile Virus *221*
23.3.1 Susceptibility of Japanese Mosquitoes Against West Nile Virus *221*
23.3.2 Transmission of Japanese Mosquitoes Against West Nile Virus *223*
23.4 Conclusion *223*
23.5 Acknowledgments *224*
23.6 References *224*

24 **Life Science Applications of Fukui Functions** *227*
Michael E. Beck, Michael Schindler
24.1 Introduction *227*
24.2 Theoretical Background *228*
24.2.1 Some Results from Conceptual DFT [19–23] *228*
24.2.2 Why Fukui Functions May be Related to Sites of Metabolism *230*
24.3 Methods *230*
24.4 Example Applications *231*
24.4.1 Parathion and Chlorpyrifos: Fukui Functions Related to Biotic Degradation *231*
24.4.2 Selective Thionation of Emodepsid *233*
24.4.3 Fukui Functions Reveal the Nature of the Reactive Species in Cytocrome P450 Enzymes *234*
24.5 Conclusion and Outlook *235*
24.6 Acknowledgments *235*
24.7 References *236*

IV **Formulation and Application Technology** *239*

25 **Homogeneous Blends of Pesticide Granules** *241*
William L. Geigle, Luann M. Pugh
25.1 Introduction *241*
25.2 Granule Blend Products *242*
25.2.1 General Theory of Segregation *242*
25.2.2 Sampling of Granule Blends *242*
25.2.3 Manufacture of Granule Blends *244*
25.2.4 Regulatory Requirements for Granule Blends *244*
25.2.5 Measurement of Homogeneity *245*
25.2.6 Advantages of Granule Blends *246*
25.3 Conclusion *246*
25.4 References *247*

26 **Sprayable Biopesticide Formulations** *249*
Prem Warrior, Bala N. Devisetty
26.1 Introduction *249*
26.2 The Biopesticide Market *250*
26.3 Biological Pesticides *250*
26.4 Factors Affecting Biopesticide Use *251*
26.4.1 The Living System *251*
26.4.2 The Production System *252*
26.4.3 Biological Activity *253*
26.4.4 Stabilization *253*
26.4.5 Quality Control *253*
26.4.6 Delivery *254*

26.5 Role of Formulations in Sprayable Biopesticides *254*
26.6 Future Outlook and Needs *256*
26.7 References *257*

V Mode of Action and IPM *259*

27 Molecular Basis of Selectivity of Neonicotinoids *261*
Kazuhiko Matsuda
27.1 Introduction *261*
27.2 Interactions with Basic Residues Induce a Positive Charge in Neonicotinoids which Mimics the Quaternary Ammonium of Acetylcholine *261*
27.3 Exploring Structural Features of nAChRs Contributing to the Selectivity of Neonicotinoids Employing the α7 nAChR *263*
27.4 Homology Modeling of nAChRs has Assisted in the Identification of Key Amino Acid Residues Involved in the Selective Interactions with Neonicotinoids of Heteromeric Nicotinic Acetylcholine Receptors *265*
27.5 Conclusion *268*
27.6 Acknowledgments *269*
27.7 References *269*

28 Target-Site Resistance to Neonicotinoid Insecticides in the Brown Planthopper *Nilaparvata lugens* *271*
Zewen Liu, Martin S. Williamson, Stuart J. Lansdell, Zhaojun Han, Ian Denholm, Neil S. Millar
28.1 Introduction *271*
28.2 Identification of Target-Site Resistance in *Nilaparvata lugens* *271*
28.3 Characterization of a nAChR (Y151S) Mutation in *N. lugens* *272*
28.4 Discussion *273*
28.5 Conclusion *274*
28.6 References *274*

29 QoI Fungicides: Resistance Mechanisms and Its Practical Importance *275*
Karl-Heinz Kuck
29.1 Introduction *275*
29.2 Resistance Risk Assessments Before Market Introduction *275*
29.3 Resistance Mechanisms of QoI Fungicides in Field Isolates *276*
29.3.1 Mutation Upstream Complex I *276*
29.3.2 Metabolization by Fungal Esterases *277*
29.3.3 Target Mutation G143A *277*
29.3.4 Target Mutation F129L *277*
29.4 Practical Importance of Individual Resistance Mechanisms to QoIs *278*

29.5 Resistance Management *280*
29.6 Perspectives *280*
29.7 Conclusion *282*
29.8 References *282*

30 Chemical Genetic Approaches to Uncover New Sites of Pesticide Action *285*
Terence A. Walsh
30.1 Introduction *285*
30.2 The Chemical Genetic Approach *285*
30.3 Components of a Chemical Genetic Process *287*
30.3.1 Chemical Libraries *287*
30.3.2 Phenotype Screens *287*
30.3.3 Target Site Identification *288*
30.4 Three Examples of Chemical Genetic Target Identification *288*
30.4.1 NP-1, a Complex Natural Product *289*
30.4.2 ATA-7, a Bleaching Phenotype *290*
30.4.3 DAS534, a Picolinate Auxin *291*
30.5 Key Learnings *293*
30.6 Conclusion *293*
30.7 Acknowledgments *294*
30.8 References *294*

31 The History of Complex II Inhibitors and the Discovery of Penthiopyrad *295*
Yuji Yanase, Yukihiro Yoshikawa, Junro Kishi, Hiroyuki Katsuta
31.1 Introduction *295*
31.2 Discovery of Penthiopyrad (MTF-753) *297*
31.3 Biological Attributes *298*
31.3.1 Target Site of Penthiopyrad *299*
31.3.2 Mode of Action *299*
31.3.3 Effect on Resistant Strains of Other Fungicides *300*
31.3.4 The Risk of Occurrence of Resistance to Penthiopyrad *301*
31.4 Conclusion *302*
31.5 Acknowledgments *302*
31.6 References *302*

32 The Costs of DDT Resistance in *Drosophila* and Implications for Resistance Management Strategies *305*
Caroline McCart and Richard ffrench-Constant
32.1 Introduction *305*
32.2 Global Spread of DDT Resistance *305*
32.3 Lack of Fitness Cost *307*
32.4 Single Genes in the Field and Many in the Laboratory *309*
32.5 Implications for Resistance Management *309*

32.6 Conclusion *310*
32.7 References *311*

VI Human Health and Food Safety *313*

33 New Dimensions of Food Safety and Food Quality Research *315*
James N. Seiber
33.1 Introduction *315*
33.2 New Analytical Methods for Identification and Source Tracking *316*
33.3 Methods for Reducing Aflatoxins in Foods *318*
33.4 Molecular Biology and Food Safety *319*
33.5 Healthy Food Constituents *320*
33.6 Conclusion *321*
33.7 References *321*

34 Impact of Pesticide Residues on the Global Trade of Food and Feed in Developing and Developed Countries *323*
Jerry J. Baron, Robert E. Holm, Daniel L. Kunkel, Hong Chen
34.1 Introduction *323*
34.2 Potential Solutions *324*
34.2.1 The IR-4 Model and Other Minor Use Programs *325*
34.2.2 Tools for Harmonization *326*
34.2.2.1 Crop Grouping *326*
34.2.2.2 Work Sharing *328*
34.2.2.3 Rationalized Global Data Requirements *329*
34.3 References *330*

35 Pesticide Residue Assessment and MRL Setting in China *331*
Yibing He, Wencheng Song
35.1 Introduction *331*
35.2 Regulations and National Standards for Pesticide Residue Management in China *331*
35.2.1 Key Components of Pesticide Management Regulation of China for Pesticide Residue *331*
35.2.2 Key Components of National Standards for Pesticide Residue Management in China *332*
35.3 Summary of Date Requirements of Pesticide Registration *332*
35.4 Residue Data Requirements *333*
35.4.1 The Protocol of Field Trials *333*
35.4.2 Residue Analysis Method *333*
35.4.3 Experimental Results *333*
35.5 Procedures for Establishing MRLs and Setting Up PHI in China *334*
35.6 Examples of MRL Setting in China *335*
35.7 Perspective *339*

36 Harmonization of ASEAN MRLs, the Work towards Food Safety and Trade Benefit *341*
Nuansri Tayaputch
36.1 Introduction *341*
36.2 Role of Codex MRLs in Regulating Food Quality *341*
36.3 Pesticide Residues in Developing Countries *342*
36.4 Issues on Minor Crops *343*
36.5 The Work of ASEAN Expert Working Group on Pesticide Residues *343*
36.6 Principles of Harmonization *344*
36.7 Several Observations Made During the Process of Harmonization from 1998 to 2005 *345*
36.8 Future Outlook *346*
36.8.1 ASEAN MRLs with Quality Data Conducted at Regional Levels on Tropical Crops Should be Established as International Standards *346*
36.8.2 Member Countries Should Have Comprehensive Knowledge on MRLs' Establishment Consistent with International Guidelines *346*
36.9 Conclusion *347*
36.10 References *347*

37 Possible Models for Solutions to Unique Trade Issues Facing Developing Countries *349*
Cecilia P. Gaston, Arpad Ambrus, and Roberto H. González
37.1 Introduction *349*
37.2 Possible Solution to the Lack of Analytical Facilities and Expertise on Developing Data to Support Establishment of MRLs *350*
37.3 A Solution to the Lack of MRLs on Spices *351*
37.3.1 Rationale for an Alternative Approach to Setting MRLs for Spices *351*
37.3.2 Codex MRLs for Spices *352*
37.3.3 Codex MRLs for Dried Chili Peppers *352*
37.4 Difficulties of Complying with Unharmonized MRLs, Including 'Private' MRLs *355*
37.4.1 Program to Facilitate Exports of Chilean Fruits *356*
37.4.2 Generating Pre-Harvest Interval Data *356*
37.4.3 An Example of a Supervised Trial Model in the "Pesticide Agenda" *357*
37.5 Conclusion *357*
37.6 References *358*

38 Genetically Modified (GM) Food Safety *361*
Gijs A. Kleter, Harry A. Kuiper
38.1 Introduction *361*
38.2 General Principles of GM Food Safety Assessment *362*
38.3 General Data *364*
38.4 Molecular Characterization of the Introduced DNA *364*

38.5 Comparison of the GMO with a Conventional Counterpart *364*
38.6 Potential Toxicity of Introduced Foreign Proteins *365*
38.7 Potential Toxicity of the Whole Food *365*
38.8 Potential Allergenicity of the Introduced Foreign Proteins *365*
38.9 Potential Allergenicity of the Whole Food *366*
38.10 Potential Horizontal Gene Transfer *366*
38.11 Nutritional Characteristics *367*
38.12 Potential Unintended Effects of the Genetic Modification *367*
38.13 Pesticide Residues *369*
38.14 Research into the Safety of GM Crops *369*
38.15 Conclusion *370*
38.16 Acknowledgment *370*
38.17 References *370*

39 Toxicology and Metabolism Relating to Human Occupational and Residential Chemical Exposures *373*
Robert I. Krieger, Jeff H. Driver, John H. Ross
39.1 Introduction *373*
39.2 Pesticide Handlers *374*
39.3 Harvesters of Treated Crops *376*
39.4 Residents Indoors *376*
39.5 Estimates of Human Exposure *377*
39.6 Exposure Biomonitoring *378*
39.7 Conclusion *380*
39.8 References *380*

40 Bioavailability of Common Conjugates and Bound Residues *383*
Michael W. Skidmore, Jill P. Benner, Cathy Chung Chun Lam, James D. Booth, Terry Clark, Alex J. Gledhill, Karen J. Roberts
40.1 Introduction *383*
40.2 Literature Search *384*
40.3 Experimental Phase *385*
40.3.1 Conjugates *386*
40.3.2 Chemical and Enzymatic Hydrolysis *386*
40.3.3 Prediction of Permeability *388*
40.3.4 Bound Residues *389*
40.3.5 Characterization of the Bound Residues *390*
40.3.6 Chemical and Enzymatic Hydrolysis *390*
40.3.7 Bioavailability of Bound Residues *391*
40.4 Conclusion *392*
40.5 Acknowledgment *392*
40.6 References *393*

41 **Multiresidue Analysis of 500 Pesticide Residues in Agricultural Products Using GC/MS and LC/MS** *395*
Yumi Akiyama, Naoki Yoshioka, Tomofumi Matsuoka
41.1 Introduction *395*
41.2 Multiple Residue Analysis *395*
41.3 Monitoring Results *398*
41.4 Conclusion *399*
41.5 References *399*

VII **Environmental Safety** *401*

42 **Current EU Regulation in the Field of Ecotoxicology** *403*
Martin Streloke
42.1 Introduction *403*
42.2 Regulatory Process *403*
42.3 Standard Risk Assessment *404*
42.4 Refined Risk Assessments *406*
42.4.1 Refined Risk Assessments for Birds and Mammals *407*
42.4.2 Persistent Compounds in Soil *407*
42.4.3 Use of Probabilistic Risk Assessment Methods for Regulatory Purposes *407*
42.4.4 Microcosm/Mesocosm Testing with Aquatic Organisms *409*
42.4.5 Data from Monitoring Studies *410*
42.4.6 Endocrine Disruption *410*
42.5 Risk Mitigation Measures *411*
42.6 Conclusions *412*
42.7 References *412*

43 **A State of the Art of Testing Methods for Endocrine Disrupting Chemicals in Fish and Daphnids** *415*
Satoshi Hagino
43.1 Introduction *415*
43.2 Fish Testing Methods for Sex Hormones *415*
43.3 S-rR Strain Medaka and Sex Reversal Test *416*
43.4 Effects of Pesticides Listed in SPEED '98 *419*
43.5 Advantages and Disadvantages of the Endpoints Selected *419*
43.6 Consistency of the Results Obtained Between Sex Reversal Assay, PLC, and FLC *421*
43.7 Development of Test Method for Thyroid Hormone *422*
43.8 Endocrine Disrupting Effect of JH Mimics to Daphnids *422*
43.9 Conclusion *423*
43.10 References *424*

44 Pesticide Risk Evaluation for Birds and Mammals – Combining Data from Effect and Exposure Studies *425*
Christian Wolf, Michael Riffel, Jens Schabacker
44.1 Introduction *425*
44.2 Principles of the Risk Assessment within the EU *426*
44.3 Refined Risk Assessment *426*
44.4 Higher-Tiered Studies *426*
44.5 Case Study for Combining Effects and Exposure Studies *427*
44.6 Conclusion *428*
44.7 Reference *429*

45 Bioassay for Persistent Organic Pollutants in Transgenic Plants with Ah Receptor and GUS Reporter Genes *431*
Hideyuki Inui, Keiko Gion, Yasushi Utani, Hideo Ohkawa
45.1 Introduction *431*
45.2 Dioxins *432*
45.3 Dioxin Bioassays *432*
45.4 The AhR *433*
45.5 POP Bioassay Using Transgenic Plants *435*
45.6 Prospects *437*
45.7 Acknowledgments *437*
45.8 References *437*

46 Recent Developments in QuEChERS Methodology for Pesticide Multiresidue Analysis *439*
Michelangelo Anastassiades, Ellen Scherbaum, Bünyamin Taşdelen, Darinka Štajnbaher
46.1 Introduction *439*
46.2 Reagents *440*
46.3 Apparatus *441*
46.4 Procedure *442*
46.5 Discussion *446*
46.5.1 Improving the Recoveries of Certain Pesticides *446*
46.5.2 Improving Selectivity *450*
46.5.3 Expanding the Commodity Spectrum Covered by QuEChERS *453*
46.6 Measurement *456*
46.7 Validation *457*
46.8 Conclusion *457*
46.9 Acknowledgment *458*
46.10 References *458*

47 Summary of Scientific Programs in 11th IUPAC International Congress of Pesticide Chemistry *459*
Hisashi Miyagawa, Isao Ueyama
47.1 Introduction *459*

47.2 Plenary Lectures *459*
47.3 Session Lectures and Special Workshops *460*
47.4 Poster Session *472*
47.5 Luncheon and Evening Seminars *474*
47.6 Other Scientific Programs *477*
47.7 Acknowledgments *477*

Appendix *479*

Author Index *485*

Subject Index *489*

List of Contributors

Yumi Akiyama
Hyogo Prefectural Institute of Public Health and Environmental Sciences
2-1-29 Arata-cho
Hyogo-ku
Kobe, 652-0032
Japan

Arpad Ambrus
Hungarian Food Safety Office
Gyáli út 2–6
1097 Budapest
Hungary

Michelangelo Anastassiades
EU-Community Reference Laboratory for Pesticide Analysis using Single Residue Methods hosted at the Chemisches und Veterinäruntersuchungsamt Stuttgart
Schaflandstr. 3/2
70736 Fellbach
Germany

Tadao Asami
RIKEN, Discovery Research Institute
2-1 Hirosawa
Wako
Saitama, 351-0198
Japan

Luke D. Bainard
Faculty of Land and Food Systems
University of British Columbia
Vancouver, BC, V6T 1Z4
Canada

Jerry J. Baron
IR-4 Project Headquarters
500 College Road East
Suite 201W
Princeton, NJ 08540
USA

Michael E. Beck
Bayer CropScience AG
BCS-Research-Discovery
Alfred-Nobel-Str. 40
40789 Monheim a. Rh.
Germany

Cheryl A. Bellin
DuPont Crop Protection
Stine-Haskell Research Center
1094 Elkton Road
Newark, DE 19711
USA

Eric A. Benner
DuPont Crop Protection
Stine-Haskell Research Center
1094 Elkton Road
Newark, DE 19711
USA

Pesticide Chemistry. Crop Protection, Public Health, Environmental Safety
Edited by Hideo Ohkawa, Hisashi Miyagawa, and Philip W. Lee

ISBN: 978-3-527-31663-2

Jill P. Benner
Syngenta
Jealott's Hill
International Research Centre
Bracknell
Berkshire, RG42 6EY
UK

James D. Booth
Syngenta
Jealott's Hill
International Research Centre
Bracknell
Berkshire, RG42 6EY
UK

Peter Boutsalis
Syngenta Crop Protection AG
Schwarzwaldallee 215
4002 Basel
Switzerland

Hans-Georg Brunner
Syngenta Crop Protection AG
Schwarzwaldallee 215
4002 Basel
Switzerland

Timothy Caspar
DuPont Crop Protection
Stine-Haskell Research Center
1094 Elkton Road
Newark, DE 19711
USA

Fredrik Cederbaum
Syngenta Crop Protection AG
Schwarzwaldallee 215
4002 Basel
Switzerland

Hong Chen
IR-4 Project Headquarters
500 College Road East
Suite 201W
Princeton, NJ 08540
USA

Cathy Chung Chun Lam
Syngenta
Jealott's Hill
International Research Centre
Bracknell
Berkshire, RG42 6EY
UK

Christopher E. Clark
DuPont Crop Protection
Stine-Haskell Research Center
1094 Elkton Road
Newark, DE 19711
USA

Terry Clark
Syngenta
Jealott's Hill
International Research Centre
Bracknell
Berkshire, RG42 6EY
UK

James C. Collins, Jr.
DuPont Crop Protection
P.O. Box 80705
Wilmington, DE 19880-0705
USA

Daniel Cordova
DuPont Crop Protection
Stine-Haskell Research Center
1094 Elkton Road
Newark, DE 19711
USA

Derek Cornes
Syngenta Crop Protection AG
Schwarzwaldallee 215
4002 Basel
Switzerland

Andrew J. Corran
Syngenta, Bioscience
Jealott's Hill
International Research Station
Bracknell
Berkshire, RG42 6EY
UK

Ian Denholm
Rothamsted Research
Harpenden, AL5 2JQ
UK

Bala N. Devisetty
Valent BioSciences Corporation
6131 RFD Oakwood Road
Long Grove, IL 60047
USA

Jeff H. Driver
infoscientific.com
Manassas, VA 20111
USA

Christine M. Dubas
DuPont Crop Protection
Stine-Haskell Research Center
1094 Elkton Road
Newark, DE 19711
USA

Stuart J. Dunbar
Syngenta, Bioscience
Jealott's Hill
International Research Station
Bracknell
Berkshire, RG42 6EY
UK

Ulrich Ebbinghaus-Kintscher
Bayer CropScience AG
Alfred-Nobel-Str. 50
40789 Monheim
Germany

Yuki Eshita
Department of Infectious Diseases
Faculty of Medicine
Oita University
Oita, 879-5593
Japan

Richard ffrench-Constant
Centre for Ecology and Conservation
University of Exeter
Penryn
Cornwall, TR10 9EZ
UK

Lindsey Flexner
DuPont Crop Protection
Stine-Haskell Research Center
1094 Elkton Road
Newark, DE 19711
USA

John H. Freudenberger
DuPont Crop Protection
Stine-Haskell Research Center
1094 Elkton Road
Newark, DE 19711
USA

Adrian A. Friedmann
Syngenta Crop Protection AG
Schwarzwaldallee 215
4002 Basel
Switzerland

Shinsuke Fujioka
Nihon Nohyaku Co., Ltd.
345, Oyamada-cho
Kawachi-Nagano
Osaka, 586-0094
Japan

Akio Fukuzawa
Hokkaido Tokai University
Sapporo, 005-8601
Japan

Takashi Furuya
Nihon Nohyaku Co., Ltd.
345, Oyamada-cho
Kawachi-Nagano
Osaka, 586-0094
Japan

Cecilia P. Gaston
Exponent, Inc.
1730 Rhode Island Avenue, NW
Washington, DC 20036
USA

William L. Geigle
DuPont Crop Protection
Stine-Haskell Research Center
Newark, DE 19714-0030
USA

Keiko Gion
Research Center for
Environmental Genomics
Kobe University
Nada-ku
Kobe, 657-8501
Japan

Alex J. Gledhill
Syngenta
Central Toxicological Laboratories
Alderley Park
Macclesfield
Cheshire, SK10 4TJ
UK

Jutta Glock
Syngenta Crop Protection AG
Schwarzwaldallee 215
4002 Basel
Switzerland

Roberto H. González
University of Chile
Casilla 1004
Santiago
Chile

Steve Gutteridge
DuPont Crop Protection
Stine-Haskell Research Center
1094 Elkton Road
Newark, DE 19711
USA

Satoshi Hagino
Sumika Technoservice Corporation
2-1, Takatsukasa 4-chome
Takarazuka-City
Hyogo, 665-0051
Japan

Zhaojun Han
Nanjing Agricultural University
Nanjing, 210095
China

Yibing He
Institute for the Control of
Agrochemicals
Ministry of Agriculture (ICAMA)
Beijing
China

Takashi Hirooka
Nihon Nohyaku Co., Ltd.
1-2-5 Nihonbashi
Chuo-Ku
Tokyo, 103-8236
Japan

Urs Hofer
Syngenta Crop Protection AG
Schwarzwaldallee 215
4002 Basel
Switzerland

Stephen Hole
Syngenta Crop Protection AG
Schwarzwaldallee 215
4002 Basel
Switzerland

J. Gary Hollingshaus
DuPont Crop Protection
Stine-Haskell Research Center
1094 Elkton Road
Newark, DE 19711
USA

Robert E. Holm
IR-4 Project Headquarters
500 College Road East
Suite 201W
Princeton, NJ 08540
USA

Qingchun Huang
Shanghai Key Lab of
Chemical Biology
School of Pharmacy
East China University of
Science and Technology
P.O. Box 544
Shanghai 200237
China

Kenneth A. Hughes
DuPont Crop Protection
Stine-Haskell Research Center
1094 Elkton Road
Newark, DE 19711
USA

Naoya Ichimaru
Division of Applied Life Sciences
Graduate School of Agriculture
Kyoto University
Kyoto, 606-8502
Japan

Hideyuki Inui
Research Center for
Environmental Genomics
Kobe University
Nada-ku
Kobe, 657-8501
Japan

Murray B. Isman
Faculty of Land and Food Systems
University of British Columbia
Vancouver, BC, V6T 1Z4
Canada

Tomonori Iwasaki
Agricultural Chemicals Research
Laboratory
Sumitomo Chemical Co., Ltd.
4-2-1 Takatsukasa
Takarazuka
Hyogo, 665-8555
Japan

Timothy C. Johnson
Dow AgroSciences LLC
9330 Zionsville Road
Indianapolis, IN 46268
USA

Hiroyuki Katsuta
Mitsui Chemicals, Inc.
1144, Togo, Mobara-shi
Chiba, 297-0017
Japan

Jack Kiser
Pioneer Hi-Bred International
700A Bay Road
Redwood City, CA 94063
USA

Junro Kishi
Mitsui Chemicals, Inc.
1144, Togo, Mobara-shi
Chiba, 297-0017
Japan

Nobutaka Kitahata
RIKEN, Discovery Research Institute
2-1 Hirosawa
Wako
Saitama, 351-0198
Japan

Alexander Klausener
Bayer CropScience AG
BCS-Research
Alfred-Nobel-Str. 50
40789 Monheim
Germany

Gijs A. Kleter
RIKILT – Institute of Food Safety
Wageningen University and
Research Center
PO Box 230
6700 AE Wageningen
Holland

Hiroki Kodama
Nihon Nohyaku Co., Ltd.
345, Oyamada-cho
Kawachi-Nagano
Osaka, 586-0094
Japan

Hiroshi Kodama
Nihon Nohyaku Co., Ltd.
345, Oyamada-cho
Kawachi-Nagano
Osaka, 586-0094
Japan

Narumon Komalamisra
Faculty of Medicine
Mahidol University
Bangkok 10400
Thailand

Robert I. Krieger
Personal Chemical Exposure Program
Department of Entomology
University of California
Riverside, CA 92521
USA

Karl-Heinz Kuck
Bayer CropScience
Alfred-Nobel-Str. 50
40789 Monheim
Germany

Harry A. Kuiper
RIKILT – Institute of Food Safety
Wageningen University and
Research Center
PO Box 230
6700 AE Wageningen
Holland

Daniel L. Kunkel
IR-4 Project Headquarters
500 College Road East
Suite 201W
Princeton, NJ 08540
USA

Ichiro Kurane
National Institute of
Infectious Diseases
Tokyo, 162-8640
Japan

George P. Lahm
DuPont Crop Protection
Stine-Haskell Research Center
1094 Elkton Road
Newark, DE 19711
USA

Stuart J. Lansdell
Department of Pharmacology
University College London
Gower Street
London, WC1E 6BT
UK

Zhong Li
Shanghai Key Lab of
Chemical Biology
School of Pharmacy
East China University of
Science and Technology
P.O. Box 544
Shanghai 200237
China

Zewen Liu
Nanjing Agricultural University
Nanjing, 210095
China

Peter Luemmen
Bayer CropScience AG
Alfred-Nobel-Str. 50
40789 Monheim
Germany

Cristina M. Machial
Faculty of Land and Food Systems
University of British Columbia
Vancouver, BC, V6T 1Z4
Canada

Thomas Maetzke
Syngenta Crop Protection AG
Schwarzwaldallee 215
4002 Basel
Switzerland

Rick K. Mann
Dow AgroSciences LLC
9330 Zionsville Road
Indianapolis, IN 46268
USA

Timothy P. Martin
Dow AgroSciences LLC
9330 Zionsville Road
Indianapolis, IN 46268
USA

Takao Masaki
Nihon Nohyaku Co., Ltd.
345 Oyamada-cho
Kawachi-Nagano
Osaka, 586-0094
Japan

Kazuhiko Matsuda
Department of Applied
Biological Chemistry
School of Agriculture
Kinki University
Nara, 631-8505
Japan

Noritada Matsuo
Agricultural Chemicals Research Laboratory
Sumitomo Chemical Co., Ltd.
4-2-1 Takatsukasa
Takarazuka
Hyogo, 665-8555
Japan

Tomofumi Matsuoka
Hyogo Prefectural Institute of Public Health and Environmental Sciences
2-1-29 Arata-cho
Hyogo-ku
Kobe, 652-0032
Japan

Tadahiro Matsunaga
Agricultural Chemicals Research Laboratory
Sumitomo Chemical Co., Ltd.
4-2-1 Takatsukasa
Takarazuka
Hyogo, 665-8555
Japan

Caroline McCart
Department of Biology and Biochemistry
University of Bath
Bath, BA2 7AY
UK

Neil S. Millar
Department of Pharmacology
University College London
Gower Street
London, WC1E 6BT
UK

Saber Miresmailli
Faculty of Land and Food Systems
University of British Columbia
Vancouver, BC, V6T 1Z4
Canada

Hisashi Miyagawa
Division of Applied Life Sciences
Graduate School of Agriculture
Kyoto University
Kyoto, 606-8502
Japan

Hideto Miyoshi
Division of Applied Life Sciences
Graduate School of Agriculture
Kyoto University
Kyoto, 606-8502
Japan

Kenji Mori
The University of Tokyo
1-20-6-1309 Mukogaoka
Bunkyo-ku
Tokyo, 113-0023
Japan

Tatsuya Mori
Agricultural Chemicals Research Laboratory
Sumitomo Chemical Co., Ltd.
4-2-1 Takatsukasa
Takarazuka
Hyogo, 665-8555
Japan

Michel Muehlebach
Syngenta Crop Protection AG
Schwarzwaldallee 215
4002 Basel
Switzerland

Masatoshi Murai
Division of Applied Life Sciences
Graduate School of Agriculture
Kyoto University
Kyoto, 606-8502
Japan

René Mutti
Syngenta Crop Protection AG
Schwarzwaldallee 215
4002 Basel
Switzerland

Takeshi Nakano
RIKEN, Discovery Research Institute
2-1 Hirosawa
Wako
Saitama, 351-0198
Japan

Hayami Nakao
Nihon Nohyaku Co., Ltd.
345, Oyamada-cho
Kawachi-Nagano
Osaka, 586-0094
Japan

Thierry Niderman
Syngenta Crop Protection AG
Schwarzwaldallee 215
4002 Basel
Switzerland

Tetsuyoshi Nishimatsu
Nihon Nohyaku Co., Ltd.
1-2-5 Nihonbashi
Chuo-Ku
Tokyo, 103-8236
Japan

Hideo Ohkawa
Research Center for
Environmental Genomics
and
Research Center for Green Science
Fukuyama University
Gakuencho 1
Fukuyama
Hiroshima, 729-0292
Japan

Shivaji Pandey
FAO
Viale delle Terme di Caracalla
00100 FAO
Rome
Italy

Prabhu Pingali
FAO
Viale delle Terme di Caracalla
00100 FAO
Rome
Italy

Luann M. Pugh
DuPont Crop Protection
Stine-Haskell Research Center
Newark, DE 19714-0030
USA

Xuhong Qian
Shanghai Key Lab of
Chemical Biology
School of Pharmacy
East China University of
Science and Technology
P.O. Box 544
Shanghai 200237
China

Marco Quadranti
Syngenta Crop Protection AG
Schwarzwaldallee 215
4002 Basel
Switzerland

Kenneth D. Racke
Dow AgroSciences
9330 Zionsville Road
Building 308/2E
Indianapolis, IN 46268
USA

James J. Ragghianti
DuPont Crop Protection
Stine-Haskell Research Center
1094 Elkton Road
Newark, DE 19711
USA

Klaus Raming
Bayer CropScience AG
BCS-Research
Alfred-Nobel-Str. 50
40789 Monheim
Germany

James J. Rauh
DuPont Crop Protection
Stine-Haskell Research Center
1094 Elkton Road
Newark, DE 19711
USA

Daniel F. Rhoades
DuPont Crop Protection
Stine-Haskell Research Center
1094 Elkton Road
Newark, DE 19711
USA

Michael Riffel
RIFCon GmbH
Breslauer Str. 7
69493 Hirschberg a. d. B.
Germany

Karen J. Roberts
Syngenta
Central Toxicological Laboratories
Alderley Park
Macclesfield
Cheshire, SK10 4TJ
UK

John H. Ross
infoscientific.com
Carmichael, CA 95608
USA

Matthew D. Sacher
DuPont Crop Protection
Stine-Haskell Research Center
1094 Elkton Road
Newark, DE 19711
USA

Jens Schabacker
RIFCon GmbH
Breslauer Str. 7
69493 Hirschberg a. d. B.
Germany

Ellen Scherbaum
EU-Community Reference
Laboratory for Pesticide Analysis
using Single Residue Methods
hosted at the Chemisches und
Veterinäruntersuchungsamt Stuttgart
Schaflandstr. 3/2
70736 Fellbach
Germany

Michael Schindler
Bayer CropScience AG
BCS-Research-Discovery
Alfred-Nobel-Str. 40
40789 Monheim a. Rh.
Germany

Klaus-Jürgen Schleifer
BASF Aktiengesellschaft
Computational Chemistry and
Biology
Bldg. A 30
67065 Ludwigshafen
Germany

Anita Schnyder
Solvias AG
Klybeckstr. 191
4002 Basel
Switzerland

James N. Seiber
United States Department of Agriculture
Agricultural Research Service
Western Regional Research Center
800 Buchanan Street
Albany, CA 94710
USA

Thomas P. Selby
DuPont Crop Protection
Stine-Haskell Research Center
1094 Elkton Road
Newark, DE 19711
USA

Akira Seo
Nihon Nohyaku Co., Ltd.
1-2-5 Nihonbashi
Chuo-Ku
Tokyo, 103-8236
Japan

Xusheng Shao
Shanghai Key Lab of Chemical Biology
School of Pharmacy
East China University of Science and Technology
P.O. Box 544
Shanghai 200237
China

Jinliang Shen
Shanghai Key Lab of Chemical Biology
School of Pharmacy
East China University of Science and Technology
P.O. Box 544
Shanghai 200237
China

Yoshinori Shono
Agricultural Chemicals Research Laboratory
Sumitomo Chemical Co., Ltd.
4-2-1 Takatsukasa
Takarazuka
Hyogo, 665-8555
Japan

Michael W. Skidmore
Syngenta
Jealott's Hill
International Research Centre
Bracknell
Berkshire, RG42 6EY
UK

Ben K. Smith
DuPont Crop Protection
Stine-Haskell Research Center
1094 Elkton Road
Newark, DE 19711
USA

Rejane M. Smith
DuPont Crop Protection
Stine-Haskell Research Center
1094 Elkton Road
Newark, DE 19711
USA

Wencheng Song
Institute for the Control of Agrochemicals
Ministry of Agriculture (ICAMA)
Beijing
China

Jeffrey S. Sopa
DuPont Crop Protection
Stine-Haskell Research Center
1094 Elkton Road
Newark, DE 19711
USA

Darinka Štajnbaher
Public Health Institute
Prvomajska 1
2000 Maribor
Slovenia

Klaus Stenzel
Bayer CropScience AG
BCS-Research
Alfred-Nobel-Str. 50
40789 Monheim
Germany

Thomas M. Stevenson
DuPont Crop Protection
Stine-Haskell Research Center
1094 Elkton Road
Newark, DE 19711
USA

André Stoller
Syngenta Crop Protection AG
Schwarzwaldallee 215
4002 Basel
Switzerland

Martin Streloke
Federal Office of Consumer Protection and Food Safety
Division of Plant Protection Products
Messeweg 11/12
38104 Braunschweig
Germany

Masayo Sugano
Agricultural Chemicals Research Laboratory
Sumitomo Chemical Co., Ltd.
4-2-1 Takatsukasa
Takarazuka
Hyogo, 665-8555
Japan

Tomohiko Takasaki
National Institute of Infectious Diseases
Tokyo, 162-8640
Japan

Ikuo Takashima
Graduate School of Veterinary Medicine
Hokkaido University
Sapporo, 060-0818
Japan

Yong Tao
DuPont Crop Protection
Stine-Haskell Research Center
1094 Elkton Road
Newark, DE 19711
USA

Bünyamin Taşdelen
EU-Community Reference
Laboratory for Pesticide Analysis
using Single Residue Methods
hosted at the Chemisches und
Veterinäruntersuchungsamt Stuttgart
Schaflandstr. 3/2
70736 Fellbach
Germany

Nuansri Tayaputch
Laboratory Center for Food and
Agricultural Products (LCFA)
Paholyothin Road
Jatujak
Bangkok 10900
Thailand

Zhongzhen Tian
Shanghai Key Lab of
Chemical Biology
School of Pharmacy
East China University of
Science and Technology
P.O. Box 544
Shanghai 200237
China

Masanori Tohnishi
Nihon Nohyaku Co., Ltd.
345, Oyamada-cho
Kawachi-Nagano
Osaka, 586-0094
Japan

Kenji Tsubata
Nihon Nohyaku Co., Ltd.
345, Oyamada-cho
Kawachi-Nagano
Osaka, 586-0094
Japan

Isao Ueyama
Yuki Research Center
Bayer CropScience
Ibaraki 307-0001
Japan

Kazuya Ujihara
Agricultural Chemicals Research
Laboratory
Sumitomo Chemical Co., Ltd.
4-2-1 Takatsukasa
Takarazuka
Hyogo, 665-8555
Japan

Hiroshi Ushijima
Graduate School of Medicine
The University of Tokyo
Tokyo, 113-0033
Japan

Yasushi Utani
Graduate School of
Science and Technology
Kobe University
Rokkodaicho 1-1
Kobe
Hyogo, 657-8501
Japan

Terence A. Walsh
Dow AgroSciences
Discovery Research
9330 Zionsville Road
Indianapolis, IN 46268
USA

Yanli Wang
Shanghai Key Lab of
Chemical Biology
School of Pharmacy
East China University of
Science and Technology
P.O. Box 544
Shanghai 200237
China

Prem Warrior
Valent BioSciences Corporation
6131 RFD Oakwood Road
Long Grove, IL 60047
USA

Sebastian Wendeborn
Syngenta Crop Protection AG
Schwarzwaldallee 215
4002 Basel
Switzerland

Jean Wenger
Syngenta Crop Protection AG
Schwarzwaldallee 215
4002 Basel
Switzerland

Martin S. Williamson
Rothamsted Research
Harpenden, AL5 2JQ
UK

Christian Wolf
RIFCon GmbH
Breslauer Str. 7
69493 Hirschberg a. d. B.
Germany

Lihong Wu
DuPont Crop Protection
Stine-Haskell Research Center
1094 Elkton Road
Newark, DE 19711
USA

Takashi Yamamoto
Pioneer Hi-Bred International
700A Bay Road
Redwood City, CA 94063
USA

Yuji Yanase
Mitsui Chemicals, Inc.
1144, Togo, Mobara-shi
Chiba, 297-0017
Japan

Yong Zhen Yang
Institute for the Control of
Agrochemicals
Ministry of Agriculture (ICAMA)
Beijing
China

Noriaki Yasokawa
Nihon Nohyaku Co., Ltd.
345 Oyamada-cho
Kawachi-Nagano
Osaka, 586-0094
Japan

Yukihiro Yoshikawa
Mitsui Chemicals, Inc.
1144, Togo, Mobara-shi
Chiba, 297-0017
Japan

Naoki Yoshioka
Hyogo Prefectural Institute of Public
Health and Environmental Sciences
2-1-29 Arata-cho
Hyogo-ku
Kobe, 652-0032
Japan

I
Keynote and Plenary Lectures

Pesticide Chemistry. Crop Protection, Public Health, Environmental Safety
Edited by Hideo Ohkawa, Hisashi Miyagawa, and Philip W. Lee

ISBN: 978-3-527-31663-2

1
Challenges and Opportunities in Crop Production Over the Next Decade

James C. Collins, Jr.

1.1
Meeting Society's Agricultural Needs

There has never been a more complex and challenging business environment in global agriculture. The world we live and work in is changing every single day, and nowhere is it more evident than in the agrochemical industry.

In agriculture, there are few single solutions that will be capable of addressing the problems that farmers and the growing world population face today and in the future. Instead, combinations of solutions and collaborations across the public and private sectors will be required. Neither biotechnology nor chemistry has all the answers.

The intersection of biology, chemistry, and environmental stewardship/ sustainability has created unique opportunities to meet societal needs for food, feed, fiber, and fuel. New products are safer, better, and more effective in meeting these needs. Productivity improvements will remain the most important factor in determining the characteristics of global markets, especially as the population grows and available land for agriculture is reduced.

This industry will always be, I believe, about new technology. New technology is driving change and will win in the marketplace. It is going to be the recipe for us to continue to succeed in the face of the challenges ahead. So as companies at this conference who are actively investing in not only chemistry but also biotechnology, I believe we're investing in the right place.

1.2
Global Trends and Uncertainties

There are a number of trends and uncertainties, over which we have no control, that have a significant impact on the agriculture industry. The global population is increasing. Energy demand is increasing, and we are all aware of the price of oil. In some parts of the world, there are shortages of critical resources, such as

Pesticide Chemistry. Crop Protection, Public Health, Environmental Safety
Edited by Hideo Ohkawa, Hisashi Miyagawa, and Philip W. Lee

ISBN: 978-3-527-31663-2

land, water, and capital. There are fantastic emerging agricultural markets in India, Brazil, Argentina, and parts of southeast Asia. But access to these critical resources is a constraining factor. There is tremendous uncertainty about what is going to happen to agricultural subsidies and global trade policies. Exchange rates have a huge impact on our industry, as does global unrest and terrorism, the economic balance of power, natural disasters, the regulatory environment.

Global population growth is about 1.3% every year and is expected to reach 7 billion by 2015 and 9 billion by 2050. Forecasters have estimated that the number of cities greater than one million people will double to 60 over the next decade. These new cities will be in developing countries, placing significant pressure on the food supply infrastructure. World income is rising annually at 1.3%. More disposable income impacts people's diets as they demand more protein.

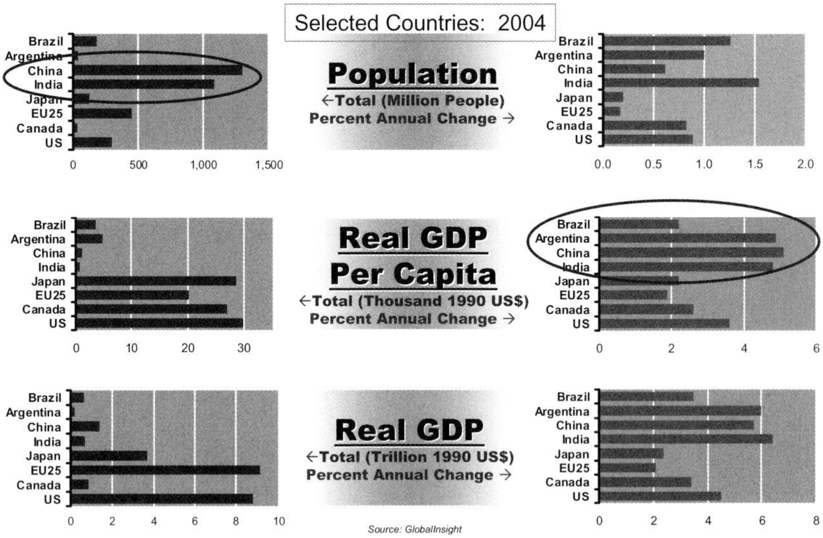

Looking at the above chart, everyone knows that China and India have big populations, but it is interesting to note that countries with large percentage changes in population also include Brazil and Argentina.

A similar situation exists with per capita Gross Domestic Product. The United States, Canada, Japan, and Europe are the leaders as expected. But the percent change in GDP is highest in places like Argentina, China, and India where we are seeing a wealth buildup. Likewise, the percentage change in total GDP in the developing world is astounding. That money is being spent on food and infrastructure and is placing unique demands on our industry. So, in thinking about where to target business growth, it will not be the traditional countries. Similarly, we can probably expect the same scenario over the next ten years in Africa as the political environment begins to settle down.

1.3
Grain Stocks

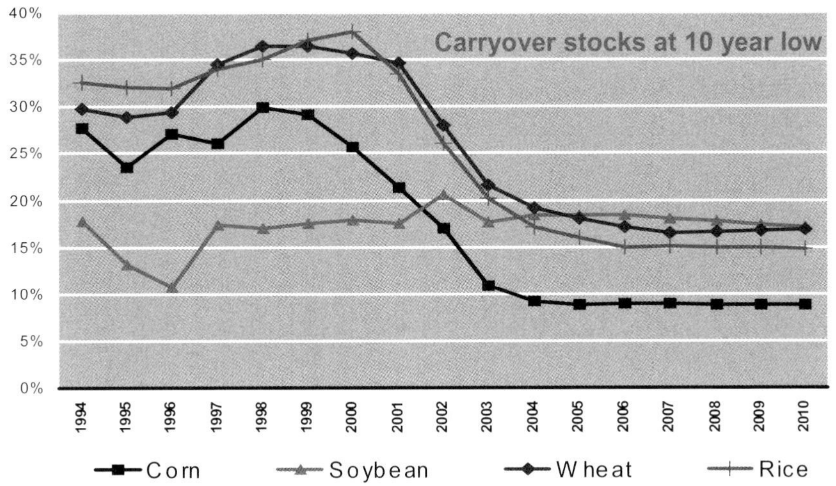

Another leading metric that we look at very closely is carryover stocks of grain. In 1999–2000, the industry had high stocks for traditional crops, like rice, wheat, corn, and soybeans. By 2005, productivity had gone up. A good example is Brazil. Brazil is producing five times the amount of soybeans they were producing just a few years ago. During that time frame, the amount of grain that was left over at the end of consumption went down every single year. We can expect these carryover stocks to continue to decline, which complicates the logistics of moving limited supplies to the place where they are needed most to feed the world's growing population. Companies and countries will upgrade the efficiency of their distribution systems to address this issue.

As Brazil's soybean production grew, U.S. production remained flat. You might expect that Brazil flooded the market with all these new soybeans and that the prices fell dramatically because of the supply glut. The fact is, however, that the additional supply was directed to China. China made a fundamental decision that it is better at other activities than producing soybeans and partnered with Brazil to set up a bilateral trade agreement. It has become a fantastic agricultural partnership, and similar agreements between other countries can be expected.

1.4 Exchange Rates

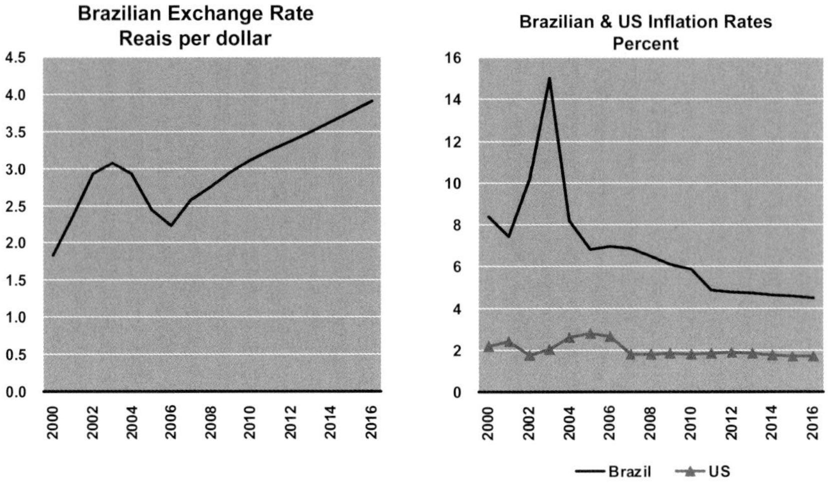

Exchange rates certainly affect our view of the world. Consider what is going on with the reais in Brazil compared to the dollar. It has created tremendous fluctuations in profitability because Brazilian farmers sell very little of their soybeans in Brazil so when that exchange rate fluctuated, growers in Brazil were only able to recover about 70% of what they invested in the crop. Next year, Brazilian farmers might decide to give up on production or plow a crop under because they will make more from government subsidies than by selling on the open market. If we adjust the price of soybeans to take inflation into account, there isn't a great deal of difference between the profitability of a U.S. and a Brazilian farmer. The government has stepped in and proposed some major financing proposals to help the farmers with these liquidity issues.

Exchange rates can also mask revenue ups and downs. In 2002, many were predicting the collapse of the crop protection business because a number of new active ingredients and products had come along and extracted most of the value. Investing in new technology needed to be re-thought as a priority. If we look at the change from 1997 through 2002, the market was down 14% – a very significant drop. But when you analyze the situation more carefully, it is clear that the real reason the market was down was the exchange rate. It just depends on whether we counted dollars, euros, or yen. During that time frame, product hectares were up 3%.

The same scenario occurred between 2002 and 2003, it had a 6% growth. It was a return to profitability, we thought, but the reason for the improvement was how we define the dollar's exchange rate. In reality, we treated absolutely no new acres during that time frame.

The crop protection market is essentially flat at about 32 billion USD. It has grown somewhat over the last few years so some companies have benefited from

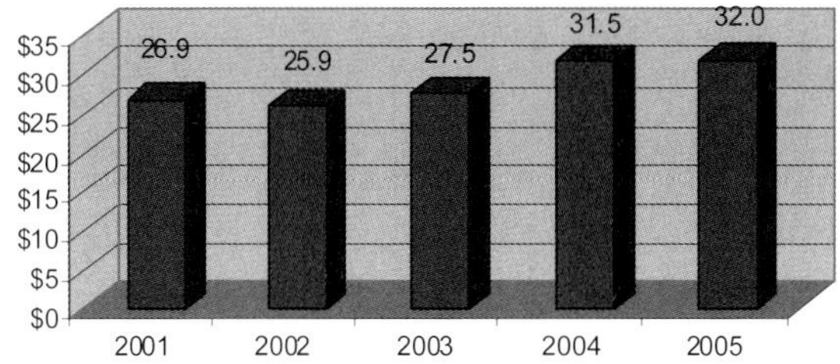

Phillips McDougall	$31,190
Cropnosis	$32,867
Average	$32,029

Average of Cropnosis and Phillips McDougall data

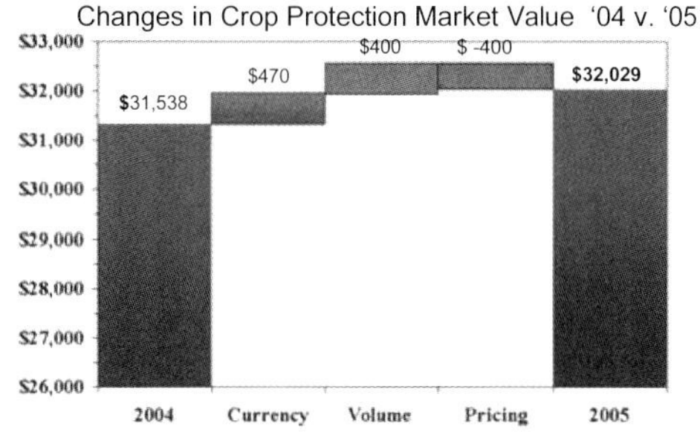

market share increases. The conventional seed market is in the 15 billion USD range. The genetically modified seed business is about 5 billion USD.

If we consult with Cropnosis or Phillips McDougal and ask them about the future, we get two completely different answers. Phillips McDougal says the market is going up by 1.3% and Cropnosis says that there is going to be a 1.5% decline. So you draw those two curves and you realize they are estimating our future is somewhere between 28 and 36 billion USD. We all need to determine how to navigate through those estimates because the answer is not clear cut.

We belong to a classic cyclical industry. What goes up one year will go down the next and vice versa. The problem is the magnitude of the cycle is getting much larger every year. So, in past years, we would have been up or down 1–2%; now we are seeing 4–7% swings. The other problem is the cyclical swings are occurring virtually annually, instead of in two- or three-year cycles. Principal factors impacting these cycles are changes in exchange rates, carryover grain stocks, and protein consumption.

We have products that were invented and launched in the 1950's that are still being sold today. Any product developed before the mid-1970's is a commodity. Then there is a group of products that we call generic. They have come off patent but are still available. Finally, there are products still under patent protection that were launched in the late 1980's or afterwards. Market share for each product category is about one-third.

If we analyze product use on global high-performance acres, the percent of proprietary herbicides, fungicides, and insecticides being applied to those acres has increased over the last few years. While the perception is that the market is becoming increasingly generic and more commodity-based, the reality is it is becoming more proprietary on the high-production acres. The percent of value represented by proprietary products has increased most dramatically in fungicides.

1.5 Biofuels

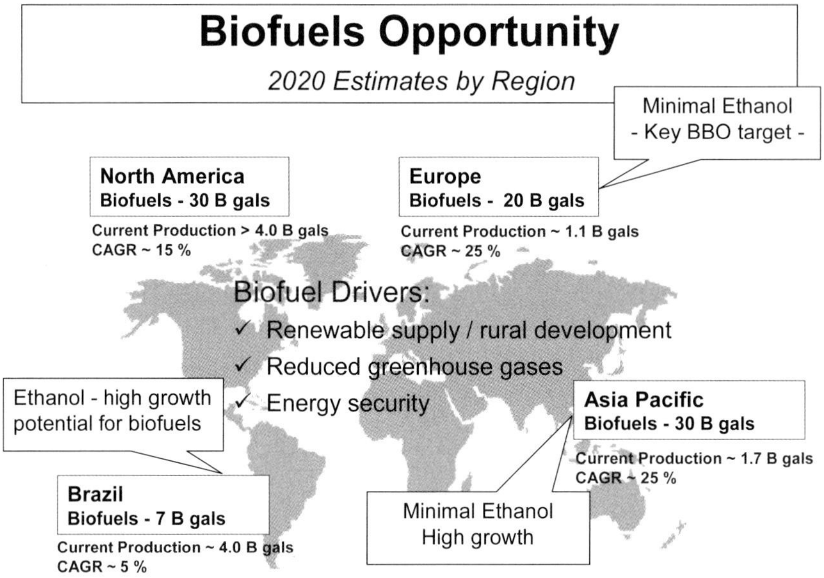

Biofuels Opportunity
2020 Estimates by Region
Minimal Ethanol - Key BBO target -
North America
Biofuels - 30 B gals
Current Production > 4.0 B gals
CAGR ~ 15 %
Europe
Biofuels - 20 B gals
Current Production ~ 1.1 B gals
CAGR ~ 25 %
Biofuel Drivers:
✓ Renewable supply / rural development
✓ Reduced greenhouse gases
✓ Energy security
Ethanol - high growth potential for biofuels
Asia Pacific
Biofuels - 30 B gals
Current Production ~ 1.7 B gals
CAGR ~ 25 %
Brazil
Biofuels - 7 B gals
Current Production ~ 4.0 B gals
CAGR ~ 5 %
Minimal Ethanol
High growth

Biofuels are a huge opportunity, we believe. The concepts of renewable supply, reduced greenhouse gases, and energy security are driving biofuels opportunities globally.

Today, about 1.6 billion bushels of corn are being used in ethanol production. We expect that to approach 4 billion bushels over the next decade. Most counties in the midwestern U.S. have one to two ethanol plants that have received building permits, and this is not even counting the facilities already in existence. It will be interesting to watch the price of corn as it moves into new markets.

We think that there are about 30 billion gallons of potential in North America, with only about 4 billion gallons being produced today. In Brazil, we are probably only looking at a potential of about 7 billion gallons and we are already producing four of that. So we have already seen the explosion in Brazil using sugarcane. Other areas with significant market expansion potential are Europe and Asia-Pacific.

1.6
Counterfeit Products

An issue that is significantly damaging the industry is the prevalence of counterfeit products. There is an estimated 500 billion USD impact to global GDP based on counterfeit goods, including an economic impact of 35–40 million USD annually to DuPont's sulfonylurea product line. So, counterfeiting has become an area we are going to address aggressively over the next few years. We hope to collaborate with CropLife and the rest of the industry to undertake a joint and concerted effort. Counterfeit products can be found in any region of the world so it impacts all of us.

It can be very hard to distinguish counterfeit from actual products. I saw two product containers where the registered trademark on the real product was about 2-mm wide. The counterfeit one was about 1-mm wide. There were some wording changes on the label itself. However, the only way you could really tell that one was counterfeit was to take this product home and apply it to your corn. In fact, the product had the wrong active ingredient in it and did substantial damage to crops of some farmers in Europe.

A number of our regulatory partners, especially in China, the U.S., and India have been very proactive at policing this practice, but they just do not have the manpower or product knowledge to identify where it is occurring. This is an emerging area, and we have a lot of work to do to help educate not only consumers and regulators, but also ourselves on how to better police some of these issues. It is critical we take on this responsibility.

List	No. of Products	Products of Commercial Significance*	Accepted into Annex 1	Re-registration admissible/ pending	Not accepted/ Not supported
Existing Products					
1	90	90	53	8	29
2	148	114	12	38	64
3	389	263		135	128
4	204	11		9	2
Total	**831**	**478**	**65**	**190**	**223**
New Active Ingredients			55	47	7
Total Existing + New a.i.s.			**120**	**237**	
			357 (120 + 237)		

* as active ingredients for crop protection

The re-registration procedure that is going on in the EU is difficult. The industry has about 470 products right now in that system. These are commercial products that have useful roles to play for farmers. Today, there are only about 350 of those that we can say are either going to be accepted or have a pending admissible label. So, there are some 200 products that will not be available to customers.

1.7 Product Commercialization

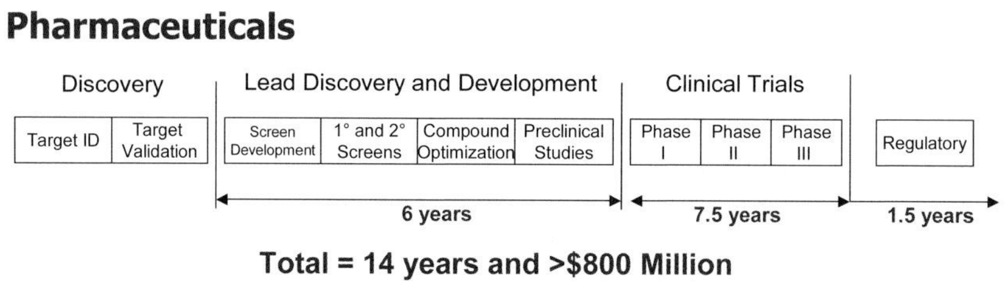

Source: Tufts University study as reported in *Chemical & Engineering News*, Dec 2005

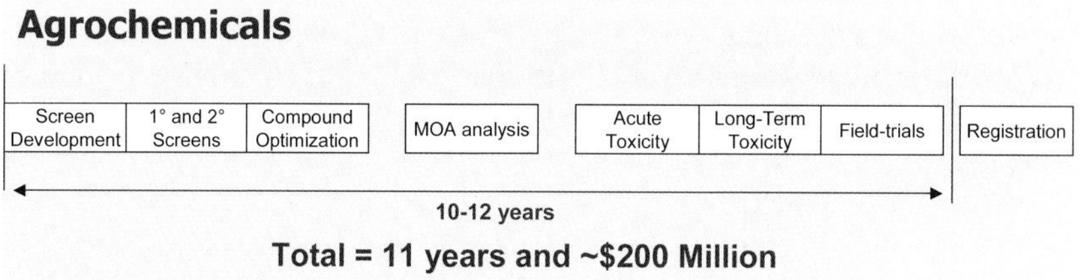

Source: Phillips McDougall Study on Ag R&D, Dec 2005

Commercialization is a major issue for our industry. We're going to be developing a lot of new products, which is very good news. The bad news is that we estimate it costs about 200 million USD to bring a new product to market. The time frame is ten to 12 years. So, there is a lot of investment to be made up front to bring in new products. At DuPont, we decided that unless a product had 150 million USD sales potential, we would not pursue it.

Once a company receives a product registration and starts to sell, it takes at least six years to break even and get to the point that we are actually making money for the shareholders. This is certainly an interesting business proposition when you consider the initial investment and lengthy time required to recover investment cost and to generate profile. How many companies have that financial ability? If we lose time in the regulatory submission cycle due to adding a field trial or researching additional data, we run into an even greater delay.

If you take a snap shot of the major industry players that are above the 500 million USD range that existed back in 1990, you can tell that of the 13, there are less than half still here today due to industry consolidation. So, while that can be daunting, it can also be a great opportunity, because these companies have come together to really improve the efficiency and effectiveness of their research and development programs.

As an industry, we have gone through fundamental changes in how we approach product discovery and development. From the 1950's to the 1990's, we did work the old-fashioned way – screening 10,000–50,000 compounds a year, mostly in-house. We knew that we needed to decrease cycle time because of the magnitude of the investment and a declining success rate. We are now looking at 250,000 products a year from multiple chemistry sources including brokers, vendors, and universities. Samples are only 2.5 mg and cost 10–40 USD apiece. The screening system for new active ingredients handles a greater volume, is more efficient, and has been targeted around areas of chemistry we know have activity.

One area that has been very interesting for us is the composition of the final product. In 2001, DuPont received about 25 registrations globally, primarily single actives. In 2005, we had 135 registrations and did not include a single active. These are all mixtures, which are combinations of products that come together to make unique leads. We generated five times the number of new registrations but moved away from a single molecule focus.

A good example of the value of mixtures is glyphosate. The amount of straight glyphosate that has been used in the North America corn and soybean market has actually been declining over the last few years. The reason is farmers and our retail partners are looking for ways to add other products into the mix. So, rather than have tank mixes, companies are trying to create these mixtures ahead of time, which is what is driving this regulatory explosion for unique products. Customers do not want to become chemists. They do not want to have to mix five products together in a tank in order to get the control they need. They want to be able to put a single product into the tank and know it will do the job.

1.8 Convergence of Factors

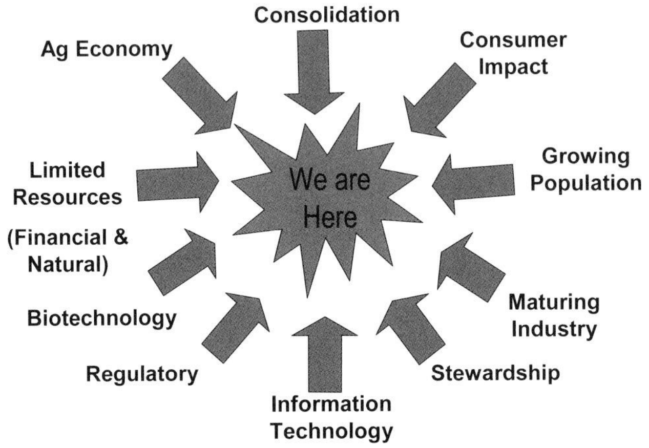

To summarize, our industry is facing an amazing convergence of factors that up until now have been relatively independent of each other. This complicated environment where we work has made it virtually impossible to find a single technology solution and highlights the critical intersection of biology, chemistry, and sustainability.

There has never been a more complex and challenging time in our industry. But, I also believe there has never been a more exciting time for chemistry. I am very encouraged by the fact that there are so many industry members at this conference. I know that you will continue to invest your time, energy, talent, and creativity into meeting the constantly emerging customer needs. I am convinced it is going to be a very collaborative process because, otherwise, the work will not get done.

I will end my presentation with a quote by Paul Anderson, a prolific science fiction writer, who said: "The only thing certain about the future is that we are going to be surprised." I would add to that a comment of my own: "I believe that we are also going to have a lot of fun as the future unfolds."

2
Searching Environmentally Benign Methods for Pest Control: Reflections of a Synthetic Chemist

Kenji Mori

2.1
Introduction

I have worked for nearly half a century as a synthetic chemist in the field of natural products chemistry relating to pesticide science. This review comprises four topics: (1) pesticides and our daily life, (2) contributions to pesticide discovery by Japanese scientists, (3) my lab's past natural product syntheses in relation to pesticide science, and (4) our past efforts in pheromone synthesis. In other words, Sections (1) and (2) are the general treatise on the historical description of human endeavor for searching environmentally benign methods for pest control, while Sections (3) and (4) are reviews of works undertaken in my own laboratory.

2.2
Pesticides and Our Daily Life

From 1945 to 1950 after World War II, Japan experienced very poor sanitary conditions due to the destruction caused by the war. Various kinds of infectious diseases were prevalent. In order to control the vector arthropods of insect-borne diseases, about 4 million Tokyo citizens were treated with DDT. Indeed, DDT saved the lives of many Japanese people after World War II.

In that period after World War II, Japan citizens also suffered from shortages of food, especially rice. So as to increase the rice production, synthetic pesticides were used extensively in Japan. For example, parathion was used to control the Asiatic rice borer, *Chilo suppressalis*. Unfortunately, this phosphorus insecticide was highly toxic, and a number of farmers were killed by poisoning. Another notorious pesticide was phenylmercuric acetate, which was used against rice blast disease. Because it contained mercury, the public could not accept it.

The reason why the public disliked mercury was the result of a pollution problem caused by the chemical industry. Minamata disease was first reported in 1953, and became the biggest pollution problem in Japan at that time. The

Pesticide Chemistry. Crop Protection, Public Health, Environmental Safety
Edited by Hideo Ohkawa, Hisashi Miyagawa, and Philip W. Lee

ISBN: 978-3-527-31663-2

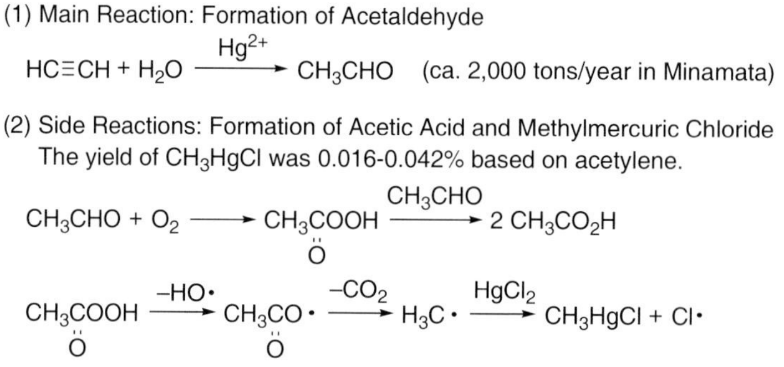
(1) Main Reaction: Formation of Acetaldehyde
HC≡CH + H2O —Hg2+→ CH3CHO (ca. 2,000 tons/year in Minamata)
(2) Side Reactions: Formation of Acetic Acid and Methylmercuric Chloride
The yield of CH3HgCl was 0.016-0.042% based on acetylene.
CH3CHO + O2 → CH3COOH —CH3CHO→ 2 CH3CO2H
CH3COOH —–HO•→ CH3CO• —–CO2→ H3C• —HgCl2→ CH3HgCl + Cl•

Figure 1. Hg^{2+}-catalyzed hydration of acetylene and its side reactions.

disease was caused by the toxicological effect of methylmercuric chloride to the brain tissues. This organomercuric compound was produced by a free-radical side reaction in the course of the mercuric ion-catalyzed hydration of acetylene to yield acetaldehyde.

Shin-Nippon-Chisso Co. manufactured about 2,000 tons per year of acetaldehyde by the process described in Figure 1, and the side reaction yielded 0.016–0.042% of methylmercuric chloride based on acetylene. Fish and shellfish in the Minamata Bay were contaminated with methylmercuric chloride. Then, those who consumed the seafoods became extremely ill, and some died. By knowing the Minamata disease, the Japanese public became much more aware and deeply concerned about pollution caused by chemical industries.

What causes pollution problems? Firstly, scientific theories are often based on approximation, and one may neglect side reactions. Minamata disease was due to the by-products generated from the side reaction. Secondly, careless mistakes or human errors may take place as exemplified in Bhopal and Chernobyl. The concern for safety should always be of first priority to the chemical industry.

In 1962, Rachel Carson published her seminal book "Silent Spring" to emphasize the reality of environmental problems caused by the excessive and improper use of persistent synthetic pesticides. The book was translated into Japanese, and became quite influential among Japanese intellectuals including myself as a young Ph. D. chemist. Knowledge about the toxicity of pesticides as well as the concern about potential environmental disasters made researchers in the agrochemical industry strive to develop environmentally benign pesticides.

2.3
Contributions in Pesticides Discovery by Japanese Scientists

Since the 1950's, Japanese pesticide scientists developed a number of new pesticides in pursuit of less toxic ones. Figure 2 summarizes their endeavors until now. A monograph was recently published, in which the development of agrochemicals in Japan was discussed in detail [1].

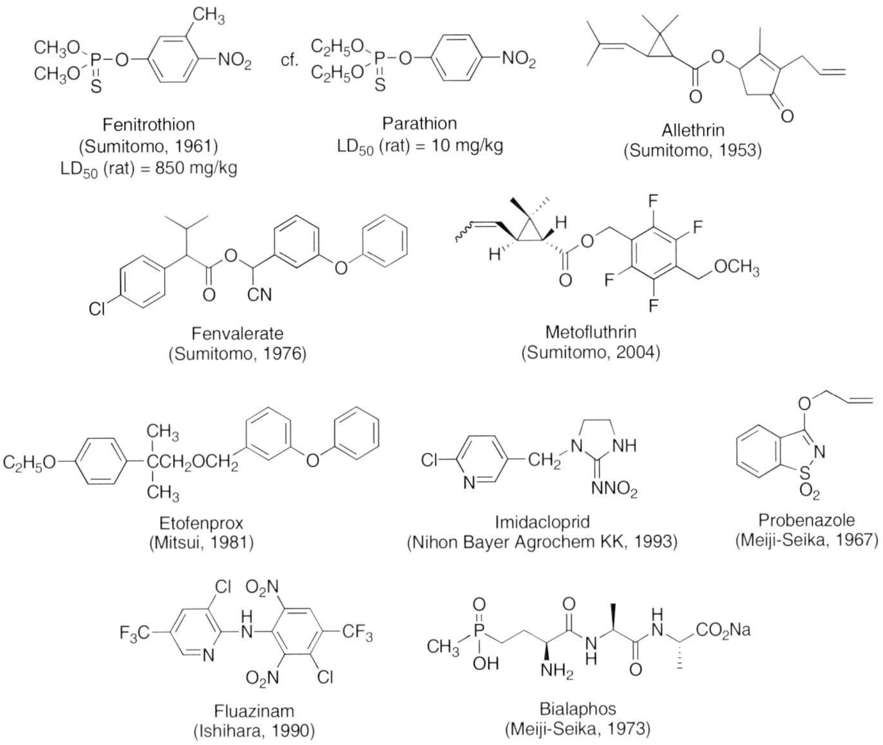

Figure 2. Japan's contribution in pesticide discovery and manufacturing.

Fenitrothion (Sumithion) was developed by Sumitomo as a potent phosphorus insecticide with low toxicity. Sumitomo developed the industrial synthesis of allethrin in the early 1950's, and continued to invent effective pyrethroids such as fenvalerate and metofluthrin. Mitsui's etofenprox is a unique pyrethroid possessing no stereogenic carbon atom. The discovery of photo-stable synthetic pyrethroids such as fenvalerate and etofenprox enabled their widespread use in outdoor agriculture.

Imidacloprid was discovered by Nihon Bayer Agrochem K.K. in 1993, and found to be a very effective systemic insecticide. Meiji-Seika Co. developed probenazole, a unique compound against rice blast disease. It activates the resistance capability of rice plants, although it has almost no fungicidal activity itself. Ishihara Sangyo developed in 1990 a new fungicide fluazinam by using novel fluorination technology. Some agricultural antibiotics were also developed and used in Japan. Bialaphos is a naturally occurring contact herbicide isolated by Meiji-Seika Co. in 1973 as a metabolite of *Streptomyces hygroscopicus*.

2.4 Natural Products Synthesis and Pesticide Science

My research group has been active for nearly 50 years on the synthesis of bioactive natural products of plant, insect, and microbial origins. A brief summary of our work is presented in Figure 3 [2].

Gibberellins are phytohormones first discovered at the University of Tokyo in 1938. In my earlier research career, I worked for 9 years since 1959 to synthesize gibberellin A_4 in Prof. M. Matsui's laboratory at the University of Tokyo, in the Department of Agricultural Chemistry. When I finished the work in 1968 [3], a famous microbiologist Prof. K. Arima of the same Department said to me, "Congratulations, Dr. Mori on the completion of the gibberellin synthesis. But you spent 9 years of your life to do it. Don't forget that the fungus *Gibberella fujikuroi* makes the gibberellins within a couple of days." This criticism made me think

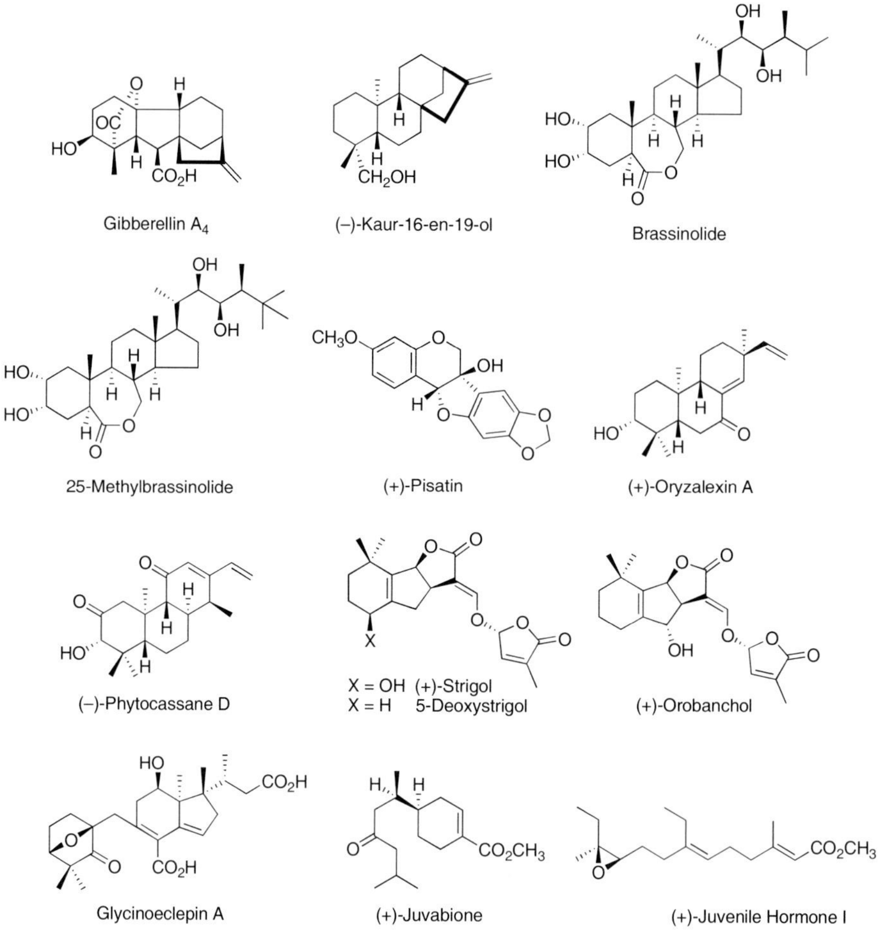

Figure 3. Examples of bioactive natural products synthesized by Mori and co-workers.

that we chemists can be respected by biologists only when we synthesize those compounds which are difficult to prepare by biological systems. In connection with my gibberellin work, I synthesized the racemate of (–)-kaur-16-en-19-ol, a precursor of gibberellin biosynthesis and a plant-growth promotor itself [4]. The racemate was only 50% as active as the natural (–)-alcohol. I decided that I should synthesize only the bioactive enantiomer in the future.

From 1980 to 1988, we worked on the synthesis of brassinosteroids, another important group of phytohormones. Brassinolide was isolated in 1979 from *Brassica napus* in a small amount (4 mg from 40 kg of pollen) as a plant-growth promotor. We synthesized brassinolide [5], and also prepared non-natural 25-methylbrassinolide [6]. The latter was more bioactive than the former. These brassinosteroids were very easily metabolized by plants, and could not be used practically. Thus, natural products often remain only as prototypes of practically useful man-made products.

Another target of ours in plant science was the synthesis of phytoalexins. (–)-Pisatin, isolated in 1960, is the best known phytoalexin of *Pisum sativum*. In 1989, we synthesized both the enantiomers of pisatin [7]. Dr. G. Russel in New Zealand kindly bioassayed them. Both of them showed antifungal activity, however, they were weaker than that of the commercial fungicides. Our synthesis of the enantiomers of oryzalexin A, a phytoalexin from *Oryza sativa*, also showed both of them to be bioactive [8]. Phytocassanes are more complicated phytoalexins isolated in 1995 from *Oryza sativa* infected with *Magnaporthe grisea*. We synthesized (–)-phytocassane D, and determined its *ent*-cassane stereochemistry [9]. Phytocassanes share in common the same *ent*-diterpene skeleton as those of the gibberellins and oryzalcxins.

We then worked on strigolactones such as (+)-strigol and (+)-orobanchol. The former was isolated from cotton plants in 1972 as a stimulant for the germination of *Striga*, while the latter was isolated in 1998 from *Trifolium pratense* as a stimulant for the germination of *Orobanche*. Our synthesis made (+)-strigol readily available, and the structure of orobanchol could be determined [10]. Recently in 2005, 5-deoxystrigol was shown to be the branching factor for arbuscular mycorrhizal fungus.

Cyst nematode is a serious pest in agriculture. In 1985, Masamune *et al.* isolated glycinoeclepin A (1 mg) as a hatching stimulus for the soybean cyst nematode by extracting the dried roots (1 ton) of the kidney bean. We synthesized it (220 mg), and examined its bioactivity [11]. Although glycinoeclepin A showed very strong hatch-stimulating activity at 10^{-12}–10^{-13} g/mL *in vitro*, it showed no nematicidal effect in laboratory pots and in a soybean field as tested by Sumitomo and Novartis.

In addition to pheromones (vide infra), we were interested in juvenile hormones (JHs), and synthesized (±)-juvabione [12] and (+)-juvabione [13–14]. JH mimics were later found to be useful as practical insect growth regulators (IGRs). We synthesized (±)-JH I [15], (+)-JH I [16] and unnatural (–)-JH I [17]. The naturally occurring (+)-JH I was 1.2×10^4 times more active than (–)-JH I. Chirality plays an important role at JH receptor sites.

I learned three things by synthesizing these bioactive natural products related to pesticides. Perhaps every pesticide chemist is familiar with the following points. First of all, natural products are often too complicated and fragile to be used as practical pesticides. Nevertheless, they can be the prototypes or lead compounds of new pesticides as in the cases of pyrethroids and juvenile hormones. Secondly, chirality plays a key role in the binding of these bioactive molecules at the active site. Thirdly, bioactivity observed *in vitro* or *in vivo* in lab tests may not always be reproduced in the field.

2.5
Enantioselective Pheromone Synthesis and Pesticide Science

After reading “Silent Spring”, I became interested in insect pheromones, because its application may provide us with a new and environmentally benign method of pest control. I was also interested in the evolving field of asymmetric synthesis. Accordingly, I started my enantioselective pheromone synthesis in 1973. The first work was the determination of the absolute configuration of the dermestid beetle pheromone [18]. By synthesizing the (*S*)-(+)-enantiomer of the pheromone from (*S*)-2-methyl-1-butanol, the levorotatory natural pheromone was shown to be the (*R*)-isomer (Figure 4).

Subsequently in 1974, I synthesized both the enantiomers of *exo*-brevicomin, the pheromone of the western pine beetle [19], and only the (+)-isomer was bioactive. In the case of frontalin, the (–)-isomer was the bioactive for te southern pine beetle [20]. Both the enantiomers of sulcatol [21], the pheromone of the ambrosia beetle *Gnathotrichus sulcatus*, were totally inactive. Their mixture, however, showed strong pheromone activity.

Recently in 2005, we synthesized 10 g of (+)-*endo*-brevicomin, the minor component of the pheromone of the male southern pine beetle, *Dendroctonus frontalis* [22–24]. We used lipase AK in this synthesis to desymmetrize the prochiral diol. Dr. B. T. Sullivan at the U.S. Forest Service is currently studying the practicality of the pheromone traps with a mixture of (+)-*endo*-brevicomin, frontalin and α-pinene.

In 2006, I converted (*S*)-perillyl alcohol into (*R*)-cryptone, which afforded (1*S*,4*R*)-2-menthen-1-ol, the male-produced aggregation pheromone of *Platypus quercivorus* [25]. This ambrosia beetle is the vector of an ambrosia fungus (*Raffaelea* sp.), which causes the dieback of deciduous oak (*Quercus crispula*) in northern Japan. The pheromone may be useful in monitoring the population of *Platypus quercivorus*. Synthesis of the pure enantiomers of pheromones allowed us to examine the relationships between stereochemistry and pheromone activity as shown in Figure 5. The relationships turned out to be complicated and diverse. Chirality was thus shown to be of key importance in pheromone perception [26–29]. Pheromones are now used practically in communication disruption among pest insects, and also in monitoring their population. Although pheromone technology is still weak, this method of insect control will be fully developed in the future.

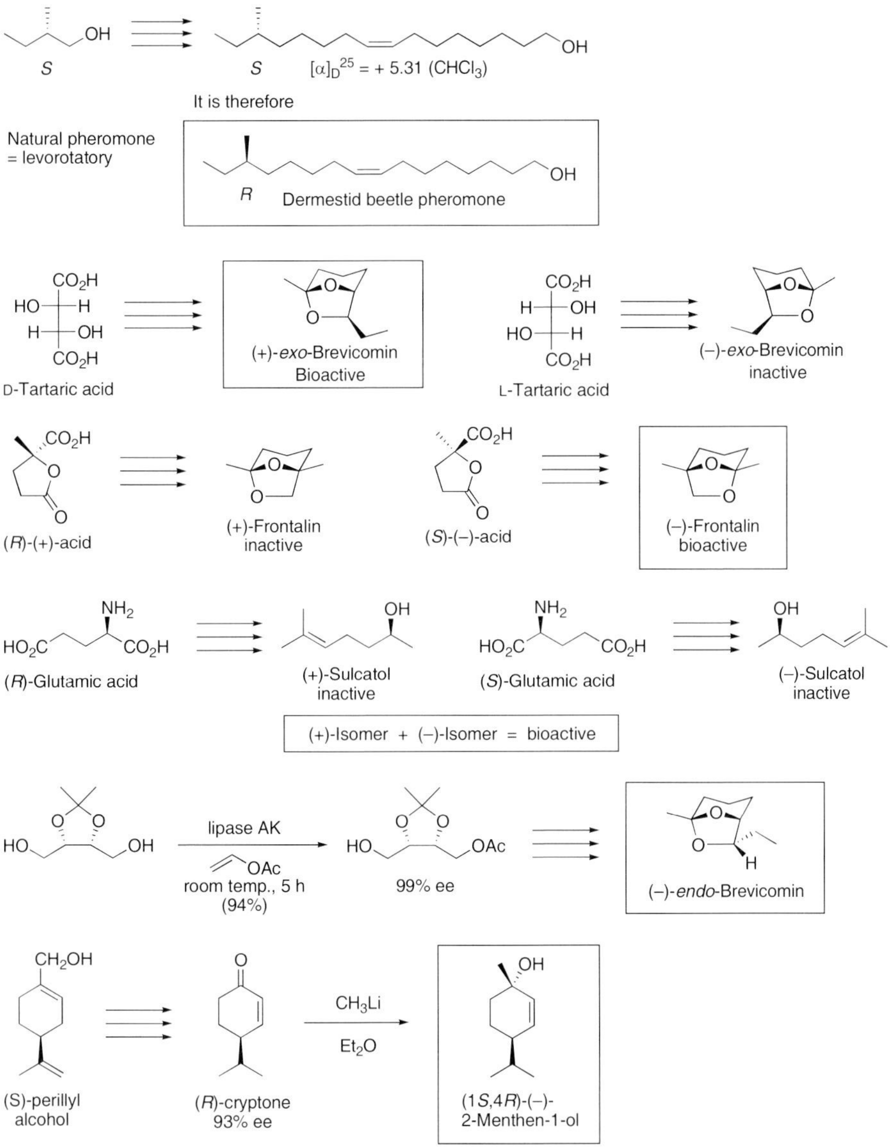

Figure 4. Enantioselective synthesis of insect pheromones.

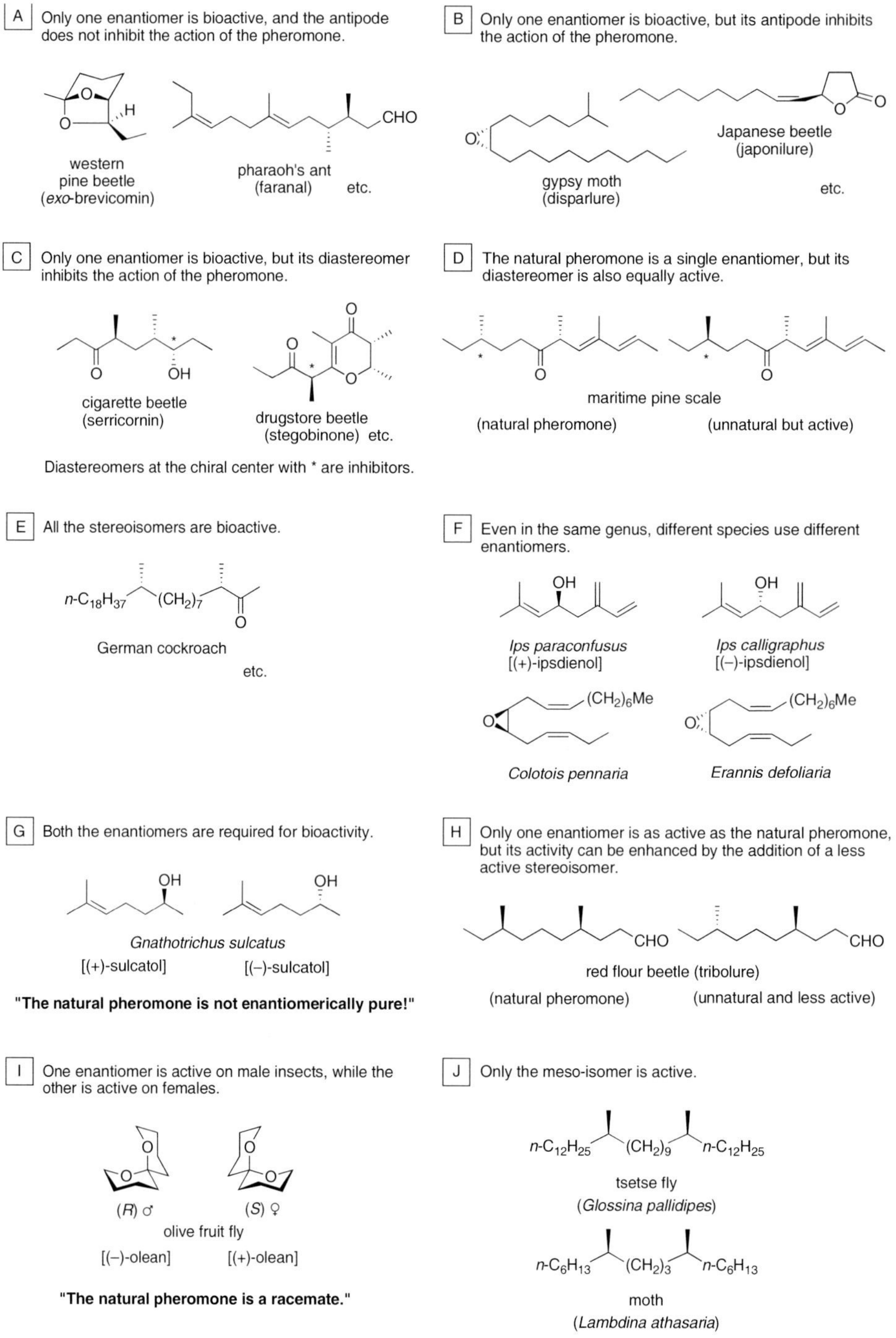

A
Only one enantiomer is bioactive, and the antipode does not inhibit the action of the pheromone.
H
O
O
CHO
western pine beetle (exo-brevicomin)
pharaoh's ant (faranal)
etc.
B
Only one enantiomer is bioactive, but its antipode inhibits the action of the pheromone.
O
O
O
Japanese beetle (japonilure)
gypsy moth (disparlure)
etc.
C
Only one enantiomer is bioactive, but its diastereomer inhibits the action of the pheromone.
O
OH
O
O
O
cigarette beetle (serricornin)
drugstore beetle (stegobinone) etc.
Diastereomers at the chiral center with * are inhibitors.
D
The natural pheromone is a single enantiomer, but its diastereomer is also equally active.
O
O
maritime pine scale
(natural pheromone)
(unnatural but active)
E
All the stereoisomers are bioactive.
n-C18H37
(CH2)7
O
German cockroach
etc.
F
Even in the same genus, different species use different enantiomers.
OH
OH
Ips paraconfusus [(+)-ipsdienol]
Ips calligraphus [(–)-ipsdienol]
O
(CH2)6Me
O
(CH2)6Me
Colotois pennaria
Erannis defoliaria
G
Both the enantiomers are required for bioactivity.
OH
OH
Gnathotrichus sulcatus
[(+)-sulcatol]
[(–)-sulcatol]
"The natural pheromone is not enantiomerically pure!"
H
Only one enantiomer is as active as the natural pheromone, but its activity can be enhanced by the addition of a less active stereoisomer.
CHO
CHO
red flour beetle (tribolure)
(natural pheromone)
(unnatural and less active)
I
One enantiomer is active on male insects, while the other is active on females.
O
O
O
O
(R) ♂
(S) ♀
olive fruit fly
[(–)-olean]
[(+)-olean]
"The natural pheromone is a racemate."
J
Only the meso-isomer is active.
n-C12H25
(CH2)9
n-C12H25
tsetse fly (Glossina pallidipes)
n-C6H13
(CH2)3
n-C6H13
moth (Lambdina athasaria)

Figure 5. Relationships between stereochemistry and pheromone activity.

2.6 Conclusion

The following three points can be made through my past synthetic studies on bioactive natural products of agricultural interest. Firstly, I am convinced that further studies on small molecules such as hormones, pheromones, and other bioregulators will provide keys to developing environmentally benign pesticides. Secondly, chirality of a biomolecule is very important for the expression of bioactivity. Use of the bioactive enantiomer can reduce the environmental exposure by at least half. Thirdly, pursuing soft or 'green' chemistry instead of persistent nonselective compounds will be key in future pest control practices. We should not be too selfish and narrow-sighted to assume that everything is here only for us without regard for the environment. In this context, I highly appreciate the words of Kenji Miyazawa (1896–1933), "Let us seek the happiness of all the creatures." He was a poet, a storyteller and a sincere Buddhist, who always desired the happiness of all the creatures. He was originally a soil scientist, and a 1918 graduate of Morioka Agricultural College, Department of Agricultural Chemistry. In conclusion, I pray "Give us this day our daily bread", believing that pesticide science will continue to be useful in assisting to fulfill this prayer.

2.7 References

1. M. Sasaki, N. Umetsu, H. Saka, K. Nakamura, K. Hamada (Eds.), *Development of Agrochemicals in Japan*, Pesticide Science Society of Japan, Tokyo, **2003** (in Japanese).
2. K. Mori, *Acc. Chem. Res.*, **2000**, *33*, 102–110.
3. K. Mori, M. Shiozaki, N. Itaya, M. Matsui, Y. Sumiki, *Tetrahedron*, **1969**, *25*, 1293–1321.
4. K. Mori, M. Matsui, *Tetrahedron*, **1968**, *24*, 3095–3111.
5. K. Mori, M. Sakakibara, K. Okada, *Tetrahedron*, **1984**, *40*, 1767–1781.
6. K. Mori, T. Takeuchi, *Liebigs Ann. Chem.*, **1988**, 815–818.
7. K. Mori, H. Kisida, *Liebigs Ann. Chem.*, **1989**, 35–39.
8. K. Mori, M. Waku, *Tetrahedron*, **1985**, *41*, 5653–5660.
9. A. Yajima, K. Mori, *Eur. J. Org. Chem.*, **2000**, 4079–4091.
10. K. Hirayama, K. Mori, *Eur. J. Org. Chem.*, **1999**, 2211–2217.
11. H. Watanabe, K. Mori, *J. Chem. Soc.*, Perkin Trans., **1991**, *1*, 2919–2934.
12. K. Mori, M. Matsui, *Tetrahedron*, **1968**, *24*, 3127–3138.
13. E. Nagano, K. Mori, *Biosci. Biotechnol. Biochem.*, **1992**, *56*, 1589–1591.
14. H. Watanabe, H. Shimizu, K. Mori, *Synthesis*, **1994**, 1249–1254.
15. K. Mori, *Tetrahedron*, **1972**, *28*, 3447–3456.
16. K. Mori, M. Fujiwhara, *Tetrahedron*, **1988**, *44*, 343–354.
17. K. Mori, M. Fujiwhara, *Liebigs Ann. Chem.*, **1990**, 369–372.
18. K. Mori, *Tetrahedron Lett.*, **1973**, 3869–3872; *Tetrahedron*, **1974**, *30*, 3817–3820.
19. K. Mori, *Tetrahedron*, **1974**, *30*, 4223–4227.
20. K. Mori, *Tetrahedron*, **1975**, *31*, 1381–1384.
21. K. Mori, *Tetrahedron*, **1975**, *31*, 3011–3012.
22. T. Tashiro, K. Mori, unpublished work based on [23] and [24].

23 K. Mori, Y.-B. Seu, *Tetahedron*, **1985**, *41*, 3429–3431.

24 K. Mori, H. Kiyota, *Liebigs Ann. Chem.*, **1992**, 989–992.

25 K. Mori, *Tetrahedron: Asymmetry*, **2006**, *17*, 2133–2142.

26 K. Mori, *Biosci. Biotechnol. Biochem.*, **1996**, *60*, 1925–1932.

27 K. Mori, *Chem. Commun.*, **1997**, 1153–1158.

28 K. Mori, *Chirality*, **1998**, *10*, 578–586.

29 K. Mori, *Eur. J. Org. Chem.*, **1998**, 1479–1489.

Keywords

Brassinosteroids, Brevicomin, Chirality, Enantioselective Synthesis, Gibberellins, Glycinoeclepin A, Juvenile Hormones, Minamata Disease, Pheromones, Phytoalexins, Strigolactones

3
The Current Status of Pesticide Management in China

Yong Zhen Yang

3.1
Introduction

Dramatic progress has been made in China's agriculture over the last 25 years. Today, China is emerging as the first in pesticide production and second in usage. There are about 2,800 manufacturers, 200,000 distributors, and more than 400 million small-scale farmers. More than 600 active ingredients (approximate 1 million tons) and 22,000 products (approximately 1.4 million tons) were registered and produced last year. Pesticide management is critical to meet China's increasing requirements on food quality/safety, environmental safeguard, and international trades for pesticide products, and agricultural production. This chapter introduces the current status and future direction of pesticide management in China.

3.2
The Current Direction of Pesticide R&D in China

As the largest country in pesticide production, China has been moving from a reformulation/manufacture base to a new R&D platform. This R&D platform is still at the early stage of development with the synthesis capability of new compounds at about 30,000 and the screening capability of about 20,000 compounds. More than 20 proprietary compounds (flumorph, phenazino-L-carboxylic acid, etc.) have been discovered in the last 5 years and about 2–3 new products have been introduced into the market each year. Pesticide regulation and trade in accordance with international standards, and consumer health/safety and environmental protection are some of the many challenges facing China. The regulation of pesticide manufacturing is getting more and more restrictive after the new policy issued by the State Development and Reform Committee (SDRC) in 2003, which restricts issuance of the manufacturing permit of "Three High" (high toxic, high pollution, and high energy cost) products. A new regulation issued by SDRC in July 2006 stipulated that the pesticides listed in Rotterdam Convention or Stockholm

Pesticide Chemistry. Crop Protection, Public Health, Environmental Safety
Edited by Hideo Ohkawa, Hisashi Miyagawa, and Philip W. Lee

ISBN: 978-3-527-31663-2

Convention are no longer to be produced in China; for example, the five highly toxic OP insecticides (methamidofos, parathion, methyl parathion, monocrotophos, and phosphamidon) are prohibited from use on fruits and vegetables. They will be forbidden to be produced, distributed, and used on January 1, 2007.

There are four internal focuses in China at present: (1) refinement of regulatory process and requirements, especially on registration and post-registration supervision; (2) strengthening the extension services and support to local farmers (a very big task due to a large number of small-scale farmers); (3) establishing the guidelines or standards for preventing the environment and human health from hazards of pesticides; (4) improving people training and promote public awareness.

On the other hand, the global issues, such as food safety, risk assessment, and harmonization of data requirements greatly impact China on pesticide production, usage, and management. First, food safety is an increasingly important issue for China from both the domestic and trade point of view. MRLs as a potential trade barrier create a big influence on the export of agricultural commodities of China. It has become more and more important to China in recent years due to the increase of the trade disputes caused by pesticide residue in exported fresh agricultural products. Second, the strong international competition of pesticide market and agricultural commodities produce a great influence on China. Third, downturn trends in product development, cost, and emerging technology, global mutual acceptance of regulatory review and harmonization of data requirements, the regulation of GMO, and risk assessment process also impact China deeply.

3.2.1
The Status of Pesticide Production and Usage in China

There are more than 2,800 pesticide manufacturers in China. More than 20 big factories with the capacity of 5,000–10,000 tons every year; about 300 factories can produce technical.

There are more than 600 active ingredients registered and about 22,000 products (or formulations) to be registered up to 2005. The production amount of technical materials was more than 1 million tons last year. The usage of pesticides is about 0.28 million tons in terms of active ingredient (approximately 1.4 million tons of formulated product) in amount and 20 million ha every year. Approximately 30–40% yield loss could be avoided. There are 200,000 distributors and more than 400 million small-scale farmers.

3.2.2
Pesticide Regulation and Management Systems in China

1) Scope of Pesticide Management in China

The scope of pesticide management in China is to regulate a substance or a mixture of substances chemically synthesized or originating from biological and other natural substances and the formulations made from these substances

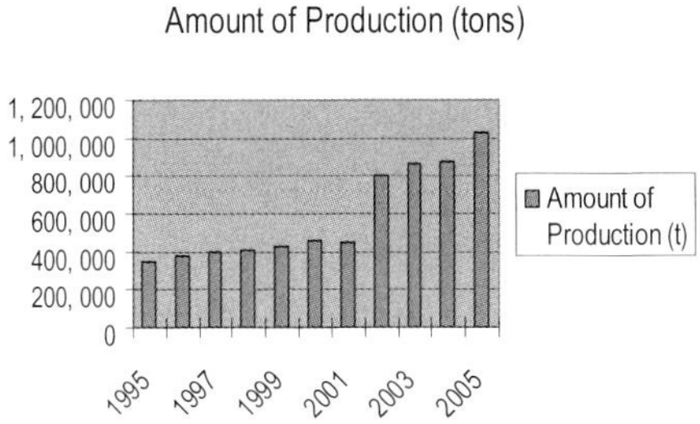

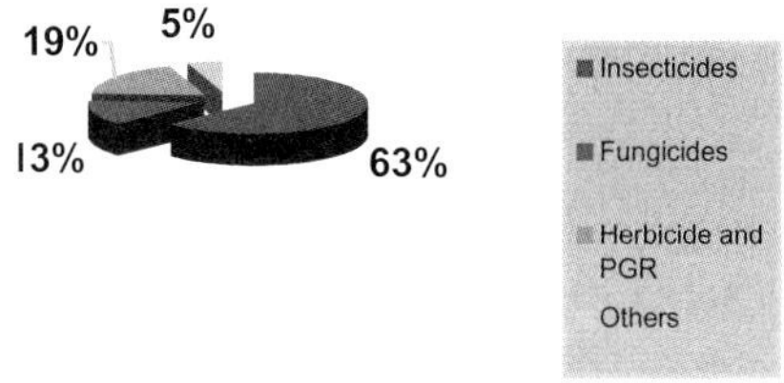

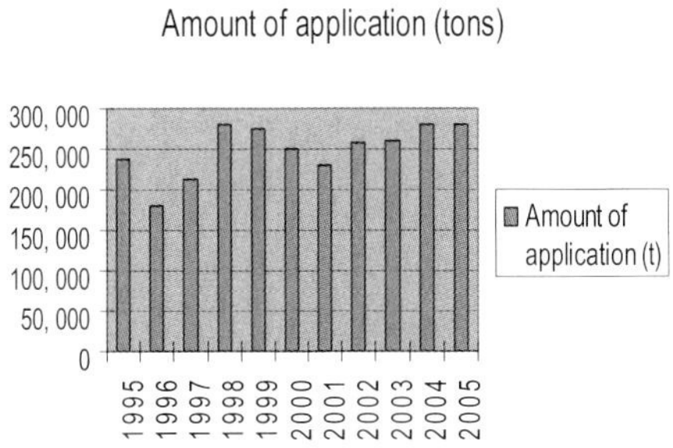

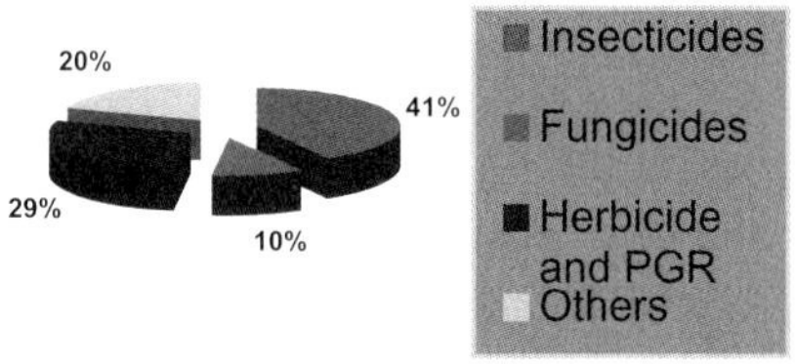

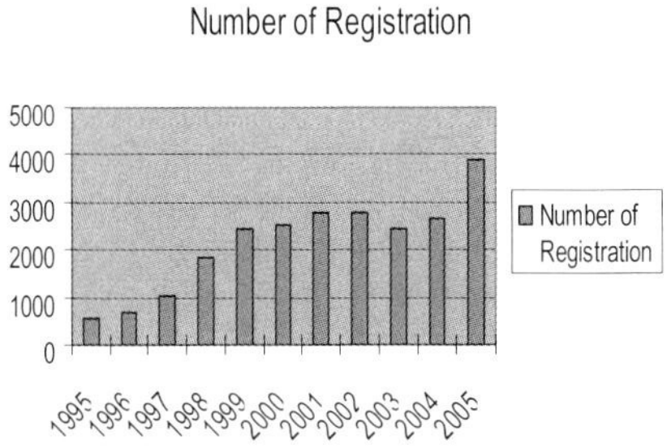

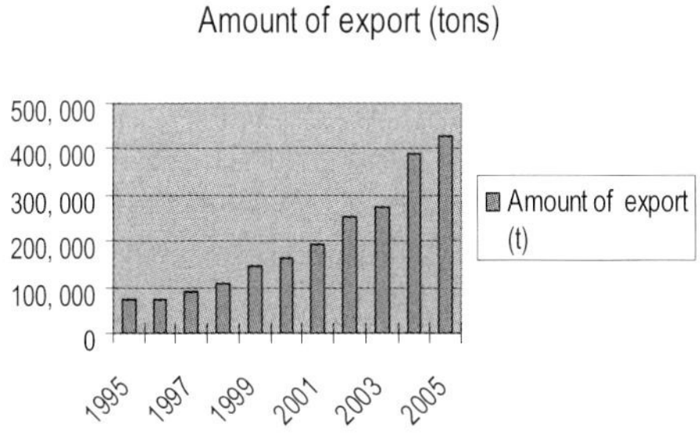

used for (1) preventing, destroying or controlling diseases, pests, weeds, and other harmful organisms detrimental to agriculture, forestry, and public health; (2) regulating the growth of plants and insects, such as insecticides, antiseptics, herbicides, plant growth regulators, rodenticides, hygiene pesticide; (3) GMO products and natural enemies.

There are four main goals of pesticide management of China: (1) controlling crop pests to ensure agricultural production; controlling disease-bearing insects to prevent breakout of epidemics; (2) minimizing the adverse effects to protect human health and environmental safety; (3) building a fair and competitive environment for the pesticides market; (4) promoting import and export of pesticide (international trade).

2) The History of Pesticide Management in China

The development of Chinese pesticide management can be divided into five stages: (1) in the 1950s-1960s, focus on preventing actual poisoning and product quality control; (2) in the 1970s, emphasis on the safe use of pesticides; (3) in the 1980s, establishment of the registration system (1982); (4) in 1990s, production premising system set up; distribution premising system regulatory residue monitoring system, import and export management system, Regulation on Pesticide Administration issued in 1997; (5) 21st century: priority in safety management.

3.3
China's Policies in Pesticide Regulation and Management

The legislation of pesticide management at the state level consists of "Regulation on Pesticide Administration", promulgated by Decree No. 216 of the State Council of the People's Republic of China on May 8, 1997, amended in accordance with the Decision of the State Council on Amending the Regulations on Pesticide Administration on November 29, 2001. "Implementation of Regulation on Pesticide Administration" was issued by MOA in 1999. There are 21 provincial regulations in provincial level. The regulations, guidelines, and technical standards related to the pesticide registration are well-developed. They are mainly listed as follows:

- Guidelines on Pesticide Field Trial
- Guidelines on Pesticide Environment Safety Test and Evaluation

Activates	System	Responsible	Corporation
Trial and registration	Registration	MOA	MOH, SEPA, SDRC
Manufacture	Permit/license standardization	SDRC/SQIQA AQSIQ (SA)	MOA, SDRC
Distribution	Permit/license	MOA, SAIC	
Application	Permit for use of special product	MOA	MOH
Advertisement	Permit	MOA, SAIC	

Legend:
MOA – Ministry of Agriculture
MOH – Ministry of Health
SDRC – State Development and Reform Committee
SEPA – State Environemental Protection Agency
AQSIQ – General Adeministrate of Quality Supervision, Inspection and Quarantine
SA – Standardization Administration
SAIC – State Administration for Industry & Commerce

- Guidelines on Pesticide Toxicity Test
- Pesticide Toxicity Evaluation Procedures
- Guidelines on Pesticide Labeling
- General Standard of Pesticide Packaging
- Standard of Pesticide Toxicity Classification
- Rules of Safety Handling in Pesticide Storage and Transportation
- Maximum Residue Limit of Pesticides on Food, etc.

3.4 The Regulatory Infrastructure within China in the Regulation of Pesticide

At the national level, ICAMA is the administration authority under the Ministry of Agriculture, which was first established on October 7, 1963, and reinstated on September 20, 1978. It is organized into 12 divisions with more than 100 staff. The responsibilities of ICAMA are the pesticide registration and post-registration management, with the cooperation of MOH, SEPA SDRC, SQIQA, AMSDC, SSPAB, and SICAB. At the provincial level, there are 30 ICAs and 1,600 administrative organizations, with more than 20,000 pesticide inspectors. Responsibilities of the provincial ICA include drafting or making regulations and rules on pesticide control within local administrative divisions and organizing to carry them out; assisting ICAMA in preparatory registration; and being in charge of safe and suitable use, supervision, quality control, and residue supervising of the pesticide act.

3.5 Key Administrative Actions on Pesticide Management

Current administrative activities on pesticide management include: (1) restricting the production/uses of highly toxic and persistent pesticides. More than 700 products containing 5 highly toxic organo-phosphorus compounds were cancelled in 2005; (2) encouraging new pesticide development and the introduction of safer pesticides in China, especially the new technology from overseas. The period for patent protection has been changed from 15 years to 20 years and the registration data protection system has been improved with a period of 10 years; (3) improving regulatory infrastructure; (4) harmonizing technical standards on quality control, registration evaluation, and MRL; (5) monitoring of pesticide residue in agricultural commodities, food and feeds through full process control from field to market; (6) carrying out a "pollution free" food action plan and traceable system on pesticide safety. Monitor pesticide residue in raw agricultural commodities and processed food (farm gate to dinner table). The compliance of pesticide residues detected in fresh agricultural commodities (such as vegetable, fruit, and tea) below the Codex MRL, reaches 96–98%.

3.6
Future Direction of Pesticide Regulation in China

Philosophy and emphasis of pesticide management has changed in China. The priority of pesticide management in China is changing from quality control to safety management, also from supervision only to service and guidance also. The main tasks are (1) implementation of legal framework to improve regulations on safety management such as strengthening marketing permits; (2) strengthening legal infrastructure, rigorously enforcing registration rules and regulations, enhancing safety requirements thus strengthening legislative and safety management; (3) improving risk assessment evaluation procedures for more scientific, fair transparence; (4) participating in international activities, especially on the international harmonization of pesticide management system and requirements. After 25 years of efforts, China has made great progress on pesticide management. As a large and diverse country of pesticide production and usage, China still faces many challenges to establish pesticide regulatory system that promotes agricultural development, and adequate protection to the environment, farmers, and consumers. China will continually collaborate with international organizations in the enforcement of international conventions such as PIC, POPs, IPPC, Montreal, FAO Code, and the activities of JMPR, CAC, FAO/WHO specification, GLP, MAD, etc; continually conduct bilateral collaboration in GLP with the USA, in MLHD (Minimum Lethal Herbicide Dose) technology with The Netherlands; in disposal of obsolete pesticide with Germany, in efficacy tests of pesticide with Japan.

The principle of China's pesticide management in the future is to fulfill the following five requirements: (1) requirement of reforming agricultural economical structure; (2) requirement of enhancing pesticide quality/safety and human healthy; (3) requirement of promoting pesticide industry progress; (4) requirement of developing sustainable agriculture; (5) requirement to increase China's competitiveness.

3.7
Conclusion

China has made significant progress in the regulation and administration of pesticide usage. With the increasing demand of food consumption, food quality and safety, and the continued economic growth of this country, China faces new challenges in meeting human health protection and environmental stewardship demands. Harmonization of risk assessment procedures, international MRLs, and the strengthening of China's pesticide research and development platform are some of the high priority short-term objectives. China will play a significant role in the international pesticide regulatory community.

4
Pesticide Residues in Food and International Trade: Regulation and Safety Considerations

Kenneth D. Racke

4.1
Introduction

In their efforts to supply a safe and abundant food supply, the world's farmers must cope with a variety of production challenges. To face threats posed by insect pests, weeds, and fungi during the growing season and post-harvest, a variety of tactics, including the use of pesticides, may be necessary as part of an integrated pest management (IPM) approach. If it is necessary to use pesticides, the potential presence of trace concentrations of pesticide residues in food commodities at harvest and after processing poses a dilemma. Consumers generally would prefer to eat food free of pesticide residues, yet pesticides are often integral components of IPM programs. To resolve this situation, a "food-chain compromise" has been reached in practice to meet the needs of both farmers and consumers. This compromise assumes that pesticides may be used in the production and storage of food, but only under the conditions that (1) no more pesticide is used than is necessary to be effective, and (2) the residues which may remain on food are not harmful to human health. The food-chain compromise is embodied in the definition of good agricultural practice (GAP) recognized by regulatory authorities and industry via the FAO Code of Conduct [1] as "... the officially recommended or nationally authorized uses of pesticides under actual conditions necessary for effective and reliable pest control ... applied in a manner which leaves a residue which is the smallest amount practicable." This chapter will reexamine the food-chain compromise in light of today's increasingly global trade environment, review the regulation and monitoring of residues in food, and summarize recent trends influencing the effective future management of residues in food on a worldwide basis.

Pesticide Chemistry. Crop Protection, Public Health, Environmental Safety
Edited by Hideo Ohkawa, Hisashi Miyagawa, and Philip W. Lee

ISBN: 978-3-527-31663-2

4.2
Globalization of the Food Chain

Farmers and consumers today are increasingly part of a global food chain. Substantial growth in the volume and variety of agricultural commodities traded globally, particularly fruits and vegetables, has occurred since the 1980's. This growth has been fueled by a number of factors including improved personal incomes and increased demand for year-round access to fresh fruits and vegetables, reduced transportation costs and improved handling technology, and international trade agreements [2]. By 2001, fruits and vegetables comprised 17% of the value of world agricultural trade, up from the 11% of 40 years earlier. Table I lists some of the major fruit and vegetable commodities moving in international trade. Significant increases in export volumes occurred between 1989 and 2001 for such commodities as frozen potatoes (11%), orange juice (14%), mangoes (13%), chillies and peppers (7%), bananas (4%), and melons (8%) [2].

Three Northern Hemisphere trading regions, each dominated by one key food exporter and one key food importer, drive worldwide trade in fruits and vegetables. These include the Europe/Africa/Middle East region dominated by the European Union (EU) for both imports and exports, the North American Free Trade Agreement (NAFTA) area dominated by the U.S. for both imports and exports, and the Asia-Pacific area dominated by Japan as the key importer and China as the key exporter [2]. Much of the existing trade occurs between countries within each region, but significant inter-regional trade also exists. In addition, an important flow of trade between regions involves global South-to-North movement of fruits and vegetables due to the counter-cyclical seasons of the two hemispheres. Key Southern Hemisphere exporters include Argentina, Australia, Brazil, Chile, New Zealand, and South Africa. The banana-exporting countries of Colombia, Costa Rica, Cote d'Ivoire, Ecuador, Guatemala, Honduras, and Philippines are also important partners in interregional trade.

The globalization of the food chain means that consumers and the farmers who supply them may reside in different regions separated by great distances and political boundaries. For any given meal, a fresh banana or apple or mango may have been grown half a world away. This brings up the question of how pesticide residues in food and the food-chain compromise are regulated at the international level.

Table I. Shares of total fruit and vegetable value moving in world trade (1999 to 2001) [2].

Commodity	Share (%)	Commodity	Share (%)
Bananas	6.3	Grapes	3.5
Citrus Fruits	5.6	Nuts	2.3
Potatoes	5.0	Cucurbits	2.3
Pome Fruits	4.9	Peppers/Chillies	2.3
Tomatoes	4.3	Stone Fruits	1.2

4.3
Regulation of Pesticide Residues in Food

The primary regulatory standard employed to control pesticide residues in food is the maximum residue limit or MRL. The MRL has been defined as "the maximum concentration of pesticide residue that is legally permitted or recognized as acceptable on a food, agricultural commodity, or animal feed" [3]. The MRL is intended primarily as a check that use of pesticide is occurring according to authorized labels and GAP. Detection of residues at or below the MRL implies that label directions and GAP have been properly followed. MRLs for pesticides are established by national authorities or advisory bodies primarily based on field residue trials that provide insight as to the levels of residues that may occur under the worst-case scenario (i.e., maximum permitted application rate and number of applications, minimum permitted pre-harvest interval). MRLs are not set on the basis of toxicology data, but once proposed based on GAP they must be evaluated for safety. This is generally accomplished through a risk assessment process that compares dietary intakes estimated from expected residue concentrations in food(s) consumed with the relevant health-related regulatory endpoints, the acceptable daily intake (ADI) and the acute reference dose (ARfD). A detailed discussion of the basis for MRL establishment and dietary intake assessment of pesticide residues is beyond the scope of this chapter, but several excellent overviews are available [4–6].

4.3.1
The World Food Code and Codex MRLs

The Codex system was designed to promulgate a set of voluntary, globally relevant standards related to food commodities which may move in international trade. More than 170 countries now ascribe as Codex members. Food standards elaborated by Codex include harmonized MRLs for pesticide residues in food. These standards are developed through activities coordinated by the Codex Committee on Pesticide Residues (CCPR). The scientific evaluations upon which Codex MRLs are based result from the FAO/WHO Joint Meeting on Pesticide Residues (JMPR), active since 1963. As part of the JMPR, a WHO panel reviews pesticide toxicology data to estimate the ADI and the ARfD. A FAO panel reviews pesticide GAP and residue chemistry data to estimate MRLs. Following adoption of the JMPR recommendation by CCPR, the Codex Alimentarius Commission (CAC) formally promulgates the MRLs as Codex standards.

The importance of Codex standards is that they offer a globally harmonized, unbiased and authoritative source of MRLs that take into account the various national GAP for a particular pesticide-commodity as well as available residue trial data. The authoritative nature of Codex MRLs has, in fact, been recognized and agreed in principle (if not always in practice) by the majority of important trading countries. The World Trade Organization (WTO), through a 1995 agreement on the Application of Sanitary and Phytosanitary Measures (SPS), identified Codex

MRLs as the official reference for food safety issues which affect international food trade and the basis for resolution of trade disputes. Thus, it would appear that with respect to management of residues and MRL issues associated with global trade, the mechanism for preempting potential national differences in GAP is neatly in place. Indeed, Codex MRLs are quite useful as reference points for many countries which do not establish their own national MRLs (e.g., Algeria, Chile, Colombia, Pakistan, Philippines) or may defer to Codex MRLs when they are available (e.g., Brazil, China, India, Israel, Korea).

Two primary factors, however, have served to retard the universal implementation of Codex MRLs for worldwide regulation of pesticide residues on food moving in international trade. The first is that Codex MRLs have not been established for all important pesticides and crops. Although more than 700 pesticide active ingredients are authorized in one country or another on a worldwide basis, as of 2006, Codex MRLs had only been established for around 180 pesticides. The primary causes for this incomplete set of Codex MRLs include the historically slow nature of the Codex standard elaboration process (e.g., 3–6 years), limited JMPR resources to complete evaluations and failure of members to submit sufficient field residue trials at GAP for some crops. Thus, farmers may use many pesticides on crops for which no Codex MRLs are available. The second factor hindering the effective regulation of pesticide residues in world food trade by Codex MRLs is the incomplete recognition of their applicability for trade by several major food-importing regions including the EU, Japan, and the U.S. In these regions, legislation mandates the development of a specific set of national or regional pesticide MRLs based primarily on locally approved GAP. Although Codex MRLs may be considered in the development of such MRL systems, in practice the MRL promulgation process strongly favors local GAP as the basis for standard-setting. As might be expected, this approach leads to national/regional MRLs which may differ in some cases from Codex MRLs (Table II).

Thus, the promise of Codex MRLs offering a single, harmonized listing of globally applicable MRLs to facilitate world trade has not yet been fully realized due to internal problems with the Codex process and also the divergent interests

Table II. Example comparison of Codex and national/regional MRLs (mg/kg) for grapes.

Pesticide	Codex	EU	Japan	U.S.
Captan	None	3	5	50
Chlorpyrifos	0.5	0.5	1	0.5
Dimethoate	1	0.02	1	1
Endosulfan	1	0.5	1	2
Fludioxonil	None	None	5	1
Myclobutanil	1	1	2	1
Spinosad	0.5	None	0.5	0.5
Tebuconazole	2	None	2	5

and MRL lists of several influential food-importing regions. Let's examine the underlying policies and characteristics of three increasingly influential MRL systems.

4.3.2
U.S. Tolerances

U.S. MRLs, referred to as "tolerances", are established by the U.S. Environmental Protection Agency (EPA) under auspices of the Federal Food, Drug, and Cosmetic Act (FFDCA). Tolerances are established on raw agricultural commodities (RAC) and also on processed commodities (i.e., food additive tolerance) if the residue level in the process fraction will be greater than that for the RAC. Tolerances for more than 300 active ingredients have been established by EPA. Enforcement of U.S. tolerances is the responsibility of the U.S. Food and Drug Administration (FDA). In the absence of a specific tolerance, residues must be below detectable levels. Based on modifications to FFDCA mandated by the Food Quality Protection Act (FQPA) of 1996, several new elements were introduced to the EPA tolerance process. These include the need to consider the special sensitivity of infants and children, the potential exposure via multiple routes of exposure (i.e., aggregate exposure from dietary and non-dietary sources), and the potential for exposure to other pesticides and chemicals with a common mechanism of toxicity (i.e., cumulative exposure). Under FQPA, the EPA was also required to complete a reevaluation of all existing tolerances during a 10-year period. Domestically established MRLs apply also to imported commodities, but there is an established (if somewhat slow) process for evaluation of residue data from other countries in support of import tolerances.

The importance of U.S. tolerances stems from both the stringency of the scientific evaluation process upon which they are based and the important role of the U.S. in international food trade. Some countries actually defer to U.S. tolerances in lieu of their own legislation for export purposes (e.g., Costa Rica, Mexico). It is expected that U.S. tolerances will be highly influential in future development of MRL policy within the NAFTA countries, particularly as regional harmonization efforts may some day lead to a system of NAFTA MRLs. The FQPA-mandated tolerance reassessment process has resulted in many changes in U.S. tolerances, particularly the loss or reduction of tolerances for some older products, and these changes have the potential to impact use and food export practices in U.S. trading partners.

4.3.3
Japan MRLs

Japan MRLs are established by the Ministry of Health, Labor, and Welfare (MHLW) in consultation with the Food Safety Committee (FSC) under auspices of the Food Sanitation Law. Until recently, specific with-holding limits (WHLs) were set under the Agricultural Chemical Control Law to govern residue limits associated with approved GAP, but these were applicable only for domestically grown agricultural commodities. For some pesticides, MRLs were also established by MHLW to govern the residue levels on both domestic and imported food commodities. This incomplete listing of MRLs meant that, for many crop/pesticide combinations, no specific regulation of pesticide residues was practiced. This situation become unpopular due to several food safety controversies (e.g., BSE), and thus in 2003 amendments to the Food Sanitation Law dramatically revised the way Japan regulates pesticide residues in food. In place of the old system, Japan activated on May 29, 2006, a so-called "positive list" system of MRLs aimed at prohibiting the distribution of foods that contain agricultural chemicals unless MRLs for them are established under the Food Sanitation Law. The Japan positive list legally applicable as of May 29, 2006, includes two sections. The first section includes the permanent MRLs for around 230 pesticides previously established by MHLW under the old system. The second section includes newly established, provisional MRLs for some 758 pesticides. These provisional MRLs were set based upon either (1) national registration WHLs, (2) Codex MRLs, or (3) the mean value of MRLs of key OECD trading partners (Australia, Canada, EU, New Zealand, U.S.). Provisional MRLs will be evaluated with respect to dietary intake and revised/adopted as permanent MRLs during the next several years. For pesticide/crop combinations not found on either section of the positive list, or on a short list of exempted active ingredients, a default level of 0.01 mg/kg will apply. An import MRL process to revise or add MRLs to the Japan system based on approved GAP in other countries is available, but timeliness is uncertain in light of requirements for a toxicology and ADI reevaluation by the FSC.

The importance of Japan MRLs stems from the highly influential role of Japan as a major food importer from neighboring countries within Asia as well as the broader Pacific Rim and beyond. The recent move to adopt a comprehensive set of "positive list" MRLs will greatly increase the importance of Japan MRLs for world trade, and exporting nations whose GAP was not specifically considered in development of the positive list (e.g., China, Korea, Thailand) may be most impacted. Another factor which increases the impact of Japan MRLs is the strict system of compliance monitoring and enforcement which is implemented by the MHLW and local government. In addition to random monitoring, targeted and mandatory monitoring of 50 or 100% of certain commodities may be required following one or two MRL violations, respectively. Continued violations may result in targeted import bans for the problem commodities.

4.3.4
EU MRLs

At present, complete harmonization of MRLs across the European Union (EU) member states has not yet been accomplished. Thus, most regulation of pesticide residues in food is based on MRLs established by national legislation in each member state. These un-harmonized MRLs may reflect different GAP and thus may differ between members. A program for creation of a harmonized set of EU

MRLs applicable across all member states has been making slow but significant progress since the early 1990's. Harmonized EU MRLs are established by the European Commission. Harmonized EU MRLs have so far been established for around 150 pesticide active ingredients. New legislation was approved by the European Parliament during 2005 which established an accelerated program for achieving a single, harmonized set of EU-wide MRLs for all crop/pesticide combinations. This harmonized listing will be based on (1) existing EU MRLs, (2) MRLs currently in force within the 25 members' states, and (3) Codex MRLs. Promulgation of a complete set of EU MRLs may take several years to occur based on the complexities of selecting the most appropriate value to reflect differences in GAP among member states and requirements for safety determination via dietary intake assessment. The new European Food Safety Agency (EFSA) is expected to play a major role in implementation of the accelerated EU MRL process. A process for establishment of an EU import MRL based on overseas GAP and data will also be available in the future.

EU MRLs are important in light of both the major trading relationships which exist among EU member states and also with countries in other regions, including those in Africa and the Middle East. A significant amount of trade also occurs with the U.S., several Latin American countries, and such Pacific area countries as New Zealand. It should be mentioned that the process for establishment of EU MRLs is a precautionary one that has been characterized by a tendency to set lower MRLs than Codex, the U.S., and other non-EU countries. The new legislation will also implement a default MRL of 0.01 mg/kg for all pesticide/crop combinations not covered by a harmonized EU MRL. A major issue to be resolved will be the possibility of establishing or maintaining EU MRLs for the hundreds of pesticides withdrawn from the EU Review process under Directive 91/414 due either to minor commercial interests or risk evaluation concerns.

4.4 Disharmonized MRLs, Monitoring, and Consumer Safety

The existing world situation of partially harmonized regulation of pesticide residues in food, with influential MRL systems including those of Codex, the EU, Japan, and the U.S., has led to negative consequences for growers and consumers alike. First, mismatches between GAP and applicable MRLs of food-exporting and food-importing regions may lead to the creation of trade barriers and irritants [7]. Farmers in one country may not be able to employ authorized GAP for certain crops and pesticides in their own country because of such discrepancies due to fears of or actual import violations. Such fears may be theoretical because in many cases the actual residues present in food moving in international trade are much lower than established MRLs (e.g., dissipation of residues during processing, storage, and transport; application of reduced rates or fewer sprays than the maximum allowed). Such trade-related concerns are often high for minor crops, which may lack specific data or grouping with major crops for purposes of MRL establishment [8]. A second consequence of MRL disharmony is the retarded adoption by farmers of many of the newer, reduced risk pesticides which may take several years to achieve worldwide approvals and all required MRLs. Although the evaluation policies of key organizations such as U.S. EPA and Codex have accelerated the introduction of new pesticides with more favorable human health and environmental safety profiles, farmers in food-exporting countries may be forced to continue to use older pesticides while they await establishment of all applicable MRLs. Third, there may be significant economic impacts which may result from disharmonized MRL standards among trading partners. A World Bank case study of divergent banana MRLs indicated that a 1% increase in regulatory stringency for one key pesticide could decrease world banana imports by 1.6%. If the lowest existing national MRL rather than the higher Codex MRL had to be observed by all banana growers, who might not know the destination of their harvest, an estimated 5.5 billion USD in lost exports would be annually predicted [9]. Finally, discordant MRLs and the trade violations they may yield have spawned sensational and inaccurate publicity regarding pesticide residues in food and decreased consumer confidence.

Actual pesticide monitoring programs from key food-importing countries indicate that in many instances no pesticide residues are detected and in the vast majority of instances where detectable residues of pesticides occur, these levels are well below established MRLs. For example, monitoring of domestic and imported foods in the U.S. during 2003 by the U.S. Department of Agriculture Pesticide Data Program found that 43% of fresh fruits and vegetables had detectable residues. In 0.3% of the samples, U.S. tolerances were exceeded and, in 1.6% of the samples, residues were detected for which no U.S. tolerance existed [10]. Similarly, monitoring of foods in the UK by the Pesticide Residues Committee during 2004 found that 31% of food commodities had detectable residues and the established MRL was exceeded in 1% of the samples [11]. Compliance monitoring in Japan has revealed similar levels of detection and MRL violation rates < 1%, although implementation of the positive list system of comprehensive MRLs has been predicted to increase the violation rate by 5- to 6-fold [12].

What about the dietary intake and consumer safety relevance of detected residues and the low incidence of MRL violations? First, for residues at or below the MRL it should be mentioned that dietary intake assessments are conducted in setting the MRL to ensure that cumulative residues which may be present in all food sources are below toxicological endpoints. The endpoints employed for the dietary risk assessments, such as the ADI, are conservative in nature and generally established at levels 100-fold lower than those found to cause no adverse effects in test animals. Thus, human exposures many times the level of the MRL would be required to reach even those levels which may have minimal biological impacts. Although the MRL is not a health-based or toxicological standard, the favorable comparison of estimated food intake containing residues at the MRL level gives confidence that food with residues at or below the MRL poses no human health concern. Second, for residues present above the established MRL, it must be kept in mind that this is only an indication that either GAP has not been followed or, for food imports, that GAP in the country of origin may differ from that of the importing country. Thus, an MRL exceedence should be considered as a trade violation and not as a human safety risk. In fact, the vast majority of MRL violations constitute a negligible level of exposure and health risk despite news media headlines to the contrary.

4.5 Recent Trends

4.5.1 Improvements in the Codex Sytem

There have been several changes in practice which have accelerated adoption of Codex MRLs for newer, reduced risk pesticides. New pesticides which qualify as reduced risk products based on human health considerations are now accorded a high priority in the JMPR schedule. An accelerated process adopted by CCPR in 2006 will provide the opportunity, in cases where there are no dietary intake concerns, for Codex MRLs to be established less than one year after the JMPR recommendation. The adoption of work-sharing practices to make use of existing or ongoing technical evaluations by major regulatory authorities promises to alleviate some of the delays in JMPR evaluations. The recent adoption of Codex MRLs for spices based on monitoring data and of revisions to the Codex crop classification to accommodate minor crops are also promising developments reflecting a renewed will and flexibility for establishing a full set of Codex MRLs for all major pesticides. The Codex system still offers the only promising option for a harmonized approach to global MRL regulation.

4.5.2 Regionalization of MRL Policies

The increased prominence and influence of EU MRLs, Japan MRLs, and U.S. tolerances has been associated with significant changes in food safety legislation and MRL processes of the past decade. In light of the dominant position in regional and world trade played by these three regions, it is safe to say that these MRL systems now rival Codex with respect to global significance. Much of the ongoing effort related to pesticide food regulation within the EU is directed toward harmonization of MRLs among all 25 member states. Within the NAFTA countries of Canada, Mexico, and the U.S., harmonization efforts are in progress concerning field residue trials, labeling, and MRLs. It remains to be seen whether the consolidation of influential, regional MRL systems represents a step on the road to a broader global harmonization or whether such regionalization signals a move toward a persistent Balkanization of MRL policy. The future enjoyment of worldwide free trade in agricultural products may depend on the ability of the “big three” regional MRL systems to find creative ways to move toward a harmonized approach, and also the foresight of the rapidly evolving regulatory systems of such influential nations as Brazil and China (new host country for the CCPR).

4.5.3
Growth of Private Standards

A complication and threat to implementation of harmonized MRLs in support of world trade is represented by the increase in private pesticide residue standards and policies. A traditional example would involve organic agriculture, for which certifications generally specify not only a lack of detectable synthetic pesticides at harvest, but a complete lack of their use during the production process. A more recent example would be the actions by food retailers in the UK, Germany, and some other EU member states to establish specific policies on pesticide use and residues at harvest. For example, some of these programs set more stringent limits for pesticide residues than are reflected in relevant international or national standards (e.g., 1/2 or 1/3 of the MRL), or prohibit the presence of specific pesticide residues or the use of certain pesticides in the production process (i.e., even if residues are not present at harvest) [13]. Such approaches are not based on scientific principles but represent instead marketing campaigns designed to play to the unfounded fears of consumers. Implementation of such private standards may undermine consumer confidence and usurp the rightful role of the MRL, a science-based trade standard.

4.5.4
Communication of MRL Information

It had often been difficult for farmers and exporters to readily locate authoritative and updated listings of Codex and regional MRLs. Fortunately, increased use of the internet as a worldwide communication tool has spurred better information sharing practices by the major MRL-setting organizations. For example, web sites are now readily available which authoritatively list Codex MRLs, U.S. tolerances, EU MRLs, and Japan MRLs among others (Table III). Along with these official listings, a variety of unofficial, multi-national MRL databases have also arisen to meet perceived needs of specific stakeholders. These include large, multi-commodity public databases such as the one operated by the U.S. Department of Agriculture Foreign Agriculture Service and private, subscription-based systems such as Homologa. Commodity-focused MRL databases are also increasingly common (e.g., for U.S. fruit exports, for Australia grape/wine exports), but the primary challenge for such secondary sources of information will be maintenance and accuracy. Overall, there could be significant benefits from cooperation of the owners of secondary databases in pooling resources to develop a unified database of MRLs and avoid proliferation of disharmonized MRL databases. It must be remembered, however, that all private or multi-country MRL databases contain secondary, derived information and the authoritative listings published by Codex and regulatory authorities must remain the primary reference points.

Table III. Web-based databases and listings of pesticide MRL information.

Scope	Sponsor
Codex MRLs	Codex Alimentarius Commission
http://www.codexalimentarius.net (Codex home page) http://www.codexalimentarius.net/mrls/pestdes/jsp/pest_q-e.jsp (MRL database)	
EU MRLs	European Commission
http://europa.eu/index_en.htm (European Commission home page) http://ec.europa.eu/food/plant/protection/pesticides/index_en.htm (MRL listing)	
Japan MRLs	Ministry of Health, Labor, and Welfare
http://www.mhlw.go.jp/english/index.html (MHLW home page) http://www.mhlw.go.jp/english/topics/foodsafety/positivelist060228/index.html	
U.S. Tolerances	Environmental Protection Agency
http://www.epa.gov/pesticides/regulating/tolerances.htm (EPA residues home page) http://www.access.gpo.gov/nara/cfr/waisidx_05/40cfr180_05.html (2005 listing)	
International MRL Database	U.S. Department of Agriculture
http://www.mrldatabase.com	
Homologa MRL Database*	Agrobase-Logiram
http://www.homologa.com	

* Subscription required

4.5.5
Adoption of Practices to Preempt or Mitigate Residue Issues

In some countries, recommended practices for minimizing residue and MRL concerns in exported foods have been developed for farmers and exporters. For example, it has been the practice of the Chilean Exporters Association jointly with the University of Chile to develop an information system for growers based on supervised trials to determine "withholding periods" (i.e., preharvest intervals) to meet the MRL standards recognized by the authorities in major export markets [14]. Likewise, the Australia Pesticides and Veterinary Medicines Authority has encouraged data generation by the registrants for the purposes of establishing an advisory "export interval", which is the minimum time between pesticide application and harvesting of the crop commodity for export to fall within the MRLs of trading partners [15]. Such measures will likely be required for the foreseeable future to deal with MRL disharmony among key export destinations. It will be important for the agrochemical industry and other stakeholders to cooperate in the development of residue data supporting such practices.

4.6 Conclusion

The food-chain compromise practiced by farmers and consumers assumes that pesticides are used according to good agricultural practice (GAP) in a manner which minimizes residues in harvested food and does not adversely impact human health. The maximum residue limit (MRL) is a regulatory standard that reflects GAP and allows control of pesticide use and residues in food. Increased international trade of fruits and vegetables has increased the complexity of the food chain compromise. The world food code as promulgated by Codex has supported development of a system of internationally harmonized MRLs and represents the best hope for ensuring fair trade and consumer protection on a worldwide basis. The increased prominence and influence of EU MRLs, Japan MRLs, and U.S. tolerances associated with recent changes in food safety legislation has created a regionalized approach to regulation of trade. Continued disharmony of MRLs between major food-importing regions may result in trade barriers and irritants, retarded adoption of new, reduced risk pesticides, and decreased consumer confidence. To address this latter development, some food associations and retailers have adopted private standards and food policies which may undermine science-based approaches. Accurate food safety information and effective risk communication are required to prevent this from occurring. Creative approaches must be adopted to help develop a more harmonized international approach toward regulation of pesticide residues in food if the benefits of global free trade are to be realized. Encouraging developments include recent improvements in the Codex system which have accelerated MRL promulgation, cooperative evaluation approaches being pursued by several of the OECD countries, availability to farmers of practical recommendations for proactively managing residues in exported commodities, and more accurate communication of MRL standards and pesticide residue information.

4.7 Acknowledgments

The author is indebted to stimulating ideas on this topic exchanged by members of the IUPAC Advisory Committee on Crop Protection Chemistry, and in particular for the thought-provoking insights of Denis Hamilton and Roberto Gonzalez. Special thanks are also in order for Graham Roberts, who provided very useful peer review comments.

4.8
References

1 FAO International Code of Conduct on the Distribution and Use of Pesticides, Rome, Italy, **2002**.
2 S. W. Huang, *Global Trade Patterns in Fruits and Vegetables*, U.S. Department of Agriculture, Agriculture and Trade Report No. WRS-04-06, Washington, DC, USA, **2004**.
3 P. Holland, *Pure Appl. Chem.*, **1996**, *68*, 1167–1193.
4 D. J. Hamilton, P. T. Holland, B. Ohlin, W. J. Murray, A. Ambrus, G. C. deBaptista, J. Kovacicova, *Pure Appl. Chem.*, **1997**, *69*, 1373–1410.
5 D. Hamilton, A. Ambrus, R. Dieterle, A. Felsot, C. Harris, B. Petersen, K. Racke, S.-S. Wong, *et al.*, *Pest Management Sci.*, **2004**, *60*, 311–339.
6 D. Hamilton, S. Crossley, *Pesticide Residues in Food and Drinking Water: Human Exposure and Risks*, John Wiley and Sons, New York, USA, **2004**.
7 W. L. Chen, Agrolinks, *CropLife Asia*, **2003**, December, 12–14.
8 R. E. Holm, J. J. Baron, D. L. Kunkel, *Proc. Brit. Crop Prot. Council*, **2005**, 31–40.
9 J. S. Wilson, T. Otsuki, *Food Policy*, **2004**, *29*, 131–146.
10 Pesticide Data Program, Annual Summary Calendar Year 2003, U.S. Dept of Agriculture, Washington, DC, USA, **2005**.
11 Annual Report of the Pesticide Residue Committee 2004, UK Pesticide Residue Committee.
12 M. Uno, *Food Sanit. Res.*, **2006**, *56*.
13 Joint Food-Chain Briefing on Non-Regulatory Residues Targets for Plant Protection Products (Pesticides), Freshfel, June **2006**.
14 R. H. Gonzalez in *Pesticide Chemistry and Bioscience*: The Food-Environment Challenge (G. T. Brooks, T. R. Roberts, Eds.), Royal Society of Chemistry, London, UK, **1998**, 386–401.
15 Australia Pesticides and Veterinary Medicines Authority, in *Manual of Requirements and Guidelines*, **2006**, Part 5B, Edition 3.

Keywords

Pesticide Residues, International Trade, Food Safety, Dietary Intake, Maximum Residue Limits, Codex

5
Hunger and Malnutrition Amidst Plenty: What Must be Done?

Shivaji Pandey, Prabhu Pingali

5.1
Introduction

While the world has been successful in producing enough food to meet the additional demand created by rising incomes and population growth, more people were hungry in 2000–2002 (852 million) than in 1990–1992 (800 million). Poverty and hunger are tightly inter-linked and their alleviation seems to rest not only in increasing availability of food but also in enhancing the financial capacity of the poor to purchase it. Today, several countries are either unable to produce enough food to meet their demand or lack the financial resources to buy food or both. The world community set for itself the noble goal of reducing hunger and extreme poverty in developing countries by half by 2015, considering 1990 as baseline. Since we are slightly more than halfway through, it is useful to examine the progress being made.

Based on the analysis of FAO (2005), the results appear to be mixed. On the positive side, in the developing countries there has been an increase in food production, productivity, and income and decrease in population growth, on the average, and some countries and regions have made greater progress toward the goal than expected. On the negative side, however, several countries are lagging behind. The goal of halving the number of hungry from 800 million in 1990–1992 to 400 million in 2015 appears difficult to achieve. With the projected increase in the population of two billion people between 1990 and 2015, the number of hungry in 2015 would be at least 600 million. Only some countries in East Asia and Latin America are likely to meet their targets. Sub-Saharan Africa, for example, would need about 100 more years to reduce its malnutrition from 33% in 1990 to the stated goal of 18% in 2015, at the current rate of progress. There appears to be greater progress in addressing the poverty problem. Most regions of the developing countries are expected to meet their goal of halving poverty (defined as the income of < 1 USD a day) between 1990 (29% of the population) and 2015 (12% of the population). Unfortunately, the absolute number of poor is not likely to be halved. It will decline from 1.27 billion in 1990 to 0.75 billion in

Pesticide Chemistry. Crop Protection, Public Health, Environmental Safety
Edited by Hideo Ohkawa, Hisashi Miyagawa, and Philip W. Lee

ISBN: 978-3-527-31663-2

2015. In sub-Saharan Africa (SSA) both the % of poor and their actual numbers may in fact increase.

Since 80% of the world's poor live in rural areas and largely depend on agriculture for their livelihoods, it is inconceivable to reduce hunger and malnutrition, without agricultural development. Poor are hungry and hungry will remain poor because of the impact of hunger on human health and productivity. Access to improved and appropriate technologies by the poor, well-planned and executed trade policies, and investment in agricultural research and development will help reduce the problem. Unfortunately, both the developing country governments and the international community have not fulfilled their promises of supporting and investing in agricultural development. Private sector and international investors are also reluctant to invest in many poor countries due to their non-democratic governments, non-transparent governance, political instability, and non-conducive social environment (lack of peace and harmony). While not all issues can be addressed at the same time, addressing one or more will be a good step toward meeting the set goal and will help create an appropriate environment for the implementation of other options in the future.

5.2
The Current Situation

The number of malnourished in the world during 2000/2002 was 852 million, about 815 million of them in the developing countries. About 61% of the malnourished were in Asia, 24% in SSA, and the rest in Latin America and elsewhere. About 33% of the population in SSA, 16% in Asia and Pacific, 10% in Latin America and the Caribbean, and 10% in Near East and North Africa were malnourished in 2000/02 [1].

Hunger and malnutrition are responsible for 6 million child deaths each year resulting not from starvation but from diseases. They affect people's behaviour, weaken their bodies and immune systems contributing to 5 million new infections from HIV/AIDS, 8 million new infections from tuberculosis, and 300 million new infections from malaria each year. These diseases together kill some 6 million people each year as well. Hunger does not just affect people; it affects their environment as well. In want of food, people plough forests and marginal lands causing further degradation of natural resources. The annual loss of 9.4 million ha of forests is in major part also caused by hunger and poverty.

If that were not enough, hunger and poverty are linked. Poor are hungry and hungry will always be poor: Hunger adversely affects health, labour productivity and investment choices, and perpetuates poverty. FAO has estimated that hunger costs developing countries a loss of about 500 billion USD a year (SOFI, 2004). Investment in hunger reduction is neither charity nor welfare; it's an investment that generates high return!

The Millennium Development Goal (MDG) Number 1, that used 1990/92 as its base line, wished for halving global hunger and poverty by 2015. In 1990/92,

20% (824 million) of the developing country population was undernourished. In 2000/02, 17% (815 million) of it was undernourished and the proportion is poised to go down to 11% by 2015. However, regional data paint a more telling picture. The proportion of malnourished in Asia went down from 16% in 1990/92 to 11% in 2000/02 and is projected to be 8% in 2015. The corresponding percentages for South Asia are 26%, 22%, and 12% and for Latin America 13%, 10%, and 6.5%. While progress in Asia is significant, it is worth noting that 61% of the malnourished of the developing countries are in Asia, most of them in India. Other regions of developing countries are not likely to achieve this goal without significantly greater investments and efforts. In SSA, for example, about 35% of its population was undernourished in 1990/92, which went down to only 33% in 2002/02 and is expected to be 23% in 2015 unless the pace of efforts to achieve it is accelerated. In fact, the number of malnourished in SSA increased from 92 million in 1969/71 to 204 million in 2000/02 [2].

Reducing the number of malnourished from about 824 million in 1990/92 to 412 million by 2015 must take into consideration an additional issue. World population would have increased by another 2 billion people during 1990 and 2015. So, even if we were able to reduce the number of malnourished to about 400 million, about 600 million people in developing countries will still suffer from hunger in 2015. To reach the goal of reducing the number of hungry and malnourished to about 400 million in 2015, the proportion of undernourished must be reduced not by half but by two-thirds.

People are hungry in spite of the fact that there is more food produced now than ever before. So, chronic hunger is not a question of absolute shortages or too low production. It is the result of the inability to obtain food through work or income. During the second half of the 20th century, per capita food production increased by 25% as the global population doubled. The productivity of agricultural land has doubled and it now takes two times less water to produce a kilo of wheat than 40 years ago. About 37% of the people living in the developing countries were malnourished (consumed < 2200 kcal a day) in 1969/71 compared to 17% in 2000/02. But, once again, the progress is not uniform. While per capita food production in Asia nearly doubled between 1970 and 2004, it declined by 20% in SSA. Part of the explanation for this disparity lies in low agricultural productivity in the SSA. For example, cereals today yield about 1 ton of grain a hectare there; that was the cereal yield in England 2000 years ago when Christ walked the earth.

Unfortunately, countries that are unable to meet their food needs today are predicted to be able to do so even less in the future. In the early 1960's, developing countries had an agricultural trade surplus of over 6 billion USD; by 2030 they are expected to have an agricultural trade deficit of about 30 billion USD, the deficit being the highest in the countries considered "least developed". This is in part due to lower competitiveness of developing country agriculture, due to apathy of their governments toward investing in agriculture, to cheaper imports from countries subsidizing agriculture, to rising populations, and to rising urbanization.

5.3 What Must be Done?

About 2.6 billion people make their living in rural areas where 75% of the world's hungry live off agriculture. So, agricultural growth is the key to alleviating hunger and poverty in developing countries. And, without reduction in hunger, other MDGs will be difficult to achieve. A World Bank study in India found that growth in agriculture had much greater impact on reducing poverty than urban or industrial growth. In India, prevalence of hunger decreased during 1980s when the agriculture sector grew and the national economy was stagnant. But progress in reducing hunger stalled during the second half of the 1990s when agricultural growth stagnated and national GDP took off. Similarly for hunger reduction, a study conducted by FAO showed that only those countries reduced hunger where the agriculture sector grew.

Following are some specific actions which must be taken if we wish to alleviate hunger and poverty:

5.3.1 National Commitment and Good Governance

Perhaps the most important missing action to alleviate hunger and poverty is for the involved governments to commit themselves to doing it. Poorest countries have not put alleviation of hunger and poverty on their national agenda. The countries with highest indices of hunger and poverty have been investing the least in agriculture. This reduces the ability of their farmers to produce and compete and directly affects food security. Aware of this fact, African Heads of State adopted the Maputo Declaration on Agriculture and Food Security in July 2003, to allocate 10% of their national budgets to agriculture within 5 years. None is even close.

Poverty and hunger contribute to conflicts and insecurity and divert national resources away from developmental activities. Peace and stability are necessary for reducing hunger and poverty: Conflicts and wars affect rural areas and agriculture first, exacerbating the problem of food security. And food insecurity is known to cause wars and conflicts.

Corruption is highest in the poorest and most food-insecure countries. A few years ago, a Prime Minister of one of these countries estimated that 78% of the aid money received in the country did not reach the people it was meant to help. Lack of transparency and accountability discourages external assistance and private investments. The war on hunger and poverty will not be won until the affected governments themselves make it their own war and fight it honestly and bravely with assistance of their allies. Others may help but will not win the war for them.

5.3.2
Investment in Rural Infrastructure

This is critical to improving access of rural poor and small farmers to inputs and markets. Roads, energy, storage, and markets all are needed to facilitate farmer participation in national development. Road density in the SSA today is $1/6^{th}$ of the road density in India in 1950 and needs urgent intervention. Improved rural infrastructure also increases income by contributing to livelihoods' diversification on the farm and increasing non-farm income of rural dwellers, which is significant in most areas of the world (42% in Africa, 32% in Asia, and 40% in Latin America). With more income-generating opportunities in rural areas, farmers will have less need to migrate to urban centers. Several countries in Asia and Latin America have benefited from public-private collaboration in improving their rural infrastructure.

5.3.3
Improving Irrigation Infrastructure

Irrigated agriculture obtains yields three times higher than rainfed agriculture. In Asia and Latin America, much of the irrigation infrastructure developed in the 1960s and 1970s is now inefficient. In SSA, only 7% of the total cultivated land is irrigated at present, against 38% for Asia and 14% for Latin America. In particular, a group of 30 countries, most of them in Africa, experiences difficulty both in producing enough food for their own population and in generating sufficient resources for importing necessary goods unavailable within their borders and achieve national food security. These countries are highly dependent on agriculture and would greatly benefit from improvement in their irrigation infrastructure.

5.3.4
Improving Soil Fertility

It has been estimated that over 132 million tons of nitrogen, 15 million tons of phosphorus and 90 million tons of potassium have been removed from the soils of 37 least developed countries in the SSA during the last 30 years. This has been costing the countries some 11 billion USD a year in loss of productivity. To make things worse, only about 10 kg of fertilizer is applied to each hectare of arable land in the SSA compared to 144 kg in Asia. Among the main reasons for low application of fertilizers is the cost of fertilizers. One tons of urea costs 90 USD in Europe and between 400 and 700 USD in SSA. Given the low purchasing power of African farmers and the inability of their governments to provide subsidies, fertilizer use in SSA is low which lowers crop yields and perpetuates poverty. Soil fertility can be enhanced by appropriate crop rotations, use of green manures, organic manure, and fertilizers. Appropriate technologies and policies can facilitate restoration of the fertility of African soils.

5.3.5
Improved Agricultural Technologies

Between 2000 and 2030, production of food crops in developing countries must be increased by 67%. Yield increase and higher cropping intensities will fill 80% of this need, and expansion of agricultural land the remaining 20%. Agricultural research is critical to providing farmers the tools to produce more and in better quality under changing climatic and biotic pressures. Lately, however, indicators show a decrease in the growth rate of productivity of the three primary cereals – rice, wheat, and maize – especially in the intensively cultivated lowlands of Asia.

In 2000, the world spent about 37 billion USD on agricultural research, about $1/3^{rd}$ of which was spent by private sector. Considering public and private funding together, only about $1/3^{rd}$ of total research dollars are spent in developing countries. The public sector provided about 45% of the research investment in the developed world and over 90% in developing countries. The private sector spent over 90% (12.6 billion USD) of its research funds in developed countries and only about 8% of it in developing countries. So, the private sector does not yet contribute much to agricultural research in developing countries; this service and support must continue to be provided by the public sector [3]. Unfortunately, public sector funding for agricultural research and development is on the decline: In Africa, for example, it has fallen from 0.8% of agricultural GDP in the 1980s to 0.3% in the 1990s.

Technologies friendly to smallholder farmers, mostly women, that help them produce more at lower costs and increase their income and competitiveness, are needed. Extra production also lowers food prices and helps urban poor. Technologies must also be friendly to natural resources, so farmers' capacity to produce more in the future is protected. Technologies developed in collaboration with farmers in their own environment are likely to be more relevant and have a low transactions cost for adoption. Among the most effective low-cost technologies are improved crop varieties with higher yield potential and nutritional quality and greater tolerance to drought, insects, diseases, and weeds. Tools of biotechnology can help accelerate development of appropriate technologies but most of their products must be supported by conventional plant breeding and some by regulatory processes and intellectual property laws.

5.3.6
Energy Supply Needs to be Improved

No country has been able to alleviate its poverty and achieve economic development without substantially investing in energy. However, energy is either not available to the majority of the poor or it is too expensive for them. It has been said that poor are only poor because their time and talents have low or no value. So, the key to alleviating poverty lies in increasing the value of the time and talents of the poor. The poor spend up to 5 hours a day collecting fuelwood to cook and sometimes an entire day to get a bucket of water. They spend huge amounts of time plowing their land, harvesting and threshing their produce, and pounding their millet. For buying and selling, they must spend an entire day or more to go to a market. With access to energy, these activities can be accomplished in a fraction of the time, and the saved time can be used to either do other things to increase the value of their time or to learn a new trade and skills to enhance their talent. So, access to energy directly increases the value of the time and talent of the poor which alleviates poverty. With growing interest of the world community in bio-energy, rural poor cannot only be the users of energy but also producers of energy.

5.3.7 Development Assistance is Needed

Over the past 20 years, official development assistance (ODA) for agriculture and the rural sector has declined by more than 50%, from an average of 5.14 billion USD per year to 2.2 billion USD. External assistance to agriculture in SSA has gone down from 43 USD per agricultural worker in 1982 to 9 USD per worker in 1994. In Latin America and the Caribbean, this assistance also plummeted from 98 USD per agricultural worker in 1983 to 29 USD per worker in 2002; and external assistance to agriculture in the Asia and the Pacific region and in the Near East and North Africa region are today 4 USD and 9 USD per worker, respectively. Today, nations invest some 975 billion USD a year in military spending and spend about 10% of that in aid to reduce hunger and poverty that breed conflict! Last year, the Commission for Africa declared that "agriculture is the key to Africa" and promised billions in aid. Much of the slightly increased assistance has gone to rebuild one country; little money has reached Africa. The United Nations Millennium Project has stated that "the global epicentre of extreme poverty is the smallholder farmer". So, there is no shortage of goodwill or slogans; what is lacking is money and action. Recent decision of major donors to increase aid and forego loans of the poor is a step in the right direction but it needs to be implemented.

5.3.8 Trade Helps Rural Poor

With proper support and policies, small farmers can participate in domestic and international trade, which helps raise their income. This support involves investment in infrastructure and policies and may also mean protecting the farmers from competition from subsidized agriculture from abroad. It also involves reduction of transactions costs for small farmers to increase their competitiveness. This is best done by helping them to produce and sell in groups (farmers' associations), better linking them to markets (through contract farming) and facilitating partnerships between farmers and the private sector.

However, participation in global trade requires adherence to certain standards for safety and quality of products. Producers, processors, and marketers must learn about policies, procedures, and standards of such conventions and commissions as IPPC, Rotterdam Convention, CODEX, EUREPGAP, and others. This requires capacity building of farmers, scientists, and public officials.

5.3.9
Implementing Policies that Promote Protection of Natural Resource Base

Input subsidies sometimes promote inappropriate use of inputs which has a negative effect on natural resources. Such policies should be discouraged or closely monitored. Discouraging agriculture in unsuitable marginal lands and their use for environment-friendly purposes, conservation agriculture, and use of input-use efficient technologies are some of the options to help protect natural resources.

5.3.10
Preparing for the Future

Income growth, urbanization, and globalization are leading to diet diversification and homogenization. Even in relatively poor countries now, super-marketization is beginning to occur. Local traditional markets are being replaced with more sophisticated markets which demand higher standards in quality and greater punctuality in the delivery of products. Intellectual Property Regulations are assuming greater importance in resource-poor countries now. There is greater awareness of environmental and health issues which determine which pesticides are used, how much, and when. While resource-poor farmers and their governments must work hard to meet the food needs of their people today, they must also prepare themselves for these new circumstances for tomorrow.

While not all issues can be addressed at the same time, addressing one or more will be a good step toward meeting the set goal and will help create appropriate environment for the implementation of other options in the future. A twin-track approach is needed that provides investments in safety nets and direct support to poor in the short term and that develops policies and infrastructure to provide for long-term development. Farmers are at the centre of any process of change. They need to be encouraged and guided to produce more and better and to conserve natural ecosystems and their biodiversity to minimize the negative impacts of agricultural development. This goal will only be achieved if the appropriate policies and technologies are in place that increase farmers' capacity to produce more to feed her family and market her extra produce in the globalized market in a competitive way.

5.4
References

1 FAO, *The State of Food Insecurity in the World*, FAO, Rome, **2005**.

2 P. Pingali, K. Stamoulis, R. Stringer, *Eradicating Extreme Poverty and Hunger*, ESA Working Paper No. 06-01, FAO, Rome, **2006**.

3 P. G. Pardey, J. M. Alston, R. R. Piggott, Shifting Ground: Agricultural R&D Worldwide. In *Agricultural R&D in the Developing World: Too Little Too Late?* P. G. Pardey, J. M. Alston, R. R. Piggott (Eds.), International Food Policy Research Institute, Washington, D.C., **2006**.

Keywords

Poverty Alleviation, Agricultural Production, Agricultural Productivity, Agricultural Technologies, Income Generation, MDGs

II
New Chemistry

Pesticide Chemistry. Crop Protection, Public Health, Environmental Safety
Edited by Hideo Ohkawa, Hisashi Miyagawa, and Philip W. Lee

ISBN: 978-3-527-31663-2

6
Modern Tools for Drug Discovery in Agricultural Research

Alexander Klausener, Klaus Raming, Klaus Stenzel

6.1
Introduction

Global food supply poses a continuous challenge to agriculture today and will continue to do so in the future. The growing world population and an increasing demand for higher quantity and quality of food are not compensated by an equivalent increase of available farmland or other resources, such as for instance, water. In contrast, the area of farmland per head of population has decreased dramatically during the last 50 years. This trend is continuing and even accelerating in some areas of the world. In addition, crop losses due to pest, disease, and weed damage are still as high as 50% overall. This situation calls for the systematic use of more intensive and sustainable crop production methods in order to feed the world's growing population and to satisfy both their basic and developing needs.

Today's crop protection chemistry has to fully meet the requirements of modern societies, both from an ecological and economical standpoint. Although increasing food and feed quality and quantity may be the primary focus, modern crop protection products must also be environmentally friendly, for example they must offer a high margin of safety to beneficial and non-target organisms. New compounds should have favorable toxicological properties, as well as acceptable degradation behavior in the environment. From an economical point of view, they should offer solutions for existing as well as upcoming problems and open new opportunities in the market-place. New active ingredients for agrochemical use will only be successful if they are broadly applicable and easy to use for the farmer and if they show a favorable cost/benefit ratio. The answer to all these challenges is, and will remain to be, *via* innovation. The capability to create innovation will be the key success factor within the agrochemical business in the coming years.

Pesticide Chemistry. Crop Protection, Public Health, Environmental Safety
Edited by Hideo Ohkawa, Hisashi Miyagawa, and Philip W. Lee

ISBN: 978-3-527-31663-2

6.2 Tools and Their Integration in the Drug Discovery Process

Innovation in pesticide chemistry is often driven by the discovery and identification of potential new modes of action within the biochemical pathways of the target organism or moreover by the exploitation of compounds which express their activity *via* such previously uncommercialized (new) modes of action. Discovery of such new active compounds can open up new possibilities for broad spectrum control of target diseases, pests or weeds and thus may often offer new business opportunities for R&D-based companies. Additionally, due to resistance development of many target organisms, especially within the world of fungal pathogens, new active ingredients with novel modes of action are of very high interest.

However, when classifying the presently available crop protection products of substantial commercial importance, one finds only a comparatively low number of biochemical modes of action for which compounds have been commercialized. Altogether, four modes of action account for more than 75% of the current insecticide sales (Figure 1). In the fields of herbicides and fungicides, the situation is similar. Here six different modes of action dominate each market. Of these, several represent modes of action that have been commercialized during the last decade. Further classes of compounds demonstrating other modes of action which have been identified in the past in all three indications have not gained major market shares from an economical point of view.

Reflecting this, the characterization of modes of action and the elucidation of novel targets are of high importance for R&D-driven companies in order to be able to focus activities and resources on novel, innovative compound classes and development candidates. More than ever before, the last decade has seen a significant development of new technologies in life science research. New tools for target identification and mode of action characterization have entered the drug discovery processes in both the pharmaceutical and the agrochemical industries.

A comprehensive technology portfolio and a multidisciplinary approach are keys for success (Figure 2). Conventional chemical, biological and biochemical technologies as well as molecular modeling, formulation, toxicology, and ecotoxicology are well established at the core of the R&D process. Modern and newly developed tools including functional genomics, transcriptomics, proteomics, bioinformatics, high-throughput screening (HTS), automated synthesis, and ADME studies complement the more traditional areas to accelerate the discovery of new active ingredients.

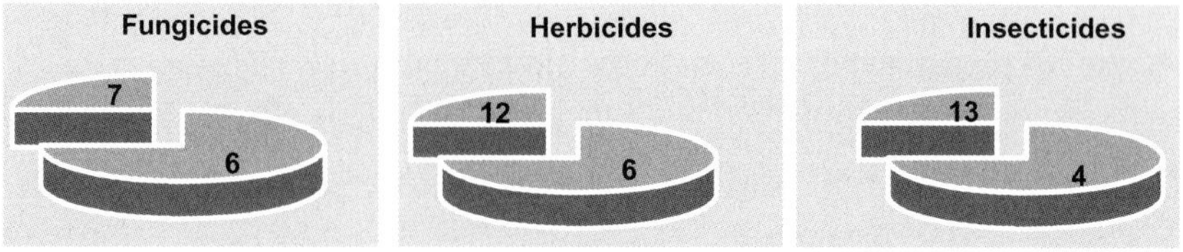

Figure 1. Number of modes of action and respective market share by segments.

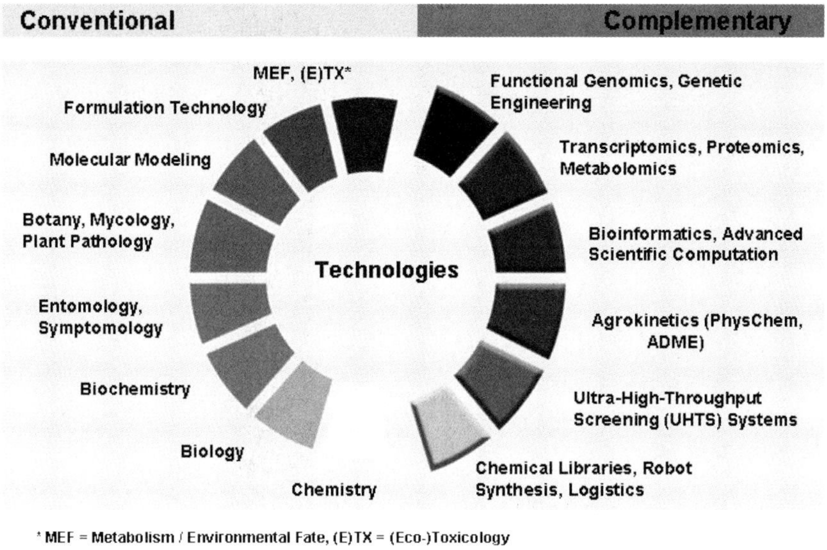

Figure 2. Conventional and complementary technologies as parts of a comprehensive technology portfolio.

Successful agrochemical companies today have fully integrated both conventional and advanced technologies into the active ingredient discovery process according to the needs of each phase of the R&D process (Figure 3). Within the Bayer CropScience discovery platform, HTS (high-throughput screening) systems support the identification of chemical inhibitors at the biochemical target level as well as playing an important role in the search for molecules showing activity

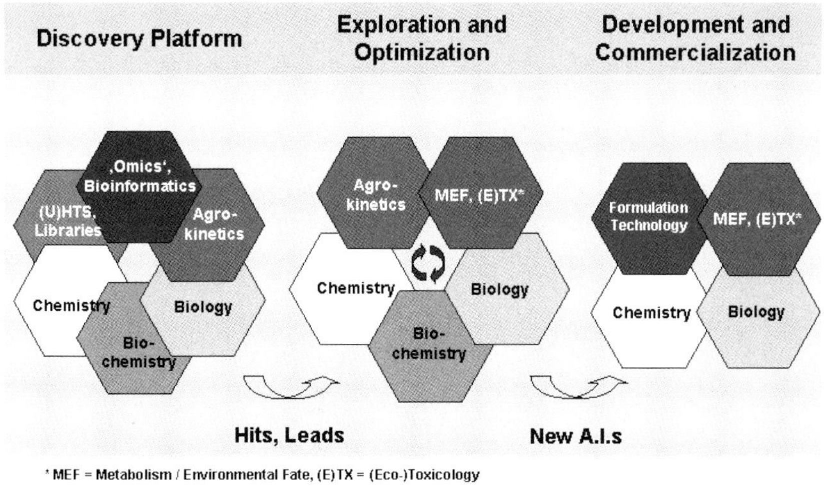

Figure 3. Integration of new complementary technologies into the drug discovery process.

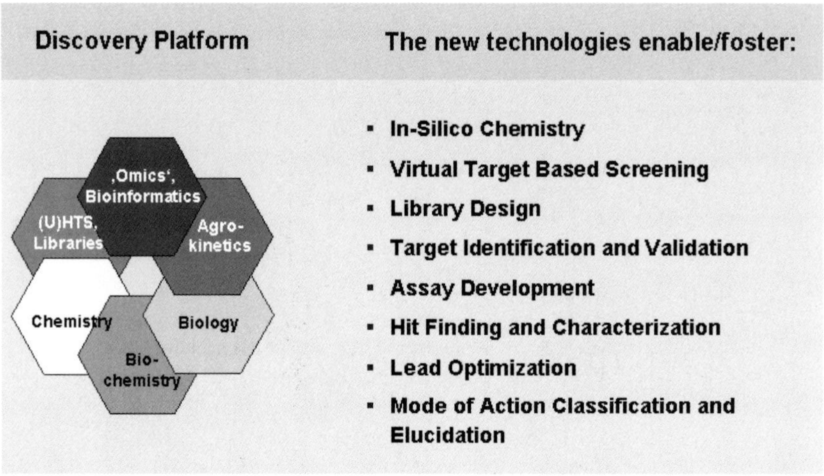

Discovery Platform
The new technologies enable/foster:
‚Omics', Bioinformatics
(U)HTS, Libraries
Agro-kinetics
Chemistry
Biology
Bio-chemistry
In-Silico Chemistry
Virtual Target Based Screening
Library Design
Target Identification and Validation
Assay Development
Hit Finding and Characterization
Lead Optimization
Mode of Action Classification and Elucidation

Figure 4. Contribution of new technologies to the early research phase.

against fungi, weeds, and insects in *in-vivo* biological screens. In addition, the new 'omics'-technologies are powerful instruments to foster the design of chemical libraries, the identification of new biochemical targets, the identification and characterization of new hits and leads and the classification and elucidation of modes of action of new chemical entities (Figure 4). The study of agrokinetics facilitates a better understanding of the physico-chemical properties of hit and lead compounds, their efficacy related to the target organisms, and their toxicological and metabolic behavior. This toolbox is of extremely high value for the exploration and optimization of lead structures to afford development candidates which fulfill the relevant biological and physiological requirements and which allow a successful translation of laboratory and greenhouse activity into the commercially relevant field situation.

The later phases of the R&D process (optimization cycles, development, and commercialization) are still primarily dominated by conventional technologies.

6.3
Mode of Action Elucidation – An Example for the Integration of New Technologies

The example of the mode of action elucidation of flubendiamide is described below to illustrate the successful integration and application of the different disciplines and cutting edge technologies into the R&D process at Bayer CropScience.

Flubendiamide is a promising new insecticide which is particularly active against lepidopteran pest species and is currently being co-developed by Nihon Nohyaku and Bayer CropScience. It is the first member of a new chemical class of insecticides named phthalic acid diamides (Figure 5) to be developed. It has been shown to be extremely potent against lepidopterous pests including those resistant

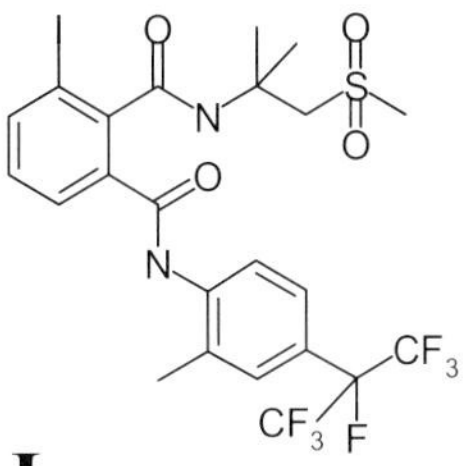

I

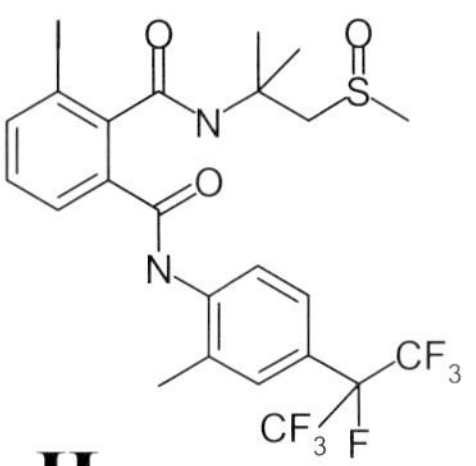

II

Figure 5. Structures of the phthalic acid diamides used in the MoA study.
I: Flubendiamide [3-iodo-*N*-(2-methanesulfonyl-1,1-dimethyl-ethyl)-*N'*-[2-methyl-4-(1,2,2,2-tetrafluoro-1-trifluoromethyl-ethyl)-phenyl]-phthalamide]
II: Flubendiamide sulfoxide [3-iodo-*N*-(2-methanesulfinyl-1,1-dimethyl-ethyl)-*N'*-[2-methyl-4-(1,2,2,2-tetrafluoro-1-trifluoromethyl-ethyl)-phenyl]-phthalamide]

to different classes of established insecticides and will be launched in 2007 for foliar application in many crops, including vegetables, fruits and cotton [1–2].

The mode of action of this important new chemical class was unknown during initiation of the development. Therefore scientists from Bayer CropScience and Nihon Nohyaku applied several different technologies in order to further elucidate the source of the biological activity of these compounds on a molecular level [3–4]. To get a first idea of the mechanism of action, *Spodoptera frugiperda* larvae were treated with flubendiamide and the resulting macroscopic activity was carefully analyzed. The larvae showed unique symptoms of poisoning resulting in complete contraction paralysis. By applying the compound in a gut muscle assay, further evidence was observed supporting the hypothesis that the target was most likely to be located in the neuromuscular system. To characterize the target on the cellular level, isolated individual *Heliothis* neuronal cells were used to measure the intracellular calcium concentration as indicated by fluorescence probes. It could be shown that flubendiamide caused a significant, transient increase of the intracellular calcium concentration (Figure 6).

In further experiments, it was shown that flubendiamide induced a Ca^{2+} release from internal stores. Flubendiamide evoked Ca^{2+} transients not only under standard conditions but also when Ca^{2+} was not included in the application solution.

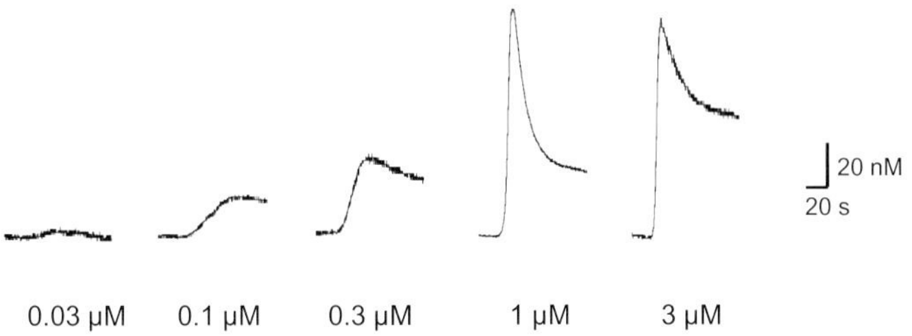

Figure 6. Transient increase of the cytosolic Ca^{2+} concentration $[Ca^{2+}]$ induced by application of different concentrations of flubendiamide in single FURA2-AM loaded *Heliothis* neurons.

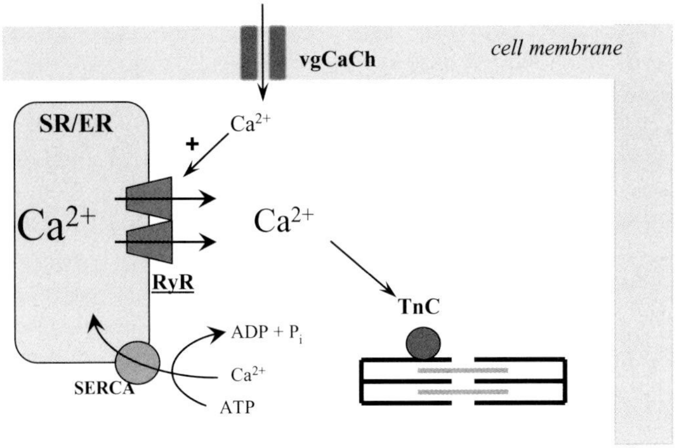

Figure 7. Illustration of the localization and function of the ryanodine receptor (RyR).

Consistent with this, the flubendiamide responses were suppressed after incubation with 10 μM thapsigargin, a known inhibitor of the endo(sarco)plasmatic Ca^{2+}-ATPase which causes depletion of endoplasmic Ca^{2+} stores. The conclusion of these experiments was that the molecular target was localized in the endo(sarco)plasmic reticulum. In principle, calcium release from the endo(sarco)plasmic reticulum is mediated by two distinct but structurally related channels, the ryanodine-sensitive calcium release channel and the inositol-1,4,5-triphosphate receptors (IP3R). Interestingly, the calcium transients induced by flubendiamide were almost completely inhibited by addition of ryanodine, but not by the IP3R inhibitor xestospongine C. These inhibitor studies indicated that the calcium release was mediated by the activation of ryanodine receptors. Ryanodine receptors (RyR) are intracellular Ca^{2+} channels responsible for the rapid and massive release of Ca^{2+} from intracellular stores, which is necessary for excitation-contraction (EC) coupling in muscle cells (Figure 7).

This hypothesis was further supported using molecular biology and cellular tools where the insect ryanodine receptor gene was heterologously expressed in appropriate cells. In untransfected CHO cells, application of flubendiamide sulfoxide (a better soluble analogue) did not cause an $[Ca^{2+}]$ increase (Figure 8). In CHO cells transfected with the RyR from *Drosophila* (CHO-RyR), flubendiamide sulfoxide induced Ca^{2+} responses with similar kinetic responses to those found in *Heliothis* neurons (Figure 8).

Therefore it was concluded that flubendiamide acts as a selective activator of the insect ryanodine receptor, inducing ryanodine-sensitive cytosolic Ca^{2+} transients. Furthermore, radioligand binding studies using microsomal membranes from *Heliothis* flight muscles demonstrated that flubendiamide allosterically increased the ryanodine affinity. Flubendiamide was found to bind to *Heliothis* microsomal membranes with an apparent K_D of 4.7 nM. Known ryanodine receptor ligands such as cyclic ADP-ribose, caffeine, ryanodine, and dantrolene did not interfere

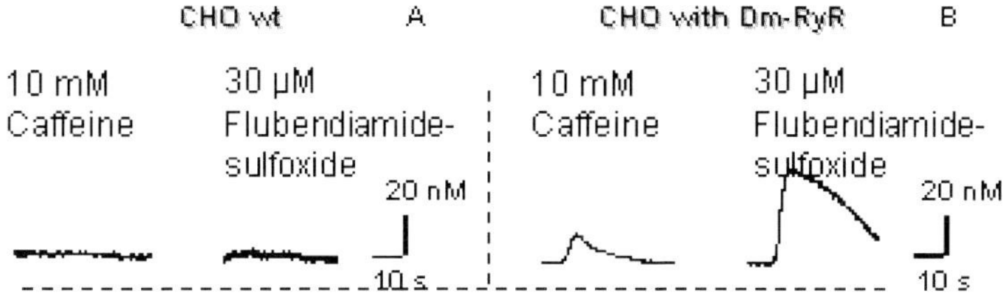

Figure 8. Flubendiamide activated the *Drosophila* RyR expressed in CHO cells. Un-transfected control CHO cells did not respond to caffeine or flubendiamide sulfoxide with an increase of [Ca^{2+}], in contrast to CHO cells which were transfected with a full-length cDNA of the *Drosophila* ryanodine receptor. Both compounds induced Ca^{2+} responses with similar kinetics as to those found in *Heliothis* neurons.

with flubendiamide binding, indicating that flubendiamide interacts with an alternative binding site on the ryanodine receptor complex. Of particular note was the finding that the number of flubendiamide binding sites was almost four times higher than for ryanodine.

Last but not least, it was also shown that flubendiamide and its sulfoxide are specific to insect ryanodine receptors and do not affect mammalian ryanodine receptors. Even high concentrations of flubendiamide sulfoxide applied on differentiated mouse muscle C2C12 cells which express the muscle Subtype I and Subtype III did not either elicit Ca^{2+} signals nor did they prevent the Ca^{2+} transients elicited by caffeine (Figure 9). Therefore, we conclude that flubendiamide and related compounds do not affect mammalian RyR Type I and III. These observations provide a good explanation for the excellent toxicological profile observed in the case of flubendiamide.

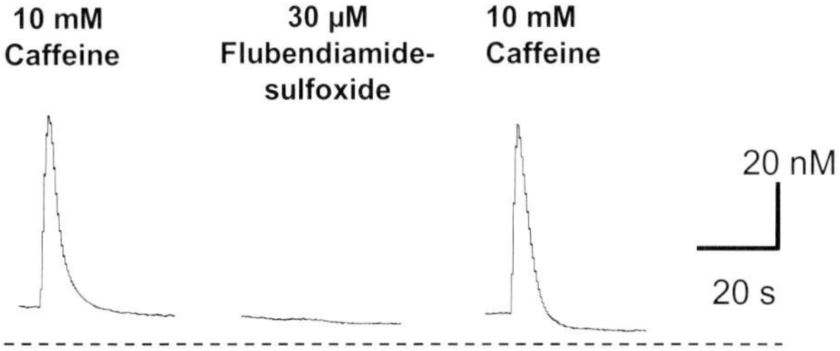

Figure 9. Effect of caffeine and phthalic diamides on mouse muscle cell line C2C12. representative [Ca^{2+}] traces of a Fura2-AM-loaded C2C12 cell during application of caffeine, flubendiamide sulfoxide, and again caffeine.

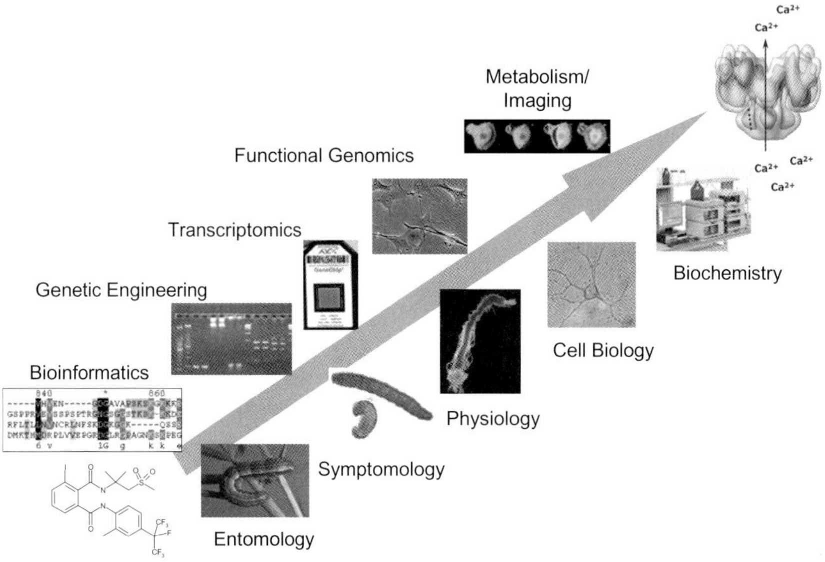

Figure 10. The Experimental strategy of the elucidation of the novel mode of action of flubendiamide is a good example for a successful integration and application of different disciplines as well as of cutting edge technologies in the research process.

6.4 Conclusion

Innovation in crop protection is essential for the sustainability of agriculture and global food production. A continuous evaluation of chances and limitations of new and established technologies is necessary in order to build up and to run modern and innovative platforms in the area of crop protection research. Modern tools, complementing conventional ones, have to be integrated into the drug discovery process to create a comprehensive technology portfolio.

Successful agricultural research has to resolve a diverse spectrum of challenges due to the broad diversity of target organisms, the stringent requirements for specificity of crop protection products and the broad range of scientifically valuable technological approaches. A flexible application of these tools to solving specific problems and a consequent follow-up of promising experimental strategies by a team of experts representing all areas of expertise are key factors of success for an efficient research process.

By bringing innovative active ingredients with novel modes of actions to the market makes it possible to offer new and attractive solutions to today's challenges in the agribusiness.

Flubendiamide has been presented in this talk as an example for this approach. This molecule is the first representative of a new chemical class of insecticides with extremely potent activity against lepidopterous pests including those resistant to already established classes of insecticides. The new mode of action (ryanodine receptor agonist), was elucidated in a coordinated team approach combining several cutting edge technologies (Figure 10). This novel mode of action combined with the favorable toxicological profile provides excellent opportunity to introduce a new, innovative insecticide into the market.

Several products out of our very recent development pipeline have shown new modes of action. This fact alone and the fast and successful elucidation of the modes of action are excellent examples for the successful integration and application of different disciplines and cutting edge technologies into the research of Bayer CropScience.

6.5 Acknowledgments

The evaluation of the mode of action of flubendiamide is a result of the joint efforts of Nihon Nohyaku and Bayer CropScience researchers: Ulrich Ebbinghaus-Kintscher, Rüdiger Fischer, Peter Lümmen, Klaus Raming (BCS) and Takao Masaki, Noriaki Yasokawa, Masanori Tohnishi (Nihon Nohyaku).

6.6 References

1 T. Nishimatsu, H. Kodama, K. Kuriyama, M. Tohnishi, D. Ebbinghaus, J. Schneider, International Conference on Pesticides 2005, Kuala Lumpur, Malaysia, *Book of Abstracts*, **2005**.

2 M. Tohnishi, H. Nakao, T. Furuya, A. Seo, H. Kodama, K. Tsubata, S. Fujioka, H. Kodama, *et al.*, *J. Pestic. Sci.*, **2005**, *30*, 354–360.

3 P. Luemmen, U. Ebbinghaus-Kintscher, N. Lobitz, T. Schulte, C. Funke, R. Fischer, Abstracts of Papers, 230th ACS National Meeting, Washington D.C., United States, Aug. 28 – Sept. 1, AGRO-025, **2005**.

4 U. Ebbinghaus-Kintscher, P. Luemmen, N. Lobitz, T. Schulte, C. Funke, R. Fischer, T. Masaki, N. Yasokawa, *et al.*, *Cell Calcium*, **2006**, *39*, 21–33.

Keywords

Agricultural Research, Technology Portfolio, Tools, Discovery Platform, Mode of Action, Flubendiamide, Ryanodine Receptor

7
Target-Based Research: A Critical Review of Its Impact on Agrochemical Invention, Focusing on Examples Drawn from Fungicides

Stuart J. Dunbar, Andrew J. Corran

7.1
Introduction

Although target-based research is core to the development of new pharmaceuticals [1], it has not (yet) realized its theoretical potential in the agrochemical arena. In agriculture, there are no products currently on the market, or visibly in development, where a clear target-based approach has been successfully applied. Is this true and, if so, why is this the case? This chapter sets out to investigate the issue, illustrating the pitfalls and opportunities encountered along the journeys involved in target-based research, focusing on examples drawn from fungicide invention in Syngenta and other companies.

The development of high-throughput technologies in the 1980's and 1990's drove the implementation of target-based research across bioscience-based industries. Take-up in the agricultural industry was, perhaps, slower than in other industries because we have a real advantage over pharmaceutical research; we can test at high-throughput on our target organisms [2]. However, the need for novel modes of action and the fact that the invention success rate was slowing down, measured by the number of compounds screened to deliver a new compound into the marketplace [3], made the implementation of a different paradigm based on knowledge of a target protein attractive. This was especially so in fungicide research where the availability of model species' genomes alongside the tractability of transformation and expression profiling technologies made hypothesis testing easier than in insecticides or herbicides. Nevertheless, even in fungal research, doing transformation in real pest species has only recently become routine [4–6]. In parallel to these molecular advances, developments in protein expression and crystallization, homology modeling and computational chemistry techniques bridged the gap between molecular biology, biochemistry, and synthetic chemistry.

What has been the impact of all of these technological advances on fungicide invention? Several different models of target-based research were implemented across the industry. These ranged from high-throughput screening of a target

Pesticide Chemistry. Crop Protection, Public Health, Environmental Safety
Edited by Hideo Ohkawa, Hisashi Miyagawa, and Philip W. Lee

ISBN: 978-3-527-31663-2

protein *in vitro* [2] to so-called "rational design" using structural knowledge to inform synthetic chemistry [7–8]. Projects often involved combinations of approaches [9]. These different models will be discussed, for example, where target-based approaches have been applied at the front end of invention looking for novel hits linked to a new mode of action as starting points for synthesis. Alternatively, target-based approaches can be applied to advanced optimization projects where the mode of action may be known [10] and *in vitro* screening and/or structure-based design is used to inform chemical synthesis. The success of *in vitro* (or *in vivo*) screening depends entirely on the nature of the chemical inputs into the screen. This critical issue has been recently discussed elsewhere [11] and, although outside the scope of this review, the issues involved in generating novel chemistry inputs are central to a successful implementation of target-based research in all industries. We hope to highlight the challenges involved in target-based research in fungicide discovery, ranging from selection of a valid or druggable target, screening issues, protein expression and crystallization, linkage of an *in vivo* effect to an *in vitro* target, and *in vitro* to *in vivo* translation, critically reviewing successes and failures along the way. An excellent recent review of high-throughput screening in agrochemical research in general sets the scene for this discussion [12].

7.2
Selection of Targets for *In Vitro* Screening

Opinions on what constitutes a good antifungal target for *in vitro* screening vary from one that is well chemically validated, i.e., one where there is good evidence that specific inhibitors exist and this inhibition leads directly to the inhibition of growth of a fungus, to one where a target protein has been validated only genetically. Typically, genetic validation requires that a gene knockout has either a lethal or potentially a non-pathogenic phenotype. Taken to extremes, the former view is limiting in that this restricts the choice of targets to those that are well-defined and therefore potentially lacking in novelty, i.e., there may be compounds already being marketed as fungicides with this mode of action. These targets are truly well chemically validated but clearly are not novel. On the other hand, the latter opinion suggests that there may be many hundreds of potentially effective antifungal targets but this view takes no account of other factors such as "drugability", and whether inhibition of this protein is likely to give a sustained fungistatic or fungicidal effect in a wide range of commercially relevant agrochemical fungi. Decades of research into the discovery of novel antifungal agents indicates there are likely to be relatively few good targets that can deliver such an effect. The challenge is therefore to find those potentially few effective targets that have yet to be discovered. However, even today, there is relatively little reverse genetic information about the importance of genes from plant pathogens and so we turn to model organisms such as *Saccharomyces cerevisiae* where there has been a concerted, systematic effort over many years to gain a better understanding of its genes and the proteins they encode. This valuable source of information is available *via* the Yeast Protein Database [13] and the *Saccharomyces* Genome Database [14]. More recently, a number of fungal genomes of relevance to agrochemical research have been sequenced with more being completed every year. Many of these genomes are either being sequenced by the Joint Genome Institute or the Broad Institute. These fungal genomes, when annotated, can be compared with yeast and other model systems such as *Arabidopsis* to select targets that are, for example, unique to fungi or to study gene families such as G-protein coupled receptors [15–18]. In addition, molecular tools such as efficient homologous recombination and RNA interference are becoming increasingly available to probe the role of a particular gene in the life cycle of fungus of relevance to the agrochemical industry [4–6, 19–26]. These techniques will have a major impact over the next decade on our knowledge and understanding processes that are critical to the life cycle of plant pathogenic fungi.

The strategy for the selection of targets for *in vitro* screening that has been adopted by Syngenta is one where targets are selected if the gene is essential, present as a single copy, and found in a broad range of phytopathogenic fungi. To maximize the opportunity for selectivity, the target protein should be significantly different, or absent, in non-target species such as plants and man. A good example of such a target is *AUR1* which encodes inositol phosphorylceramide synthase in fungal and plant sphingolipid biosynthesis. Sphingolipids are essential membrane components and mammals and fungi share a common pathway up to the formation of sphinganine after which the pathway diverges and inositol phosphorylceramide synthase is found in the fungal-specific branch [27]. Fungi does not absolutely require cytochrome b in the bc_1 complex of mitochondrial electron transport, however, it has been shown to be the target site for the commercially important strobilurin class of fungicides [28]. This protein is highly conserved between fungi and man. In the 1990's, a number of different papers were published describing natural products with potent inhibitory activity against inositol phosphorylceramide synthase and reasonable antifungal activity [29–31]. These publications provided key information on the "drugability" of inositol phosphorylceramide synthase, validating the target chemically and complimenting the existing genetic validation. As a result, inositol phosphorylceramide synthase has been suggested as a good target for the development of potent and selective antifungal agents [32].

Further limitations on the traditional approach to *in vitro* screening, and therefore target selection, is the requirement for a supply of functionally active protein (unless a cell-based assay is employed) as well as an assay method that is amenable to plate-based, high-throughput screening formats. Inositol phosphorylceramide synthase is an integral membrane protein and is therefore difficult to produce efficiently in a suitable expression system. In addition, the substrates and products of this enzyme-catalyzed reaction are complex lipids making assay design challenging [33]. Published assay methods typically employ a post-assay separation technique followed by either fluorescence or radiochemical detection [29, 31, 34], making high-throughput screening for this particular protein target impractical.

In summary, there are many difficulties in the selection of targets that need to be addressed, both in terms of a lack of detailed knowledge especially on the commercially relevant phytopathogens as well as a more practical problem of being able to develop robust *in vitro* assays that can be used to screen thousands of potential inhibitors quickly and cost-effectively.

7.3
Assay Design and Implementation

Syngenta's *in vitro* high-throughput screens for anti-fungal targets have included both enzyme as well as cell-based assays such as a reporter assay for ergosterol biosynthesis based in *S. cerevisiae* [35]. Cell-based screens have the advantage that they allow screening for targets that would otherwise be too technically challenging. In addition, multiple targets can be assayed at the same time, for example, the entire ergosterol biosynthetic pathway [35]. However, there are also a number of disadvantages of this approach such as the need for follow-up assays to determine the site and potency of inhibition and the potential for false positives or negatives due to cellular uptake, variable responses from different sites of inhibition or cellular toxicity. In addition, there is the potential for the model not to fully represent the disease system. Fenhexamid, for example, is a novel inhibitor of 3-keto reductase step in ergosterol C4-demethylation in *Botryotinia fuckeliana* [36]. However, this compound has no activity in the *S. cerevisiae* ergosterol reporter assay and neither does it inhibit the growth of *S. cerevisiae*, suggesting that *S. cerevisiae* is not a good model system to discover novel inhibitors of 3-keto reductase.

The Syngenta strategy on *in vitro* high-throughput screening is to fix the assay format, as far as possible, in order to simplify the logistics of compound presentation, data capture, and managing the output from the screens. As such, test compounds are provided in 1 μL of 100% DMSO in 384-well microtiter plates that are stamped out from 'mother plates' which are prepared by solubilizing compounds in the company collection in DMSO to one of a number of fixed concentrations. Typically, compounds are tested in batches of 200–250 K. In the initial screen, compounds are tested at one concentration and as one replicate and standard inhibitors and suitable controls are employed on every plate for consistency and quality control. At the end of each batch of compound testing, the data is analyzed using in-house data analysis software and 'hits' are selected for 'Tier 2' screening where multiple rates are used to generate IC_{50} values. Novel 'targeted input' chemistry can also be included in Tier 2 testing to supplement hits obtained by random screening. This targeted input is generated by searching commercially available databases for structures with similarity to known types of inhibitor or to reaction intermediates and as a result targeted input has proved to be a rich source of both hits and new leads compared to random input. Compounds that have an IC_{50} below a certain cut-off value are selected as leads. The value selected as the cut-off will vary according to the nature of the screen, the potency of the standard inhibitors, and the hit rate. However, typically for an enzyme assay, a cut-off value of 1–5 μM would be chosen whereas for a cell-based screen, leads may be selected that are less potent provided the signal of activity is sufficiently robust.

Tier 3 tests are normally used a try to confirm the mode of action or to generate additional information that will add value to the lead. A good example of this is in the use of strains of *S. cerevisiae* that have been engineered to over-express a protein of interest. Provided the compound has inhibitory activity against *S. cerevisiae*, then a strain over-expressing the target protein would be expected to be less susceptible to growth inhibition than the parental strain [37]. Another approach, where sufficient information about the target is known, is in the generation of site-specific resistance, for example, the glycine to alanine (G143A) point mutation in cytochrome b which when introduced into *S. cerevisiae* confers strobilurin resistance [38]. This resistant strain has been used to screen bc_1 complex inhibitors for molecules that overcome this particular mode of resistance. For example, the compound shown in Figure 1 was discovered as a potent inhibitor of mitochondrial electron transport when using beef heart mitochondria in a high-throughput assay to discover novel inhibitors of respiration. Further tests showed it to inhibit the ubiquinone:cytochrome c oxidoreductase complex and to inhibit *Plasmopara viticola* (downy mildew on grapes) in fungicide *in-planta* tests. However, at the time of this project we were trying to discover molecules that would inhibit ubiquinone:cytochrome c oxidoreductase at novel sites, i.e., that would not be cross-resistant to azoxystrobin, so we tested this molecule in our Tier 3 screen to determine whether it would control the growth of the G143A strain of *S. cerevisiae*. The results indicated that this molecule was not cross-resistant with azoxystrobin as it was equally active against the wild-type and G143A strain of *S. cerevisiae* (Figure 2).

A further example where such Tier 3 assays are useful is in the inhibition of cell wall biosynthesis. Fungicidal inhibitors of chitin or glucan biosynthesis cause a weakening of the cell wall leading to a susceptibility to osmotic shock. This can be exploited to generate cell-based assays [39–40] that provide a linkage between inhibition of cell wall biosynthesis to the observed inhibition of fungal growth. Such data provides confidence that the selected biochemical target is responsible for the inhibition of the fungus, rather than a secondary, unrelated site of inhibition and allows the assay to be used to optimize *in vitro* potency further.

Compounds that have *in vitro* IC_{50} values < 1–5 μM (depending on the assay), and have been demonstrated to be linked to effects on whole cells, are regarded

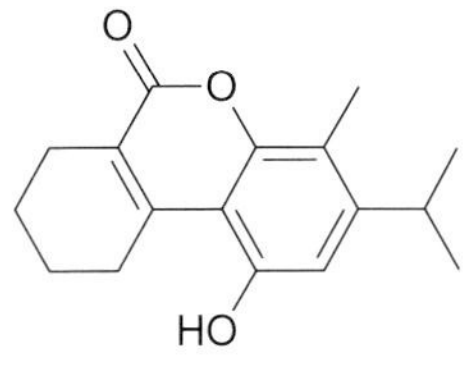

Figure 1. Molecule discovered as an inhibitor of electron transport using beef heart mitochondria.

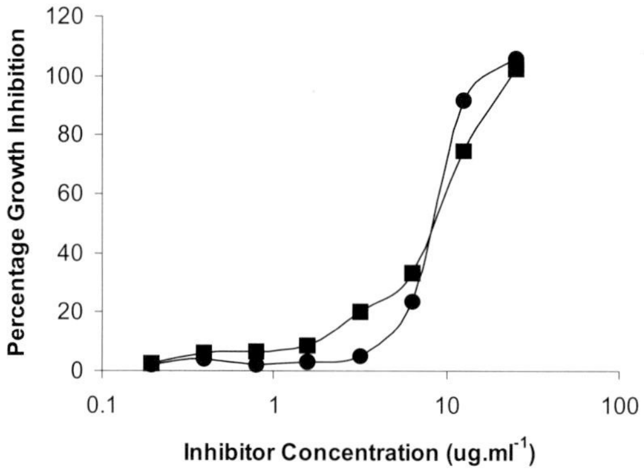

Figure 2. The effect of the novel inhibitor of ubiquinone:cytochrome c oxidoreductase on the growth of the wild-type (■) and G143A (●) strains of *S. cerevisiae* using lactic acid as the carbon-source.

as the best candidates for development as new fungicide leads. One advantage of these leads compared to those generated by traditional *in vivo* screening methods, is that the mode of action is known and can now be used to guide synthetic chemistry. The strategy adopted by Syngenta is one of lead exploration whereby selected leads are used to generate more targeted input for the next batch of compounds for the screen. Commercial databases are again searched for compounds that contain molecules similar to, or contain the same substructure as, the original lead. In this way, a certain amount of exploration of the chemical space around the lead can take place very rapidly and before valuable synthetic chemistry effort is employed.

7.4
In Vitro to *In Vivo* Translation

We define *in vitro* to *in vivo* translation as the processes whereby an *in vitro* lead is developed into a fungicidally active *in vivo* lead, where the biological activity is linked to potency at the *in vitro* target. It is very common to discover leads with target binding potency at levels we believe should be sufficient to give biological activity (μM and below) but which display no, or only weak, activity on biological screens. Converting these *in vitro* hits into *in vivo* actives has been the key issue that has prevented target-based approaches from delivering robust leads and products in agricultural invention. Why is this and what strategies have been adopted in attempts to overcome it?

We have used biokinetic approaches (defined as the study of uptake, movement and metabolism of compounds within target organisms) to try to address *in vitro* to *in vivo* translation in Syngenta. Studies have included uptake into the fungus, often using non-radiolabeled, LC-MS analytical techniques at this early stage. Metabolism, tracking loss of the parent compound, has been applied to understand the rate of loss which can be important in the kinetics involved in delivering biological activity, alongside studies to determine if the compound is stored or excreted. Biokinetics has been invaluable in identifying key issues limiting biological activity and has led to hypotheses for chemical synthesis to overcome them. An example of this approach helping decision-making in Syngenta was in the bc_1 project. The compound in Figure 1 was discovered to be glycosylated at the hydroxyl site. Unfortunately the -OH moiety was also critical to *in vitro* activity and chemical strategies to replace it were not successful and the area was dropped. However, whilst there are examples whereby we have succeeded in identifying the critical biokinetic issue limiting *in vivo* potency, they are few in number when the starting point was an *in vitro* only hit. Greater success has been achieved where a weak biological signal was associated with the hit in the first place and we have used biokinetics to improve this. Checks are also made to determine whether or not any biological activity we do see is linked back to the target protein of interest. These linkage tests are discussed above and often involve genetically modified strains and/or SAR analyses to track both *in vivo* and *in vitro* potency levels together.

7.5 Structure Based Design

Structure Based Design (SBD) has been a very powerful tool to inform synthesis within projects once an initial lead has been discovered, either *via* a target-based screen or from an *in vivo* hit where the Mode of Action (MOA) has been elucidated. In Syngenta it has been particularly successful in projects where there is an *in vitro* screen used to generate novel chemical scaffolds or to provide biochemical potency to help derive the structural model. The detailed challenges and opportunities associated with SBD in agrochemical research have been well documented elsewhere [8] and are outside the scope of this short review; however, we will briefly illustrate the application of SBD in fungicide discovery using pharmaceutical and agrochemical examples. One of the early published examples using SBD in pharmaceutical discovery was *N*-myristoyltransferase (NMT) [41–44]. Initial genetic and biochemical studies determined that NMT1 was an essential, single copy gene and that the fungicidal activity of a range of peptidomimetics could be linked to NMT activity [41]. Further studies by workers at Roche [7, 42–45] elegantly developed novel fungicidal benzofurans using a combination of *in vivo* and *in vitro* assays alongside knowledge of the crystal structure. In agrochemical discovery, elegant studies by Jordan's group (see [46] on scytalone dehydratase and other enzymes in the melanin biosynthesis pathway (discussed in [8]) have successfully been used to design novel fungicidal inhibitors.

One of the greatest challenges associated with application of SBD to all discovery processes is the production of protein and subsequent crystallization, particularly when the protein itself is membrane-bound. Many validated fungicidal targets unfortunately fall into the membrane-bound group and this has inhibited the universal application of SBD approaches in the industry. Greater examples of successful SBD approaches can be found for soluble enzymes in herbicides (see [8]). However, recent developments in elucidating the crystal structure of membrane bound targets indicate that there are exciting opportunities ahead for agrochemical design. For example, the publication of the structure of the bc_1 complex in the respiratory chain [47–49] and recent studies on Complex II [50–51] open up new avenues for the design of novel agrochemicals at these well-validated sites, particularly when focused on design to overcome resistance (e.g., at the strobilurin binding, Q_o, site in the bc_1 complex).

These are among the first pioneering examples of structural studies on membrane bound proteins but new high-throughput crystallography techniques, novel methodologies for packing membrane bound crystals, coupled with the high energy X-ray sources such as the new Diamond laboratory in the UK (see http://www.diamond.ac.uk/default.htm) could herald a new era in the application of SBD approaches to fungicide invention.

7.6
Conclusion and a Forward Look

It is clear that, measured by products in development or in the marketplace, target-based approaches to fungicide invention have not succeeded in delivering a rich vein of novel modes of action to our businesses. However, the science is still in its infancy. The tools needed are still being developed, for instance the sequencing and (most importantly) functional annotation of key pest species' genomes remain to be completed [52]. Even in the best characterized fungal genome, *S. cerevisiae*, *ca.* 20% of the genes known to encode proteins are unannotated and the function of essential genes is still being discovered, more than a decade after the genome was first sequenced [53]. How these genes and their gene products interact is also uncertain and the evolving science of Systems Biology [54] will open up new insights into the difficult challenges associated with fungal control, targeting pathways and systems that are essential to life.

New techniques to overcome the blocking hurdles associated with *in vitro* to *in vivo* translation are also just becoming available. The move towards cell-based systems where at least some metabolism is incorporated into the screen is an indication of this, although the associated higher false positive rate is a consideration that needs to be incorporated into the screen design. A further development is the use of engineered organisms in target-based research. Here an organism is engineered to report activity on an enzyme, pathway or system and the whole organism is used in the screen. The advantage is that this can report activity on target proteins *in vivo*, bypassing *in vitro* to *in vivo* translation issues. When used in parallel with biochemistry, this approach is particularly powerful. There remain difficulties in achieving this routinely in our pest species but the pace of the development of transformation tools for filamentous fungi indicates that the day when this will be routine is not that far ahead.

The past 10 years has been the first phase of target-based research, although the paradigm of high-throughput screening and combinatorial chemistry has not delivered its promise, the next phase of the science will be radically different. We believe that the integration of target-based concepts into the discovery process, alongside biology and chemistry is the future, selecting robust, chemically validated targets/pathways and incorporating strategies to report/bypass metabolic *in vitro* to *in vivo* translation issues. Scientific developments including structural studies on membrane-bound targets, genetic transformation of pest species, and understanding the function of proteins and genes using Systems Biology [55] will herald an exciting future which will, we believe, begin to deliver on the potential of the science.

7.7 References

1 P. Veerpandian, *Structure-Based Drug Design*, Marcel-Dekker, New York, **1997**.
2 A. J. Corran, A. Renwick, S. J. Dunbar, *Pestic. Sci.*, **1998**, *54*, 338–344.
3 M. Pragnell in *Chemistry of Crop Protection*, G. Voss, G. Ramos (Eds.), Wiley-VCH, Weinheim, **2003**.
4 K. Adachi, G. H. Nelson, K. A. Peoples, S. A. Frank, M. V. Montenegro-Chamorro, T. M. DeZwaan, L. Ramamurthy, J. R. Shuster, *et al.*, *Curr. Genet.*, **2002**, *42*, 123–127.
5 P. J. Balint-Kurti, G. D. May, A. C. Churchill, *FEMS Microbiol. Lett.*, **2001**, *195*, 9–15.
6 J. Kamper, Z. An, M. L. Farman, A. Budde, S. Taura, S. A. Leong, *Mol. Genet. Genomics*, **2004**, *271*, 103–110.
7 S. Sogabe, M. Masubuchi, K. Sakata, T. A. Fukami, K. Morikami, Y. Shiratori, H. Ebiike, K. Kawasaki, *et al.*, *Chem. Biol.*, **2002**, *9*, 1119–1128.
8 M. W. Walter, *Nat. Prod. Rep.*, **2002**, *19*, 278–291.
9 S. Gutteridge, S. O. Pember, L. Wu, Y. Tao, M. Walker in *New Discoveries in Agrochemicals: ACS Symposium Series 892*, J. Marshall Clark, H. Ohkawa (Eds.), American Chemical Society, Hawaii, **2005**.
10 J. J. Steffens, D. A. Kleier in *Modern Selective Fungicides*, H. Lyr (Ed.), Fischer, New York, **1995**.
11 S. C. Smith, J. S. Delaney, M. P. Robinson, M. J. Rice, *Comb. Chem. High Throughput Screen*, **2005**, *8*, 577–587.
12 K. Tietjen, M. Drewes, K. Stenzel, *Comb. Chem. High Throughput Screen*, **2005**, *8*, 589–594.
13 M. C. Costanzo, M. E. Crawford, J. E. Hirschman, J. E. Kranz, P. Olsen, L. S. Robertson, M. S. Skrzypek, B. R. Braun, *et al.*, *Nucl. Acids Res.*, **2001**, *29*, 75–79.
14 S. Weng, Q. Dong, R. Balakrishnan, K. Christie, M. Costanzo, K. Dolinski, S. S. Dwight, S. Engel, *et al.*, *Nucl. Acids Res.*, **2003**, *31*, 216–218.
15 J. R. Xu, Y. L. Peng, M. B. Dickman, A. Sharon, *Annu. Rev. Phytopathol.*, **2006**, *44*, 337–366.
16 R. A. Dean, N. J. Talbot, D. J. Ebbole, M. L. Farman, T. K. Mitchell, M. J. Orbach, M. Thon, R. Kulkarni, *et al.*, *Nature*, **2005**, *434*, 980–986.
17 D. M. Soanes, N. J. Talbot, *Mol. Plant Pathol.*, **2005**, *7*, 61–70.
18 M. Axelson-Fisk, P. Sunnerhagen, *Topics Curr. Genet.*, **2006**, *15*, 1–15.
19 S. B. Goodwin, C. Waalwijk, G. H. J. Kema, *Appl. Mycol. Biotechnol.*, **2004**, *4*, 315–330.
20 R. Beffa, *Pflanzenschutz-Nachrichten Bayer*, **2004**, *57*, 46–61.
21 S. J. Assinder, *Appl. Mycol. Biotechnol.*, **2004**, *4*, 137–160.

22 J. E. Hamer, *Advances in Fungal Biotechnology for Industry, Agriculture, and Medicine*, J. S. Tkacz, L. Lange (Eds.), Kluwer Academic/Plenum Publishers, New York, **2004**, 31–39.
23 C. B. Michielse, P. J. J. Hooykaas, C. A. M. J. J. Hondel, A. F. J. Ram, *Curr. Genet.*, **2005**, *48*, 1–17.
24 R. E. Cardoza, J. A. Vizcaino, M. R. Hermosa, S. Sousa, F. J. Gonzalez, A. Llobell, E. Monte, S. Gutierrez, *Fungal Genet. Biol.*, **2006**, *43*, 164–178.
25 H. Nakayashiki, S. Hanada, B. Q. Nguyen, N. Kadotani, Y. Tosa, S. Mayama, *Fungal Genet. Biol.*, **2005**, *42*, 275–283.
26 T. M. DeZwaan, K. D. Allen, M. M. Tanzer, K. Adachi, L. Ramamurthy, S. Mahanty, L. Hamer, *Mycology Series*, **2003**, *18*, 163–201.
27 R. Dickson, R. Lester, *Biochim. Biophys. Acta*, **1999**, *1426*, 349–357.
28 B. C. Baldwin, J. M. Clough, C. R. A. Godfrey, J. R. Godwin, T. E. Wiggins. In *Modern Fungicides and Antifungal Compounds*, H. Lyr, P. E. Russell, H. D. Sisler (Eds.), Intercept, Andover, UK, **1996**, 69–77.
29 M. M. Nagiec, E. E. Nagiec, J. A. Baltisberger, G. B. Wells, R. L. Lester, R. C. Dickson, *J. Biol. Chem.*, **1997**, *272*, 9809–9817.
30 S. M. Mandala, R. A. Thornton, M. Rosenbach, J. Milligan, M. Garcia-Calvo, H. G. Bull, M. B. Kurtz, *J. Biol. Chem.*, **1997**, *272*, 32709–32714.
31 S. M. Mandala, R. A. Thornton, J. Milligan, M. Rosenbach, M. Garcia-Calvo, H. G. Bull, G. Harris, G. K. Abruzzo, *et al.*, *J. Biol. Chem.*, **1998**, *273*, 14942–14949.
32 Y. Sugimoto, H. Sakoh, K. Yamada, *Curr. Drug Targets Infect. Disord.*, **2004**, *4*, 311–322.
33 P. A. Aeed, A. E. Sperry, C. L. Young, M. M. Nagiec, A. P. Elhammer, W. Zhong, D. J. Murphy, N. H. Georgopapadakou, *Biochemistry*, **2004**, *43*, 8483–8493.
34 W. Zhong, D. J. Murphy, N. H. Georgopapadakou, *FEBS Lett.*, **1999**, *463*, 241–244.
35 G. Dixon, D. Scanlon, S. Cooper, P. Broad, *J. Steroid Biochem. Mol. Biol.*, **1997**, *62*, 165–171.
36 D. Debieu, J. Bach, M. Hugon, C. Malosse, P. Leroux, *Pest. Manag. Sci.*, **2001**, *57*, 1060–1067.
37 J. Rine, W. Hansen, E. Hardeman, R. W. Davis, *Proc. Natl. Acad. Sci. USA*, **1983**, *80*, 6750–6754.
38 N. Fisher, B. Meunier, *Pest. Manag. Sci.*, **2005**, *61*, 973–978.
39 J. M. Evans, P. G. Zaworski, C. N. Parker, *J. Biomol. Screen.*, **2002**, *7*, 359–366.
40 D. J. Frost, K. D. Brandt, D. Cugier, R. Goldman, *J. Antibiot.* (Tokyo), **1995**, *48*, 306–310.
41 J. K. Lodge, E. Jackson-Machelski, M. Higgins, C. A. McWherter, J. A. Sikorski, B. Devadas, J. I. Gordon, *J. Biol. Chem.*, **1998**, *273*, 12482–12491.
42 H. Ebiike, M. Masubuchi, P. Liu, K. Kawasaki, K. Morikami, S. Sogabe, M. Hayase, T. Fujii, *et al.*, *Bioorg. Med. Chem. Lett.*, **2002**, *12*, 607–610.
43 M. Masubuchi, K. Kawasaki, H. Ebiike, Y. Ikeda, S. Tsujii, S. Sogabe, T. Fujii, K. Sakata, *et al.*, *Bioorg. Med. Chem. Lett.*, **2001**, *11*, 1833–1837.
44 K. Kawasaki, M. Masubuchi, K. Morikami, S. Sogabe, T. Aoyama, H. Ebiike, S. Niizuma, M. Hayase, *et al.*, *Bioorg. Med. Chem. Lett.*, **2003**, *13*, 87–91.
45 S. A. Weston, R. Camble, J. Colls, G. Rosenbrock, I. Taylor, M. Egerton, A. D. Tucker, A. Tunnicliffe, *et al.*, *Nat. Struct. Biol.*, **1998**, *5*, 213–221.
46 G. S. Basarab, D. B. Jordan, T. C. Gehret, R. S. Schwartz, *Bioorg. Med. Chem.*, **2002**, *10*, 4143–4154.
47 S. Iwata, J. W. Lee, K. Okada, J. K. Lee, M. Iwata, B. Rasmussen, T. A. Link, S. Ramaswamy, *et al.*, *Science*, **1998**, *281*, 64–71.
48 C. A. Yu, D. Xia, H. Kim, J. Deisenhofer, L. Zhang, A. M. Kachurin, L. Yu, *Biochim. Biophys. Acta*, **1998**, *1365*, 151–158.
49 C. A. Yu, H. Tian, L. Zhang, K. P. Deng, S. K. Shenoy, L. Yu, D. Xia, H. Kim, *et al.*, *J. Bioenerg. Biomembr.*, **1999**, *31*, 191–199.
50 F. Sun, X. Huo, Y. Zhai, A. Wang, J. Xu, D. Su, M. Bartlam, Z. Rao, *Cell*, **2005**, *121*, 1043–1057.

51 R. Horsefield, V. Yankovskaya, G. Sexton, W. Whittingham, K. Shiomi, S. Omura, B. Byrne, G. Cecchini, *et al.*, *J. Biol. Chem.*, **2006**, *281*, 7309–7316.

52 U. Güldener, G. Mannhaupt, M. Munsterkötter, D. Haase, M. Oesterheld, V. Stümpflen, H. W. Mewes, G. Adam, *Nucl. Acids Res.*, **2006**, *34*, D456–458.

53 J. E. Hirschman, R. Balakrishnan, K. R. Christie, M. C. Costanzo, S. S. Dwight, S. R. Engel, D. G. Fisk, E. L. Hong, *et al.*, *Nucl. Acids Res.*, **2006**, *34*, D442–445.

54 R. Mustacchi, S. Hohmann, J. Nielsen, *Yeast*, **2006**, *23*, 227–238.

55 E. C. Butcher, E. L. Berg, E. J. Kunkel, *Nat. Biotech.*, **2004**, *22*, 1253–1259.

Keywords

Target-Based Research, Agrochemical Invention, High-Throughput Technologies, *In Vitro* Screening, *In Vitro* to *In Vivo* Translation, Structure Based Design

8
Virtual Screening in Crop Protection Research

Klaus-Jürgen Schleifer

8.1
Introduction

The aim of all R&D activities in crop protection companies is to discover new active ingredients with ideal properties at lowest cost in the quickest time. This simple principle led to the introduction of a multitude of new technologies designed to accelerate the classical research process, e.g., combinatorial chemistry and high-throughput screening. However, not each chemical needs to be purchased or synthesized to test its potency in expensive field trials. Very often, knowledge derived from previous experiments and an expert's intuition are sufficient to judge the potential of compounds with reasonable quality. Increased capacity of chemicals, however, resulting from new high-throughput techniques, prevents a rating by hand. Therefore, virtual screening strategies are increasingly applied to separate "good from bad" compounds at a very early stage. The following chapter provides an overview of virtual screening strategies applied in the different stages of BASF's crop protection research process.

8.2
General Lead Identification Strategies

Potential lead structures are identified following two general screening strategies. First, chemicals may be tested directly on harmful organisms, e.g., weed, and relevant phenotype modifications such as bleaching, etc., are rated. This so-called random screening or organism–based approach indicates biological effects without knowledge of the addressed mode of action (MoA). Lead optimization strategies have to consider that several MoA's may be involved and that the observed effect reflects both bioavailability and intrinsic target activity.

The second, so-called mechanism–based approach tries to optimize target activity. This procedure relies on a suitable biochemical assay to study target protein function in the presence of screening compounds. The challenge for

Pesticide Chemistry. Crop Protection, Public Health, Environmental Safety
Edited by Hideo Ohkawa, Hisashi Miyagawa, and Philip W. Lee

ISBN: 978-3-527-31663-2

target-optimized leads is the transfer of activity from the biochemical assay to the biological system, i.e., plant, fungus or insect. Usually physicochemical features, octanol-water partition coefficient, or metabolic stability must be adjusted in order to allow the active ingredients to reach their molecular targets.

8.3 Virtual Screening Based on 1-D and 2-D Descriptors

Only a few compounds screened in early lead identification phases are synthesized in-house. More flexible and cost effective is to purchase chemicals from external suppliers. Most vendors provide lists of some ten to hundred thousand chemicals on compact discs and guarantee delivery within days to weeks. To explore this huge amount of data with the aid of computers, chemical information is transformed to computer-readable strings, e.g., smiles code, and different descriptors are determined. 1-dimensional (1-D) descriptors encode chemical composition and physicochemical properties, e.g., molecular weight, stoichiometry ($C_mO_nH_k$), hydrophobicity, etc. 2-D descriptors reflect chemical topology, e.g., connectivity indices, degree of branching, number of aromatic bonds, etc. 3-D descriptors consider 3-D shape, volume or surface area.

In order to avoid redundant molecules, the external library has to be cleansed of chemicals that are already in the corporate compound repository. For this purpose, hash coding procedures may be applied that thoroughly eliminate identical compounds [1]. A hash code is typically a highly compressed encoding of a data structure with a fixed value range and therefore a fixed bit/byte length. Applying a hierarchy of hash codes allows establishing the topological identity of atoms, bonds, molecules, and ensembles of molecules from a basic connection table. Descriptors used for hashing may be the number of bonded heavy and hydrogen neighbors of an atom, system number of elements, molecule size (number of atoms in molecule) and (formal) charge in system [1].

Alternative to hash coding, USMILES (Unique Simplified Molecular Input Line Entry System) codes may be used that even allow restoring the chemical structures [2].

Akin to Lipinski's "rule of five" [3] that predicts a poor absorption of orally administered pharmaceuticals in case of exceeding more than one of four particular molecular properties, i.e., mass, clogP, number of hydrogen bond donors and acceptors, Briggs presented his "rule of 3" for agrochemical compounds [4]. Thus, bioavailability is likely to be poor when three or more of his limits are exceeded. Tice summarized rules for insecticidal or post-emergence herbicidal activities [5] and Clarke-Delaney [6–7] suggested in their "guide of 2" that the probability of lead progression correlates with specific physicochemical features (Table I).

Although such agro relevant filters reduce the total number of compounds in question, even the decreased quantity may lead to a budget overrun. Therefore, further compound selection may be necessary by retaining characteristics of

Table I. Bioavailability enhancing properties of agrochemicals according to Briggs [4], Tice [5], and Clarke-Delaney [6–7] compared to pharmaceuticals based on Lipinsky's "Rule of Five" [3].

Authors	Briggs	Tice	Clarke-Delaney	Lipinsky
molecular weight	~ 300	150–500	200–400 (500 I)	≤ 500
$\log P_{oct}$*	≤ ~ 3		1–5 (7 I)	
mlogP		≤ 3.5 (H), 0–5 (I)		≤ 4.15
alogP		≤ 3.5 (H), 0–6.5 (I)		
clogP				≤ 5
$\Delta \log P_{(oct-alk)}$	< 3		0.5–4 (3 H, 5 I)	
H-bond donors	≤ 3	≤ 3 (H), ≤ 2 (I)	0–1	≤ 5
H-bond acceptors		2–12 (H), 1–8 (I)	0.7–2	≤ 10
melting point	< 300 °C		< 200 °C	

* exact method not indicated; mlogP [16]; alogP [17]; clogP [18];
(H) indicates postemergence herbicides; (I) indicates insecticides.

the entire data set. In order to yield the most information, each of the selected representatives should be maximally diverse compared to all others. To achieve this, different strategies may be applied: (1) Cluster sampling methods, which first identify a set of compound clusters followed by the selection of several compounds from each cluster [8–11]; (2) Grid based sampling, which places all the compounds into a low-dimensional descriptor space divided into many cells, and then chooses a few compounds from each cell [12]; (3) Direct sampling methods, which try to obtain a subset of optimally diverse compounds from an available pool by directly analyzing the diversity of the selected molecules [13–15].

All methods require compound characterization by multiple molecular descriptors and appropriate dissimilarity scoring functions must be used. The purpose of the diversity selection can be formulated as follows: select a subset S_0 of n_0 representative compounds from a database S containing n compounds, which is "the most diverse" in terms of chemical structure. The key to each of the different methods is the mathematical function that measures diversity. Since each molecule is represented by a vector of molecular descriptors, it is geometrically mapped to a point in a multidimensional space. The distance between two points, such as Euclidian distance, measures the dissimilarity between two molecules. Thus, the diversity function should be based on all pairwise distances between molecules in the subset.

An additional requirement for a diversity function is that, after the diversity value has been maximized by choosing different subtypes of molecules, the final subset that corresponds to the maximum function value is most diverse, i.e., from different clusters of points in the descriptor space. This property (distance *D*) may be described as the summation over all pairwise distances between the selected molecules (d_{ij}) [19]. In accord with Equation I, smaller values represent a more diverse and representative sampling. Exponent *a* may be set to values between 1 and 6 (Equation I).

$$D = \frac{1}{\sum_{i}^{m-1} \sum_{j>i}^{m} d_{ij}^{a}} \qquad \text{Equation I}$$

The diversity level may be individually adjusted to reach the desired number of representatives. The selected subset of compounds may now be purchased and tested in biological assays. If one of them shows activity, i.e., is a hit, a further evaluation is performed by testing similar compounds in order to identify an experimentally validated hit cluster of structurally related compounds, i.e., hit validation. For this similarity search, indicating chemicals in the multidimensional descriptor space that are close to a hit, a multitude of methods may be applied. Holliday *et al.* [20] published 22 different similarity coefficients for searching databases of 2-D fragment bit-strings. To demonstrate the principle, one example is given applying a prominent method, the Tanimoto coefficient (Equation II) [21]. This pairwise comparison counts the number of bits on (a and b) representing the presence of special features, e.g., a hydroxyl group, in molecule A and B relative to the sum of all common bits on in both molecules (c). Using the same bit strings, the Tanimoto coefficient may also be applied as a measure of diversity (i.e., $d_{Tan} = 1 - S_{Tan}$).

$$S_{Tan} = \frac{c}{a + b - c} \qquad \text{Equation II}$$

The following example indicates the bit strings (0/1 means feature absent or present) of two molecules (A and B) with four common features yielding Tanimoto similarity coefficients of 0.5.

A: (0 1 0 1 1 1 1 1 0...0 0)
B: (1 0 1 0 1 1 1 1 0...0 0)

$$S_{Tan} = \frac{4}{6 + 6 - 4} = 0.5 \quad \text{(4 common features)}$$

The virtual screening procedure described above applying 1-D and 2-D descriptors is illustrated in Figure 1.

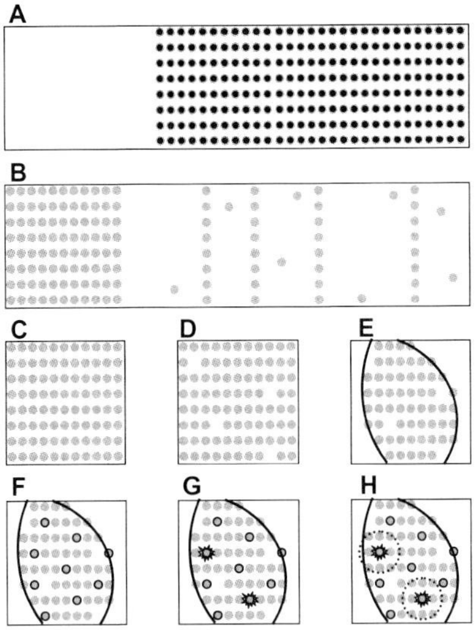

Figure 1. A: Characterization of the chemical space described by all compounds of the corporate library; **B:** Chemical space of the compounds offered by an external vendor; **C:** Unique compounds from the vendor not present in the corporate library; **D:** Elimination of toxic and reactive chemicals (white patches); **E:** Exclusion of not "agro-like" chemicals ("agro-like" space is defined by lines); **F:** Selection of diverse subsets representing the entire "agro-like" space (dark circles); **G:** Detection of two experimental hits (explosions); **H:** Selection of the hits nearest neighbors for experimental validation of the active chemotypes (dashed circles).

8.4 Virtual Screening Based on 3-D Descriptors

1-D and 2-D descriptors are fast and inexpensive to calculate and therefore ideal for screening of large libraries with millions of compounds. Sometimes, however, geometrical features like shape or a particular localization of functional groups in 3-D space are necessary to describe molecules unambiguously. In this situation, it is necessary to use conformers of active (and non-active) compounds as reference templates for a ligand-based screening strategy. A second, even more promising concept may be applied in the presence of the 3-D coordinates of the molecular target protein. This so-called structure-based ligand design uses the binding site as a lock in order to find virtually the best fitting key, i.e., new ligand.

8.4.1 Ligand-Based Screening Strategies

Output from biochemical high-throughput screenings is a list of active compounds with indicated activity values, e.g., IC_{50}. On a three-dimensional (3-D) level, relevant conformers of active molecules may be superimposed in order to locate similar

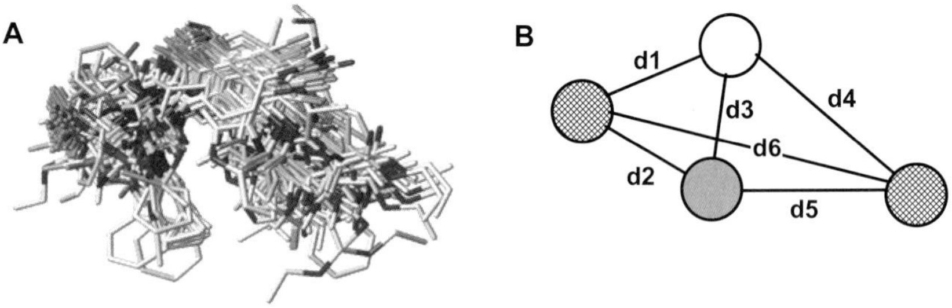

Figure 2. A: Agrophore model derived from 318 *Protox* inhibitors. B: Hypothetical 4-point agrophore model based on four molecular key functions (circles) and relevant spatial interfunction distances (d1–d6).

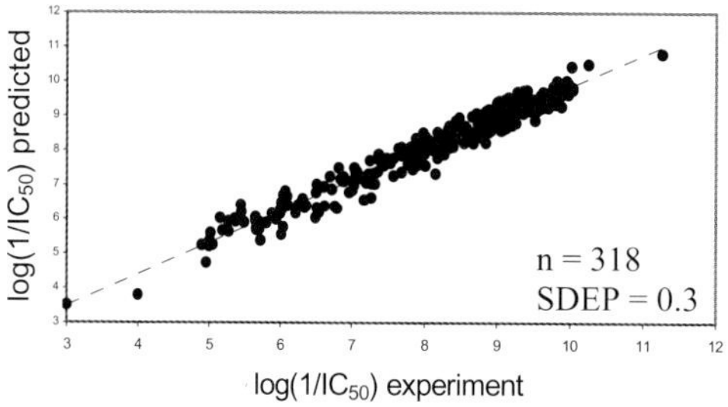

Figure 3. Statistics derived from a comparative molecular similarity indices analysis (CoMSIA) [26] for the agrophore model shown in Figure 2A. The graph depicts the predictive power of a leave-one-out cross-validation procedure for 318 *Protox* inhibitors (SDEP = standard error of prediction).

molecular functions crucial for biochemical activity in the same 3-D space. The resulting agrophore model (Figure 2A), synonymous with the pharmacophore model for drugs, should be able to differentiate active and non-active compounds by the presence or absence of particular functions in specific regions. Based on the spatial distribution of molecular key functions, 3-D pharmacophore fingerprints may be extracted and used to detect similar molecules in 3-D compound libraries (Figure 2B) [22–25].

While 3-D pharmacophore/agrophore searches indicate best matches of relevant key functions, i.e., (semi)qualitative result, a linear regression derived from a statistical analysis of spatial features and measured activities yields quantitative structure-activity relationships (3-D QSAR). This may be extremely helpful for a detailed interpretation of existing results and the activity prediction of new or hypothetical compounds (Figure 3).

In addition to statistical results, favorable and unfavorable volumes for specific features, e.g., steric or electrostatic, may be visualized to interpret the results in a more intuitive way (Figure 4).

Ligand-based approaches are generally used to accurately predict the activity of derivatives with optimized substitution pattern. In some cases, however,

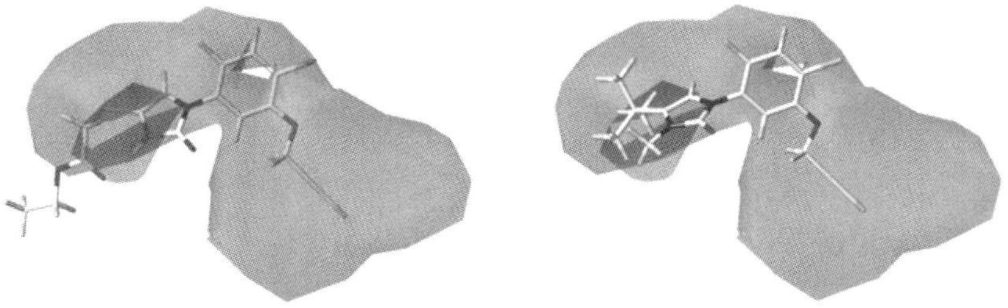

Figure 4. Sterically favorable volume derived from a 3-D QSAR study for *Protox* inhibitors. Molecules totally enclosed by the volume are, with respect to this particular property, more active (left) than derivatives with protruding residues (right).

especially the visual analysis is helpful for a scaffold-hopping procedure [27–28] that seeks to modify the molecular framework (e.g., phenyl *vs.* pyrimidine) while retaining critical residues at the same 3-D position. Benefits might be a more suitable class of compounds, e.g., with higher stability, or a broader intellectual property position.

8.4.2 Structure-Based Screening Strategies

Virtual screening concepts are most successful if coordinates of the molecular target protein are available. In this situation, the binding site of the target protein (*the* structure) is used as a lock in order to find the best fitting keys (ligands). Since this approach is independent of any further ligand information, it allows an unbiased probing of new scaffolds (Figure 5).

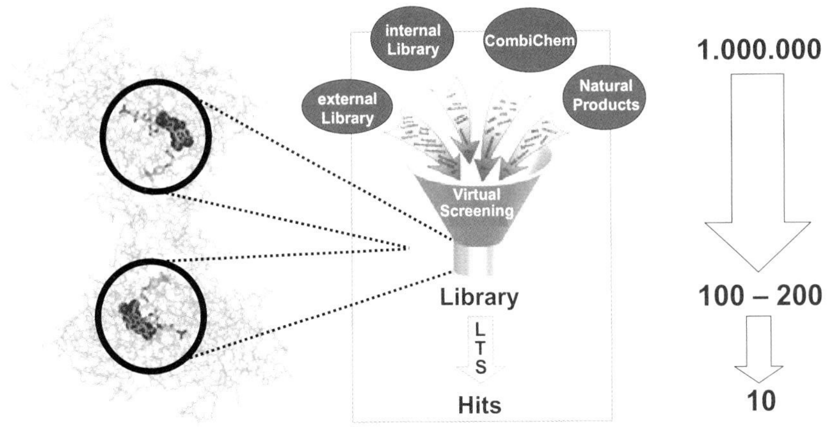

Figure 5. General procedure of a structure-based screening approach: 3-D coordinates of relevant binding sites are chosen as filters to reduce the number of compounds to a reasonable quantity.

A structure-based virtual screening starts with the docking of each compound into the binding site pocket. Based on complementary features, best fitting poses are searched for each ligand. In general, three principal algorithmic approaches are used to dock small molecules into macromolecular binding sites [29–30]. The first one separates the conformer generation of the ligands from the placement in the binding site. Subsequently, all relevant low-energy conformations are rigidly placed in the binding site (e.g., GLIDE (Schrödinger, Inc.) and FRED (OpenEye Scientific Software)). Only the remaining six rotational and translational degrees of freedom of the rigid conformer have to be considered. A second class of algorithms aims at optimizing the conformation and orientation of the molecule in the binding pocket simultaneously. The tremendous complexity of this combined optimization problem needs stochastic algorithms such as genetic algorithms or Monte Carlo simulations. The third class of docking algorithms exploits the fact that most molecules contain at least one small, rigid fragment that forms specific, directed interactions with the receptor site. Such so-called base fragments are docked rigidly at various initial positions. An incremental construction process explores the torsional conformational space adding now new fragments (e.g., FlexX [31]). Some programs apply further force field minimizations in order to optimize the actual pose of the fragments (e.g., Glide XP, eHits, ICM).

Subsequent to the detection of appropriate binding poses, scoring functions are used to estimate the free energy of ligand binding for each conformer. Commonly used scoring functions may be divided into three general categories [32–35]. Most important are empirical scoring functions. They approximate the free energy of binding as a weighted sum of terms, each term being a function of the ligand and protein coordinates. Each term describes different types of interactions such as lipophilic contacts or hydrogen bonds between receptor and ligand. The weighting factors are derived by multiple linear regression to experimental binding affinities or by approximate first principle considerations. A second class of scoring functions is based on molecular force fields, summing up the electrostatic and van der Waals interaction energies between receptor site and ligand. Knowledge-based scoring functions are derived from statistical analyses of experimentally determined protein-ligand X-ray structures. The underlying assumption is that inter-atomic contacts occurring more frequently than average are energetically favorable. Knowledge-based scoring functions are sums of many atom-pair contact distributions for protein and ligand atom type combinations.

The accuracy of docking procedures may be evaluated in two ways. First, by detecting known ligand-binding site poses of crystallized complexes, and second, by the enrichment of known active compounds of a mixed library, i.e., enrichment factor.

To illustrate the first feature, a typical re-docking result of the co-crystallized pyrazol (*INH*) *Protox* inhibitor is shown (Figure 6). In this study, complex formation of ligand and binding site is mainly driven by a directed hydrogen bond interaction of the carboxyl group of *INH* to a critical arginine residue (Arg) at the entrance of the binding pocket.

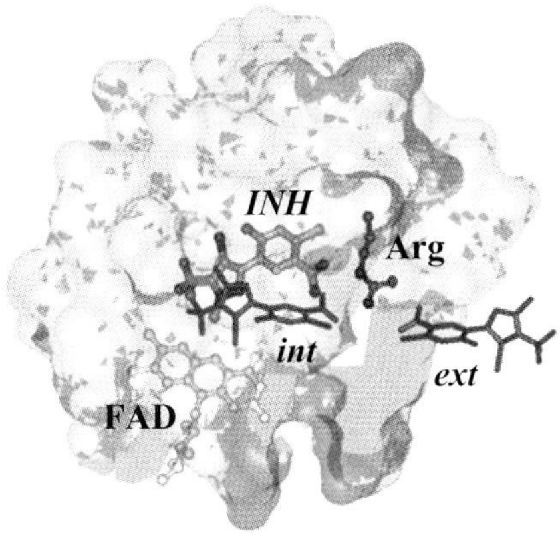

Figure 6. Binding site region of Protoporphyrinogen IX Oxidase 2 (*Protox*, pdb code 1SEZ) with the co-crystallized inhibitor *INH* (ball and sticks representation), the critical amino acid residue arginine (Arg) and the co-factor FAD. Re-docked solutions are indicated inside (int) and outside (ext) of the detected binding pocket.

This re-docking result obviously overestimates the hydrogen bond, since the energetically most favorable docking solutions locate *INH* outside the pocket in order to optimize this particular interaction. Only lower-ranked solutions are positioned in the detected binding site, although not congruent with the X-ray structure. In order to eliminate obviously false solutions, post-processing procedures or other scoring functions balancing hydrophobic and hydrophilic interactions more thoroughly may be applied.

A reasonable result is shown for the BASF *Protox* inhibitor *BPI* (Figure 7). In this solution, the acid function of *INH* is mimicked by the nitrogen atom of the benzothiazole moiety of *BPI* forming the crucial hydrogen bond to arginine. This facilitates a good match for the trifluoromethyl-pyrazol of *INH* and the trifluoromethyl-pyrimidinedione of *BPI*.

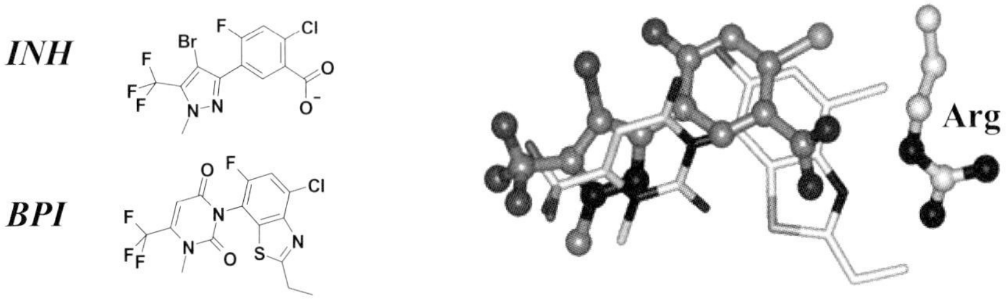

Figure 7. Docking solution for BASF's inhibitor *BPI* to the *Protox* binding site. For clarity, *INH* and *BPI* are superimposed indicating the relative orientation and mimicking functions for complexation with the critical arginine residue (Arg).

The crucial metric for assessing the performance of docking codes is the extent to which a dataset of compounds can be enriched such that only a much smaller subset needs to be prepared and assayed to identify hits or leads. Following Pearlman and Charifson [36], the enrichment factor can be calculated according to Equation III.

$$EF = \{N_{\text{total}} / N_{\text{sampled}}\} \{Hits_{\text{sampled}} / Hits_{\text{total}}\} \qquad \text{Equation III}$$

Thus, if only 10% of the scored and ranked database (i.e., $N_{\text{total}} / N_{\text{sampled}} = 10$) needs to be assayed to recover all of the active hits ($Hits_{\text{total}}$), the enrichment factor would be 10. But if only half the total number of known actives are found in the first 10% (i.e., if $Hits_{\text{sampled}} / Hits_{\text{total}} = 0.5$), the effective enrichment factor would be 5.

Enrichment factors may also be illustrated in a graphical manner as accumulation curves that show how the fraction of actives recovered varies with the percent of the database screened (Figure 8). While the diagonal shows results expected by chance, the dark line indicates the results obtained by use of a structure-based screening protocol. An enrichment factor of 3.5 is yielded when 20% of the database are screened (i.e., 69% actives are recovered). If 50% of the database are assayed, the enrichment factor is decreased to 1.8.

The crucial step to judge structure-based approaches is the careful visual inspection of all top scored ligands. An expert should appraise the results with respect to plausible poses of the hits, the presence of key interactions with the binding site residues and different docking solutions for the same ligand. If these and some other requirements are fulfilled, the virtual screening results would have been proved to be extremely valuable.

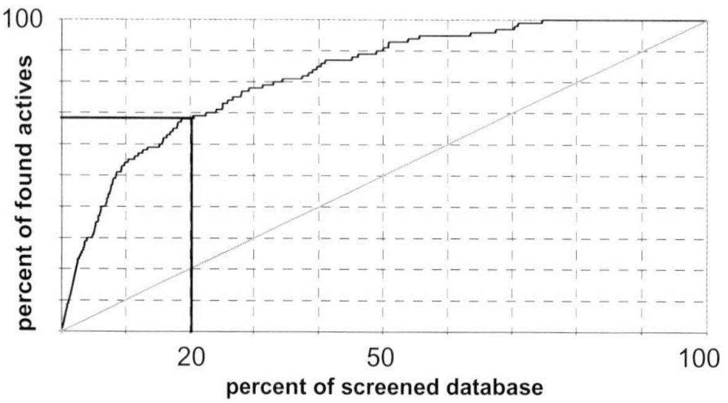

Figure 8. Graph illustrating the enrichment of found actives by a virtual screening protocol (dark line) relative to results expected by chance (grey diagonal). Assaying 20% of the database recovers 69% of all actives (enrichment factor = 3.5).

8.5 Conclusion

Virtual screening is a successful tool to prioritize representative compounds for testing. However, compared to orally administered drugs, only few (if any) rules are universally applicable to all crop protectants so that a thorough descriptor selection (1-D, 2-D or 3-D) is crucial and virtual and experimental findings have to be constantly compared and adjusted. Furthermore, an open dialog between the bench chemist and computational chemist is necessary in order to gain a common understanding of chemical feasibility, thus preventing rejection of successfully predicted hits in ongoing optimization cycles.

Although the aim of virtual screening was originally to discriminate active from non-active ingredients, other relevant characteristics like absorption, distribution, metabolism, excretion, and toxicological (ADME-Tox) behavior may nowadays be considered [37]. The quality of the computational predictions is presently limited by the lack of sufficient experimental data for crop protection relevant plants, fungi or insects. The huge potential of this cheap and fast technology, however, was recognized and is reflected by an extensive integration in classical research processes of all innovative crop protection companies.

8.6 References

1 W. D. Ihlenfeldt, J. Gasteiger, *J. Comp. Chem.*, **1994**, *15*, 793–813.
2 D. Weininger, A. Weininger, J. L. Weininger, *J. Chem. Inf. Comput. Sci.*, **1989**, *29*, 97–101.
3 C. A. Lipinski, F. Lombardo, B. W. Dominy, P. J. Feeney, *Adv. Drug Delivery Rev.*, **1997**, *23*, 3–25.
4 C. C. Briggs, "*Uptake of Agrochemicals & Pharmaceuticals*", *Predicting Uptake & Movement of Agrochemicals from Physical Properties*, SCI Meeting, **1997**.
5 C. M. Tice, *Pest Manag. Sci.*, **2001**, *57*, 3–16.
6 E. D. Clarke, J. S. Delaney in *Designing Drugs and Crop Protectants: Processes, Problems and Solutions* (M. Ford, D. Livingstone, J. Dearden, H. V. de Waterbeemd (Eds.), Blackwell Publishing, Malden, **2003**.
7 E. D. Clarke, J. S. Delaney, *Physical and Molecular Properties of Agrochemicals: An Analysis of Screen Inputs, Hits, Leads and Products*, Chimia (Aarau), **2003**, *57*, pp. 731–734.
8 P. Willet, V. Winterman, D. Bawden, *J. Chem. Inf. Comput. Sci.*, **1986**, *26*, 109–118.
9 R. G. Lawson, P. C. Jurs, *J. Chem. Inf. Comput. Sci.*, **1990**, *30*, 137–144.
10 L. Hode, *J. Chem. Inf. Comput. Sci.*, **1989**, *29*, 66–71.
11 R. D. Brown, Y. C. Martin, *J. Chem. Inf. Comput. Sci.*, **1996**, *36*, 572–584.
12 R. S. Pearlman, *Network Science*, **1996**.
13 E. J. Martin, J. M. Blaney, M. A. Siani, D. C. Spellmeyer, A. K. Wong, W. H. Moos, *J. Med. Chem.*, **1995**, *38*, 1431–1436.
14 D. K. Agrafiotis, *J. Chem. Inf. Comput. Sci.*, **1997**, *37*, 841–851.
15 M. Hassan, J. P. Bielawski, J. C. Hempel, M. Waldman, *Mol. Divers.*, **1996**, *2*, 64–74.
16 I. Moriguchi, S. Hirono, Q. Liu, I. Nakagome, Y. Matsushita, *Chem. Pharm. Bull.*, **1992**, *40*, 127–130.
17 V. N. Viswanadhan, A. K. Ghose, G. R. Revankar, R. K. Robins, *J. Chem. Inf. Comput. Sci.*, **1989**, *29*, 163–172.

18 A. J. Leo, Chem. Rev., **1993**, *93*, 1281–1304.

19 W. Zheng, S. J. Cho, C. L. Waller, A. Tropsha, *J. Chem. Inf. Comput. Sci.*, **1999**, *39*, 738–746.

20 J. D. Holliday, C.-Y. Hu, P. Willet, *Comb. Chem. HTS*, **2002**, *5*, 155–166.

21 H. Matter, *J. Med. Chem.*, **1997**, *40*, 1219–1229.

22 J. W. Godden, L. Xue, J. Bajorath, *J. Chem. Inf. Comput. Sci.*, **2000**, *40*, 163–166.

23 B. R. Beno, J. S. Mason, *Drug Discov. Today*, **2001**, *6*, 251–258.

24 J. S. Mason, A. C. Good, E. J. Martin, *Curr. Pharm. Des.*, **2001**, *7*, 567–597.

25 T. Langer, R. D. Hoffmann (Eds.), *Pharmacophores and Pharmacophore Searches, Methods and Principles in Medicinal Chemistry (Band 32)*, Wiley-VCH, Weinheim, **2006**.

26 G. Klebe, U. Abraham, T. Mietzner, *J. Med. Chem.*, **1994**, *37*, 4130–4146.

27 J. L. Jenkins, M. Glick, J. W. Davies, *J. Med. Chem.*, **2004**, *47*, 6144–6159.

28 Q. Zhang, I. Muegge, *J. Med. Chem.*, **2006**, *49*, 1536–1548.

29 I. Muegge, M. Rarey in *Reviews in Computational Chemistry*, K. B. Lipkowitz, D. B. Boyd (Eds.), Wiley-VCH, New York, **2001**.

30 A. J. Orry, R. A. Abagyan, C. N. Cavasotto, *Drug Discov. Today*, **2006**, *11*, 261–266.

31 M. Rarey, B. Kramer, T. Lengauer, G. Klebe, *J. Mol. Biol.*, **1996**, *261*, 470–489.

32 A. Ajay, M. A. Murcko, *J. Med. Chem.*, **1995**, *38*, 4953–4967.

33 H.-J. Boehm, M. Stahl, *Med. Chem. Res.*, **1999**, *9*, 445–462.

34 H. Gohlke, G. Klebe, *Curr. Opin. Struct. Biol.*, **2001**, *11*, 231–235.

35 J. R. H. Tame, *J. Comput. Aided Mol. Design*, **1999**, *13*, 99–108.

36 D. A. Pearlman, P. S. Charifson, *J. Med. Chem.*, **2001**, *44*, 502–511.

37 O. Roche, W. Guba, *Mini Rev. Med. Chem.*, **2005**, *5*, 677–683.

Keywords

Virtual Screening, Ligand-Based Ligand Design, 3-D QSAR, Structure-Based Ligand Design, Docking, Scoring, Enrichment Factor, ADME-Tox Prediction

9
Synthesis of Triazolo[1,5-c]pyrimidine Sulfonamides Leading to the Discovery of Penoxsulam, a New Rice Herbicide

Timothy C. Johnson, Timothy P. Martin, Rick K. Mann

9.1
Introduction

The triazolopyrimidine sulfonamide class of herbicides has been studied extensively since their discovery in the early 1980s. The first triazolopyrimidine sulfonamide was discovered while examining bioisosteres of the sulfonylureas [1]. Subsequent SAR studies led to the discovery of flumetsulam (**1**) and metosulam (**2**) (Figure 1). Flumetsulam (**1**) was developed for broadleaf weed control in maize and soybeans while metosulam (**2**) was developed for broadleaf weed control in maize and cereals. Studies have shown the triazolopyrimidine sulfonamides to be competitive with the amino acid leucine for binding to acetolactate synthase (ALS) isolated from cotton (*Gossypium hirsutum*) [2]. The same study showed similar results for the bioisosteric sulfonylurea herbicides.

Synthetic studies focused on the bicyclic heterocycle led to the discovery of a new sub-class of sulfonamides where the triazolo[1,5-a]pyrimidine ring was replaced with a triazolo[1,5-c]pyrimidine ring. Further investigations led to the development of diclosulam (**3**) and cloransulam-methyl (**4**) for broadleaf weed control in soybeans, and florasulam (**5**) for broadleaf weed control in cereals. To fully explore this sub-class of sulfonamides, an investigation was initiated to determine if reversing the sulfonamide linkage (**6**) would lead to compounds with the spectrum of activity on weeds and crop selectivity different from **3**, **4**, and **5**.

9.2
Chemistry

A general route for the preparation of compounds such as **6** required the synthesis of appropriately substituted 2-aminotriazolo[1,5-c]pyrimidines **7** (Scheme 1) [3–6]. Toward that end, appropriately substituted 4-hydrazino-2-methylthiopyrimidines **8** were reacted with cyanogen bromide to give the 3-amino-5-methylthiotriaz olo[4,3-c]pyrimidines **9**, usually as the hydrogen bromide salts. When treated

Pesticide Chemistry. Crop Protection, Public Health, Environmental Safety
Edited by Hideo Ohkawa, Hisashi Miyagawa, and Philip W. Lee

ISBN: 978-3-527-31663-2

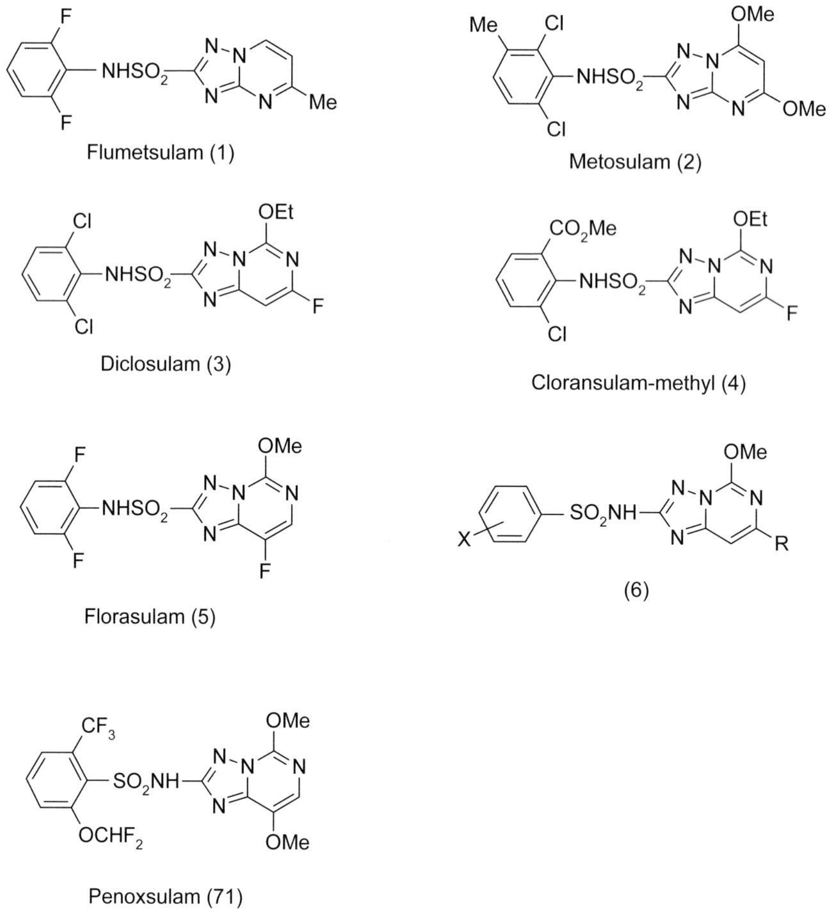

Figure 1. Commercial triazolopyrimidine herbicides and new active analogs.

with sodium methoxide, **9** undergoes a Dimroth rearrangement with loss of the methylthio group to form the desired 2-amino-5-methoxytriazolo[1,5-c]pyrimidines **7**. In this reaction, the Michael acceptor ethyl acrylate was used to consume the mercaptide by-product.

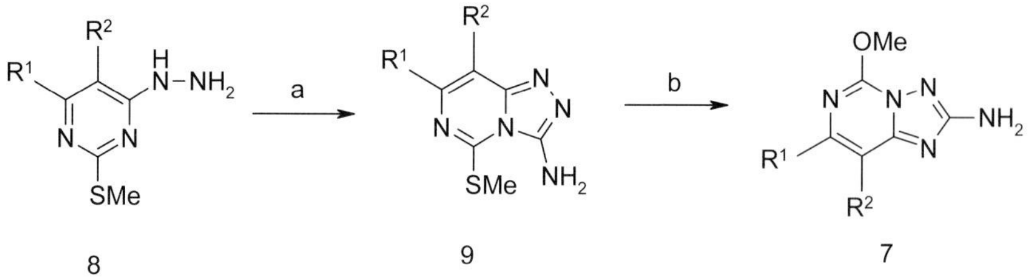

Scheme 1. Reagents: (a) BrCN, i-PrOH, (b) NaOMe, MeOH, ethyl acrylate.

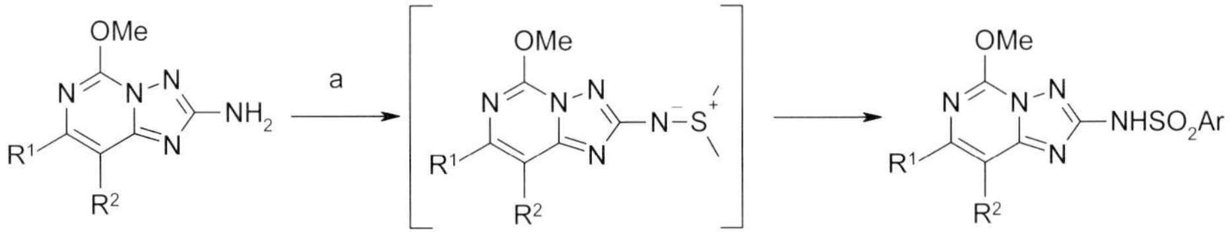

7 10 6

Scheme 2. Reagents: (a) Pyridine, DMSO (catalytic), $ArSO_2Cl$, CH_3CN.

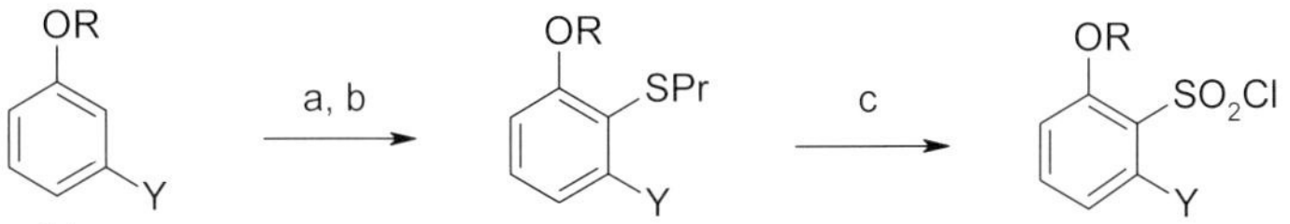

11 12 13

Y = Cl, F

Scheme 3. Reagents: (a) BuLi (1.2 eq), TMEDA, THF or Et_2O, (b) $(PrS)_2$, (c) Cl_2, H_2O, HOAc.

The target triazolo[1,5-c]pyrimidine sulfonamides **6** were prepared via coupling **7** with substituted benzenesulfonyl chlorides (Scheme 2). In these transformations, pyridine and a catalytic amount of dimethylsulfoxide (DMSO) were essential for the formation of **6**. The reactive intermediate is believed to be the *in situ* generated sulfilimine **10** which allows for the relatively non-nucleophilic amines to react with sulfonyl chlorides under mild conditions [7–8].

A number of substituted benzenesulfonamide derivatives of **6** have been investigated. With respect to crop selectivity and activity on target weeds, 2,6-disubstitued benzene rings were found to be optimal. The following section will focus on the synthesis of 2,6-disubstituted benzenesulfonyl chlorides.

Ortho-directed metalation was chosen as a general method applicable to the preparation of many 2,6-disubstituted benzenesulfonyl chlorides [3–6]. 1-Alkoxy-3-halobenzenes **11** were metalated with n-butyllithium and then quenched with dipropyl disulfide to furnish sulfides **12** (Scheme 3). Sulfides **12** were converted to the 1-alkoxy-3-halobenzenesulfonyl chlorides **13** through the use of chlorine gas in aqueous acetic acid.

A similar approach was used for the preparation of 1,3-dialkoxybenzenesulfonyl chlorides **14** (Scheme 4). 1,3-Dialkoxybenzenes **15** were sequentially treated with n-butyllithium and sulfur dioxide to furnish lithium benzenesulfinates **16**. Subsequent oxidation of **16** with sulfuryl chloride afforded 2,6-dialkoxybenzenesulfonyl

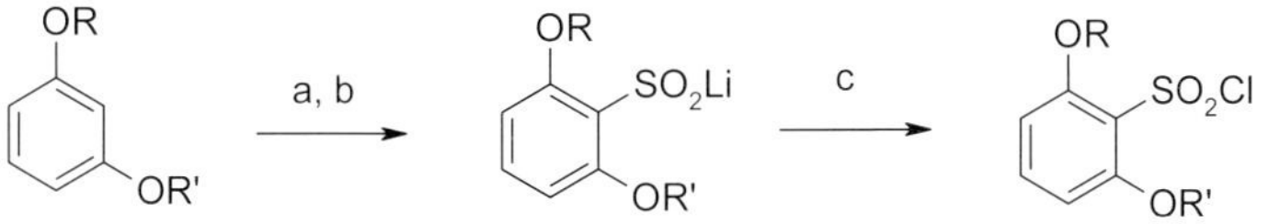

15 16 14

Scheme 4. Reagents: (a) BuLi (1.2 eq), TMEDA, THF or Et_2O, (b) SO_2, Et_2O, (c) SO_2Cl_2, hexane.

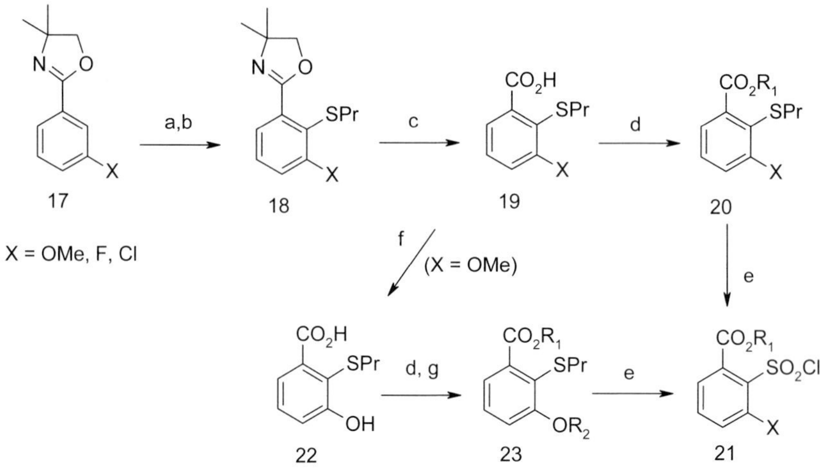

Scheme 5. Reagents: (a) BuLi (1.2 eq), TMEDA, THF or Et_2O, (b) $(PrS)_2$, (c) Cl_2, H_2O, HOAc/HCl, MeOH, (d) R_1OH, H_2SO_4, (e) Cl_2, H_2O, HOAc, (f) BBr_3, (g) NaH.

chlorides **14**. Attempts to prepare dialkoxybenzenesulfonyl chlorides by oxidation of the sulfides (not shown) resulted in ring chlorination prior to sulfonyl chloride formation.

2-Chlorosulfonyl benzoates were synthesized from 2-aryldimethyloxazolines **17** (Scheme 5). Sequential treatment with n-butyllithium and dipropyl disulfide afforded sulfides **18**. The oxazolines **18** were then hydrolyzed to the benzoic acids **19** under acidic conditions. Fisher esterification gave the esters **20**, which were oxidized to the corresponding sulfonyl chlorides **21** using chlorine and aqueous acetic acid. When methoxy-substituted benzoic acids (**19**, X = OMe) were employed, further manipulations were performed prior to oxidation. For example, phenol (**22**) was prepared by demethylation with boron tribromide, subjected to esterification, and then alkylated by a variety of electrophiles to provide **23**. As before, **23** was oxidized using chlorine and aqueous acetic acid.

Modification of the experimental procedure used for the previous ortho lithiation strategies was necessary for the preparation of trifluoromethyl- and trifluoromethoxy-substituted benzenesulfonyl chlorides. When 3-trifluoromethylanisole (**24**, PG = Me, Scheme 6) was subjected to conditions identical to those in

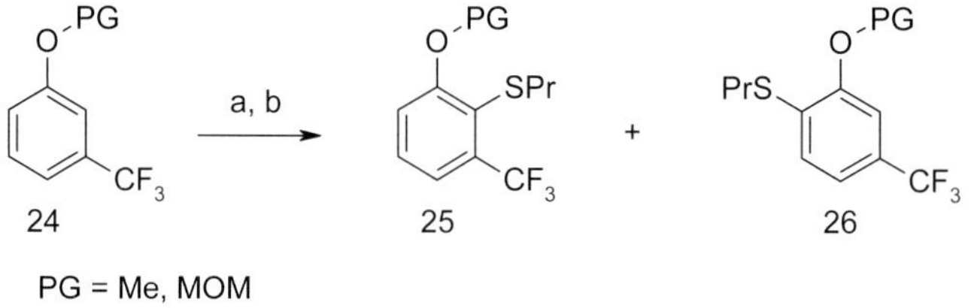

Scheme 6. Reagents: (a) BuLi (1.2 eq), THF or Et_2O, (b) $(PrS)_2$.

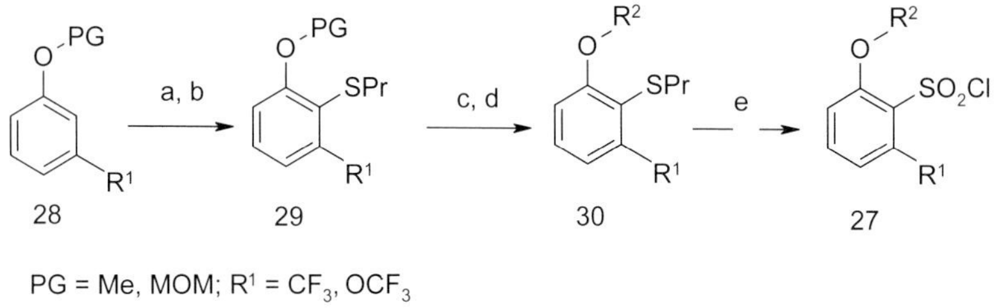

Scheme 7. Reagents: (a) BuLi (0.95 eq), TMEDA, i-Pr_2NH (10 Mol %), THF or Et_2O, (b) $(PrS)_2$, (c) PG removal, (d) alkylation, (e) Cl_2, H_2O, HOAc.

Scheme 3, the desired 1,2,3-trisubstituted benzene (**25**, PG = Me, 82%) was isolated along with inseparable 1,2,5-trisubstituted benzene (**26**, PG = Me, 10%) [4, 9–10]. Furthermore, when the protecting group was methoxymethyl (**24**, PG = MOM), the yield of undesired product increased (**26**, PG = MOM, 80%). However, when the lithiation was carried out under conditions of thermodynamic control, **25** was formed as the sole product. Thus, when **24** was treated with a catalytic amount of diisopropylamine and a substoichiometric quantity of n-butyllithium, allowed to equilibrate, and then treated with dipropyl disulfide **25** was isolated in > 99% yield [11]. This method was used to prepare a number of 2-alkoxy-6-trifluoromethylbenzenesulfonyl chlorides (**27**; $R^1 = CF_3$) and 2-alkoxy-6-trifluoromethoxylbenzenesulfonyl chlorides (**27**; $R^1 = OCF_3$) (Scheme 7). The protected phenols **28** were converted to the sulfides **29** using the above conditions, the protecting group removed, and the phenols reacted with various electrophiles to afford **30**. The previously described oxidation conditions were used to convert **30** to the sulfonyl chlorides **27**.

9.3 Biology

Unless otherwise noted, the *in vivo* greenhouse screening data presented in the following sections is a tabulation of postemergence foliar applied results and expressed as a "percent in growth reduction" (GR) for treated plants, where 0 is no effect and 100 is complete kill, as visually compared to untreated plants. The broadleaf weed activity (BW) is given as an average of 80% reduction in growth at a given concentration, expressed in parts per million (ppm), over five to eight broadleaf weeds chosen from the following: *Xanthium strumarium, Datura stramonium, Chenopodium album, Helianthus* spp., *Ipomoea* spp., *Amaranthus retroflexus, Abutilon theophrasti, Veronica heteraefolia, Ipomoea hederacea, Stellaria media*, and *Polygonum convolvulus*. The grass weed activity (GW) is averaged over five weeds chosen from *Alopecurus* spp., *Echinochloa crus-galli, Setaria fabari, Sorghum halapense, Digitaria sanguinalis*, and *Avena fatua*, and expressed in a manner similar to broadleaf weeds. The injury for a specified crop is expressed as a 20% reduction in growth compared to the untreated crop. The *in vitro* activity is given as the concentration of compound, expressed in nano molar (nM), to affect 50% inhibition (I_{50}) of the enzyme, acetolactate synthase (ALS).

Initial efforts focused on testing several analogs to determine if triazolo[1,5-c] pyrimidine sulfonamides such as **6** would have herbicidal activity comparable to their sulfonanilide counterparts (**3**, **4**, and **5**). A summary of the herbicidal activity for the first analogs prepared (**31**, **32**, and **33**) and florasulam (**5**) is presented in Table I. Test results demonstrated that **33** is more active than **5** on grass weeds and has comparable activity to **5** on broadleaf weeds. The activity of **32** on broadleaf weeds was slightly less than **5** and **31** had weak activity on both grass and broadleaf weeds. Additionally, **32** showed a trend for selectivity to wheat (*Triticum aestivum*) similar to **5**. The trends observed for *in vivo* activity correlates with the activity observed *in vitro*. The high levels of activity observed for **33** on grass weeds was interesting and warranted further investigation.

A comparison of *in vivo* and *in vitro* activity for compounds with various substituents in the 7- and 8-positions (R^7 and R^8, respectively) of the triazolo[1,5 -c]pyrimidine ring is given in Table II. In general, the *in vivo* and *in vitro* activity is greater for compounds with substitutions in the 8-position (**33**–**39**) than those with substitution in the 7-position (**40**–**42**). Compounds with substitution in the 7-position have weak activity on both grass and broadleaf weeds. The highest levels of activity on both grass and broadleaf weeds are observed with the 8-methoxy (**33**) and 8-chloro (**34**) substitution. Unfortunately, **33** and **34** also cause significant injury to rice (*Oryza sativa*). Increasing the alkoxy size from methoxy (**33**) to ethoxy (**39**) causes a loss in activity on both grass and broadleaf weeds. The 8-bromo (**35**) and 8-iodo (**36**) substituted compounds have good *in vivo* activity, but the activity on broadleaf and grass weeds is somewhat less than that observed for **33** and **34**.

An initial effort to compare the activity for various substituents on the phenyl ring was accomplished by preparing sulfonamides from commercially available ortho-substituted benzenesulfonyl chlorides. The *in vivo* and *in vitro* activity for the ortho-substituted sulfonamides is summarized in Table III. With the exception of the nitro substitution (**50**), these molecules have good activity on broadleaf weeds. However, only the methoxy (**46**), trifluoromethoxy (**47**), and methyl ester (**48**) have good levels of activity on grass weeds. Additionally, all of these sulfonamides are very injurious to rice, even those with weak levels of overall grass activity.

Disubstitutions around the phenyl ring were investigated by preparing sulfonamides from the commercially available dichlorobenzenesulfonyl chlorides. The activity for these compounds is summarized in Table IV. While all the compounds have good levels of *in vitro* activity, only the 2,6-dichloro (**52**) has good levels of activity on grass and broadleaf weeds. The activity on broadleaf weeds is better when one ortho substituent is present, which can be seen by comparing the activity of **52** and **53** to that of **54**.

In the above table, compound 52 is the same as compound 33. The biological data reported for 33 and 52 is similar but not identical.

Based on the structure-activity relationships shown in Tables I–IV, further investigations were initiated to determine if crop selectivity could be found while maintaining activity on grass weeds. These efforts focused primarily on probing

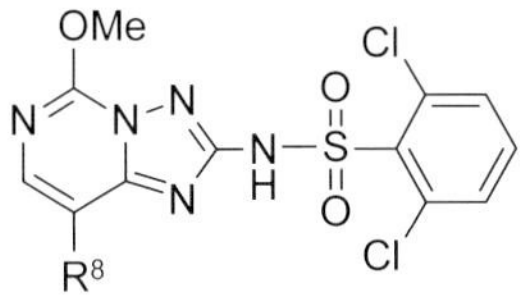

Table I. Herbicidal activity of initial analogs compared to florasulam (5).

Compound	R^8	BW GR_{80} (ppm)	GW GR_{80} (ppm)	ALS I_{50} (nM)	*Oryza sativa* GR_{20} (ppm)	*Triticum aestivum* GR_{20} (ppm)
31	H	475	> 2000	1040	500	1000
32	F	16	300	12	> 1000	> 1000
33	OMe	2	10	0.6	8	4
5		2	> 31	10	< 1	> 16

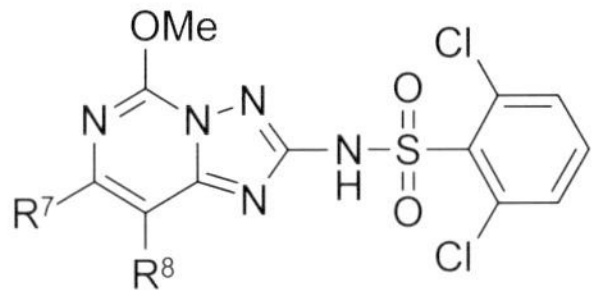

Table II. Herbicidal activity for substitutions in the 7- and 8-positions of the triazolopyrimidine ring.

Compound	R^8	R^7	BW GR_{80} (ppm)	GW GR_{80} (ppm)	*Oryza sativa* GR_{20} (ppm)	ALS I_{50} (nM)
33	OMe	H	2	10	8	0.6
34	Cl	H	1	15	4	0.6
35	Br	H	11	105	16	0.4
36	I	H	6	27	< 1	0.1
37	SMe	H	> 125	> 125	31	NA
38	Me	H	< 15	> 250	250	7
39	OEt	H	216	> 500	250	2
40	H	OMe	> 1000	> 1000	> 1000	195
41	H	F	> 250	> 250	8	49
42	H	Me	62	574	250	27

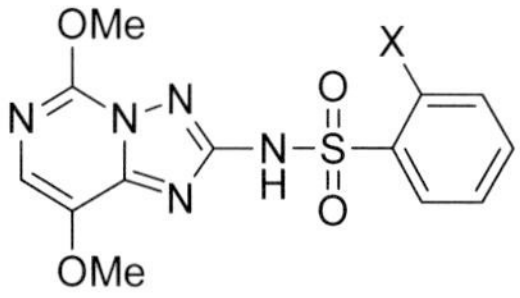

Table III. Herbicidal activity for ortho substituted phenyl analogs.

Compound	X	BW GR_{80} (ppm)	GW GR_{80} (ppm)	*Oryza sativa* GR_{20} (ppm)	ALS I_{50} (nM)
43	H	7	323	20	11
44	F	18	423	< 1	7
45	Cl	2	74	< 1	2
46	OMe	1	1	< 1	0.6
47	OCF_3	3	12	< 1	3
48	CO_2Me	14	24	< 1	13
49	CF_3	13	435	< 1	107
50	NO_2	152	> 250	1	43
51	Me	5	160	1	13

2,6-disubstitutions on the phenyl ring with combinations of substituents where at least one of the substituents was methoxy, trifluoromethoxy, or ester, the three most active ortho substituents from Table III, and keeping the most active 5,8-dimethoxy substitution pattern on the triazolo[1,5-c]pyrimidine ring. The results of these investigations are summarized for selected compounds in Table V. In general, a large number of compounds show high levels of activity on both grass and broadleaf weeds and the majority of these compounds are also very injurious to crops such as rice. However, when one of the substituents is trifluoromethyl and the other a higher alkoxy, such as ethoxy (**61**) or fluoroethoxy (**62**), the level of injury to rice is significantly less than when the substituents are methoxy and trifluoromethyl (**59**). In addition, **61** and **62** maintained good levels of activity on barnyard grass (*Echinochloa crus-galli*) even though the overall level of activity on grass weeds was poor (GW GR_{80} = 191 and 66, respectively) compared to many of the other compounds.

Based on the results shown in Table V, a number of 2-trifluoromethylphenyl analogs were prepared with various alkoxy and substituted alkoxy groups in the 6-position of the phenyl ring. A number of these molecules demonstrated trends for selectivity toward rice with activity on barnyard grass and broadleaf weeds. Table VI summarizes the activity on rice and key rice weeds for specific 2-alkoxy-6-trifluoromethylphenyl substituted analogs when applied as a water-injected treatment in the greenhouse to rice and weeds in the 1–3 leaf stage.

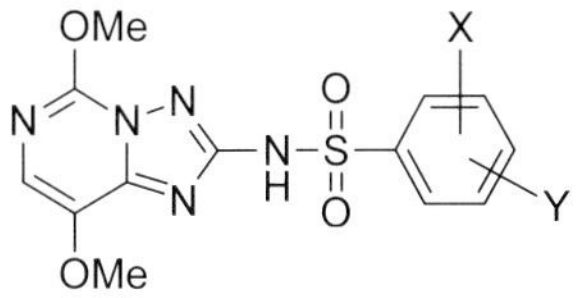

Table IV. Herbicidal activity for dichloro substitutions on the phenyl ring.

Compound	X	Y	BW GR_{80} (ppm)	GW GR_{80} (ppm)	ALS I_{50} (nM)
52	2-Cl	6-Cl	2	10	0.6
53	2-Cl	5-Cl	19	> 1000	0.4
54	3-Cl	5-Cl	126	> 1000	1

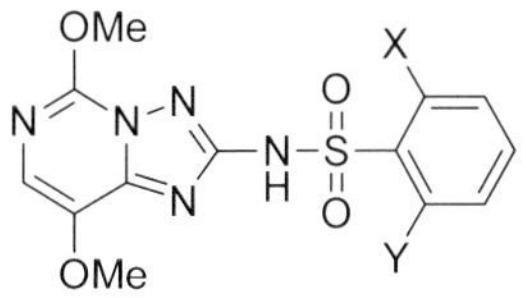

Table V. Herbicidal activity for various 2,6-disubstituted phenyl analogs.

Compound	X	Y	BW GR_{80} (ppm)	GW GR_{80} (ppm)	*Oryza sativa* GR_{20} (ppm)	*Echinochloa crus-galli* GR_{80} (ppm)
55	OMe	F	1	3	< 1	4
56	OMe	Cl	1	2	< 1	2
57	OMe	Br	0.5	2	< 1	1
58	OMe	OMe	0.5	< 0.1	< 0.5	< 0.5
59	OMe	CF_3	< 0.2	1	2	1
60	OMe	OCF_3	0.5	5	0.5	1
61	OEt	CF_3	2	191	250	4
62	$O(CH_2)_2F$	CF_3	6	66	250	1
63	OMe	CO_2Me	10	2	< 1	2
64	OEt	CO_2Me	3	15	2	1
65	OMe	CO_2Et	15	15	2	15
66	OEt	OCF_3	8	15	< 1	1
67	$O(CH_2)_2F$	OCF_3	2	4	1	2
68	OEt	OEt	2	15	< 1	4

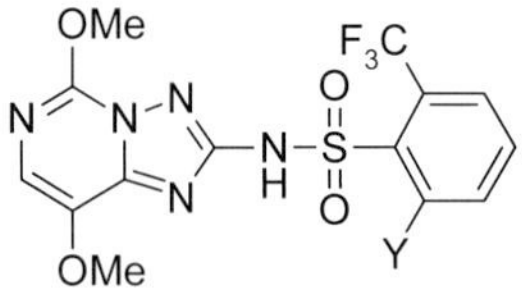

Table VI. Herbicidal activity of alkoxy trifluoromethyl phenyl analogs when applied as a water injection treatment on transplanted paddy rice and selected weeds.

Compound	Y	*Oryza sativa* GR_{20} (g ai/ha)	*Echinochloa crus-galli* GR_{80} (g ai/ha)	*Monochoria vaginalis* GR_{80} (g ai/ha)	*Scirpus juncoides* GR_{80} (g ai/ha)	*Cyperus difformis* GR_{80} (g ai/ha)
61	OEt	1	1	1	3	NA
62	OCH_2CH_2F	14	10	5.2	9	8
69	OCH_2OMe	124	16	4	15	18
70	OCH_2CF_3	51	19	5	21	36
71	OCH_2CF_2H	75	12	< 2	12	14
72	$OCH(CH_2F)_2$	140	14	1	9	31
73	O-*n*-Pr	20	15	3	7	21
74	O-*i*-Pr	66	7	3	8	23
75	O-*n*-Bu	39	25	3	23	18

In these greenhouse trials, rates were expressed as grams active ingredient per hectare (g ai/ha). Particularly noteworthy are the 2-fluoroethoxyphenyl (**62**), 2,2-difluoroethoxyphenyl (**71**), 2-(1-fluoromethyl-2-fluoroethoxy)phenyl (**72**), and 2-isopropoxyphenyl (**74**) analogs which showed high levels of activity on all weeds species, particularly barnyard grass, with good selectivity to rice.

9.4
Selection of Penoxsulam for Development

Based on the greenhouse results, several 2-trifluoromethyl-6-alkoxyphenyl analogs were tested in key rice-growing countries from 1997–1999 to characterize field performance. From these analogs, the 2,2-difluoroethoxyphenyl analog (**71**) was identified as having good rice tolerance, broad spectrum weed control, and providing good residual weed control depending on the rates applied. Other analogs tested were not selected for a number of reasons, such as being too injurious to rice, providing poor weed control, or having short residual activity, when compared to **71**. Additionally, it was discovered that **71** could be co-applied with the grass herbicide cyhalofop-butyl which cannot be tank-mixed with commercially available ALS or auxin mode of action herbicides without antagonizing the control of *Echinochloa* spp. Based on the ability to meet many of the commercial rice herbicide needs in transplanted rice, direct-seeded rice and water-seeded rice in over 25 rice countries, **71** was identified for development as a new rice herbicide with the code number DE-638 and the common name "penoxsulam" [12–20].

9.5 Conclusion

The triazolopyrimidine sulfonamide class of ALS inhibitors has grown to include a number of compounds which have demonstrated control of grass, broadleaf, and sedge weeds in several important crops. Penoxsulam is the sixth member to be successfully developed and the first to be registered for use in rice for control of grass, broadleaf, and sedge weeds. Penoxsulam was first registered and sold in Turkey in 2004. Registrations and launches in many rice-growing countries, including USA, Colombia, Argentina, Vietnam, Philippines, Indonesia, Korea, China, and Italy, continued in 2005. Further registrations are pending in additional countries.

9.6 References

1 W. A. Kleschick, M. J. Costales, J. E. Dunbar, R. W. Meikle, W. T. Monte, N. R. Pearson, S. W. Snider, A. P. Vinogradoff, *Pest. Sci.*, **1990**, *29*, 341–355.
2 M. V. Subramanian, V. Loney-Gallant, J. M. Dias, L. C. Mireles, *Plant Physiol.*, **1991**, *96*, 310–313.
3 J. V. Van Heertum, B. C. Gerwick, W. A. Kleschick, T. C. Johnson, **1992**, US 5,163,995.
4 T. C. Johnson, R. J. Ehr, R. D. Johnston, W. A. Kleschick, T. P. Martin, M. A. Pobanz, J. V. Van Heertum, R. K. Mann, **1999**, US 5,858,924.
5 T. C. Johnson, R. J. Ehr, R. D. Johnston, W. A. Kleschick, T. P. Martin, M. A. Pobanz, J. V. Van Heertum, R. K. Mann, **1999**, US 6,005,108.
6 T. C. Johnson, R. J. Ehr, T. P. Martin, M. A. Pobanz, J. V. Van Heertum, R. K. Mann, **2001**, US 6,303,814.
7 T. C. Johnson, W. A. Nasitavicus, **1993**, US 5,177,206.
8 J. W. Ringer, A. C. Scott, D. L. Pearson, A. P. Wallin, **1999**, US 5,973,148.
9 T. Hamada, O. Yonemitsu, *Synthesis*, **1986**, 852–854.
10 R. E. Rosen, D. G. Weaver, J. W. Cornille, L. A. Spangler, **1993**, US 5,272,128.
11 M. G. Smith, M. A. Pobanz, G. A. Roth, M. A. Gonzales, **2002**, US 6,462,240.
12 D. Larelle, R. K. Mann, S. Cavanna, R. Bernes, A. Duriatti, C. Mavrotas, Proc. Int. Cong. Brit. Crop Prot. Conf. – Crop Science & Technology, Glasgow, UK, **2003**, *1*, 75–80.
13 R. K. Mann, R. B. Lassiter, A. E. Haack, V. B. Langston, D. M. Simpson, J. S. Richburg, T. R. Wright, R. E. Gast, *et al.*, *Proc. Weed Sci. Soc. Am.*, **2003**, *43*, 40.
14 R. K. Mann, A. E. Haack, V. B. Langston, R. B. Lassiter, J. S. Richburg, *Proc. Weed Sci. Soc. Am.*, **2005**, *45*, 308.
15 R. K. Mann, C. Mavrotas, Y. H. Huang, D. Larelle, V. Patil, Y. K. Min, I. Shiraishi, L. Nguyen, *et al.*, *Proc. 20th Asian-Pacific Weed Sci. Soc.*, Vietnam, **2005**, pp. 289–294.
16 Y. K. Min, R. K. Mann, *Korean J. Weed Sci.*, **2004**, *24*(3), 192–198.

17 Y. K. Min, R. K. Mann, *Korean J. Weed Sci.*, **2004**, *24*(3), 199–205.
18 I. Shiraishi, *J. Pestic. Sci.*, **2005**, *30*(3), 265–268.
19 C. L. Wang, M. S. Lee, Y. W. Li, Z. W. Yao, J. N. Shieh, R. K. Mann, Y. H. Huang, *15th International Plant Prot. Cong. Abstracts*, Beijing, China, **2004**, 598.
20 R. K. Mann, Y. H. Huang, D. Larelle, C. Mavrotas, Y. K. Min, M. Morell, H. Nonino, I. Shiraishi, Proc 3rd International Temperate Rice Conf., Punta del Este, Uruguay, **2003**, abstract WD055, 68.

Keywords

Triazolopyrimidine, Triazolo[1,5-a]pyrimidine, Triazolo[1,5-c]pyrimidine, Sulfonamide, Penoxsulam

10
Discovery and SAR of Pinoxaden: A New Broad Spectrum, Postemergence Cereal Herbicide

Michel Muehlebach, Hans-Georg Brunner, Fredrik Cederbaum, Thomas Maetzke, René Mutti, Anita Schnyder, André Stoller, Sebastian Wendeborn, Jean Wenger, Peter Boutsalis, Derek Cornes, Adrian A. Friedmann, Jutta Glock, Urs Hofer, Stephen Hole, Thierry Nidermann, Marco Quadranti

10.1
Introduction

Biocidal 2-aryl-1,3-diones and their enol esters showing acaricidal and pre- and postemergence herbicidal activity were reported since the mid-1970's as exemplified by 3-hydroxy-2-arylindones from Union Carbide [1]. 3-Aryl-pyrrolidine-2,4-diones with similar biological properties were discovered more than 15 years later by Bayer scientists [2]. The herbicidal mechanism of action of this dichlorophenyl derivative was shown by Babczinski and Fischer [3] to be inhibition of the enzyme acetyl-coenzyme A carboxylase.

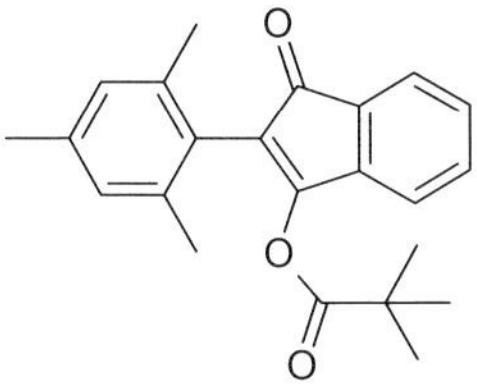

Union Carbide, 1973

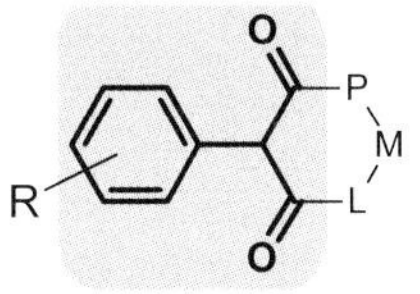

2-Aryl-1,3-diones

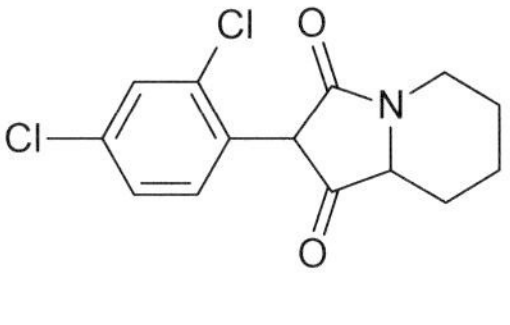

Bayer, 1990

Inspired by this background from the public literature, a research entry in this area was found by incorporating an internally available tetrahydropyridazine monoester inter-mediate (PPGO) to quickly establish CGA271312 as a first lead [4–5]. The bridge L-M-P thereby represents a simple *N-N* bond, thus involving a hydrazine.

In retrospect it turns out that, independently of one another, Cederbaum [4] and Krueger [6] almost simultaneously claimed the pre- and postemergence herbicidal action of 4-mesityl-pyrazolidine-3,5-dione CGA271312. Unawares to both parties, a race towards a proprietary subclass was taking place. A beneficial

Pesticide Chemistry. Crop Protection, Public Health, Environmental Safety
Edited by Hideo Ohkawa, Hisashi Miyagawa, and Philip W. Lee

ISBN: 978-3-527-31663-2

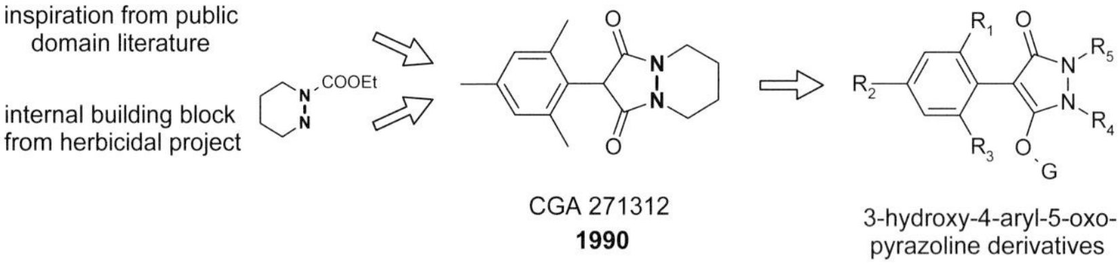

difference of only two days in the priority filing of two similar patent applications finally allowed Cederbaum and co-workers to secure IP protection in the chemical class of hydroxy-phenyl-oxo-pyrazoline derivatives. A textbook example of timely patenting!

10.2
Optimization Phase and Discovery of Pinoxaden

Three main areas of the 4-aryl-pyrazolidine-3,5-dione scaffold were modified during the optimization phase towards graminicidal activity and selectivity in small grain cereals. A major breakthrough with regard to the herbicidal activity was achieved as aryl moieties bearing ethyl, ethynyl or methoxy groups in the 2,6-positions were synthesized. The functional group R2 is also important and methyl seems optimum. Such a substitution pattern boosts the herbicidal activity and the aromatic moiety can be seen as a potency contributor.

The hydrazine region was found to be equally highly relevant. Carbocyclic hydrazines are clearly beneficial over open chain analogs, whereby the ring size (5- to 7-membered) is not of primary significance. [1,3,4]Oxadiazinane and diverse oxa-diazepane derivatives showed good activity, and most noteworthy, a considerable increase in crop tolerance was achieved, especially in barley. [1,4,5]Oxadiazepane is the preferred oxygen containing cyclic hydrazine and can be seen as a tolerance contributor.

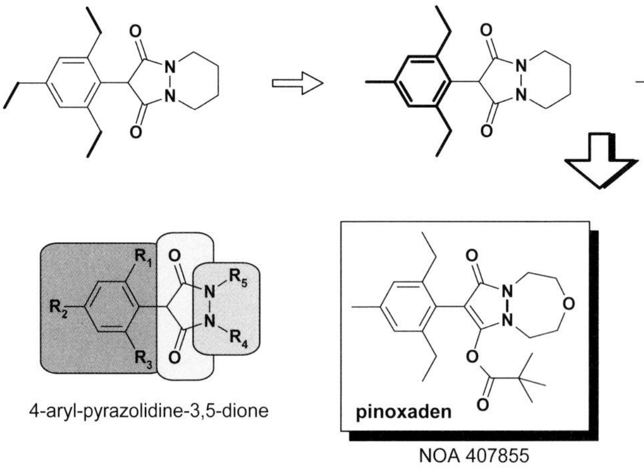

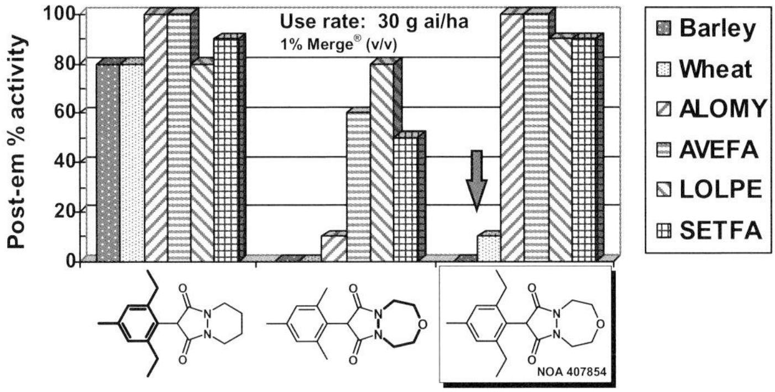

Figure 1. Summary of the aryl-pyrazolidine-dione optimization demonstrating key fragment contributions to activity level against grass weed species (aryl moiety) and tolerance in cereal crops (hydrazine moiety).

With a pKa of about 3.8 (vinylogous acid), the dione area is well suited for propesticide formation with the aim to increase leaf penetration. Various analogs were synthesized, sulfonates and pivaloates in particular were found to be ideal regarding increased herbicidal activity. The combination of all key elements ultimately led to the discovery of pinoxaden [7].

10.3 Chemistry

General access to analogs: Aryl-pyrazolidine-diones, precursors of the target molecules, were primarily prepared via a thermal condensation reaction between hydrazines and aryl malonates. While scouting for a technical synthesis of pinoxaden, it was moreover discovered that aryl malonamides will react with equal efficacy (NH3 extrusion), a reaction unprecedented in the literature [8].

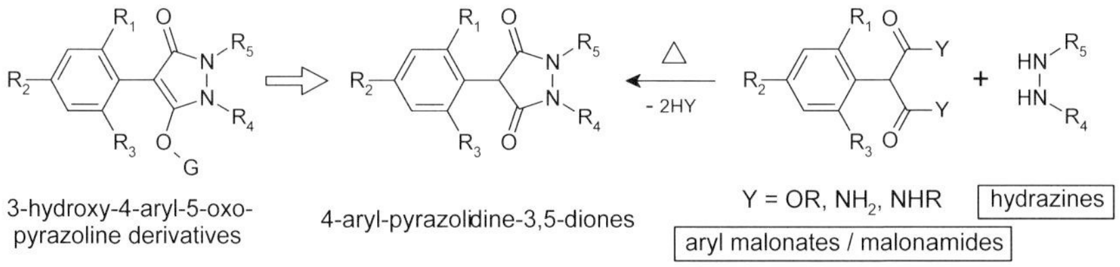

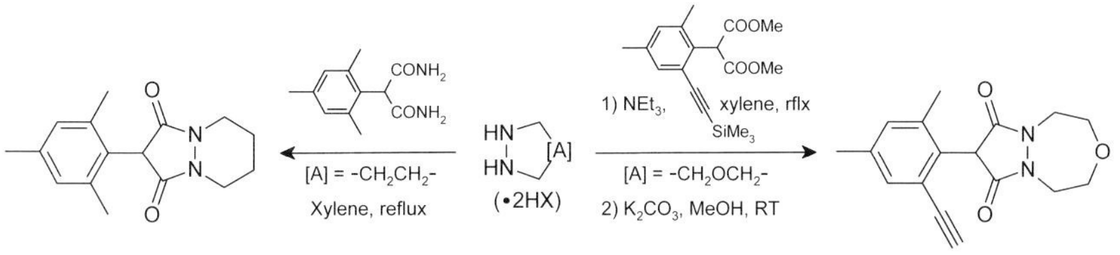

Hydrazine synthesis: 5- to 7-membered carbocyclic hydrazines are accessible through either cyclocondensation or hetero Diels-Alder cycloaddition (6-membered rings) reactions. Novel diverse oxygen containing cyclic hydrazines were found to significantly reduce phytotoxicity in cereal crops. The synthesis of these [1,4,5]oxadiazepane derivatives is simple and makes use of cheap starting materials.

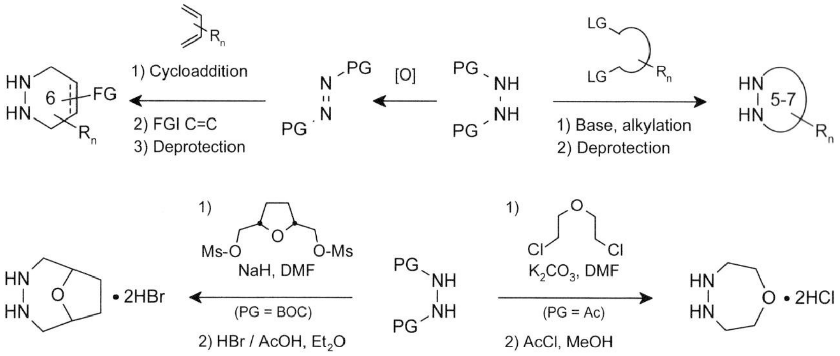

Adjusting aryl substitution: Functionalization of the aryl moiety was mainly achieved through cross-coupling and directed *ortho*-metalation strategies [9–10]. A 2,4,6-substitution pattern proved highly beneficial for herbicidal activity. Aryl moieties bearing ethyl, ethynyl or methoxy in the 2,6-positions boost the graminicidal activity decisively. Various alkyl, alkenyl, alkynyl, and (het)aryl substituents were introduced similarly using Stille, Heck, Suzuki, Kumada or Negishi type cross-coupling reactions.

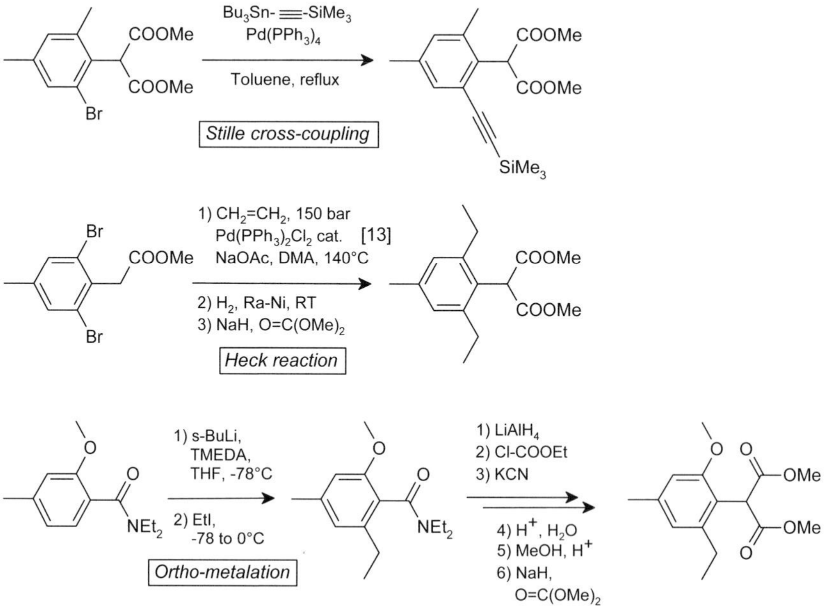

Synthesis of pinoxaden: A convergent synthetic route leads to the pyrazolidinedione NOA 407854, which is esterified in the last step with pivaloyl chloride. Pinoxaden is prepared effectively in a total of 5 (longest linear sequence) + 3 (oxadiazepane synthesis), total 8 steps. Key reactions are an efficient malononitrile arylation via cross-coupling [12] to access a sterically hindered aryl malononitrile and the unprecedented condensation of its malonamide derivative with [1,4,5]oxadiazepane [8]. This sequence fits feasibility criteria for a technical application.

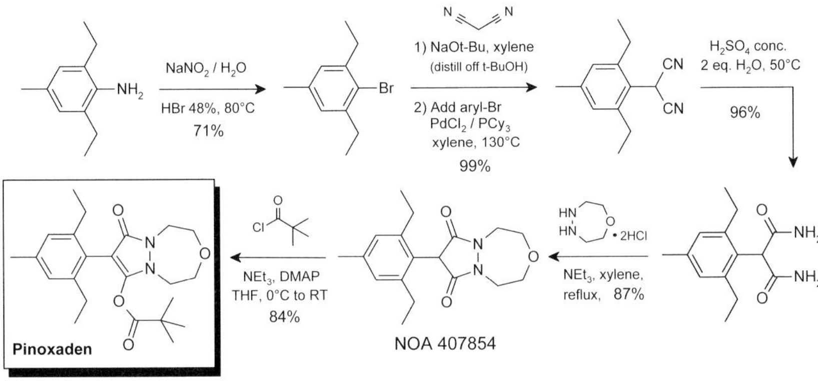

NOA 407855

10.4 Mode of Action

Pinoxaden interferes with the lipid metabolism in plant cells. It acts by inhibiting the enzyme Acetyl-CoA-carboxylase (ACCase), interrupting the synthesis of fatty acids and as a final consequence impacts the formation of biomembranes. Pinoxaden represents a unique new structure of a novel chemical class of ACCase inhibitors, the hydroxy-aryl-oxo-pyrazoline derivatives or in a more general sense the aryl-diones (ADs or den) [13]. It is different from any existing aryloxy-phenoxy-propionate (AOPP or fop) or cyclo-hexanedione (CHD or dim) herbicides, offering new properties, for example, in setting a new standard of grass control in barley.

10.5 Structure-Activity Relationships

Key fragment contributions to activity/tolerance and structure-activity relationships in the aromatic moiety as well as in the hydrazine region can be summarized as follows:

2,6-Diethyl-4-methyl is the preferred aromatic substitution pattern. The [1,4,5]oxa-diazepane ring is superior to bridged or substituted analogs, and clearly

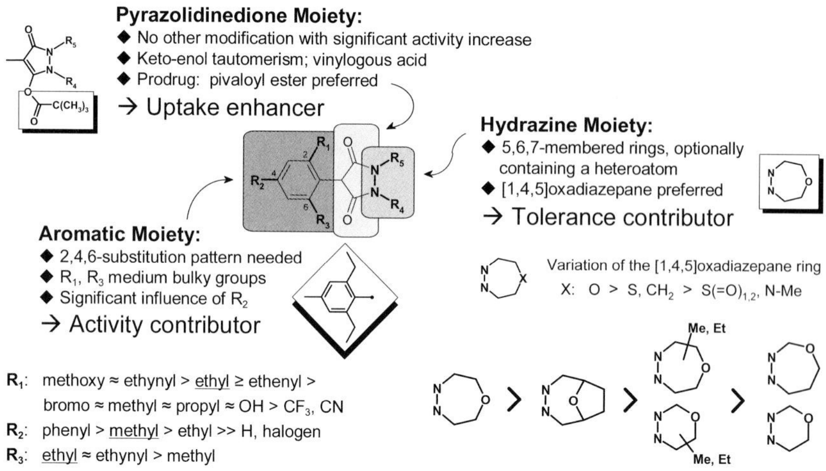
Pyrazolidinedione Moiety:
◆ No other modification with significant activity increase
◆ Keto-enol tautomerism; vinylogous acid
◆ Prodrug: pivaloyl ester preferred
→ Uptake enhancer
Hydrazine Moiety:
◆ 5,6,7-membered rings, optionally containing a heteroatom
◆ [1,4,5]oxadiazepane preferred
→ Tolerance contributor
Aromatic Moiety:
◆ 2,4,6-substitution pattern needed
◆ R1, R3 medium bulky groups
◆ Significant influence of R2
→ Activity contributor
Variation of the [1,4,5]oxadiazepane ring
X: O > S, CH2 > S(=O)1,2, N-Me
R1: methoxy ≈ ethynyl > ethyl ≥ ethenyl > bromo ≈ methyl ≈ propyl ≈ OH > CF3, CN
R2: phenyl > methyl > ethyl >> H, halogen
R3: ethyl ≈ ethynyl > methyl
Me, Et
Me, Et

better than unsubstituted [1,3,4]oxadiazepanes or [1,3,4]oxadiazinanes. A slight increase in activity over the dione form is observed with the pivaloyl enol ester propesticide.

10.6 Biological Performance

Pinoxaden is for broad spectrum grass weed management use in cereal crops [14]. It is applied postemergence at field use rates of 30–60 g ai/ha. Susceptible weed species stop growing within 48 hours of treatment, turn yellow within 1–2 weeks and are dead within 3–5 weeks. Uptake of pinoxaden is into green leaf tissue, from where it is translocated quickly within the plant.

10.6.1 Grass Weed Spectrum

The weed spectrum of pinoxaden covers a wide range of key annual grass species like *Alopecurus myosuroides* (blackgrass), *Apera spica-venti* (silky bent grass), *Avena* spp. (wild oats), *Lolium* spp. (ryegrass), *Phalaris* spp. (canary grass), *Setaria* spp. (foxtails) and other monocot weeds commonly found in cereals (Figure 2) [14].

10.6.2 Crop Tolerance in Wheat and Barley

The discovery of pinoxaden is complemented by two further key findings leading to the invention of a family of novel and innovative postemergence grass herbicides: safener efficacy and adjuvant response.

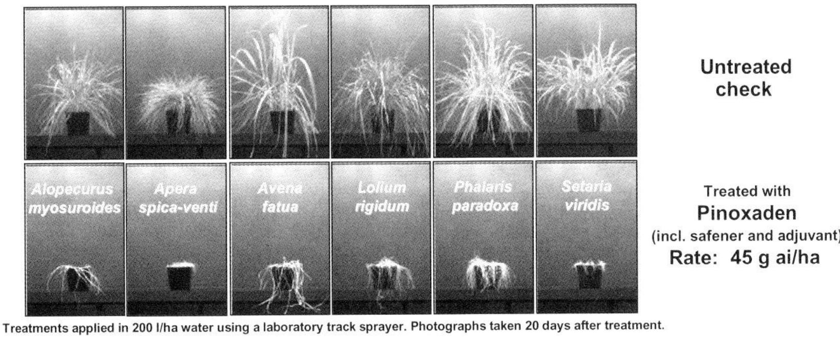

Figure 2. Performance of pinoxaden in the glasshouse against six key grass weeds at 45 g ai/ha after foliar application. Treatments applied as EC100 Formulation with 0.5% v/v Adigor®.

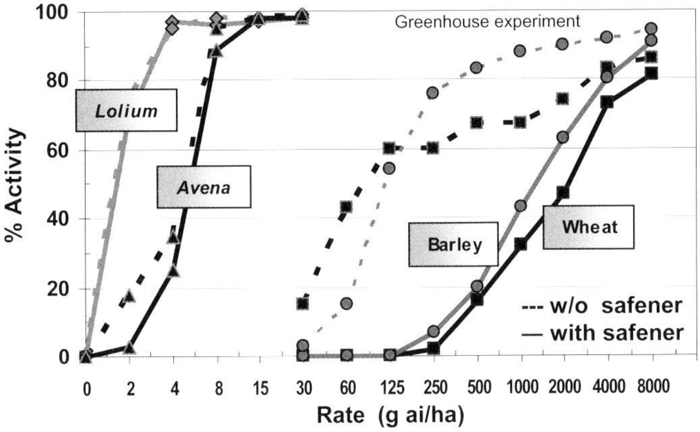

Figure 3. Dose-response curve highlighting safener efficacy and selectivity margin of the specific pinoxaden plus cloquintocet-mexyl combination (—) in contrast to treatments without safener (---).

Tolerance in key cereal crops is achieved through safener technology by inclusion of the proprietary safener cloquintocet-mexyl into the formulation. The specific pinoxaden plus safener combination provides very good tolerance in the important crops wheat and barley, without causing significant antagonism on weed control (Figure 3) [15].

10.6.3
Adjuvant Effect – Adigor®

Pinoxaden was shown to have a significant activity enhancement by certain adjuvants while still remaining safe to crops when used in combination with a safener [15]. Rationalizing the adjuvant component composition led to the discovery and development of Adigor®, a specific and optimized adjuvant for

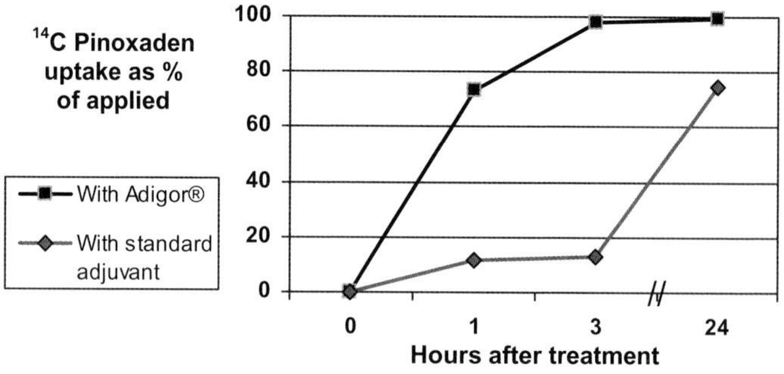

Figure 4. Effect of Adigor® on pinoxaden uptake into wild oats (*Avena fatua*) demonstrating rapid absorption into the leaf within hours.

pinoxaden. Following optimal wetting and maximal spreading, Adigor® solubilizes the leaf wax and thereby facilitates a quick uptake across the leaf cuticular waxes. This rapid leaf penetration permits translocation to the site of action in the meristematic growing tissue.

The higher rate of uptake improves important properties like rainfastness and enhances significantly reliability and robustness of the herbicide, maximizing its weed control performance.

10.6.4
Introducing Axial®

Based on the innovative chemistry of pinoxaden, Axial® is the first brand of a new family of tailor-made grass herbicide solutions introduced to growers globally. Launched by Syngenta in 2006, Axial® is for worldwide use in both wheat and barley. Tolerance and efficacy trials across the globe have confirmed the outstanding biological performance of Axial® in cereal crops. The high and consistent level of activity against six of the most important grass species tested across different parts of the world is shown in Figure 5.

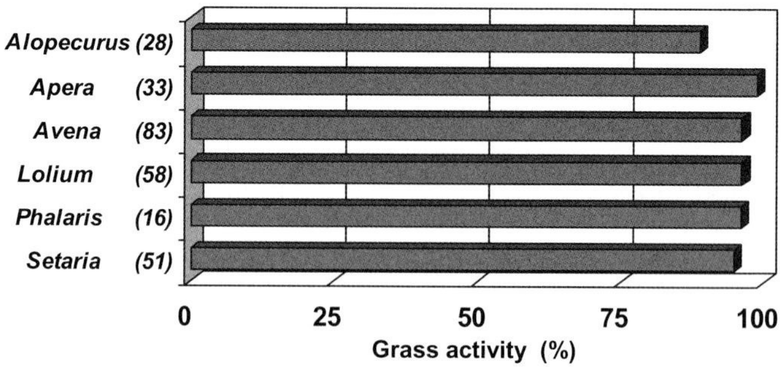

Figure 5. Field performance (% activity at 30–60 g ai/ha; number of trials in brackets) against key grass weeds.

The main features and benefits of Axial® are:
- efficacy: its high level of activity against key annual grass weeds
- selectivity: its excellent tolerance in the main cereal crops wheat and barley
- flexibility: its ease of use in terms of application timing, mixability with key partners and no rotational crop restrictions.

10.7 Conclusion

Hydroxy-phenyl-oxo-pyrazoline derivatives bearing ethyl, ethynyl or methoxy groups in the aromatic 2,6-positions and incorporating a [1,4,5]oxadiazepane ring were shown to exhibit high levels of herbicidal activity on key annual grass weed species in cereal crops. Structure-activity relationship information revealed pinoxaden (PXD™, NOA 407855) as the most promising candidate for further development. Pinoxaden is applied postemergence at low use rates and is introduced globally as Axial® into cereal markets in 2006. Axial® offers unrivaled crop safety for worldwide use in both wheat and barley.

The breakthrough unveiling the outstanding performance of Axial® is a sum of a careful optimization of many parameters in chemistry, biology, and adjuvant sciences, highlighting excellent cross-divisional research teamwork. A crucial interplay of the active ingredient (pinoxaden) with a safener (cloquintocet-mexyl) and a specially optimized adjuvant (Adigor®) proves absolutely essential to maximize broad spectrum grass control and tolerance in both wheat and barley.

10.8 Acknowledgments

The authors thank all Syngenta coworkers in Basel and around the globe involved in the discovery, research, development, and launch of pinoxaden.

10.9 References

1 A. A. Sousa, J. A. Durden, J. F. Stephen (Union Carbide), *J. Econ. Entomol.*, **1973**, 66, 584–586.
2 R. Fischer, A. Krebs, A. Marhold, H. J. Santel, R. R. Schmidt, K. Luerssen, H. Hagemann, B. Becker, *et al.*, *EP 355599*, **1990**.
3 P. Babczinski, R. Fischer, *Pestic. Sci.*, **1991**, *33*, 455–466.
4 F. Cederbaum, H. G. Brunner, M. Boeger (Syngenta, priority: 19.03.1991), WO 92/16510, **1992**.
5 M. Boeger, P. Maienfisch, F. Cederbaum, T. Pitterna, P. J. Nadkarni, V. S. Ekkundi, S. U. Kulkarni, WO 96/21652, **1996**.
6 B. W. Krueger, R. Fischer, H. J. Bertram, T. Bretschneider, S. Boehm, A. Krebs, T. Schenke, H. J. Santel, *et al.* (Bayer, priority: 21.03.1991), *EP 508126*, **1992**.

7 M. Muehlebach, J. Glock, T. Maetzke, A. Stoller, WO 99/47525, **1999**; WO 00/047585, **2000**.
8 T. Maetzke, R. Mutti, H. Szczepanski, WO 01/78881, **2001**.
9 T. Maetzke, A. Stoller, S. Wendeborn, H. Szczepanski, WO 01/17972, **2001**.
10 T. Maetzke, S. Wendeborn, A. Stoller, WO 01/17973, **2001**.
11 A. F. Indolese, H. Meier, M. Benz, F. Prüter, M. Muehlebach, internal report Syngenta.
12 A. Schnyder, WO 00/78712, **2000**; A. Schnyder, A. F. Indolese, T. Maetzke, J. Wenger, H.-U. Blaser, *Synlett*, **2006**., submitted.
13 J. Wenger, T. Niderman In *Modern Crop Protection Compounds*, W. Krämer, U. Schirmer (Eds.), Wiley-VCH, to be published.
14 U. Hofer, M. Muehlebach, S. Hole, A. Zoschke, *Journal of Plant Diseases and Protection*, **2006**, in press.
15 J. Glock, A. A. Friedmann, D. Cornes, WO 01/17352, **2001**.

Keywords

Pinoxaden, Axial®, Adigor®, Cloquintocet-mexyl, Discovery, Structure-Activity Relationship, Herbicide, Grass Weed Control, Graminicide, Postemergence, Cereals, Wheat, Barley

11
Rynaxypyr^{TM}: A New Anthranilic Diamide Insecticide Acting at the Ryanodine Receptor

George P. Lahm, Thomas M. Stevenson, Thomas P. Selby, John H. Freudenberger, Christine M. Dubas, Ben K. Smith, Daniel Cordova, Lindsey Flexner, Christopher E. Clark, Cheryl A. Bellin, J. Gary Hollingshaus

11.1
Introduction

The discovery of new insecticides that protect crops from harmful pests, and possess favorable toxicological and environmental properties, is an essential component of the agricultural industry and critical to the protection of the global food supply. In addition, the ability of insects to develop resistance makes the discovery of new crop protection chemicals, which work via new biochemical mechanisms, an important component of effective pest management strategies. We wish to report on the discovery of a new class of insecticides, the anthranilic diamides [1–2], which exhibit their action through activation of the ryanodine receptor [3–4]. Further, we describe the discovery of Rynaxypyr^{TM}, a potent ryanodine receptor (RyR) activator as the first new insecticide from this class (Figure 1). The discovery of Rynaxypyr^{TM} was based in part on the discovery of early leads related to the emerging class of phthalic diamide insecticides [5]. Studies have shown that both the anthranilic diamides and the phthalic diamides are related by their biochemical mechanism of action at the ryanodine receptor [3, 6–7].

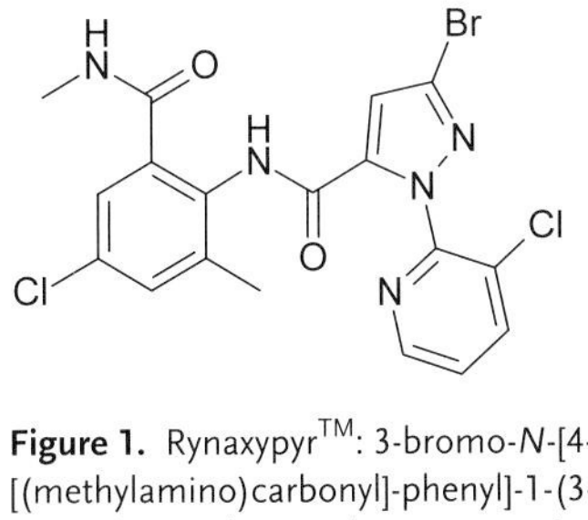

Rynaxypyr^{TM}

Figure 1. Rynaxypyr^{TM}: 3-bromo-*N*-[4-chloro-2-methyl-6-[(methylamino)carbonyl]-phenyl]-1-(3-chloro-2-pyridinyl)-1H-pyrazole-5-carboxamide. A new anthranilic diamide insecticide acting at the insect ryanodine receptor.

Pesticide Chemistry. Crop Protection, Public Health, Environmental Safety
Edited by Hideo Ohkawa, Hisashi Miyagawa, and Philip W. Lee

ISBN: 978-3-527-31663-2

11.2
Discovery of the Anthranilic Diamide Insecticides

Figure 2 illustrates the synthesis of several anthranilic diamides prepared in our early discovery program. Our initial targets consisted of derivatives including those of formula **D1–D3**. These were prepared by coupling of anthranilamide **2** with substituted benzoyl chlorides **3** in the presence of base. The anthranilamide intermediates **2** could be prepared from isatoic anhydrides by treatment with isopropylamine.

These compounds were tested against a series of Lepidoptera under standard laboratory procedures. Activity on diamondback moth (*Plutella xylostella, Px*), fall armyworm (*Spodoptera frugiperda, Sf*), and tobacco budworm (*Heliothis virescens, Hv*) was evaluated. Insecticidal potency is reported as the LC_{50} in ppm. Comparison of the insecticidal activity for compounds **D1–D3** identifies several key structural features of importance (Table I). Our initial lead compound **D1** was effective on the species *Sf* and *Px* with LC_{50} values of 68.9 and 16.8, respectively, but with no activity on *Hv*. Incorporation of a 2-methyl group on the benzamide, i.e., **D2**, significantly improved potency on *Sf* and *Hv* and showed comparable activity on *Px*. This trend was consistently observed for all anthranilic diamides, i.e., compounds containing an R_1 substituent on the benzamide were found to be significantly more active than their corresponding unsubstituted analogs. This is also consistent with the reported structure-activity profile of phthalic diamides, where *ortho*-methyl benzamides were found to be optimum [5].

The striking data of Table I come from the comparison of the 6-methyl and 3-methyl positional isomers, **D2** and **D3**. In this comparison, the 6-methyl isomer **D2** was found to be significantly more active than the corresponding 3-methyl analog **D3**. This is in direct contrast with phthalic diamide structure-activity where the R substituent is preferentially located *ortho* to the alkyl amide, as reported by Tohnishi [5], and indicates a divergent structure-activity profile for these two classes of chemistry. While R as methyl appears most preferred, we have found

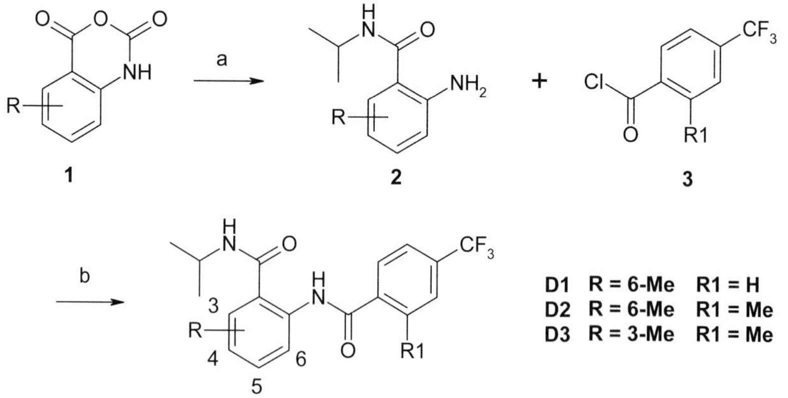

Figure 2. (a) (i) i-PrNH$_2$, THF; (b) (i-Pr)$_2$EtN, CHCl$_3$.

Table I. Insecticidal potency of anthranilic diamides against three species of Lepidoptera.[d]

Compound	Sf [a] LC_{50} ppm	Px [b] LC_{50} ppm	Hv [c] LC_{50} ppm
D1	69	17	> 500
D2	20	21	256
D3	> 500	77	> 500
D4	45	21	327
D5	70	12	102
D6	48	30	130
D7	88	35	158
D8	23	11	36
D9	48	26	354
D10	0.2	0.05	3.4

[a] Sf, Spodoptera frugiperda; [b] Px, Plutella xylostella; [c] Hv, Heliothis virescens;
[d] Insecticidal potency is reported as an LC_{50} in ppm based on larval mortality from leaves treated with serial concentrations of the experimental compounds at 96 hours post-infestation.

a variety of other groups at the 6-position to show strong activity including F, Cl, Br, I, and CF_3 among the more active.

Following this activity, we set out to prepare a series of heterocyclic derivatives of the benzamide to explore a range of physical properties including the pyridine, pyrimidine, and pyrazoles of formula **D4–D8**. Based on the emerging structure-activity trends, we also examined phenyl pyrazoles of formula **D9–D10** (Figure 3).

Comparison of the activity profiles for **D3–D5** shows the phenyl analog to be the more active on *Sf*, while the pyrimidine was best on both *Px* and *Hv*. On balance however, all three analogs **D3**, **D4**, and **D5** showed similar levels of activity. Pyrazoles **D6–D8** showed an interesting trend. Replacement of the heteroaryl group with a pyrazole containing an *N*-methyl group, as in **D6**, showed similar activity to the corresponding pyridine and pyrimidine. A slight decrease in activity was observed with the ethyl derivative **D7**. However, the 1-isopropyl pyrazole, **D8**, was the most active of the group, approaching a 5-fold increase in activity. This finding suggested that bulkier substituent groups appended to the pyrazole nitrogen may give compounds with improved activity and prompted investigation of the *N*-phenyl pyrazoles and their substituted derivatives.

The first analog from this set, the unsubstituted phenyl pyrazole **D9**, showed activity similar to pyrazoles **D6–D7**, and somewhat less than the isopropyl pyrazole **D8** especially on *Hv*. In contrast, however, the *ortho*-Cl-phenyl analog **D10** showed a remarkable improvement in activity, spanning two orders of magnitude against all three species of Lepidoptera. This discovery was the key structural breakthrough in the anthranilic diamide area and provided the framework for a large scope

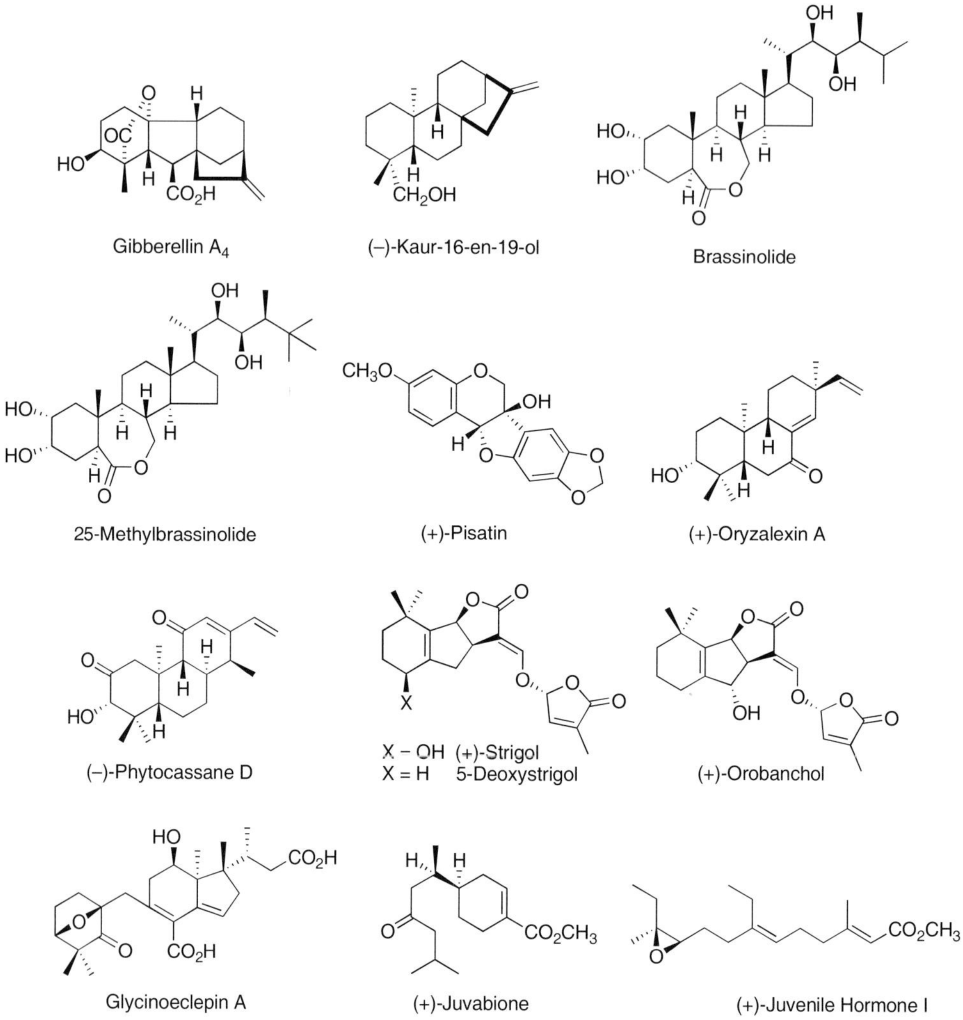

Figure 3. Heterocyclic derivatives of the anthranilic diamides.

of structurally diverse analogs with outstanding potency on all key species of Lepidoptera.

We observed that a variety of groups could be substituted for the ortho-chloro substituent and still retain high activity. Substituents including F, Br, I, CH_3, CF_3, OCF_3, CN, and CO_2CH_3 were consistently more active than corresponding unsubstituted analogs. In contrast, the isomeric meta-Cl and para-Cl analogs of D10 were several orders of magnitude less active. We speculate that the high activity for ortho-substitution is the result of a steric interaction favoring a more preferred out of plane conformation at the binding site, although we cannot rule out a direct site specific interaction of the ortho-substituent.

11.3 Discovery of Rynaxypyr™

Based on the high levels of activity obtained for **D10**, a series of *N*-pyridyl analogs were explored. We also wished to study the effect of substituents R_4 on the aryl portion of the anthranilic diamide. Figures 4–6 illustrate the routes by which these compounds were prepared [2]. The method presented in Figure 4 was used to prepare the 3-trifluoromethyl-5-pyrazolecarboxylic acid **26**. This method involves regioselective lithiation of 3-trifluoromethylpyrazole **25** followed by reaction with carbon dioxide, to afford the pyrazole carboxylic acid **26** in good yield.

3-Halo-5-pyrazolecarboxylic acids were prepared as outlined in Figure 5 as shown for the bromo derivative **32**. Treatment of **27** with n-butyllithium at –60 °C in THF followed by bromination with dibromotetrachloroethane afforded the pyrazole **28** [8]. Removal of the *N,N*-dimethylsulfamoyl protecting group afforded 3-bromopyrazole **29** directly. Conversion of **29** to 3-bromo-5-pyrazolecarboxylic acid was achieved in two steps by reaction with dichloropyridine followed by regioselective metallation with lithium diisopropylamide, and carbon dioxide quench.

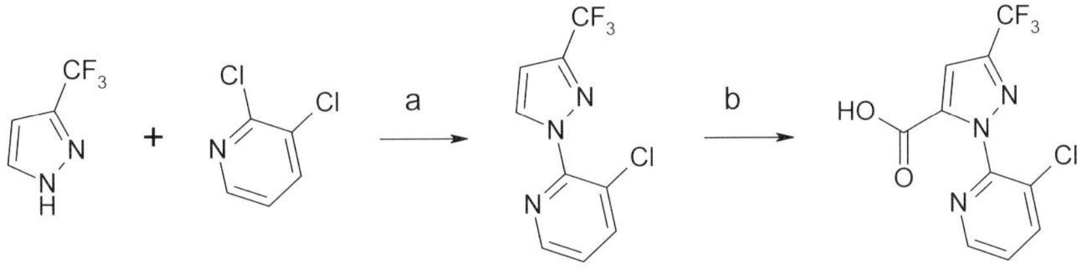

23 **24** **25** **26**

Figure 4. (a) K_2CO_3, DMF, 125 °C; (b) (i) LDA, THF, –75 °C (ii) CO_2 (iii) HCl.

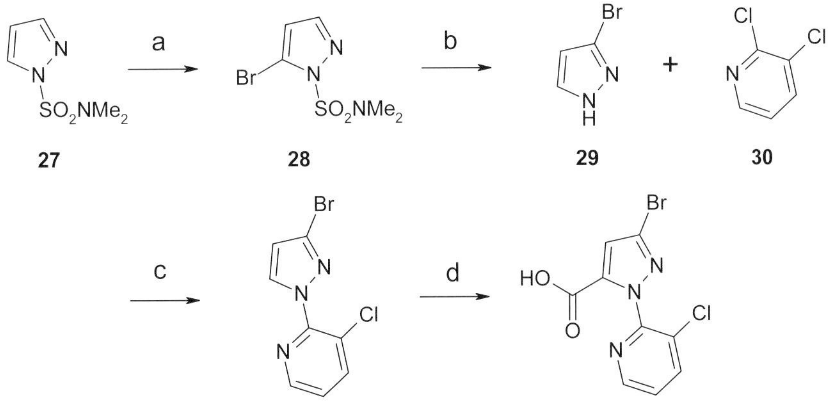

27 **28** **29** **30**

31 **32**

Figure 5. (a) (i) n-BuLi, THF, –60 °C (ii) $BrCCl_2CCl_2Br$, THF, –70 °C, (b) TFA, 25 °C, (c) K_2CO_3, DMF, 125 °C, (d) (i) LDA, THF, –78 °C (ii) CO_2 (iii) HCl.

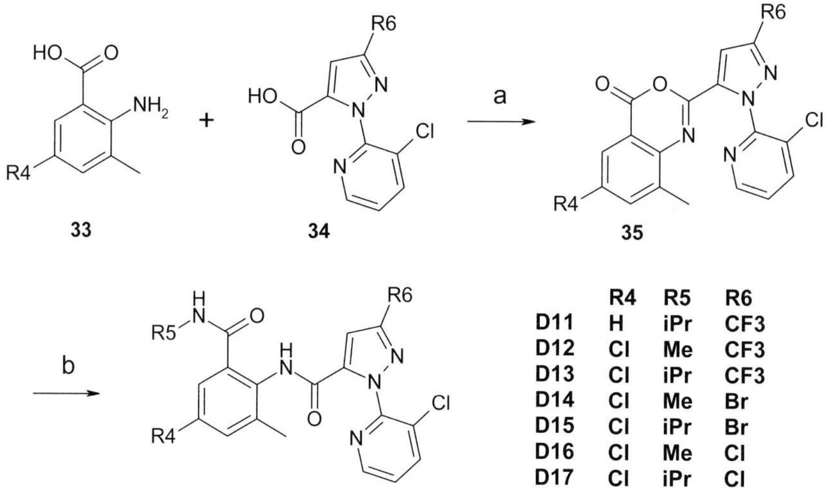

	R4	R5	R6
D11	H	iPr	CF3
D12	Cl	Me	CF3
D13	Cl	iPr	CF3
D14	Cl	Me	Br
D15	Cl	iPr	Br
D16	Cl	Me	Cl
D17	Cl	iPr	Cl

Figure 6. (a) (i) 33, $MeSO_2Cl$, Et_3N, MeCN (ii) 34 (iii) Et_3N, MeCN (iv) $MeSO_2Cl$, (b) R_5NH_2, THF.

Anthranilic diamides, **D11–D17**, were prepared in good yield by the reaction of an amine, R_5NH_2, with benzoxazinones **35** as shown in Figure 6. Synthesis of the benzoxazinones was accomplished by sequential treatment of **34** with one equivalent of triethylamine and methanesulfonyl chloride, followed by the anthranilic acid **33**, and then followed by a second equivalent of triethylamine and methanesulfonyl chloride. The intermediate 4-chloro-6-methyl-anthranilic acid **33** (R^4 = Cl) was prepared by chlorination of 6-methylanthranilic acid with *N*-chlorosuccinimide in DMF.

Insecticidal activity and the calcium mobilization threshold (CMT) values, for compounds **D11–D17** are presented in Table II. The calcium mobilization threshold value is a useful measure of ryanodine receptor activity for these compounds. Compounds **D11–D17**, containing the pyridyl pyrazole group, all display outstanding insecticidal activity with marked improvement versus the corresponding phenyl analogs. For example, **D11** was found to approach 10-fold greater activity than the corresponding phenyl analog **D10**, on the difficult to control *Hv*. Introduction of a 4-chloro substituent provided a further increase in activity as shown for compounds **D12–D17**. Comparison of **D11**, where R_4 is H, with its 4-chloro analog **D13** showed a 4- to 10-fold improvement in activity for **D13** across all three species of Lepidoptera. The structure-activity relationship for the R_6 substituents chloro, bromo and trifluoromethyl suggested the trend Br > CF_3 > Cl, although these differences were small. In laboratory insecticide screens, compound **D13** was arguably the most active analog tested. Field trials, however, showed subtle differences in performance with **D14** (Ryanxypyr™) showing consistent and outstanding results against all major species of Lepidoptera. Rynaxypyr™ was selected from this group based on a variety of properties including its outstanding insecticidal activity, exceptional performance in the field, low toxicity to mammals, and its favorable environmental profile.

Table II. Insecticidal potency and CMT values of anthranilic diamides D11–D17.

Compound	Sf LC_{50} ppm	Px LC_{50} ppm	Hv LC_{50} ppm	CMT (μM)
D11	0.12	0.07	0.41	0.30
D12	0.04	0.02	0.04	0.06
D13	0.03	0.01	0.02	0.10
D14	0.02	0.01	0.06	0.05
D15	0.04	0.03	0.02	0.03
D16	0.04	0.01	0.17	0.18
D17	0.05	0.03	0.03	0.09

11.4
Biological Attributes

Rynaxypyr™ provides outstanding crop protection against a broad spectrum of lepidopteran species. Dependent on the crop, target species, pest pressure, and other factors, typical use rates generally fall in the range of 25–75 g ai/Ha. However, use rates as low as 1–2 g ai/ha have been observed on species such as *Anticarsia gemmatalis* on soybeans. All major lepidopteran families are controlled including Noctuidae, Tortricidae, Pyralidae, and Plutellidae. Other species for which high levels of activity have been found include *Leptinotarsa decemlineata, Lyriomyza* spp., and *Lissorhoptrus*. Rynaxypyr™ provides lasting control in excess of 14 days and will be an excellent choice for IPM programs based on its control of resistant strains and minimal impact on beneficial insects.

11.5
Toxicology

Rynaxypyr™ possesses low acute mammalian toxicity with an acute oral and dermal $LD_{50} > 5000$ mg/kg in rat studies. In 90-day studies, little to no toxicity was observed following repeat dosing at doses as high as 1500 mg/kg/day. Furthermore, no developmental toxicity was observed in rats or rabbits with doses up to 1000 mg/kg/day. In a two-generation rat reproduction study, no effects on reproduction, fertility, litter size, pup survival, or developmental landmarks were observed. Moreover, Rynaxypyr™ was found to be inactive in all *in vivo* and *in vitro* mutagenicity tests. Additional studies have shown low toxicity to birds, fish, and beneficial insects. We believe the toxicology profile is extremely favorable especially in comparison to current commercial insecticides.

11.6 Mechanism of Action

To assess activity at the insect ryanodine receptor, pyridyl pyrazoles of Table II were tested in a calcium mobilization assay, using neurons from the American cockroach, *Periplaneta americana*. These studies have confirmed the mode of action to be RyR activation. Compounds **D11–D17** showed exceptional potency in this assay with activity in the range of 0.03–0.30 μM. The data shows the ability of anthranilic diamides to release internal calcium stores while failing to activate voltage-gated calcium channels. Furthermore, calcium mobilization induced by anthranilic diamides is blocked following treatment with 1 μM ryanodine, consistent with action at the ryanodine receptor.

Figure 7 shows the relationship between insecticidal activity on *Plutella xylostella* and data from the Ca^{2+} mobilization assay for a diverse structural set of anthranilic diamides.

These results demonstrate the strong linear relationship between whole insect activity and Ca^{2+} mobilization and provide further confirmation of RyR activation as the mechanism of action. A similar trend is seen in the comparison of the LC_{50} values for *Spodoptera frugiperda* and *Heliothis virescens* and their corresponding CMT's.

To determine if differential receptor selectivity is a contributing factor to the low mammalian toxicity of Rynaxypyr™, comparative studies with insect and mammalian cells were conducted (Figure 8). As the data of Figure 8 demonstrates, Rynaxypyr™ is 350-fold less potent against RyRs in the mouse cell line, C2C12, than insect RyRs. Greater selectivity is observed with the rat cell line, PC12, where

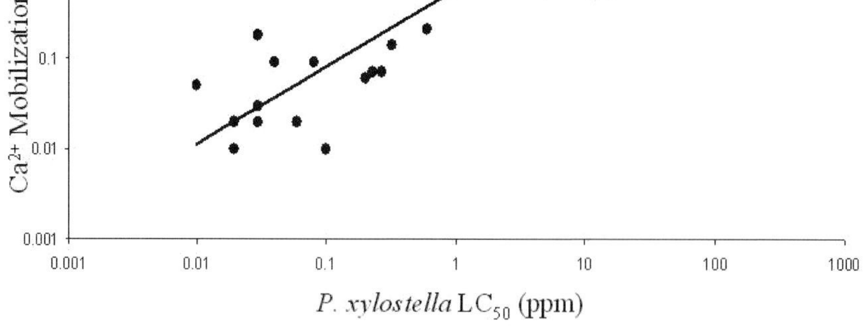

Figure 7. Anthranilic diamide LC_{50} values for *Plutella xylostella* plotted against CMT values.

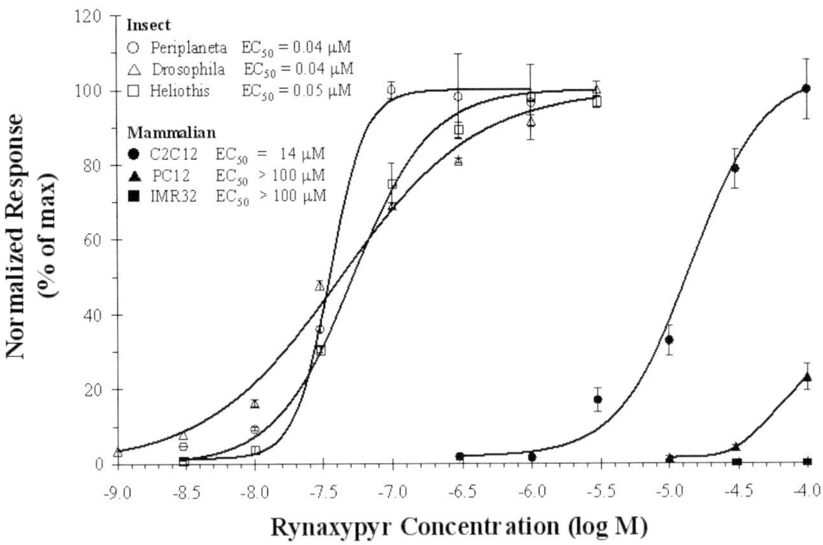

Figure 8. Comparative activity of Rynaxypyr™ against cells expressing insect and mammalian RyRs.

> 2500-fold selectivity is observed. And finally, in the human cell line IMR32, Rynaxypyr™ fails to activate RyRs when tested at concentrations up to 100 μM. This large differential selectivity toward insect RyRs is highly consistent with the observed low mammalian toxicity and almost certainly a contributing factor to the low mammalian toxicity.

11.7 Conclusion

In summary, a novel class of chemistry has been discovered with exceptional insecticidal activity against a broad spectrum of lepidoptera. These compounds have been found to exhibit their action through release of intracellular Ca^{2+} stores mediated by the ryanodine receptor. The first commercial member of this class, Rynaxypyr™, demonstrates outstanding lab and field activity on all major species of lepidoptera with lab rates in the range of 0.01–0.06 ppm. This level of activity is significantly better than current commercial standards and shows remarkable consistency across a broad insect spectrum. Rynaxypyr™ thus offers exceptional promise as a new product for crop protection based on this combination of a new mode of action with outstanding insecticidal properties.

11.8 Dedication

This paper is dedicated to the memory of our colleague and friend, J. Gary Hollingshaus, for his great enthusiasm and tireless energy during the course of this work.

11.9 References

1 G. P. Lahm, T. P. Selby, J. H. Freudenberger, T. M. Stevenson, B. J. Myers, G. Seburyamo, B. K. Smith, L. Flexner, *et al.*, *Bioorg. and Med. Chem. Lett.*, **2005**, *15*, 4898–4906.

2 G. P. Lahm, T. P. Selby, T. M. Stevenson, *PCT Int. Appl. WO 03/015519*, **2003**.

3 D. Cordova, E. A. Benner, M. D. Sacher, J. J. Rauh, J. S. Sopa, G. P. Lahm, T. P. Selby, T. M. Stevenson, *et al.*, *Pesticide Biochem. and Phys.*, **2006**, *84*, 196–214.

4 S. Gutteridge, T. Caspar, D. Cordova, J. J. Rauh, Y. Tao, L. Wu, R. M. Smith, *PCT Int. Appl. WO 2004027042*, **2004**.

5 M. Tohnishi, H. Nakao, T. Furuya, A. Seo, Hiroki-Kodama, K. Tsubata, S. Fujioka, Hiroshi-Kodama, *et al.*, *J. Pestic. Sci.*, **2005**, *30*, 354–360.

6 T. Masaki, N. Yasokawa, M. Tohnishi, T. Nishimatsu, K. Tsubata, K. Inoue, K. Motoba, T. Hirooka, *Mol. Pharmacol.*, **2006**, *69*, 1733–1739.

7 U. Ebbinghaus-Kintscher, P. Luemmen, N. Lobitz, T. Schulte, C. Funke, R. Fischer, T. Masaki, N. Yasokawa, *et al.*, *Cell Calcium*, **2006**, *39*, 21–33.

8 F. Effenberger, M. Roos, R. Ahmad, A. Krebs, *Chemische Berichte*, **1991**, *124*, 1639–1650.

Keywords

Rynaxypyr^{TM}, Anthranilic Diamide, Insecticide, Ryanodine Receptor

12
Elucidation of the Mode of Action of Rynaxypyr™, a Selective Ryanodine Receptor Activator

Daniel Cordova, Eric A. Benner, Matthew D. Sacher, James J. Rauh, Jeffrey S. Sopa, George P. Lahm, Thomas P. Selby, Thomas M. Stevenson, Lindsey Flexner, Timothy Caspar, James J. Ragghianti, Steve Gutteridge, Daniel F. Rhoades, Lihong Wu, Rejane M. Smith, Yong Tao

12.1
Introduction

There is a continuous demand for discovery of insecticides with novel modes of action to combat resistance to existing products. Interference with one of five physiological systems accounts for the mode of action of 95% of commercial insecticides [1–3]. It has been speculated that Ca^{2+} homeostasis mechanisms would offer excellent targets for insect control; however, discovery of synthetic pesticides that exploit these targets has remained elusive. Ca^{2+} signaling plays a key role in numerous biological processes, including muscle contraction, and neurotransmitter release [4]. Muscle contraction involves modulation of two distinct Ca^{2+} channels, voltage-gated channels, which regulate external Ca^{2+} entry, and ryanodine receptor channels (RyRs), which regulate release of internal Ca^{2+} stores [5]. Here we describe the mode of action of Rynaxypyr™, a new insecticide currently in development at DuPont Crop Protection, which provides unprecedented lepidopteran control through activation of insect RyRs.

12.2
Symptomology Associated with Anthranilic Diamides

Lepidopteran larvae exposed to anthranilic diamides exhibit rapid feeding cessation, general lethargy, constrictive muscle paralysis, and regurgitation. One of the earliest symptoms observed was reduction in heart rate. *Manduca sexta* larvae showed greater than a 50% decrease in heart beat frequency ten minutes following injection with Rynaxypyr™ (30 ng). Among anthranilic diamides evaluated, similar rank potency was found for this cardio-inhibitory effect and lepidopteran toxicity.

Pesticide Chemistry. Crop Protection, Public Health, Environmental Safety
Edited by Hideo Ohkawa, Hisashi Miyagawa, and Philip W. Lee

ISBN: 978-3-527-31663-2

These symptoms are unique among commercialized insecticides and pointed to a novel mode of action.

12.3
Rynaxypyr™ Stimulates Release of RyR-Mediated Internal Ca^{2+} Stores

Early studies with anthranilic diamides demonstrated a lack of activity against targets associated with commercial insecticides. However, it became clearly evident that Rynaxypyr™ and related anthranilic diamides stimulate an increase in intracellular Ca^{2+} concentration ($[Ca^{2+}]_i$). In *P. americana* neurons, a transient, dose-dependent increase in $[Ca^{2+}]_i$ is observed with Rynaxypyr™ (Figure 1). The linear relationship between such responses and lepidopteran toxicity for a series of anthranilic diamides indicates Ca^{2+} mobilization plays a key role in insect toxicity. It was subsequently determined that this Ca^{2+} mobilization is: (1) independent of external Ca^{2+}, (2) blocked by the selective RyR agent, ryanodine, and (3) absent in cells that do not endogenously express RyRs. Such findings, coupled with an absence of activity against other targets, supported RyR-mediated internal Ca^{2+} store release as the mode of action for Rynaxypyr™ and related anthranilic diamides.

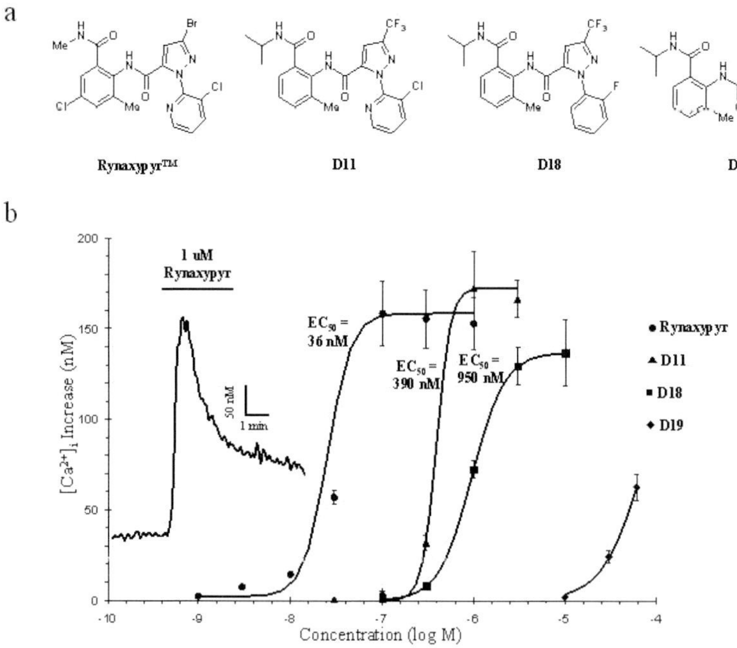

Figure 1. Ca^{2+} mobilization dose response for *P. americana* neurons challenged with anthranilic diamides shown in (a).
Rynaxypyr™ releases internal Ca^{2+} stores with an EC_{50} value of 36 nM (b).
Inset: Typical Ca^{2+} response for Rynaxypyr™.

12.4
Rynaxypyr™ Binds to a Unique Site on the RyR

The RyR is a tetrameric channel protein, which regulates release of stored intracellular Ca^{2+} (Figure 2). The name was derived from the natural plant product, ryanodine, which binds at the channel pore and alters the channel conductance state. Anthranilic diamides failed to displace or enhance ^{3}H-ryanodine binding to *P. americana* membranes, indicating that this chemistry binds to a distinct site. A radiolabeled anthranilic diamide, ^{3}H-**D18**, was prepared and found to exhibit specific, saturable binding to muscle membranes. Interestingly, under conditions of high $CaCl_2$ (500 μM), ryanodine enhances ^{3}H-**D18** binding up to 8-fold, with a $K_d = 44$ nM and $B_{max} = 9690$ fmol/mg [6]. In this preparation, Rynaxypyr™ potently displaces ^{3}H-**D18** with an IC_{50} value of 4 nM. As was observed with lepidopteran toxicity, a linear relationship was revealed between Ca^{2+} mobilization and the ability of anthranilic diamides to displace ^{3}H-**D18**.

Western blot analysis of *P. americana* membranes and photo-affinity studies using a tritiated azido-anthranilic diamide revealed that Rynaxypyr™ and related analogs bind directly to the ryanodine receptor rather than to an accessory protein.

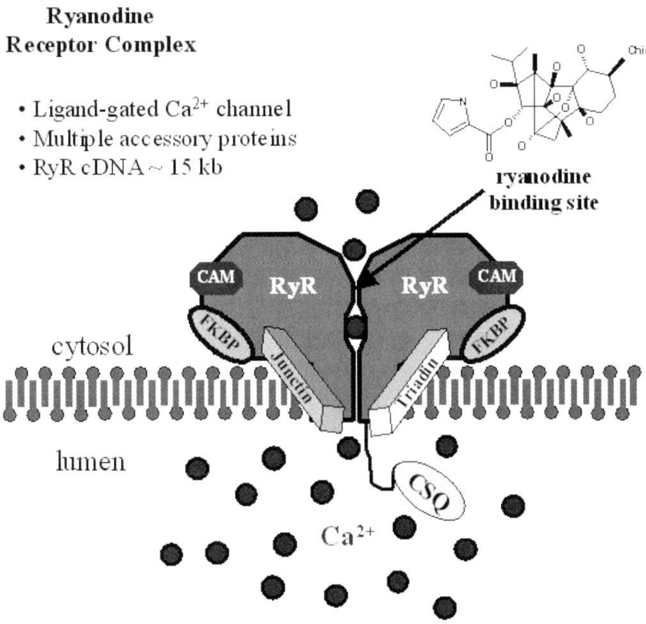

Figure 2. Diagram of the RyR complex.
Two of the four RyR subunits are shown along with accessory proteins and the location of the ryanodine-binding site.
CAM = Calmodulin, FKBP = FK506-Binding Protein, and CSQ = Calsequestrin.

12.5
Cloning and Expression of Pest Insect RyRs

Cloning and expression of multiple insect RyRs were undertaken to provide genetic validation of the mode of action of anthranilic diamides and enable high throughput screening of the insect target. Previous efforts to establish a stable cell line expressing full-length functional insect RyRs were unsuccessful [7]. Using various molecular approaches, we have successfully cloned and expressed full-length functional RyRs from dipteran, lepidopteran, and homopteran pest insects [8]. As shown in Figure 3, sensitivity to the RyR activator, caffeine, and Rynaxypyr™, is conferred upon Sf9 cells (a *S. frugiperda* cell line) expressing the *H. virescens* RyR. Sf9 cells expressing recombinant *D. melanogaster* and *H. virescens* RyRs exhibit comparable Rynaxypyr™ sensitivity to that observed with native *P. americana* neurons. Expression of such recombinant insect RyRs offers utility for target screening and is the basis of a patent application [8].

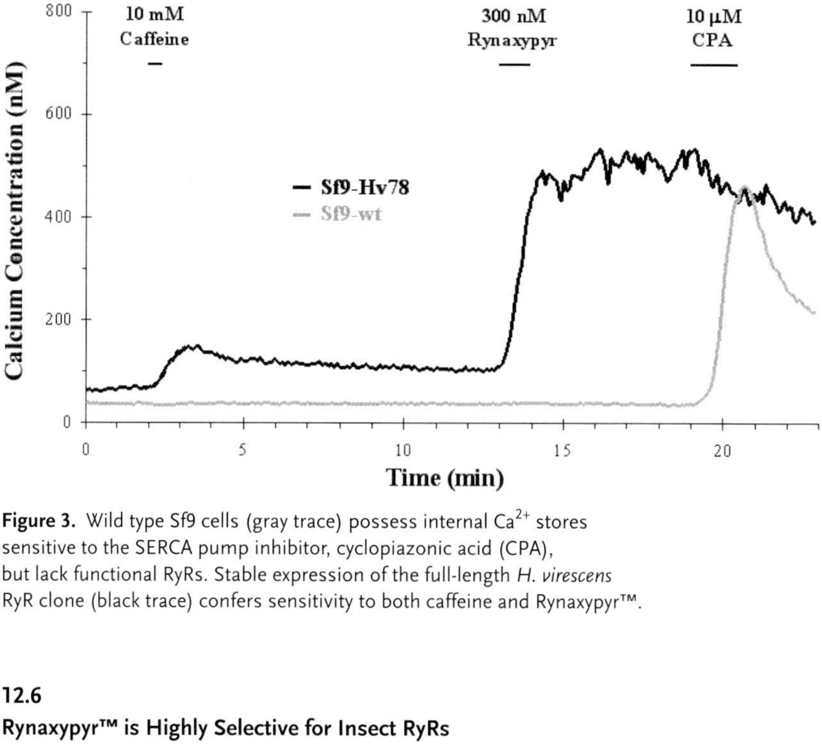

Figure 3. Wild type Sf9 cells (gray trace) possess internal Ca^{2+} stores sensitive to the SERCA pump inhibitor, cyclopiazonic acid (CPA), but lack functional RyRs. Stable expression of the full-length *H. virescens* RyR clone (black trace) confers sensitivity to both caffeine and Rynaxypyr™.

12.6
Rynaxypyr™ is Highly Selective for Insect RyRs

Mammals possess three isoforms of the ryanodine receptor; RyR1 and RyR2, distributed primarily in skeletal and cardiac muscle, respectively, and RyR3 distributed more heterogeneously. Insects, however, express a single form of the receptor, sharing only 47% sequence homology [9]. Comparative studies were conducted to determine Rynaxypyr™'s ability to activate mammalian RyRs.

Though ryanodine and caffeine show similar potency against insect and mammalian receptors (not shown), differential selectivity is observed for Rynaxypyr™. An EC_{50} value of 14 μM was determined for C2C12 cells, which express the RyR1 and RyR3 isoforms [10]. In cells expressing the RyR1 and RyR2 isoforms (undifferentiated rat PC12 cell line), Rynaxypyr™ shows even lower potency, with an EC_{50} value greater than 100 μM (limited saline solubility) [10]. Moreover, Rynaxypyr™ failed to activate RyRs in a third cell line, the human IMR32 line. Although the receptor isoforms expressed in this line have yet to be characterized, they clearly express functional RyRs as demonstrated with caffeine. Consequently, Rynaxypyr™ exhibits 350-fold to > 2500-fold differential selectivity for insect receptors over that of mammalian RyRs. This differential selectivity is almost certainly a major factor contributing to Rynaxypyr™'s mammalian safety.

12.7 Conclusion

Rynaxypyr™ is a highly potent and selective activator of insect RyRs. Activation of these receptors causes unregulated release of internal Ca^{2+} stores leading to store depletion, muscle paralysis, and ultimately insect death. Anthranilic diamides bind to a site on the RyR distinct from that of ryanodine or caffeine and appears to be impacted by the channel's state. Through cloning and expression of multiple insect RyRs we have provided genetic validation of Rynaxypyr™'s mode of action and developed a powerful tool for high-throughput screening. In addition to its superb lepidopteran potency and novel mode of action, Rynaxypyr™'s possesses strong differential selectivity for insect over mammalian RyRs. Based on these attributes, Rynaxypyr™ holds great promise for pest management strategies.

12.8 References

1 R. Nauen, T. Bretschneider, *Pesticide Outlook*, **2002**, *6*, 241–245.

2 T. Narahashi, *Mini Rev. Med. Chem.*, **2002**, *2*, 419–432.

3 M. A. Dekeyser, *Pest Manag. Sci.*, **2005**, *61*, 103–110.

4 M. J. Berridge, P. Lipp, M. D. Bootman, *Nat. Rev. Mol. Cell Biol.*, **2000**, *1*, 11–21.

5 W. Melzer, A. Herrmann-Frank, H. C. Luttgau, *Biochim. Biophys. Acta*, **1995**, *1241*, 59–116.

6 D. Cordova, E. A. Benner, M. D. Sacher, J. J. Rauh, J. S. Sopa, G. P. Lahm, T. P. Selby, T. M. Stevenson, *et al.*, *Pest. Biochem. Phys.*, **2006**, *84*, 196–214.

7 X. Xu, M. B. Bhat, M. Nishi, H. Takeshima, J. Ma, *Biophys J.*, **2000**, *78*, 1270–1281.

8 S. Gutteridge, T. Caspar, D. Cordova, J. J. Rauh, Y. Tao, L. Wu, R. M. Smith, WO 2004027042, **2004**.

9 M. Takeshima, M. Nishi, N. Iwabe, T. Miyata, T. Hosoya, I. Masai, Y. Hotta, *FEBS Lett.* **1994**, *337*, 81–87.

10 D. L. Bennett, T. R. Cheek, M. J. Berridge, H. DeSmedt, J. B. Parys, L. Missiaen, M. D. Bootman, *J. Biol. Chem.*, **1996**, *271*, 6356–6362.

Keywords

Rynaxypyr™, Anthranilic Diamide, Insecticide, Ryanodine Receptor

13
Flubendiamide, a New Insecticide Characterized by Its Novel Chemistry and Biology

Akira Seo, Masanori Tohnishi, Hayami Nakao, Takashi Furuya, Hiroki Kodama, Kenji Tsubata, Shinsuke Fujioka, Hiroshi Kodama, Tetsuyoshi Nishimatsu, Takashi Hirooka

13.1
Introduction

Resistance has often been a problem or a potential problem for insecticides and this is one of the most important reasons why the insecticides with a new mode of action have been always desired, though it is quite a difficult task to find such molecules. Flubendiamide, discovered by Nihon Nohyaku (NNC), is a novel insecticide belonging to the new chemical class of 1,2-benzenedicarboxamides or phthalic diamides, having a unique chemical structure (Figure 1) [1–3]. Flubendiamide is co-developed by NNC and Bayer CropScience globally [4]. The structure-activity relationships, the chemistry, including topics in process research, the mode of action and the biological profiles are described.

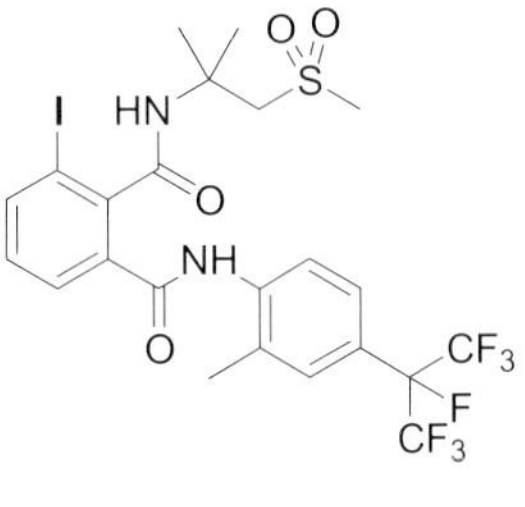

Figure 1. The chemical structure of flubendiamide, 3-iodo-*N*′-(2-mesyl-1,1-dimethylethyl)-*N*-{4-[1,2,2,2-tetrafluoro-1-(trifluoromethyl)ethyl]-*o*-tolyl}phthalamide.

Pesticide Chemistry. Crop Protection, Public Health, Environmental Safety
Edited by Hideo Ohkawa, Hisashi Miyagawa, and Philip W. Lee

ISBN: 978-3-527-31663-2

13.2
Structure-Activity Relationship

13.2.1
Lead Generation

Figure 2 shows the early phase of research for flubendiamide. In 1989, Dr. T. Tsuda, at Osaka Prefecture University in Japan, reported that some pyrazinedicarboxamide derivatives showed moderate herbicidal activity [5]. From 1990, the research for herbicide discovery was conducted at NNC Research Center. In the course of this research, a lead compound for an insecticide was discovered in 1993 from the class of benzenedicarboxamides as shown in Figure 2. This compound provided insecticidal activity on lepidoptera at the relatively high dose of 50–500 mg a.i./L. Moreover, it did not show activity against other species such as Hemiptera or Acarina. Although the level of activity was not satisfactory, this compound attracted the attention of researchers for both the novelty of its chemical structure and the characteristic insecticidal symptoms such as gradual contractions of the insect body. We therefore started the study for further optimization of this lead compound.

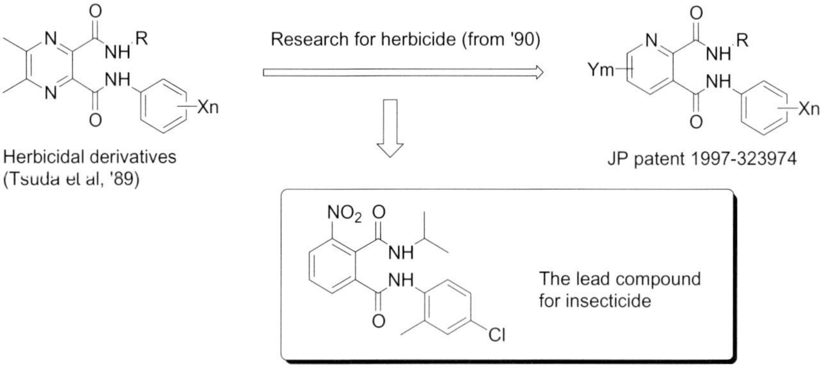

Figure 2. Optimization history of flubendiamide.

13.2.2
Lead Optimization

The weak insecticidal activity was found in the lead compound; its structure was quite new as an insecticide. However, there were various points to be improved for practical use such as increased insecticidal activity, reduced phytotoxicity to crops and instability of the compound. Two thousand derivatives were synthesized with the general formula shown in Figure 3. Many studies on the improvement of the activity were conducted, and flubendiamide was finally discovered in 1998.

The chemical structure of benzenedicarboxamides can be divided into three parts as shown in Figure 3. These are characterized by (A) the phthaloyl moiety, (B) the aliphatic amide moiety and (C) the aromatic amide moiety. A brief description of the structure-activity relationships for each part is described below.

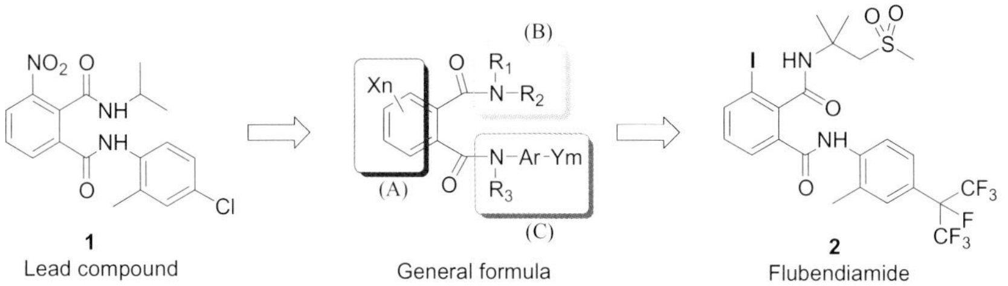

Figure 3. Lead optimization of benzenedicarboxamide derivatives.

Table I shows the insecticidal activity of 1,2-benzenedicarboxamides. Insecticidal activities are shown as EC_{50} values against common cutworm (*Spodoptera litura*) and diamondback moth (*Plutella xylostella*).

In order to improve the activity of the lead compound **1**, the nitro group was changed to other groups. Although the non-substituted derivative **3** showed similar or slightly higher activity, we found that the chloro-derivative **4** showed much stronger activity. Optimization of substituent X with chloro-derivatives at positions 3–6 resulted in the finding that the 3-position was clearly the best (Table II, compounds **4** to **7**). As for substituents at the 3-position of the phthaloyl moiety, various groups such as halogen atoms, aryl groups, haloalkyl groups, haloalkoxy groups, and haloalkylthio groups were evaluated. Among these substituents, lipophilic and bulky substituents tended to show good activity, and the iodine atom was found to be the best substituent for X. It should be noted that compounds having an iodine atom are very rare among commercial agrochemicals.

As for substituents on the aniline ring, the ortho-methyl group was fixed, because it was essential for keeping the stability of the diamide structure. The optimization of the best position with a chlorine atom as substituent Y showed that the 4-position was the best by comparison of compounds **10–12**. Other groups were introduced as substituent Y onto the aniline ring, and the results showed the tendency for a more lipophilic substituent to be preferable. Notably, the fluoroalkyl group was highly effective as exemplified with the heptafluoroisopropyl compound **15**. The heptafluoroisopropyl group has never been reported as a substituent in a commercial pesticide and is seldom used in pesticide research.

The last section shows the effect of substituents (R_1, R_2) on the aliphatic amide moiety. As for the aliphatic side chain, it was found that the alpha-branched alkyl side chain was essential for stabilizing the diamide structure. In the case of non-branched alkyl, the diamide derivatives tend to decompose to the corresponding phthalimides. A variety of substituents were examined to improve the activity. As shown in Table I, the introduction of a heteroatom or a functional group increased the insecticidal activity; especially a sulfur atom within the alkyl side chain markedly increased the activity. This sulfonylalkylamine is also novel as an amine residue in pesticide chemistry. In summary, flubendiamide has unique substituents as essential parts of the structure in three adjacent positions on the benzene ring, which characterizes the chemical structure of flubendiamide as totally novel.

Table I. Insecticidal activities of 1,2-benzenedicarboxamides against *Spodoptera litura* and *Plutella xylostella*.

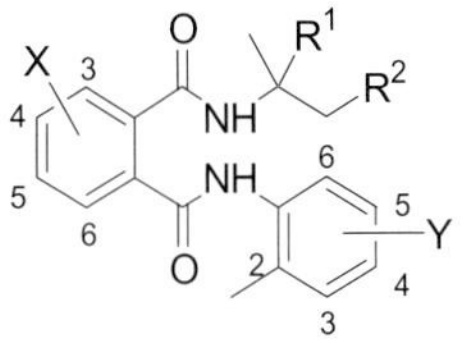

No.	X	Y	R_1	R_2	EC_{50} value (mg a.i./L)	
					S. litura	*P. xylostella*
1	3-NO_2	4-Cl	H	H	10–100	10–100
3	H	4-Cl	H	H	10–100	3–10
4	3-Cl	4-Cl	H	H	10	1–3
5	4-Cl	4-Cl	H	H	> 500	5
6	5-Cl	4-Cl	H	H	> 500	50
7	6-Cl	4-Cl	H	H	> 500	10
8	3-F	4-Cl	H	H	> 100	1–3
9	3-Br	4-Cl	H	H	10	1
10	3-I	4-Cl	H	H	3–10	0.3–1
11	3-I	3-Cl	H	H	10	3
12	3-I	5-Cl	H	H	10–100	3–10
13	3-I	4-OCH_3	H	H	30–100	10–30
14	3-I	4-OCF_3	H	H	1–3	0.3–1
15	3-I	4-$CF(CF_3)_2$	H	H	0.3–1	0.1–0.3
16	3-I	4-$CF(CF_3)_2$	CH_3	H	0.3–1	0.3–1
17	3-I	4-$CF(CF_3)_2$	CH_3	$NHCOCH_3$	0.1	not tested
2	3-I	4-$CF(CF_3)_2$	CH_3	SO_2CH_3	0.03–0.1	0.001–0.003

13.3 Chemistry

Figure 4 shows the synthetic pathway to flubendiamide employed at the early stages of discovery. 3-Iodophthalic anhydride **21** was the important intermediate, which was prepared from commercially available 3-nitrophthalic acid according to known methods via a diazonium intermediate. Phthalamic acid **22** was obtained by the reaction of **21** with thioalkylamine with high regioselectivity. Phthalamic acid **22** was treated with methyl chloroformate to give isoimide **23**, which was reacted with the corresponding aniline to afford diamide **24**. Finally flubendiamide was obtained by the oxidation of diamide **24** with hydrogen peroxide. It seemed that there was no alternative route, since a practical iodination was very limited.

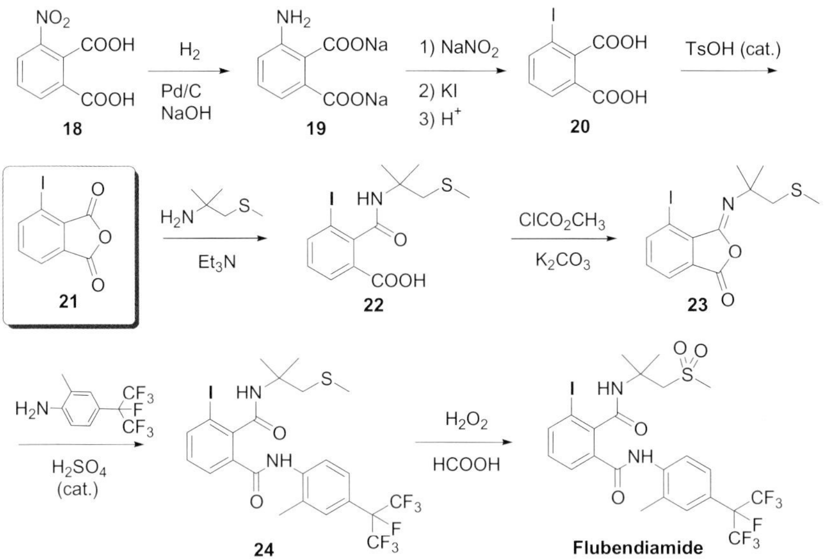

Figure 4. Synthetic pathway of flubendiamide at the early optimization stage.

Process investigation was started at a very early stage of the optimization studies before flubendiamide had been discovered, nevertheless extensive research to resolve the issues such as cost, quality, safety, and environment performance were conducted on related analogs. Since flubendiamide consisted of three characteristic building blocks, it was necessary that a regioselective introduction of the three components be found, and also an inexpensive manufacturing method for each component be established. Figure 5 shows the newly developed regioselective introduction of an iodine atom. Iodine was introduced at the ortho-position of the benzamide by a Pd(II) catalyzed reaction, which was quite novel in the area of palladium chemistry. This reaction realized direct and regioselective introduction of iodine onto the benzene ring in one step. Both the reaction yield and regioselectivity were excellent, and furthermore, the reaction itself is practical from the view point of manufacturing. It is noteworthy that the waste volume could be extremely reduced compared with the case of the former diazonium method. With these characteristic points established, this reaction may be classified as "Green chemistry".

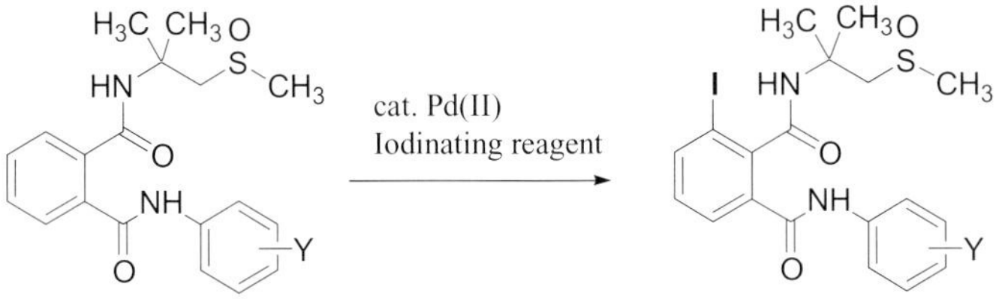

Figure 5. Regioselective introduction of an iodine atom.

13.4 Mode of Action

Flubendiamide is most effective on larvae followed by adults, but it has no ovicidal activity. In the course of extensive research on the mode of action of flubendiamide, it was determined that flubendiamide was a ryanodine receptor modulator. Flubendiamide fixes the Ca-channel of insect ryanodine receptors (RyR) in the open state, and subsequently induces calcium release from the membrane vesicle [6]. In parallel, the RyR activation by flubendiamide induces the stimulation of the Ca-pump *via* functional connection between these two components [7]. It is suggested that the effect of flubendiamide on intracellular calcium regulation is essential for the insecticidal activity. Furthermore, flubendiamide shows very little effect on the mammalian RyR isoform. This comparative study concluded that flubendiamide specifically activates insect RyR. By the binding assay using ^{3}H-flubendiamide, it was confirmed that the binding site was specific to insect RyR, and its binding site was different from those of other RyR modulators such as ryanodine. Finally, it is known that the binding site of ryanodine is located at a pore region of the RyR. Thus, we conclude that the selective action of flubendiamide is due to the specificity of the binding site.

13.5 Biological Profile

Table II shows the insecticidal activity of flubendiamide against major insect and acarina species. Flubendiamide provided high activity on all lepidopterous insect pests, and its EC_{50} values were between 0.004 and 0.58 mg a.i./L. However, flubendiamide did not show activity against other insect species. Thus, the insecticidal spectrum of flubendiamide is expected to be broad among lepidoptera pests in agriculture. Against the resistant strain of diamondback moth, flubendiamide provided the same level of activity as against the susceptible strain. This result indicates that flubendiamide will be useful for insecticide resistance management (IRM) programs.

Table III shows the activity of flubendiamide on several species of beneficial arthropods and natural enemies. Flubendiamide was inactive against beneficial arthropods (except silkworm) and natural enemies tested. This result indicates that flubendiamide should be very safe for natural enemies, and consequently will fit well into integrated pest management (IPM) programs.

Field evaluations of flubendiamide have been conducted in many areas on various crops such as vegetables, top fruits, and cotton. Flubendiamide shows excellent performance on controlling the major lepidopterous pests in the field at the recommended dose and its efficacy was better than those of standard insecticides. Furthermore, flubendiamide (20% WDG) showed no phytotoxicity to vegetables, tea and top-fruits at recommended doses [3–4].

Table II. Insecticidal spectrum of flubendiamide.

Scientific name	Common name	Tested stage	DAT	EC_{50} (mg a.i./L)
Plutella xylostella	Diamondback moth	L3	4	0.004
(Resistant strain)*		L3	4	0.002
Spodoptera litura	Common cutworm	L3	4	0.19
Helicoverpa armigera	Cotton bollworm	L3	4	0.24
Agrotis segetum	Turnip moth	L2–3	7	0.18
Autographa nigrisgna	Beet semi-looper	L3	4	0.02
Pieris rapae crucivora	Common cabbage worm	L2–3	4	0.03
Adoxophyes honmai	Smaller tea tortrix	L3	5	0.38
Homona magnanima	*Oriental tea tortrix*	L4	5	0.58
Hellula undalis	Cabbage webworm	L3	5	0.01
Chilo suppressalis	Rice stem borer	L3	7	0.01
Diaphania indica	Cotton caterpillar	L3	3	0.02
Sitophilus zeamais	Maize weevil	A	4	> 1000
Nilaparvata lugens	Brown rice planthopper	L3	4	> 1000
Myzus percicae	Green peach aphid	All	7	> 1000
Pseudococcus comstocki	Comstock mealybug	L1	7	> 100
Tetranychus urticae	Two-spotted spider mite	All	4	> 100

L2, L3, A: second, third and Adult DAT: Day(s) after treatment
* resistant strains to pyrethroids, BPUs, Ops, and carbamates

Table III. Activity of flubendiamide on beneficial arthropods and natural enemies.

Common name	Scientific name	Test stage	EC_{30} (mg a.i./L)
Honeybee	*Apis mellifera*	Adult	> 200
Horn-faced bee	*Osmia cornifrons*	Adult	> 200
Bumblebee	*Bombus ignitus*	Adult	> 200
Lady beetle	*Harmonia axyridis*	Adult	> 200
	Coccinella septempunctata bruckii	Adult	> 200
Parasite wasp	*Encarsia formosa*	Adult	> 400
	Aphidius colemani	Adult	> 400
	Cotesia glomerata	Adult	> 100
Green lacewing	*Chrysoperla carnea*	Larva	> 100
Predatory bug	*Orius strigicollis*	Adult	> 100
Predatory midge	*Aphidoletes aphidimyza*	Larva	> 100
Predatory mite	*Amblyseius cucumeris*	Adult	> 200
	Phytoseiulus persimilis	Adult	> 200
Spider	*Pardosa pseudoannulata*	Adult	> 100
	Misumenops tricuspidatus	Adult	> 200
Silkworm	*Bombyx mori*	Larva	< 50

13.6 Toxicological Properties

Flubendiamide shows low acute oral toxicity. The LD_{50} for male and female rats were both > 2000 mg/kg. The agent is slightly irritating to rabbit eyes, non-irritating to rabbit skin, non-mutagenic in the Ames test, and non-sensitizing to guinea pig skin. The acute oral LD_{50} for quail was > 2000 mg/kg. The LD_{50} for carp was > 548 mg/L. These findings suggest that flubendiamide is safe for mammals, fish, and birds.

13.7 Conclusion

Flubendiamide represents a novel class of insecticide having a unique chemical structure, and provides a new mode of action, which acts as a RyR modulator. This activity is highly selective to insect RyR, and no cross-resistance to existing insecticides is observed. Flubendiamide will also be very suitable for Insecticide Resistant Management. Furthermore, flubendiamide shows a broad insecticidal spectrum against lepidopterous insect pests, excellent efficacy in field evaluations, and excellent safety against various beneficial arthropods and natural enemies. It will be suitable for IPM programs.

13.8 Acknowledgments

The authors would like to express sincere thanks to all their distinguished colleagues involved in the discovery of flubendiamide in Nihon Nohyaku Co., Ltd. The authors also wish to acknowledge the many scientists at Bayer CropScience AG and Professor Yasuo Mori at Kyoto University during the course of the mode of action work.

13.9 References

1 M. Tohnishi, H. Nakao, E. Kohno, T. Nishida, T. Furuya, T. Shimizu, A. Seo, K. Sakata, *et al.*, *Eur. Pat. Appl.*, **2000**, EP 1006107.
2 M. Tohnishi, H. Nakao, T. Furuya, A. Seo, Hiroki Kodama, K. Tsubata, S. Fujioka, H. Kodama, *et al.*, *J. Pesticide Sci.*, **2005**, *30*, 354–360.
3 T. Nishimatsu, T. Hirooka, H. Kodama, M. Tohnishi, A. Seo, *Proceedings of the BCPC International Congress – Crop Sci. & Technology*, **2005**, *2A-3*, 57–64.
4 T. Nishimatsu, H. Kodama, K. Kuriyama, M. Tohnishi, D. Ebbinghaus, J. Schneider, *Int'l. Conf. on Pesticides, Kuala Lumpur, Malaysia, Book of Abstracts*, **2005**, 156–161.
5 T. Tsuda, H. Yasui, H. Ueda, *J. Pestic. Sci.*, **1989**, *14*, 241–243.
6 U. Ebbinghaus-Kintscher, P. Luemmen, N. Lobitz, T. Schulte, C. Funke, R. Fischer, T. Masaki, N. Yasokawa, *et al.*, *Cell Calcium*, **2006**, 39, 21–33.
7 T. Masaki, N. Yasokawa, M. Tohnishi, T. Nishimatsu, K. Tsubata, K. Inoue, K. Motoba, T. Hirooka, *Mol. Pharmacol.*, **2006**, *69*, 1733–1739.

Keywords

Flubendiamide, Benzenedicarboxamide, Insecticide, Ryanodine Receptor Modulator

14
Flubendiamide Stimulates Ca^{2+} Pump Activity Coupled to RyR-Mediated Calcium Release in Lepidopterous Insects

Takao Masaki, Noriaki Yasokawa, Ulrich Ebbinghaus-Kintscher, Peter Luemmen

14.1
Introduction

Flubendiamide is a novel insecticide possessing potent and selective activity against lepidopterous insects [1]. This insecticidal activity has been clarified to be mediated by a ryanodine-sensitive calcium release channel (RyR) [2–3]. The stabilization of RyR to open state by the compound induces robust calcium release from intracellular calcium store (Figure 1). This implies a significant impact on components involved in intracellular calcium homeostasis such as the Ca^{2+} pump, a pivotal component which reuptakes released Ca^{2+} into SR. In this study, we examined effects on the Ca^{2+} pump, as a consequence of the calcium mobilization induced by flubendiamide.

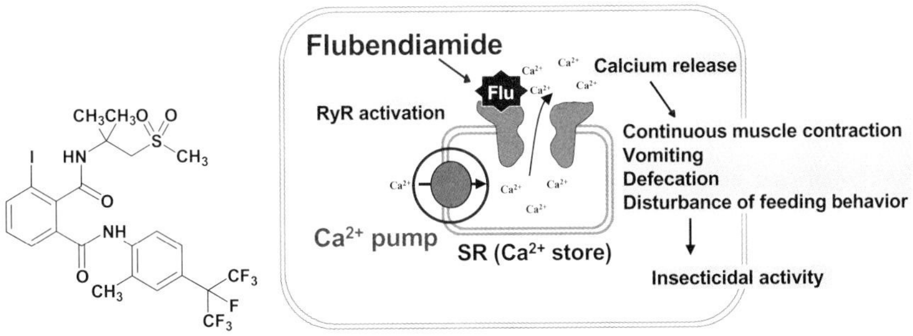

Flubendiamide

Figure 1. Proposed mode of action of flubendiamide.

Pesticide Chemistry. Crop Protection, Public Health, Environmental Safety
Edited by Hideo Ohkawa, Hisashi Miyagawa, and Philip W. Lee

ISBN: 978-3-527-31663-2

14.2
Calcium Release Induced by Flubendiamide

To reveal the functional consequence of the specific interaction between insect RyRs and flubendiamide, calcium release by flubendiamide was first examined using muscle membrane preparations from *S. litura*. As shown in Figure 2, flubendiamide induced remarkable calcium release only in the presence of the Ca^{2+} pump inhibitor, thapsigargin. Excluding thapsigargin from the assay system concealed the observable calcium release induced by flubendiamide (Figure 2A). This evidence suggests that the released Ca^{2+} was rapidly resequestered by Ca^{2+} pump which is tightly coupled to RyR activity.

14.3
Specific Stimulation of Ca^{2+} Pump

The effects of flubendiamide on calcium transport by the Ca^{2+} pump were evaluated by measuring Ca^{2+}-ATPase activity, indicative of the catalytic cycles of Ca^{2+} pump, since calcium transport is stoichiometrically coupled to hydrolysis of ATP. As shown in Figure 3, flubendiamide in the nanomolar range specifically stimulated Ca^{2+}-ATPase in a concentration-dependent manner (EC_{50} = 11 nM). The maximum velocity of Ca^{2+}-ATPase was 160% of control activity at supramaximal concentrations of flubendiamide. The potency of flubendiamide was evidently pronounced in comparison with the effects of the known RyR modulators, ryanodine and caffeine (Figure 3). Furthermore, the Ca^{2+}-ATPase stimulation was quantitatively correlated to insecticidal activity against target insect by analogous compounds. Thus the simulative effect on Ca^{2+}-ATPase should be an important process in insecticidal activity of flubendiamide.

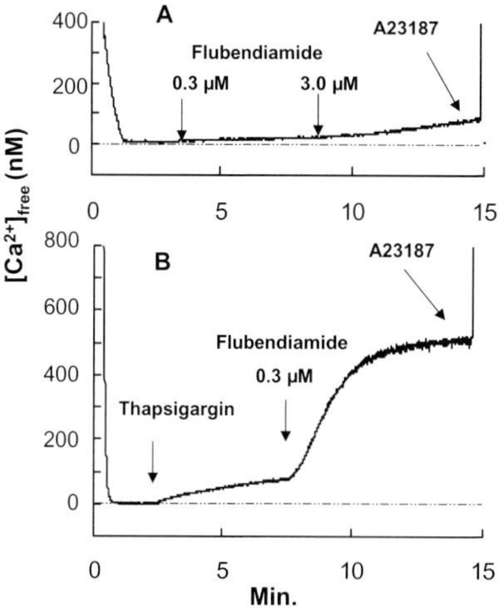

Figure 2. Calcium release from the insect membrane preparation induced by flubendiamide [3].

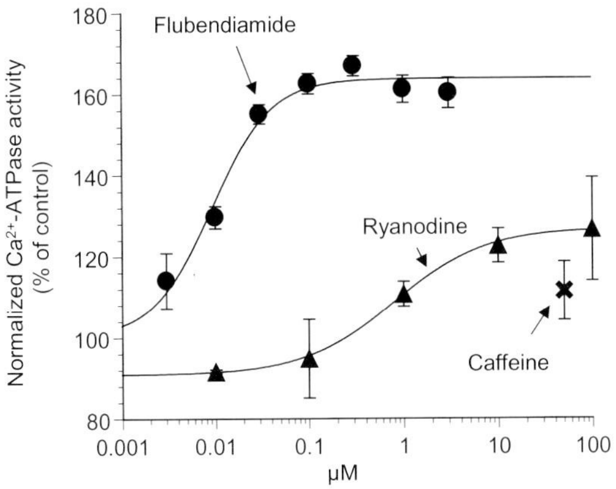

Figure 3. Effect of flubendiamide, ryanodine, and caffeine on the Ca^{2+}-ATPase activity of the membrane preparation [3].

14.4
Luminal Ca^{2+} Mediated Ca^{2+} Pump Stimulation

The calcium release rate through RyR depends on the calcium gradient across SR membrane. The limited-calcium conditions would make for a steeper transmembrane calcium gradient which can enhance the efflux of luminal Ca^{2+} through RyR. Under this condition, the maximum velocity of the flubendiamide-stimulated Ca^{2+}-ATPase activity was notably augmented (1 μM of free Ca^{2+}, as imposed by a Ca^{2+}/EGTA buffer) compared to the velocity in the presence of 50 μM Ca^{2+} [3]. Hence, the results further support the possibility that the calcium efflux rate through RyR would determine the amplitudes of the Ca^{2+}-pump stimulation by the compound (Figure 4).

The rapid calcium efflux efficiently decreases luminal calcium concentration, which induced acceleration of Ca^{2+}-pump activity due to facilitation of calcium dissociation from a low affinity calcium-binding site (luminal site). To investigate the possible involvement of luminal calcium in the Ca^{2+} pump stimulation, the luminal Ca^{2+} concentration was indirectly manipulated by a Ca^{2+}-ionophore and a calcium chelator (Figure 5). A23187 induced a five-fold increase in Ca^{2+} pump activity [3] due to an increase of calcium permeability. Under this condition, Ca^{2+} pump stimulation by flubendiamide was mostly eliminated [3], suggesting the involvement of luminal Ca^{2+} in the Ca^{2+} pump stimulation by flubendiamide.

The stimulatory effect of flubendiamide on the Ca^{2+} pump was also diminished in the calcium buffers comprised of calcium chelators with high and low calcium affinity [3]. The low affinity calcium chelator, diBr-BAPTA (Kd = 3.7 μM), evidently accelerated the catalytic cycles of Ca^{2+} pump as in the case with A23187. The result also implies importance of luminal calcium, since the low affinity of this chelator could not interrupt the calcium association with high affinity binding sites (cytoplasmic site) on Ca^{2+}-ATPase. In addition, the results also demonstrate

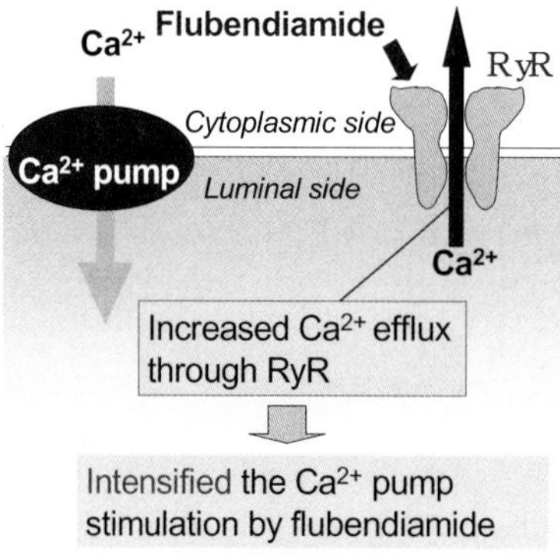

Figure 4. Increase of Ca^{2+} efflux through RyR intensified Ca^{2+} pump stimulation.

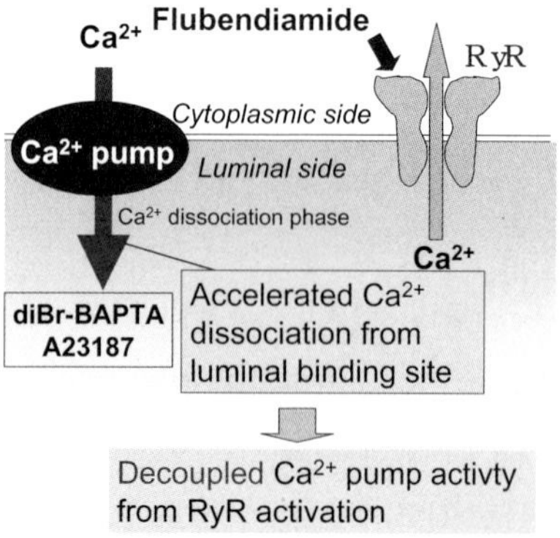

Figure 5. Ca^{2+} dissociation from luminal binding site decoupled Ca^{2+} pump activity from RyR activation.

that the elimination of effect of flubendiamide on the Ca^{2+} pump with A23187 or the chelators inferred flubendiamide has no direct effect on the Ca^{2+} pump.

14.5 Conclusion

In this study, we demonstrated that the calcium mobilization induced by flubendiamide caused sequential effects on the Ca^{2+} pump. The pronounced stimulation of the Ca^{2+} pump by flubendiamide was closely correlated to insecticidal activity. Further investigation suggested that luminal Ca^{2+} was an important mediator for the functional co-ordination of RyR and the Ca^{2+} pump. It is also suggested that this compound should provide a promising probe for understanding the functional regulations of the insect RyR.

14.6 References

1. M. Tohnishi, H. Nakao, T. Furuya, A. Seo, H. Kodama, K. Tsubata, S. Fujioka, H. Kodama, *et al.*, *J. Pestic. Sci.*, **2005**, *30*, 354–360.
2. U. Ebbinghaus-Kintscher, P. Luemmen, N. Lobitz, T. Schulte, C. Funke, R. Fischer, T. Masaki, N. Yasokawa, *et al.*, *Cell Calcium*, **2006**, *39*, 21–33.
3. T. Masaki, N. Yasokawa, M. Tohnishi, T. Nishimatsu, K. Tsubata, K. Inoue, K. Motoba, T. Hirooka, *Mol. Pharmacol.*, **2006**, *65*, 1733–1739.

Keywords

Flubendiamide, Ryanodine Receptor, Ca^{2+} Pump, Insecticide, Mode of Action

15
Novel Arylpyrazole and Arylpyrimidine Anthranilic Diamide Insecticides

Thomas P. Selby, Kenneth A. Hughes, George P. Lahm

15.1
Introduction

Calcium channels represent an attractive biological target for insect control due to the important role that they play in multiple cell functions including muscle contraction. Ryanodine receptor channels regulate intracellular calcium levels and are named after the plant metabolite ryanodine found to affect calcium release by blocking these channels in the partially open state [1–4]. Recently, substituted anthranilic diamides such as *N*-pyridylpyrazole diamide **1** were reported to be a new class of insecticides showing potent activity against a range of Lepidoptera [5–6] and causing release of intracellular stores mediated by the ryanodine receptor [7]. Work in this area has led to the discovery of Rynaxypyr™ (**2**), an exciting new product with outstanding insecticidal properties [8]. In exploring ring modifications to the heterobiaryl portion of diamide **1**, we found that related arylpyrazole and arylpyrimidine anthranilic diamides **3** and **4** also possess high insecticidal activity. Unlike the biaryl amide group on **1** and **2** where pyridine and pyrazole are attached through a carbon-nitrogen bond, the heterobiaryl rings on **3** and **4** are attached *via* a carbon-carbon bond. This chapter describes the synthesis, insecticidal activity, and structure-activity relationships observed for these types of anthranilic diamides where both rings of the biaryl amide function are attached through carbon.

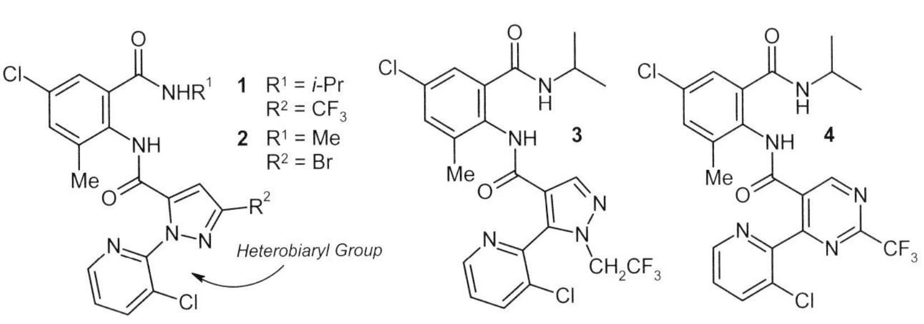

Pesticide Chemistry. Crop Protection, Public Health, Environmental Safety
Edited by Hideo Ohkawa, Hisashi Miyagawa, and Philip W. Lee

ISBN: 978-3-527-31663-2

15.2
Synthesis of Anthranilic Diamides

Figure 1 outlines a nonregioselective method for making substituted arylpyrazole esters that served as precursors to anthranilic diamides. Heating benzoylacetates **5** with *N,N*-dimethylformamide dimethyl acetal in toluene gave the corresponding ketoester enamines **6** that underwent cyclization with hydrazine to give pyrazole esters **7**. Alkylation of **7** with a variety of alkyl and haloalkyl halides gave a mixture of isomeric pyrazole esters **8** and **9** that were separated by silica gel column chromatography. The predominant isomer **8** arose from alkylation of the less sterically hindered pyrazole ring nitrogen. The regiochemical assignments for **8** and **9** were based on NOE studies involving both hydrogen and fluorine NMR.

A regioselective synthesis of arylpyrazole esters is shown in Figure 2. Cyclization of the ketoester enamines **6** with trifluoroethylhydrazine gave almost exclusive formation of pyrazole esters of formula **9a**.

The preparation of aryl pyrazole anthranilic diamides with designated "Isomer A" regiochemistry is outlined in Figure 3. Esters **8** underwent base hydrolysis to the acids **10** that were converted to the acid chlorides with oxalyl chloride and coupled with anthranilic amides **11** to afford the anthranilic diamide products of formula **12** in moderate yield.

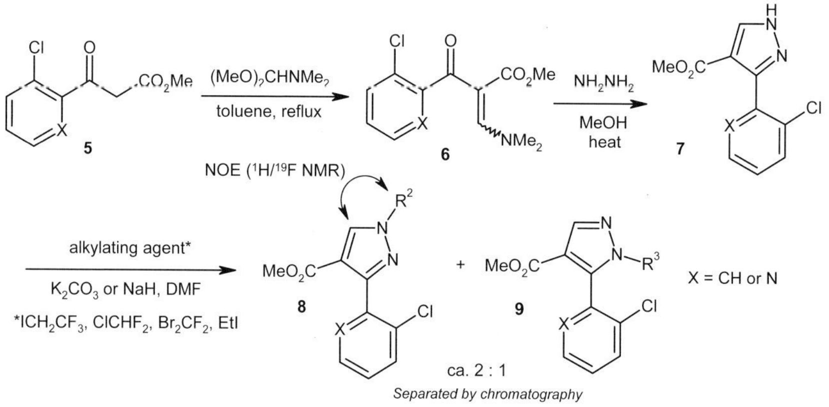

Figure 1. Nonregioselective synthesis of arylpyrazole esters.

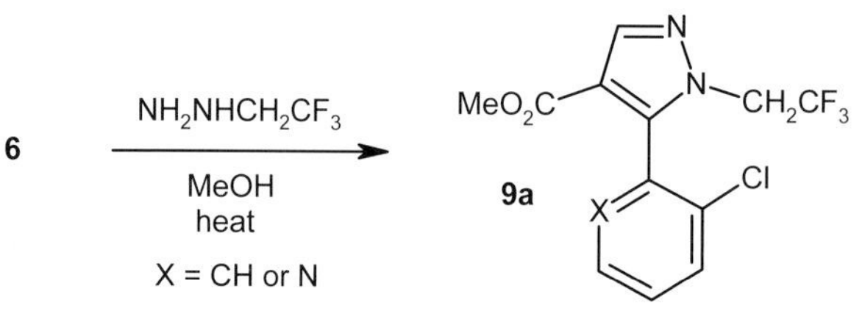

Figure 2. A regioselective synthesis of arylpyrazole esters.

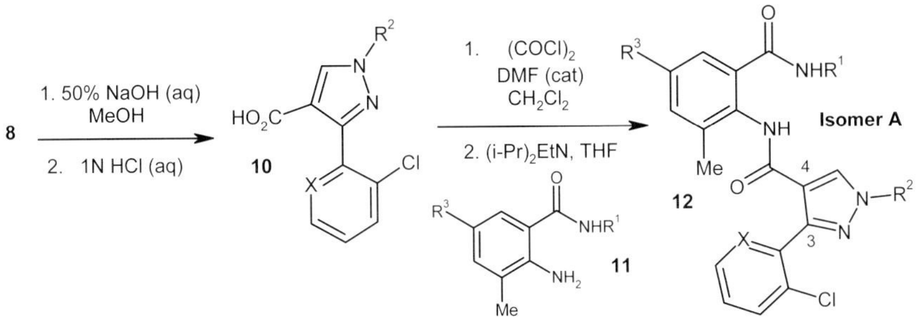

Figure 3. Synthesis of arylpyrazole diamides with "Isomer A" regiochemistry.

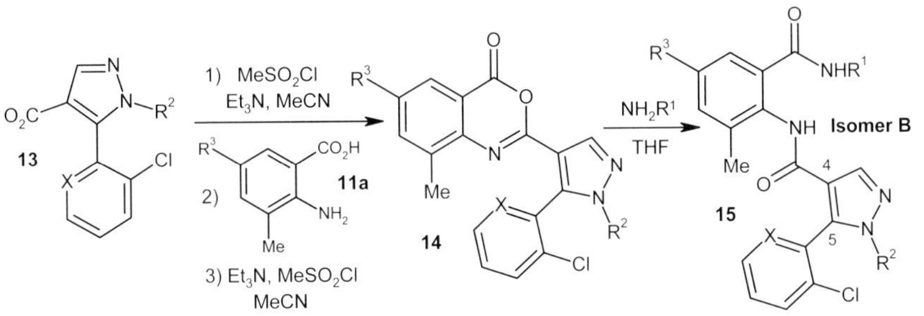

Figure 4. Synthesis of arylpyrazole diamides with "Isomer B" regiochemistry.

By the same procedure described in Figure 3, regioisomeric arylpyrazole esters **9** could also be converted to the corresponding anthranilic diamides. Alternatively, acids of formula **13**, obtained from base hydrolysis of **9**, were treated with methane sulfonyl chloride, triethylamine, and anthranilic acids **11a** (*via* the 3-step reagent addition sequence shown in Figure 4) to give benzoxazinones **14** which on ring opening with alkyl amines afforded anthranilic diamides of formula **15** ("Isomer B" regiochemistry).

Figure 5 illustrates the preparation of arylpyrimidine anthranilic diamides where two six-membered rings comprise the biaryl amide portion of the molecule versus a five- and six-membered ring (as in the case of arylpyrazole diamides). These analogs were made from the same ketoester enamine intermediates **6** used in the synthesis of arylpyrazole anthranilic diamides **12** and **15**. Heating **6** with trifluoroacetamidine in methanol afforded 4-aryl-2-trifluoromethylpyrimidine-5-carboxylates **16** that were hydrolyzed in base to the corresponding carboxylic acids **17**. Pyrimidine-5-carboxylic acids **17** were then condensed with anthranilic acids **11a** in the presence of methanesulfonyl chloride and base to yield the corresponding benzoxazinone intermediates **18**. Ring opening of the benzoxazinones **18** with alkylamines afforded arylpyrimidine anthranilic diamides of formula **19**.

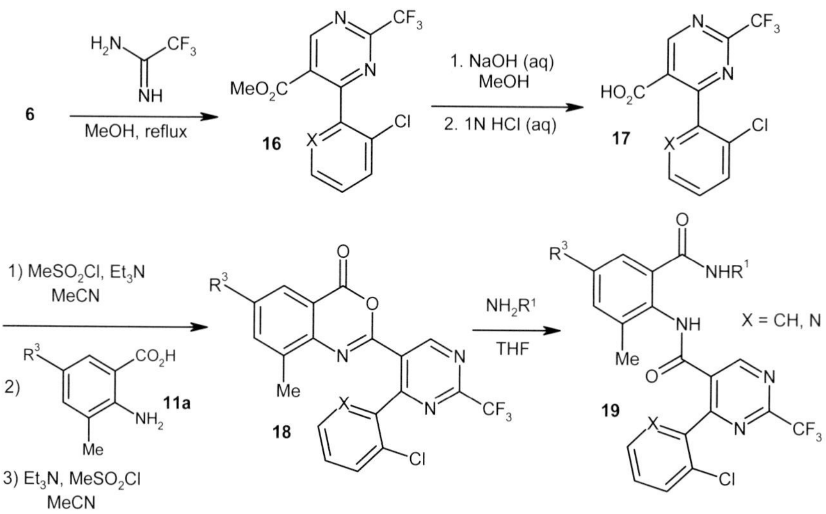

Figure 5. Synthesis of arylpyrimidine anthranilic diamides.

15.3 Insecticidal Activity

These compounds were tested against a series of Lepidoptera including *Plutella xylostella* (*Px*, diamondback moth), *Heliothis virescens* (*Hv*, tobacco budworm), and *Spodoptera frugiperda* (*Sf*, fall armyworm). Insecticidal activity is reported in Tables I–III as percent plant protection at various concentrations where reduction in plant damage generally resulted from insect mortality rather than cessation of feeding.

Table I shows that arylpyrazole anthranilic diamides of formula **12** (Isomer A regiochemistry) gave varying levels of plant protection against Lepidoptera at concentrations between 50 and 0.4 ppm. *Plutella xylostella* was generally the most sensitive insect species where some analogs (where R_1 is isopropyl or *t*-butyl, R_2 is trifluoroethyl or bromodifluoromethyl, R_3 is chlorine and X is *N*) provided excellent plant protection at 0.4 ppm. These analogs required higher concentrations for comparable control of *Heliothis virescens* and *Spodoptera frugiperda.*

As illustrated in Table II, isomeric arylpyrazole anthranilic diamides of formula **15** (Isomer B regiochemistry) usually provided higher levels of insect control than those of formula **12** with some analogs giving excellent control of all three insect species at concentrations as low as 0.4 ppm. Since diamides of formula **15** were not viewed as structurally close to *N*-pyridylpyrazole diamide **1** as analogs of formula **12**, we were quite surprised to observe high activity for these derivatives – an unexpected structure-activity finding for these types of analogs.

Compounds of formula **15** where R_1 is isopropyl or *t*-butyl, R_2 is haloalkyl (i.e., trifluoroethyl, difluoromethyl or bromodifluoromethyl), R_3 is chlorine or bromine

Table I. Insecticidal activity of "Isomer A" arylpyrazole anthranilic diamides of formula 12.

R_1	R_2	R_3	X	Conc (ppm)	% Plant Protection		
					Px [a]	*Hv* [b]	*Sf* [c]
i-Pr	CH_2CF_3	H	CH	50	100	70	10
i-Pr	Et	Cl	N	2	90	60	20
Me	CHF_2	Cl	N	2	100	20	80
i-Pr	CHF_2	Cl	N	2	100	90	90
				0.4	70	50	20
t-Bu	CHF_2	Cl	N	2	100	90	90
				0.4	100	60	20
i-Pr	$CBrF_2$	Cl	N	2	100	90	70
				0.4	100	80	40

[a] *Px, Plutella xylostella;* [b] *Hv, Heliothis virescens;* [c] *Sf, Spodoptera frugiperda*

and X is *N* provided optimum plant protection with activity comparable to that of *N*-pyridylpyrazole anthranilamide **1**.

In Table III, percent plant protection is reported for phenyl and pyridylpyrimidine anthranilic diamides of formula **19** against lepidopteran pests. A similar trend in structure-activity was observed for substitution on **19** as that for arylpyrazole anthranilic diamides **12** and **15**. Pyridylpyrimidine diamide **4** (formula **19** where R_1 is isopropyl, R_3 is chlorine and X is *N*) provided the highest level of plant protection with activity approaching that of more active pyridylpyrazoles diamides of formula **15**. However, arylpyrimidine diamides **19** tended to be less active overall than the corresponding arylpyrazole derivatives **15**.

15.4 Conclusion

The discovery of a novel class of anthranilic diamide insecticides having exceptional activity against a broad spectrum of Lepidoptera at extremely low rates of application has successfully led to the commercialization of Rynaxypyr™ (**2**). Compounds of this chemistry class exert their effect by causing release of intracellular Ca^{2+} stores in muscle cells by activation of the ryanodine receptor. As a new mode-of-action product with outstanding insecticidal properties, Rynaxypyr™ offers great promise for the marketplace.

Table II. Insecticidal activity of "Isomer B" arylpyrazole anthranilic diamides of formula 15.

R_1	R_2	R_3	X	Conc (ppm)	% Plant Protection		
					Px [a]	*Hv* [b]	*Sf* [c]
i-Pr	CH_2CF_3	H	CH	10	100	90	60
				2	80	80	30
Me	CH_2CF_3	H	N	2	100	90	60
					70	30	10
i-Pr	CH_2CF_3	H	N	2	100	90	80
				0.4	80	80	30
t-Bu	CH_2CF_3	H	N	2	100	90	90
				0.4	90	80	60
i-Pr	Et	Cl	N	2	100	90	40
				0.4	50	80	0
Me	CH_2CF_3	Br	N	2	100	90	90
				0.4	90	90	60
Me	CH_2CF_3	Cl	N	2	100	90	100
				0.4	90	90	90
i-Pr	CH_2CF_3	Br	N	2	100	100	100
				0.4	80	90	80
i-Pr	CH_2CF_3	Cl	N	2	100	90	100
				0.4	100	80	90
t-Bu	CH_2CF_3	Br	N	2	90	90	100
				0.4	90	80	90
i-Pr	CHF_2	Cl	N	2	100	90	100
				0.4	100	80	90
i-Pr	$CBrF_2$	Cl	N	2	100	100	100
				0.4	100	90	90
N-Pyridylpyrazole Diamide **1**				2	100	100	100
				0.4	100	90	70

[a] *Px, Plutella xylostella;* [b] *Hv, Heliothis virescens;* [c] *Sf, Spodoptera frugiperda*

Table III. Insecticidal activity of arylpyrimidine anthranilic diamides of formula 19.

R_1	R_3	X	Conc (ppm)	% Plant Protection		
				Px [a]	*Hv* [b]	*Sf* [c]
Et	H	CH	10	100	70	90
			2	90	10	0
i-Pr	H	CH	10	100	60	40
			2	90	50	40
t-Bu	H	CH	10	100	90	30
			2	100	10	0
Me	Br	CH	10	100	100	80
			2	90	80	70
i-Pr	Br	CH	10	100	90	60
			2	60	70	70
t-Bu	Br	CH	10	90	100	80
			2	90	60	10
i-Pr	H	N	2	100	80	90
			0.4	90	60	70
Me	Br	N	2	100	80	90
			0.4	60	40	70
i-Pr	Br	N	2	90	90	100
			0.4	90	60	90
i-Pr	Cl	N	2	100	90	100
			0.4	100	80	80
t-Bu	Br	N	2	80	90	100
			0.4	80	40	70

[a] *Px, Plutella xylostella;* [b] *Hv, Heliothis virescens;* [c] *Sf, Spodoptera frugiperda*

In further exploring the heterobiaryl portion of *N*-arylpyrazole anthranilic diamides such as **1**, we found that structurally related arylpyrazole and arylpyrimidine anthranilic diamides of formula **12**, **15**, and **19** also possess high insecticidal activity. Unlike the heterobiaryl amide group on **1** where pyridine and pyrazole are attached through a carbon-nitrogen bond, the biaryl rings on these diamides are attached *via* a carbon-carbon bond. Although some analogs showed activity close to that of Rynaxypyr™, none actually showed advantages over Rynaxypyr™ in advanced testing.

15.5 Acknowledgment

The authors express their gratitude to all those who made contributions to this effort, especially the chemists and biologists involved in the preparation and evaluation of compounds. Special thanks are due Dr. George Chiang for his early synthetic efforts in making phenylpyrazole anthranilic diamides.

15.6 References

1 E. F. Rogers, F. R. Koniuszy, J. Shavel Jr., K. Folkers, *J. Am. Chem. Soc.*, **1948**, *70*, 3086–3088.
2 E. Buck, I. Zimanyi, J. J. Abramson, I. N. Pessah, *J. Biol. Chem.*, **1992**, *267*, 23560–23567.
3 W. Melzer, A. Herrmann-Frank, H. C. Luttgau, *Biochim. Biophys. Acta*, **1995**, *1241*, 59–116.
4 R. Coronado, J. Morrissette, M. Sukhareva, D. M. Vaughan, *Am. J. Physiol.*, **1994**, *266*, 1485–1504.
5 G. P. Lahm, T. P. Selby, J. H. Freudenberger, T. M. Stevenson, B. J. Myers, G. Seburyamo, B. K. Smith, L. Flexner, *et al.*, *Bioorg. Med. Chem. Lett.*, **2005**, *15*, 4898–4906.
6 G. P. Lahm, T. P. Selby, T. M. Stevenson, *PCT Int. Appl. WO 03/015519*, **2003**.
7 D. Cordova, E. A. Benner, M. D. Sacher, J. J. Rauh, J. S. Sopa, G. P. Lahm, T. P. Selby, T. M. Stevenson, *et al.*, *Pesticide Biochem. Physiol.*, **2006**, *84*, 196–214.
8 See chapter in this book entitled: "The Discovery of Rynaxypyr™: A New Anthranilic Diamide Insecticide Acting at the Ryanodine Receptor".

Keywords

Anthranilic Diamide, Anthranilamide, Rynaxypyr™, Phenylpyrazole, Pyridylpyrazole, Phenylpyrimidine, Pyridylpyrimidine, Insecticide, Ryanodine Receptor

16
Metofluthrin: Novel Pyrethroid Insecticide and Innovative Mosquito Control Agent

Yoshinori Shono, Kazuya Ujihara, Tomonori Iwasaki, Masayo Sugano, Tatsuya Mori, Tadahiro Matsunaga, Noritada Matsuo

16.1
Introduction

Metofluthrin (SumiOne®, Eminence®) is a novel pyrethroid discovered by Sumitomo Chemical Co., Ltd. (Figure 1). It was registered in Japan in January 2005 and is under worldwide development for environmental health use. Metofluthrin has extremely high knockdown activity against various insect pests especially mosquitoes as well as high volatility and low mammalian toxicity. This chemistry is applicable to not only existing mosquito-controlling devices such as mosquito mats and coils, but also to various new formulations and devices such as paper emanators, fan-driven devices, and resin formulations. Metofluthrin is more than 40 times as active as *d*-allethrin against southern house mosquitoes (*Culex quinquefasciatus*) when used in mosquito coils. This paper describes the story behind the discovery of metofluthrin and this compound's efficacy against mosquitoes in various formulations.

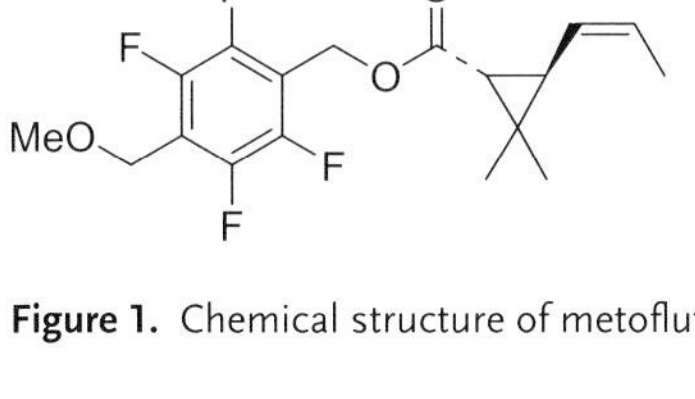

Figure 1. Chemical structure of metofluthrin (SumiOne®, Eminence®).

16.2
Discovery

At present, the main devices used for mosquito protection are mosquito coils, electric mosquito mats and liquid vaporizers, but all of these are methods that vaporize insecticides into the air using heating by means of fire or electricity to

Pesticide Chemistry. Crop Protection, Public Health, Environmental Safety
Edited by Hideo Ohkawa, Hisashi Miyagawa, and Philip W. Lee

ISBN: 978-3-527-31663-2

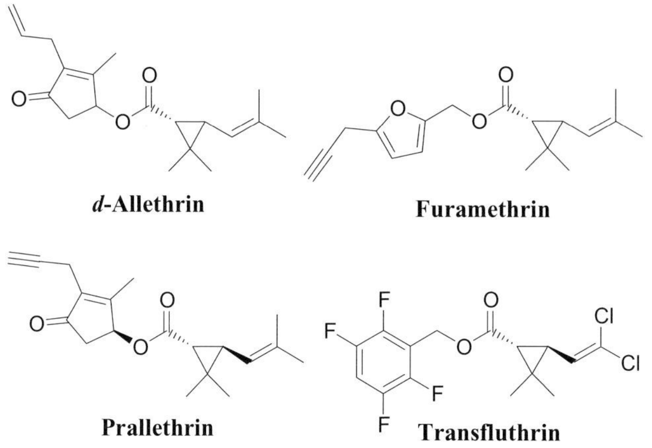

Figure 2. Typical active ingredient for mosquito-control devices.

kill the insects. Most devices contain pyrethroids as active ingredients because of their "knockdown effect," where mosquitoes are rapidly paralyzed and cannot suck blood, and their low mammalian toxicity (Figure 2). Much attention has recently been directed toward the development of non-heated formulations such as fan vaporizers because of their increased safety and ease of use, especially during outdoor activities. However, the insecticidal activity and/or the vapor pressures of the existing pyrethroids are unsatisfactory for use with such ambient-temperature devices. Therefore, we started our research to find a new pyrethroid with higher vapor activity as well as high potency against mosquitoes.

In 1924, Staudinger, *et al.*, reported a norchrysanthemic acid as the principal acidic product from the pyrolytic decomposition of chrysanthemum dicarboxylate [1] (Figure 3). The first insecticidal derivatives, 6-halo-3,4-methylenedioxy-benzyl esters were reported in 1965 [2], but their insecticidal activities were not described in the paper. In the 1970's, Ohno [3] and Elliott [4] reported their insecticidal esters. They reported that some norchrysanthemate esters showed comparable insecticidal activity to the corresponding chrysanthemates. However, further studies were discontinued at that time because they could not find any justification to develop these norchrysanthemic acid esters due to the difficulty of synthesizing norchrysanthemic acid. Despite this, we directed our attention to the norchrysanthemic acid esters because they had a lower molecular

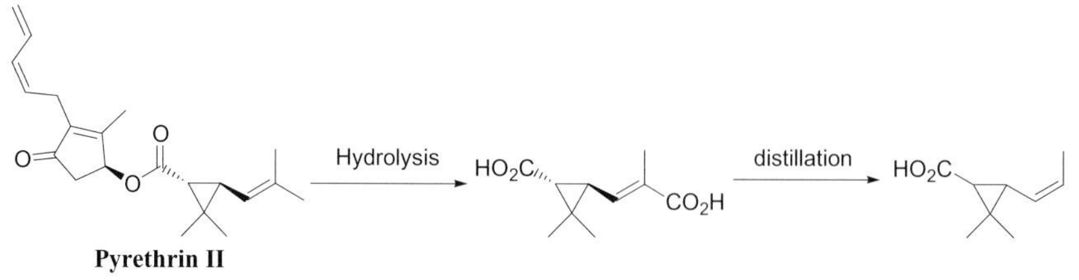

Figure 3. Origin of the norchrysanthemic acid.

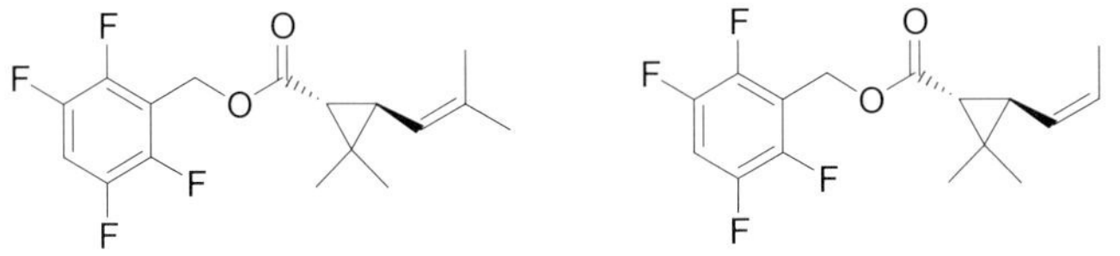

Chrysanthemate KT_{50} = 79 [min]

Norchrysanthemate KT_{50} = 55 [min]

Figure 4. Effectiveness of 2,3,5,6-tetrafluorobenzyl chrysanthemate and its norchrysanthemic analog in an ambient-temperature formulation against *Culex pipiens pallens*.

weight and showed comparable insecticidal activity to that of the corresponding chrysanthemates. We considered these features to be key properties in our search for candidate compounds.

At first, we evaluated some known norchrysanthemates in an ambient-temperature formulation but they did not provide good activity against mosquitoes. Furthermore, when we screened various alcohol esters, we found that the 2,3,5,6-tetrafluorobenzyl ester exhibited little vapor activity at room temperature. This activity did not meet our desired level, but compared with the corresponding chrysanthemate, we found that it clearly exhibited a higher knockdown efficacy (Figure 4). We then synthesized the derivatives with substituents at the 4-position on the phenyl ring.

All analogs had much higher activity against mosquitoes than compound **2** by the standard topical application method as shown in Figure 5. The relative toxicity

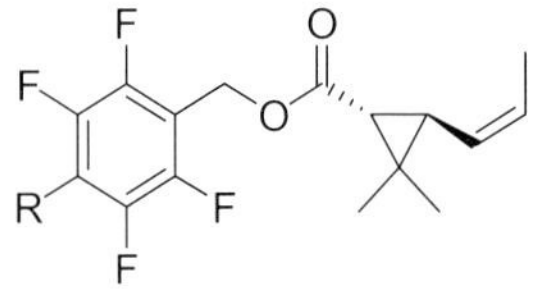

Compound	R	R.E. [a]
2	H	30
3	F	100
4	Me	200
5	Et	490
6	Pr	250
7	allyl	500
8	OMe	360
9	CH_2OMe	2500
d-Allethrin (standard)		100

[a] Relative efficacy (R.E.) against *Culex pipiens pallens* based on LD_{50} by the topical application method

Figure 5. Insecticidal effectiveness of derivatives with various substituents at the 4 position against *Culex pipiens pallens*.

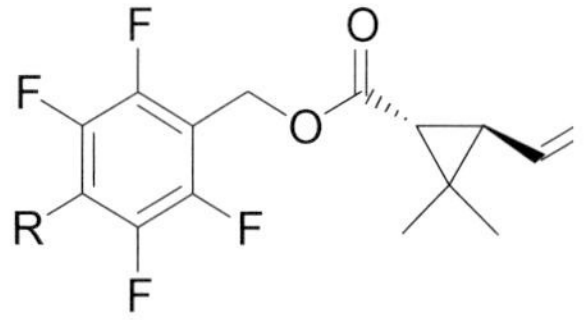

Compound	R	KT_{50} [min] [a]	KD% [b]
2	H	55	60
4	Me	38	94
8	OMe	52	70
9	CH_2OMe	27	100

a KT_{50}: Time for 50% knockdown calculated by the probit method;
b The average percentage of knocked down mosquitoes after 60 min.

Figure 6. Effectiveness of tetrafluorobenzyl derivatives in non-heated formulations against *Culex pipiens pallen.*

reached the maximum with between two and three carbon atoms at the 4-position (**2–6**). Unsaturation (**7**) and incorporation of an oxygen atom (**8–9**) also showed substantial activity, *inter alia,* the 4-methoxymethyl derivative (**9**) exhibiting the highest lethal potencies.

From these compounds, we selected compounds methyl, methoxy, and methoxymethyl derivatives as R substituents, taking into consideration their molecular weight, basic efficacy, ease of synthesis, and other physical and chemical properties, and evaluated their vapor activities with their chrysanthemic derivatives in a non-heated formulation. The results presented in Figure 6 clearly demonstrate that the methoxymethyl derivate (**9**) shows a significantly faster action in comparison with other compounds in the non-heated formulation. As a consequence of these findings, we chose this methoxymethyl derivate to be a new synthetic pyrethroid with high vapor activity against mosquitoes [5].

16.3
Efficacy

16.3.1
Intrinsic Insecticidal Activity

The relative lethal efficacy of metofluthrin to *d*-allethrin by topical application for 4 medically important mosquito species is given in Table I. The lethal efficacy of metofluthrin is 19 to 49 times as high as *d*-allethrin and also superior to that of permethrin.

Metofluthrin in particular has an extremely high lethal efficacy against mosquitoes.

Table I. Lethal efficacy of metofluthrin against 4 mosquito species.

	Culex pipiens	*Culex quinque-fasciatus*	*Aedes aegypti*	*Aedes albopictus*
Metofluthrin	0.0015[a] (25)[b]	0.00041 (32)	0.00037 (19)	0.00047 (49)
Permethrin	0.0028 (14)	0.0010 (13)	0.00056 (13)	0.0012 (19)
d-Allethrin	0.038 (1)	0.013 (1)	0.0071 (1)	0.023 (1)

[a] LD_{50} (μg/female adult) by topical application method; [b] Relative efficacy (*d*-allethrin = 1)

16.3.2 Activity in Devices

The biological activity of metofluthrin was evaluated in a variety of commercial and experimental devices. These are considered below

16.3.2.1 Heated Formulations

We carried out a detailed investigation into the performance of metofluthrin in mosquito coils and liquid vaporizers, the most commonly used forms of heated mosquito control devices.

Mosquito Coil

Mosquito coils, which were first invented in Japan in the 19th Century, are widely used throughout the world. In Southeast Asia, in particular, they are the most popular method used to protect against biting mosquitoes. To evaluate the efficacy of metofluthrin coils against various species of mosquitoes, we carried out evaluations using a large 28-m^3 chamber, containing free-flying mosquitoes.

Southern house mosquito (*Culex quinquefasciatus*) is widely distributed in tropical and subtropical areas worldwide and the most important target for mosquito coils. Mosquito coils containing 0.005% metofluthrin exhibited an efficacy exceeding that of coils containing 0.2% *d*-allethrin against this species (laboratory strain), and the relative efficacy is estimated to exceed 40 times that of *d*-allethrin (Figure 7).

To determine the practical effects of coils containing metofluthrin, field tests were conducted according to the method of Yap, *et al.* [6], using private residences in Bogor, Indonesia. The results are shown in Table II. In these tests, 95% of the mosquitoes captured were *Culex quinquefasciatus*, and coils containing 0.005% metofluthrin exhibited an efficacy exceeding that of coils containing either 0.03% transfluthrin or 0.2% *d*-allethrin.

In order to confirm efficacy against field strains, eggs of *Culex quinquefasciatus* were collected in Bogor, Indonesia, and reared in the laboratory. Large chamber free-flying tests were conducted against this field strain. The results are shown in Figure 8. Since field strains of mosquito tend to have a longer knockdown time compared with laboratory strains, these tests were carried out by releasing insects into the chamber that the test coil had been allowed to burn in for 1 hour. Coils containing 0.005% metofluthrin exhibited an efficacy almost equal to that of coils containing 0.3% *d*-allethrin. The relative efficacy is therefore approximately 60 times that of *d*-allethrin, an increase in the efficacy ratio when compared with the laboratory strain.

Table II. Field evaluation of metofluthrin coils in Bogor, Indonesia.

A.I.	Conc. (% w/w)	Collected mosquitoes*		Reduction (%)
		Pre-treatment	Treatment	
Metofluthrin	0.005	210	18	93
Transfluthrin	0.03	187	26	88
d-Allethrin	0.3	188	27	88
Control		256	303	

* Predominant species was *Culex quinquefasciatus*

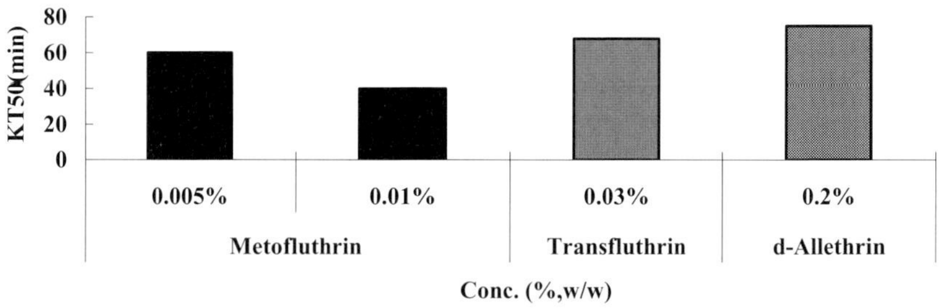

Figure 7. Knockdown efficacy of metofluthrin coil against *Culex quinquefasciatus* (laboratory strain) by large chamber free-flying method.

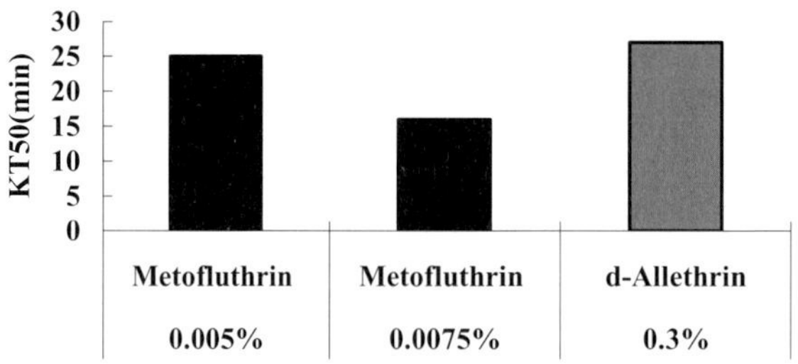

Figure 8. Knockdown efficacy of metofluthrin coils against *Culex quinquefasciatus* (Bogor, Indonesia, field strain) by large chamber free-flying method, 1-hour pre-fumigation.

Liquid Vaporizer

The efficacy of metofluthrin liquid vaporizers tested by the large chamber free-flying method (with insects released after one-hour pre-fumigation) against field strains of *Culex quinquefasciatus* from Bogor, Indonesia, is given in Figure 9. The relative efficacy ratio was estimated to be over 8 times that of prallethrin.

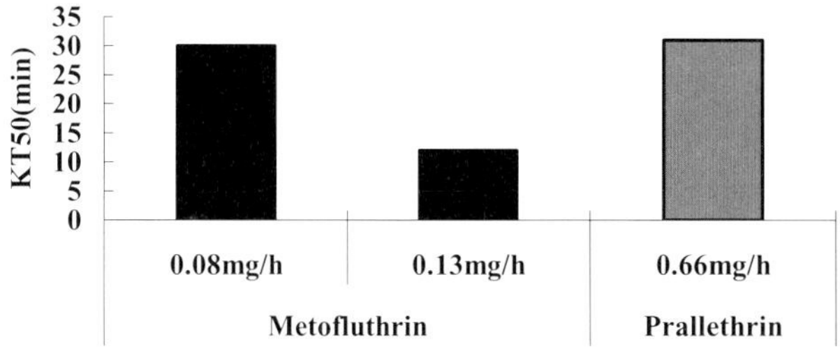

Figure 9. Knockdown efficacy of metofluthrin vaporizer liquid against *Culex quinquefasciatus* (Bogor, field strain) by large chamber free-flying method, 1-hour pre-fumigation.

16.3.2.2 **Non-heated Formulations**

One of the major characteristics of metofluthrin is its vapor activity at room temperature, which is not seen in the majority of knockdown pyrethroids, such as *d*-allethrin and prallethrin. We will describe a fan-type formulation where a motor turns a fan and the active ingredient is vaporized by the airflow from it at room temperature and a no-fan vaporizing formulation where the active ingredient is held in a paper, resin or other carrier without heating and without any motive force being provided.

Fan Emanator

Mosquito mats, liquid vaporizers, and other formulations exhibit their effects by vaporizing the active ingredient through heating. However, since these heaters require a comparatively large amount of electric power, their use has been limited to devices powered by mains electricity. On the other hand, the active ingredient in a fan vaporizer is vaporized by the airflow from a fan at room temperature, but the power required to turn the fan is much smaller than that for heating, so it is possible to utilize batteries for power. Fan vaporizers therefore have the benefit of making it possible to carry them around without the reel for electrical outlets and this makes it possible to use them outdoors.

For the purpose of determining the basic activity of metofluthrin in a fan-type formulation, we evaluated the knockdown efficacy against Culex *pipiens* (laboratory-susceptible strain) in a large chamber (28 m^3) with various evaporation rates. Transfluthrin was used as the control chemical. The results are shown in Table III. Based on the results of these tests, it was found that metofluthrin had an efficacy of more than 3 times that of transfluthrin. We also evaluated the efficacy against the Asian tiger mosquito, *Aedes albopictus* (laboratory-susceptible strain), with the large chamber free-flying method. The results are shown in Table IV. In

Table III. Knockdown efficacy of metofluthrin fan vaporizer against *Culex pipiens* (laboratory strain).

	Evaporation rate (mg/h)	KT_{50} (min.)	Linear Regression Expression
Metofluthrin	0.09	34	$\text{Log}(KT_{50}) = -0.89 \times \log$ (Evaporation rate) + 0.61
	0.18	18	
	0.26	14	
Transfluthrin	0.2	38	$\text{Log}(KT_{50}) = -0.59 \times \log$ (Evaporation rate) + 1.17
	0.36	29	
	0.54	21	

Table IV. Knockdown efficacy of metofluthrin fan vaporizer against *Aedes albopictus*.[a]

A.I.	Evaporation rate (mg/h)	KT_{50} (min) [b]
Metofluthrin	0.09	40
Transfluthrin	0.28	> 60
	0.39	55
	0.50	55

[a] Laboratory-susceptible strain; [b] Large chamber (28 m^3) free-flying method

comparisons of the efficacy based on the evaporation rate, metofluthrin exhibited an efficacy approximately 5 times that of transfluthrin.

Ambient Vaporization Devices

Ambient vaporization devices, where the active ingredient is held in paper or resin, and is vaporized without heating or use of power, are easy to use, so there are particular expectations for new developments in the field of mosquito control. Insecticides that can be used in formulations for this purpose must possess the following characteristics: vapor action of room temperature, strong efficacy; and a high level of safety to mammals. Metofluthrin meets all of these features. First of all, we investigated a device using paper as the carrier. The structure of the device is similar to that of an old Japanese toy called "denguri". In this paper, we call this device "Denguri paper strip" (Figure 10).

To confirm the efficacy of the Denguri device in a practical setting, practical tests were conducted at a home in Malaysia. The testing was carried out according to the methods of Yap *et al.* [6]. The results are shown in Figure 11. A Denguri device containing 100 mg of metofluthrin exhibited strong biting inhibitory effects against *Culex quinquefasciatus*, exceeding those of coil formulations containing 0.25% *d*-allethrin.

Practical tests using a similar Denguri formulation were conducted on Lombok Island in Indonesia and in Nagasaki, Japan. In a house on Lombok Island, a

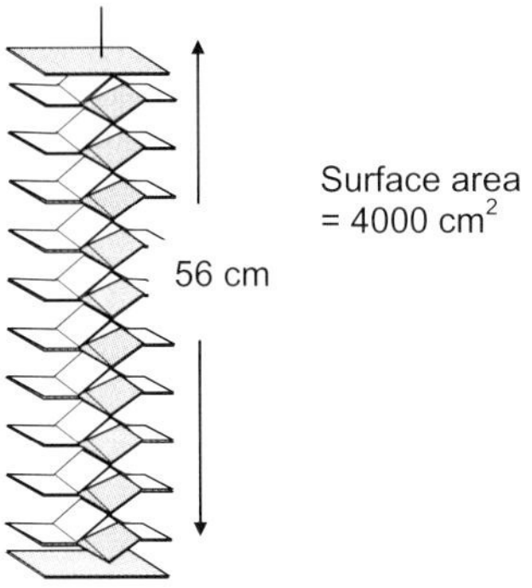

Figure 10. Denguri paper strip.

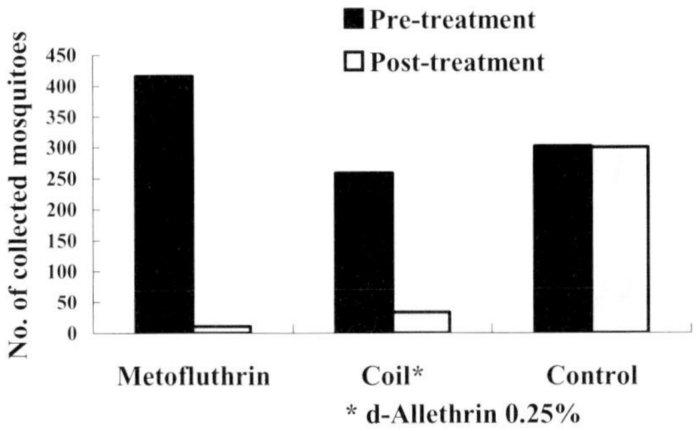

Figure 11. Field efficacy of metofluthrin denguri paper strip against *Culex quinquefasciatus* in Malaysia.

Denguri formulation impregnated with 200 mg of metofluthrin exhibited repellent effects of 80% or greater on *Culex quinquefasciatus* and anopheles mosquitoes over a period of four weeks [7]. In addition, in outdoor conditions on Lombok Island, this formulation exhibited superior repellent effects against *Culex quinquefasciatus* as well as *Anopheles balabaciensis* and *An. sundaicus*, which are vector mosquitoes for malaria [8]. On the other hand, this formulation impregnated with 200 mg of metofluthrin exhibited almost complete repellent activity against *Aedes albopictus* [9]. From these results, we were able to confirm that the Denguri formulation had excellent activity in practical settings.

Finally, we will briefly describe polymer-based resin formulations. Resins are suitable as the carriers for active ingredients, such as metofluthrin. Their durability, waterproof characteristic, and highly flexible properties make them highly suitable for both indoor and outdoor use.

Practical tests using resin formulations were conducted in Vietnam. Resin formulations containing 1 g of metofluthrin exhibited excellent spatial repellent effects against *Culex quinquefasciatus* and *Aedes aegypti* for at least six weeks, confirming the practicability of this formulation [10].

16.4 Conclusion

For more than half a century, Sumitomo Chemical Company has discovered and marketed over 20 pyrethroids with a wide range of chemical and biological characteristics. These pyrethroids have made a significant contribution to improving human health and happiness by controlling insect pests and the diseases and damage they cause in both agricultural and urban environments. We are proud to add metofluthrin to our product range and are confident it will become a major tool in the battle against mosquitoes and the diseases they transmit.

16.5 Acknowledgment

We are indebted to Mr. J. R. Lucas of Valent BioSciences Corporation, Libertyville, IL, USA, for his valuable comment on this manuscript.

16.6 References

1 H. Staudinger, L. Ruzicka, *Helv. Chim. Acta.*, **1924**, *7*, 201.
2 M. Julia, S. Julia, M. Langlois, *Bull. Soc. Chim.*, **1965**, 1007.
3 Y. Okuno, N. Itaya, T. Mizutani, N. Ohno, T. Matsuo, S. Kitamura, *Jpn. Kokai Tokkyo Koho*, **1972**, JP 470433330.
4 M. Elliott, A. Farnham, N. F. Janes, P. H. Needham, D. A. Pulman, *Nature*, **1973**, *244*, 456.
5 K. Ujihara, T. Mori, T. Iwasaki, M. Sugano, Y. Shono, N. Matsuo, *Biosci. Biotechnol. Biochem.*, **2004**, *68*, 170.
6 H. Yap, H. Tan, A. Yahaya, R. Baba, P. Loh, N. Chong, *Southeast Asian J. Med. Health*, **1990**, *21*, 558.
7 H. Kawada, Y. Maekawa, Y. Tsuda, M. Takagi, *J. Am. Mosq. Control Assoc.*, **2004**, *20*, 292.
8 H. Kawada, Y. Maekawa, Y. Tsuda, M. Takagi, *J. Am. Mosq. Control Assoc.*, **2004**, *20*, 434.
9 T. Argutea, H. Kawada, M. Takagi, *Med. Entomol. Zool.*, **2004**, *55*, 211–216.
10 H. Kawada, N. Yen, N. Hoa, T. Sang, N. Dan, M. Takagi, *Am. J. Med. Hyg.*, **2005**, *73*, 350.

Keywords

Metofluthrin, Pyrethroid, Mosquito Control Agent, Mosquito Mats, Mosquito Coils, Paper Emanotros, Fan-Driven Devices, Resin Formulations

17
Design and Structure-Activity Relationship of Novel Neonicotinoids

Xuhong Qian, Yanli Wang, Zhongzhen Tian, Xusheng Shao, Zhong Li, Jinliang Shen, Qingchun Huang

17.1
Introduction

Since imidacloprid was first reported as a potent insecticide [1], seven neonicotinoids were registered as agricultural insecticides [2–4] for crop protection and animal health usages. Neonicotinoids have excellent selectivity for insect versus mammalian nAChRs, contributing to high bioactivities against insects, and is relatively safe toward mammals and aquatic organisms [5–6]. Here, we carried out design and structure-activity relationships of novel neonicotinoids through bioinformatics, quantum chemical calculation, and chemical synthesis. That is, based on bioinformatic analysis, the selectivity mechanism of neonicotinoids was investigated, and the alternative binding model of neonicotinoids with insect nAChR was proposed through *ab initio* quantum study. On the other hand, based on chemical modification using hydro-pyridine to form the *cis* nitro configuration with *exo*-ring ether moiety, novel tetrahydro-pyridine nitromethylene derivatives were designed and synthesized, presenting good insecticidal activity. Interestingly, compound **4a** is not only active on a wide range of insects, it has lower toxicity for mammals than imidacloprid, but also has very low cross-resistance to imidacloprid.

17.2
Selectivity Mechanism and Binding Model of Neonicotinoids

17.2.1
Bioinformatic Analysis

Understanding the selectivity of neonicotinoids toward insect nAChR is essential for environment protection, human health, and insecticide resistance [7], and also a key issue for the design and structure-relationship of new derivatives.

Pesticide Chemistry. Crop Protection, Public Health, Environmental Safety
Edited by Hideo Ohkawa, Hisashi Miyagawa, and Philip W. Lee

ISBN: 978-3-527-31663-2

Researchers had proposed some residues that may contribute to the selectivity through mutation experiments [8–10]. Here, through bioinformatic and statistical analysis, residue distribution differences in the ligand binding sites between arthropods and vertebrates as well as features of nicotinic leads were studied. The specific sites that contribute to the neonicotinoids' selectivity were identified and coincide well with the known experimental data.

Data collection and process. All available neuronal nAChR sequences were retrieved from the NCBI (http://www.ncbi.nlm.nlh.gov/entrez/query.fcgi). Subunits of α5, β3, α9, α10 were excluded for their irrelevance of ligand binding [11]. Based on the ligand binding function of the subunits, sequences were divided into two sets: principal subunits (α2, α3, α4, α6, α7, and α8 subunits) and complementary subunits (β2, β4, α7, and α8 subunits). After removing the redundant sequences, 59 of arthropods' nAChR sequences and 73 of vertebrates' nAChR sequences were used for the bioinformatic analysis. Multiple sequences alignments (MSA) were performed and manually modified. To investigate the mechanism of ligand selectivity, sites in the binding domain were selected for our research. According to the physicochemical properties, residues were clustered into four non-intersecting sets: non-polar residues (NR), uncharged polar residues (PR), acidic residues with negative charge (AR), and basic residues with positive charge (BR). Based on the MSA results, residue distribution differences in sites were statistically analyzed and the specific sites that may contribute to the ligand selectivity were identified. Fisher's exact test was used to test the significant difference.

Key residues distribution. Sites with a statistically significant difference ($P < 0.01$) were: 92, 94, 95, 147, 183, 184, 186, and 189 in the principal subunits; 34, 55, 56, 57, 106, 110, 112, 160, 161, and 163 in the complementary subunits. Since the spatial locations of the residues are as important as their type, the spatial information of the sites was inspected through their analogous residues in the X-ray structure of AChBP complexed with nicotine (PDB: 1UW6). Considering that some residues may contribute to the ligand binding indirectly, residues that located close to the ligand and can interact with it directly were selected for the further analysis. These sites are 186 and 189 in principal subunit and 34, 55, 56, 57, 112 in complementary subunit, as shown in Table I.

Table I. Residue type difference in key sites of arthropods' *vs.* vertebrates' nAChRs.

Subunits	Residue Position	Arthropods nAChRs	Vertebrates nAChRs
Complementary	34, 56, 57, 112	Hydrophobic	Hydrophilic
	55, 106	Positively charged	Hydrophilic
Principal	186, 189	Hydrophobic	Negatively charged

Selectivity of neonicotinoids. The differences between arthropods' and vertebrates' nAChR are clearly revealed. The pocket environments of vertebrates' nAChR are polar and negative, and that of arthropods' nAChR are non-polar and positive, which well interpreted the species-specific mechanism of the well-known neonicotinoids insecticide: the electronegative hydrophobic group can fit more powerfully with the pocket of insect nAChR than that of homo nAChR. The results are well consistent with the known mutation experiments [8–10]. Therefore, through the sequence-based data, key sites corresponding to the ligand selectivity were identified, and may be helpful in designing the new selectivity and potency molecules for insect nAChRs.

17.2.2
Ab initio Quantum Chemical Calculation

The binding model is also a critical issue for the design and structure-activity relationship of new derivatives. Here, *ab initio* quantum chemical calculation was used for the investigation of nonbonded bindings between neonicotinoids and key residues of nAChR.

Computational model. Based on the bioinformatic analysis and the known mutation data [8–10] of insect nAChR, two key residues (Arg and Trp) were selected to build a computational model with three analogs of neonicotinoids. Considering the computational efficiency, the structures were simplified as seen in Figure 1.

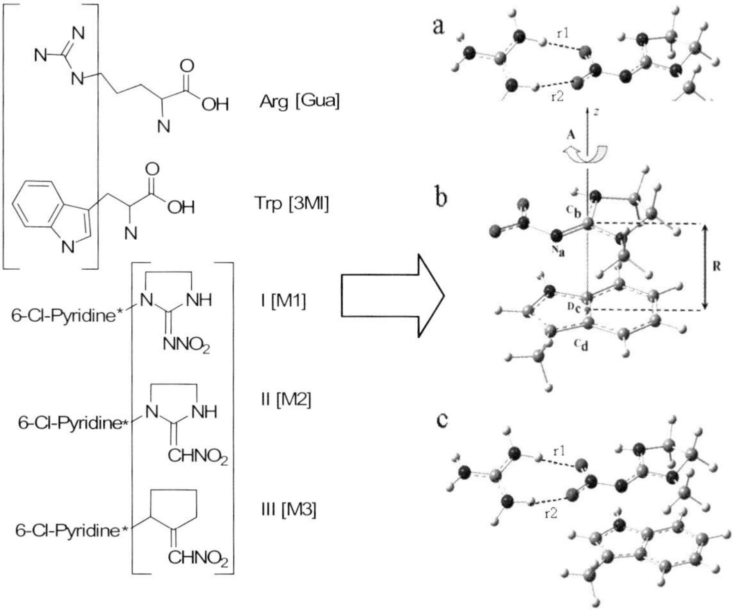

Figure 1. Chemical structures and the computational model. Left are the compound's structures of neonicotinoid derivatives I ~ III and the simplified structure Gua (protonated guanidinium), 3MI (3-methyl-indole), M1 ~ 3.

The computational models are illustrated in the right of Figure 1 using M1 as the example: a) M_x–Gua complex, b) 3MI–M_x complexes, c) 3MI–M_x–Gua complexes. The geometries of all monomers and M_x–Gua complexes were fully optimized at MP2/6-31G*. For the 3MI–M_x complexes, 3MI and M_x were parallel placed in the initial structures according to the common features of the known binding modes of neonicotinoids [12–13], and finally determined by the potential energy curves scan against the interaction distance R and the dihedral angle A (Na-Cb-Dc-Cd). For the geometries of 3MI–M_x–Gua complexes, the M_x–Gua parts were taken from the former MP2/6-31G* optimized structure. The values of R and A were obtained by the former potential energy curves scan of 3MI–M_x complexes. Taking into account the influence of Gua on the geometries of 3MI with M_x, the geometry parameters R and A were rescanned.

π-π stacking. The BSSE corrected binding energies for all complexes were listed in Table II.

Both geometries and energy data of 3MI–M_x showed that there is no obvious difference among 3MI–M_x complexes, indicating that the binding of 3MI–M_x

Table II. Calculated energies for 3MI–M_x, M_x–Gua, 3MI–M_x–Gua and 3MI–[M_x–Gua] after BSSE correction (ΔE in kcal/mol) at 6-311++G**.

	3MI–M1	3MI–M2	3MI–M3
MP2	–7.02	–6.88	–7.63
HF	7.97	8.85	6.67
	M1–Gua	M2–Gua	M3–Gua
MP2	–28.73	–33.40	–23.11
HF	–27.77	32.14	–23.70
	3MI–M1–Gua	3MI–M2–Gua	3MI–M3–Gua
MP2	–37.94	–42.90	–32.47
HF	–20.63	–24.81	–18.11
	3MI–[M1–Gua] [a]	3MI–[M2–Gua]	3MI–[M3–Gua]
MP2	–12.13	–12.68	–11.58
HF	6.05	6.21	4.69

[a] Concerning the distance and the orientation of 3MI and Gau in the three-body systems, 3MI have little interaction with Gua, so the energy value of 3MI–[M_x–Gua] (x = 1–3) is the interaction energy between 3MI and M_x, with the existence of Gua.

complexes is not sensitive to the structures difference of three analogs. M1 and M2 own two sp^2 nitrogen atoms in the five-member ring, whereas M3 does not. Therefore, the sp^2 nitrogen atom does not influence the 3MI–M_x interaction. Meanwhile, a large difference between the MP2 and HF binding energies of 3MI–M_x complexes indicates that the electronic correlation is essential for the 3MI–M_x binding. Since all analogs have a conjugate part, it can be concluded that the π-π stacking should dominate in between 3MI and M_x, instead of the π-π interaction [12], which coincides with the fact that the position of sp^2 N in the five-number ring has no impact on the activity [14]. In contrast from 3MI–M_x complexes, clear interaction energy differences were found among both M_x–Gua and 3MI–M_x–Gua complexes. The different conjugation degree between the five-number ring and the electronegative group of M_x should contribute to the different binding energies of M_x–Gua complexes, as pointed out by Tomizawa *et al.* [12]. Similar to the M_x–Gua complexes, the 3MI–M3–Gua complex is less favorable than 3MI–M1–Gua and 3MI–M2–Gua complexes. The binding energy values of 3MI–M_x–Gua complexes decrease in the sequence of $\Delta EM2 > \Delta EM1 > \Delta EM3$, and are well-correlated with the experimental binding affinities of neonicotinoid derivatives I ~ III (pIC_{50} (II) > pIC_{50} (I) > pIC_{50} (III)) [5, 14].

Cations assist the π-π interaction. In Table II, the 3MI–[M_x–Gua] stands for the interaction between 3MI and M_x with the existence of Gua. For 3MI–[M_x–Gua] complexes, the binding energies differences between the MP2 and HF level are still very large, which indicated that the dispersion interaction is significantly important for the π-π interaction between 3MI and M_x, not only in 3MI–M_x complexes but also with the existence of Gua. However, the interaction energy between 3MI and M_x increased about –3.95 ~ –5.80 kcal/mol with the existence of Gua, which indicated that the existence of Gua could strengthen the M_x-3MI interaction. Similar phenomenon has also been revealed by Reddy *et al.* [15] that the cations could assist the π-π interaction strengths greatly. The enhanced π-π interaction between 3MI and M_x should be the other important factor for the binding of neonicotinoids to nAChR.

Cooperative binding model. NPA atomic charges were calculated at the MP2/6-31G*, as shown in Table III.

For 3MI–M_x–Gua (x = 1–3) complexes, the total amount of charge transfer is the sum of the atom charges on 3MI and M_x (x = 1–3), which are 0.096, 0.111, and 0.086, respectively, for the three complexes. Similar to the energy properties, the total atom charge variations also coincide with the binding activities. But the charge transfer values between 3MI and M_x (x = 1–3) are quite small whether with or without Gua, as compared to that of M_x–Gua (x = 1–3) complexes, indicating that the hydrogen bond between M_x and Gua should be more important than the π-π stacking for the ligand binding. By the semi-empirical PM3 method, it was suggested that the nitrogen atom of imidacloprid can possess enough positive atom charge (positive by 0.11e) after interacting with Arg/Lys, and then interacts with Trp through cation-π interaction. Here, atom charge population analysis showed that

Table III. The sum of the atom charges on each monomer of the complexes (Q/e).

	3MI–M1	M1–Gua	3MI–M1–Gua
M1	0.001	0.093	0.094
3MI	–0.001	–	0.002
Gua	–	0.907	0.904
	3MI–M2	M2–Gua	3MI–M2–Gua
M2	0.001	0.108	0.109
3MI	–0.001	–	0.002
Gua	–	0.892	0.889
	3MI–M3	M3–Gua	3MI–M3–Gua
M3	–0.014	0.083	0.066
3MI	0.014	–	0.020
Gua	–	0.917	0.914

the sum of the atom charges on M1 and M2 in 3MI–M_x–Gua (x = 1,2) complexes are no more than 0.002e, meaning that the cation-π interaction could hardly exist in the binding of neonicotinoids to nAChR, which is different from the other's previous prediction [13]. From the geometry, energy and charge transfer analysis, a modified model including hydrogen bonding and cooperative π-π interaction between neonicotinoids and nAChR, was depicted here (Figure 2). The hydrogen bonds between neonicotinoids and the positively charged sidechain of Arg/Lys play essential roles in the binding, as well as the cooperative π-π interaction between the conjugate guanidine/amidine moiety and the indole ring of Trp. Our results coincide well with the experimental activities. So, the binding model proposed here might provide an alternative way close to the actual binding features of neonicotinoids.

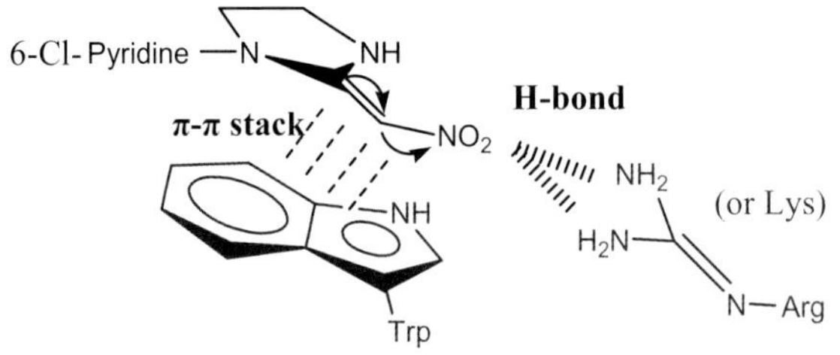

Figure 2. The alternative binding model between neonicotinoids and nAChR.

17.3
Chemical Modification for *cis* Nitro Configuration

Then we designed and synthesized a series of neonicotinoids containing *cis* nitro configuration with the help of virtual screening and the aforementioned bioinformatic analysis as well as quantum chemical calculation.

Hydropyridine fixed *cis* nitro configuration. Neonicotinoids usually have an electron-withdrawing tip (NO_2 or CN) as an important molecular feature. However, linked to C=C/C=N bond, the NO_2 or CN can be *trans* or *cis* configurations. Crystallographic and computation results showed that the *trans* configuration is absolutely dominant (**1a** and **1b**) in Figure 3) [12, 16]. Interestingly, Bay T 9992 (**1c** in Figure 3), in which the nitro group was in *cis* configuration, also showed high biological activity [17], which implied that neonicotinoids in *cis* configuration might also bind to the receptor well.

On the other hand, although 6-Cl-PMMI (**1b**) in *trans* configuration exhibits similar insecticidal activity with imidacloprid [18], photoinstability [19] and weak hydrophobicity [20] limited its use in crop protection. Herein, in order to find the diversity of nitromethylene neonicotinoids with *cis* nitro configuration, a novel neonicotinoids family was designed and synthesized by introducing a tetrahydropyridine ring into the lead compound to fix the nitro moiety in the *cis* position (**1d** in Figure 3), expecting that the new structure could not only improve its photoinstability, but also adjust hydrophobicity by *exo*-ring ether modifications.

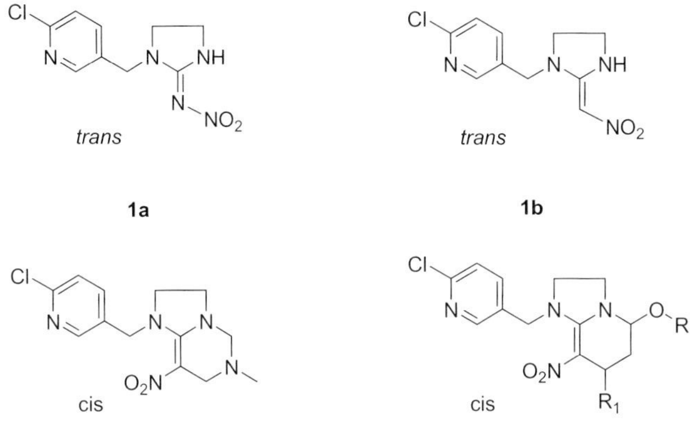

1c Hydro-pyrimidine **1d** Hydro-pyridine

Figure 3. Chemical structures of various *cis* and *trans* isomeric forms of neonicotinoids.

17.3.1
Synthesis

Starting from 2-chloro-5-chloromethylpyridine, a set of *N*′-((5-chloropyridin-2-yl)methyl) ethane-1,2-diamine and nitromethylene **1b** was synthesized following

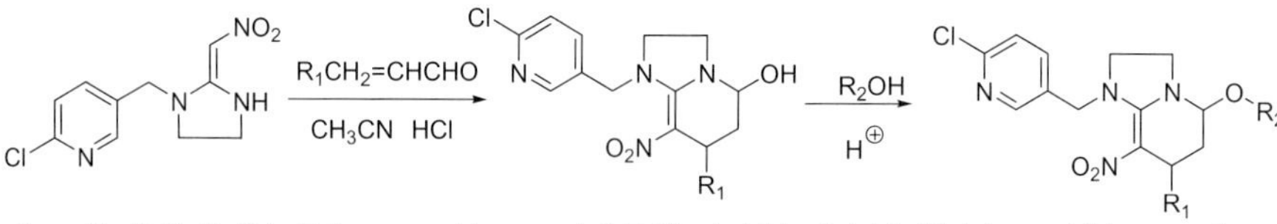

3a-n: R_1=H, R_2=H, CH_3, C_2H_5, n-propyl, iso-propyl, C_2H_5Cl, n-butyl, iso-butyl, tert-butyl, n-pentyl, iso-penpyl, benzyl, 2-Cl-5-methylpridine, propargyl
4a-c: R_1=CH_3, R_2=H, CH_3,C_2H_5

Figure 4. General synthetic scheme.

the procedure reported previously [21]. The further reaction of **1b** with acrylaldehyde or crotonaldehyde could proceed readily at 40 °C under catalysis of hydrochloride acid to give target compounds **3a** and **4a**. Compounds **3b–3n**, **4b–4c** were synthesized by the reaction of **3a** or **4a** with various alcohols in the existence of acid.

17.3.2
Biological Activity

The preliminary bioassays showed that most of them exhibited moderate insecticidal activity against pea aphids. Compound **4a** acts on a wide range of insect pests via further research, including important species, such as *Nilaparvata lugens* (LC_{50}: 7.36 µg/mL), *Myzus persicae* (LC_{50}: 2.31 µg/mL), *Bemisia tabaci* (LC_{50}: 9.18 µg/mL). Interestingly, compared to inactivity of imidacloprid, the LC_{50} value of **4a** against acarid is 18 µg/mL. On the other hand, the acute oral toxicity of **4a** to male rat has > 5000 mg/kg by LD_{50} value.

Recently, resistance to imidacloprid became a serious problem for crop protection. Comparative studies of other neonicotinoids revealed a high cross-resistance to acetamiprid and thiamethoxam against imidacloprid-resistant strains [22]. However, bioassays exhibited that **4a** has good activity against imidacloprid-resistant strains of brown planthopper (Table IV), showing very low cross-resistance to imidacloprid, as shown in Table V.

Table IV. Bioassays of IPP against sensitive and imidacloprid-resistant strains of brown planthopper (*Nilaparvata lugens*).

Insects	Chemicals	LC_{50} and 95% Confidential Level (mg a.i./L)	Index of Relative toxicity
Resistant strain	4a	20.7 (17.8 ~ 24.1)	2
	Imidacloprid	42.4 (32.7 ~ 65.0)	1
Sensitive strain	4a	7.4 (6.4 ~ 8.4)	0.026
	Imidacloprid	0.19 (0.15 ~ 0.24)	1

Table V. Cross-resistance of imidacloprid-resistant strains on 4a.

Chemicals	LC_{50} and 95% Confidential Level (mg a.i./L)		Resistant Level
	Sensitive Strain	Resistant Level	
4a	7.36 (6.44 ~ 8.43)	20.7 (17.8 ~ 24.1)	2.8
Imidacloprid	0.19 (0.15 ~ 0.24)	42.4 (32.7 ~ 65.0)	223

17.3.3 QSAR Analysis

As shown in Table VI, activity was strongly related to the group of R_1 and R_2. When R_1 is H, and CH_3, and R_2 is from H to propyl, the compounds show higher activity. When R_2 is a bulky group, such as tert-butyl or benzyl, the activity decreased. Therefore, the volume of compound may be an important factor for activity. To further explore the structural requirements for the activity of our compounds, quantitative structure-activity relationships (QSAR) analysis was performed.

Table VI. Activity (pLC_{50} mmol/L) and parameters.

Compound	Mortality (%) (500 mg/L)	Activity (pLC_{50} mmol/L)	AlogP	Dipole_Mopac
3a	> 90	6.7	2.8	11.1
4a	> 90	7.0	3.2	11.0
3b	> 90	6.9	3.2	11.3
3c	> 90	7.0	3.5	10.9
3d	> 90	7.0	4.0	11.2
3e	> 90	6.6	3.9	11.4
3f	45	–	–	–
3g	78	–	–	–
3h	76	–	–	–
3i	73	–	–	–
3j	63	–	–	–
3k	24	–	–	–
3l	> 90	6.2	4.8	11.2
3m	50	5.9	4.5	13.1
3n	47	–	–	–
4b	> 90	6.8	3.6	11.1
4c	> 90	6.7	2.8	11.1
imidacloprid	–	–	1.6	9.4

$$pLC_{50} = 11.4 - 0.22\ \text{AlogP} - 0.34\ \text{Dipole\_Mopac}$$

$$n = 10\ r = 0.87,\ XV\ r = 0.56\ \text{PRESS} = 0.771$$

The equation explained that AlogP and Dipole_Mopac are the most important factors for the activity of our compounds. The lower AlogP and Dipole_Mopac value will result in higher bioactivities.

17.4 Conclusion

In conclusion, we theoretically studied the selectivity mechanism of neonicotinoids and proposed an alternative binding model of neonicotinoids. Most importantly, the successful chemical modification of neonicotinoids resulted in a novel class of tetrahydro-pyridine nitromethylene derivatives with the *cis* position from the nitro group. Some compounds of that group, such as **4a,** showed very good insecticidal activities on a wide range of insects and have lower toxicity for mammals than imidacloprid. In summary, chemical modification using hydro-pyridine to form a *cis* configuration with the *exo*-ring ether moiety might provide potential neonicotinoids.

17.5 References

1 D. Bai, S. C. R. Lummis, W. Leicht, H. Breer, D. B. Sattelle, *Pestic Sci.*, **1991**, *33*, 197–204.

2 M. E. Schroeder, R. F. Flattum, *Pestic. Biochem. Physiol.*, **1984**, *22*, 148–160.

3 K. Kiriyama, N. Keiichiro, *Pest. Manag. Sci.*, **2002**, *58*, 669–676.

4 H. Uneme, K. Iwanaga, N. Higuchi, Y. Kando, T. Okauchi, A. Akayama, I. Minamida, *Pestic. Sci.*, **1999**, *55*, 202–205.

5 I. Yamamoto, J. E. Casida, *Nicotinoid Insecticides and the Nicotinic Acetylcholine Receptor*, Springer-Verlag, Tokyo, **1999**.

6 R. Nauen, U. Ebbinghaus-Kintscher, R. Schmuck, *Pest Manag. Sci.*, **2001**, *57*, 577–586.

7 Z. W. Liu, M. S. Williamson, S. J. Lansdell, I. Denholm, Z. Han, N. S. Millar, *Proc. Nat'l. Acad. Sci.*, **2005**, *102*, 8420–8425.

8 M. Shimomura, M. Yokota, M. Okumura, K. Matsuda, M. Akamatsu, D. B. Sattelle, K. Komai, *Brain Res.*, **2003**, *991*, 71–77.

9 M. Shimomura, H. Okuda, K. Matsuda. Komai, M. Akamatsu, D. B. Sattelle, *Br. J. Pharmacol.*, **2000**, *130*, 981–986.

10 M. Shimomura, M. Yokota, K. Matsuda, D. B. Sattelle, K. Komai, *Neurosci. Lett.*, **2004**, *363*, 195–198.

11 F. Clementi, D. Fornasari, C. Gotti, *Neuronal Nicotinic Receptors*, Springer-Verlag, **2000**.

12 M. Tomizawa, N. J. Zhang, K. A. Durkin, M. M. Olmstead, J. E. Casida, *Biochemistry*, **2003**, *42*, 7819–7827.

13 K. Matsuda, M. Shimomura, M. Ihara, M. Akamatsu, D. B. Sattelle, *Bioscience Biotechnology and Biochem.*, **2005**, *69*, 1442–1452.

14 R. S. Jérome Boëlle, P. Gérardin, B. Loubinoux, P. Maienfisch,

A. Rindlisbacher, *Pestic. Sci.*, **1998**, *54*, 304–307.
15 A. S. Reddy, D. Vijay, G. M. Sastry, G. N. Sastry, *J. Phys. Chem. B*, **2006**, *110*, 2479–2481.
16 S. Kagabu, H. Matsuno, *J. Agric. Food Chem.*, **1997**, *45*, 276–281.
17 B. Latli, M. Tomizawa, J. E. Casida, *Bioconjugate Chem.*, **1997**, *8*, 7–14.
18 S. Kagabu, H. Nishiwaki, K. Sato, M. Hibi, N. Yamaoka, Y. Nakagawa, *Pest Manag. Sci.*, **2002**, *58*, 483–490.
19 S. Kagabu, S. Medej, *Biosci. Biotech. Biochem.*, **1995**, *59*, 980–985.
20 I. Yamamoto, M. Tomizawa, T. Satio, T. Miyamoto, E. C. Walcott, W. Sumikawa. *S. Arch. Insect Biochem. Physiol.*, **1998**, *37*, 24–32.
21 S. Kagabu, K. Moriya, K. Shibuya, Y. Hattori, S. Tsuboi, K. Shiokawa, *Biosci. Biotech. Biochem.*, **1992**, *56*, 362–363.
22 A. Elbert, R. Nauen, *Pest Manag. Sci.*, **2000**, *56*, 60–64.

Keywords

Neonicotinoids, Bioinformatics, *an initio* Chemistry, Binding Mode, Nitromethene, Tetrahydro-pyridine, QSAR, Cross-Resistance

18
Synthesis and Inhibitory Action of Novel Acetogenin Mimics Δlac-Acetogenins: A New Class of Inhibitors of Mitochondrial NADH-Ubiquinone Oxidoreductase (Complex-I)

Hideto Miyoshi, Naoya Ichimaru, Masatoshi Murai

18.1
Introduction

NADH-ubiquinone oxidoreductase (Complex I) is the first energy-transducing enzyme of the respiratory chains of most mitochondria and many bacteria. It catalyzes the oxidation of NADH by ubiquinone, coupled to the generation of an electrochemical proton gradient across the membrane that drives energy-consuming processes such as ATP synthesis and flagella movement [1]. Complex I is the most complicated multisubunits enzyme in the respiratory chain; e.g., the enzyme from bovine heart mitochondria is composed of 46 different subunits with a total molecular mass of about 1 MDa [2]. Recently, the crystal structure of the hydrophilic domain (peripheral arm) of Complex I from *Thermus thermophilus* was solved at 3.3 angstrom resolution [3]. However, our knowledge about the functional and structural features of the membrane arm, such as the ubiquinone redox reaction, proton translocation mechanism, and mode of action of numerous specific inhibitors, is still very limited [4–5].

Many structurally diverse inhibitors of Complex I are known [6–8]. With the exception of rhein [9] and diphenyleneiodonium [10], which inhibit electron input into Complex I, all inhibitors act at the terminal electron transfer step of the enzyme [6, 11]. Although these inhibitors are generally believed to act at the ubiquinone reduction site, there is still no hard experimental evidence to verify this possibility. Rather, a recent photoaffinity labeling study using azidoquinone suggested that the inhibitor binding site is not the same as the ubiquinone binding site [12]. To begin with, both the number and the location of the ubiquinone binding site(s) remain controversial [3–4, 12–13]. On the other hand, mutagenesis studies using the yeast *Yarrowia lipolytica* and *Rhodobacter capsulatus* [14–16] and photoaffinity labeling studies [17–18] indicated that PSST, ND5, and 49-kDa subunits contribute to the inhibitor-binding domain. Using a strong inhibitor with intense fluorescence [6-amino-4-(4-*tert*-butylphenetylamino)quinazoline, AQ], Ino *et al.*, suggested that the apparent competitive behavior among potent Complex I inhibitors cannot be explained simply based on competition for the same binding

Pesticide Chemistry. Crop Protection, Public Health, Environmental Safety
Edited by Hideo Ohkawa, Hisashi Miyagawa, and Philip W. Lee

ISBN: 978-3-527-31663-2

region [19]. Thus it remains to be learned how binding sites of diverse Complex I inhibitors relate to each other.

18.2 Mode of Action of Δlac-Acetogenins

Acetogenins isolated from the plant family Annonaceae, such as bullatacin (Figure 1) and rolliniastatin-1, are among the most potent inhibitors of bovine heart mitochondrial Complex I [6, 8, 11]. We recently synthesized new acetogenin mimics named Δlac-acetogenins [20], which consist of the hydroxylated bis-THF ring and two hydrophobic side chains without a α,β-unsaturated γ-methylbutyrolactone ring which is a structural feature common to a large number of natural acetogenins (Figure 1).

Some Δlac-acetogenins elicit very potent inhibition of bovine Complex I at the nanomolar level despite the lack of a γ-lactone ring. An electron paramagnetic resonance (EPR) spectroscopic study on the redox state of iron-sulfur clusters indicated that the inhibition site of Δlac-acetogenins is downstream of the iron-sulfur cluster N2 [20], as is the case for other ordinary Complex I inhibitors such as rotenone and piericidin A [6, 11]. However, several lines of evidence, as summarized below, strongly suggest that the inhibition manner of Δlac-acetogenins is different from that of natural acetogenins as well as ordinary Complex I inhibitors [21]; (1) the profile of the structure-activity relationship of Δlac-acetogenins is entirely different from that of natural-type acetogenins, (2) double-inhibitor titration of steady state Complex I activity shows that the extent of inhibition by Δlac-acetogenin and bullatacin is not additive, and (3) competition tests using a fluorescent ligand (AQ) indicate that the binding site of Δlac-acetogenins does not overlap with that of other Complex I inhibitors.

Since a detailed study of these unique inhibitors might provide new insight into the terminal electron transfer step of the enzyme, we further characterized their inhibitory action using the most potent Δlac-acetogenin derivative (compound **1**). Unlike ordinary Complex I inhibitors, **1** has a dose-response curve for inhibition of the reduction of exogenous short-chain ubiquinones that was difficult to

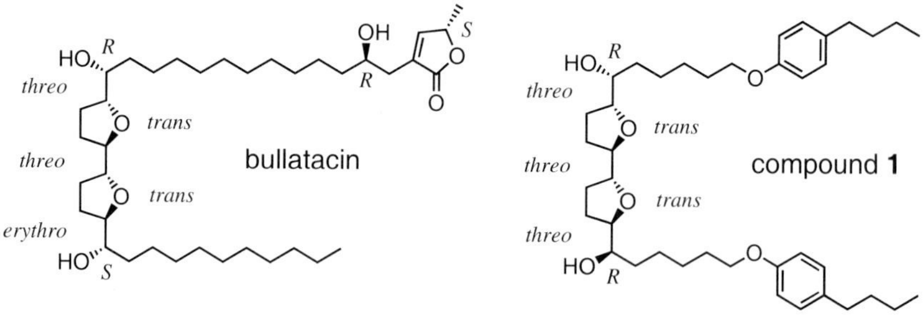

Figure 1. Structures of natural acetogenin (e.g., bullatacin) and a representative Δlac-acetogenin (Compound 1).

explain with a simple bimolecular association model. The inhibitory effect of **1** on ubiquinol-NAD^+ oxidoreductase activity (reverse electron transfer) is much weaker than that on NADH oxidase activity (forward electron transfer), indicating a direction-specific effect. These results suggest that the binding site of **1** is not identical to that of ubiquinone and the binding of **1** to the enzyme secondarily (or indirectly) disturbs the redox reaction of ubiquinone [22]. Using endogenous and exogenous ubiquinone as an electron acceptor of Complex I, we investigated the effect of **1** in combination with different ordinary inhibitors on the superoxide production from the enzyme. The results indicated that the level of superoxide production induced by **1** is significantly lower than that induced by ordinary inhibitors probably because of fewer electron leaks from the ubisemiquinone radical to molecular oxygen and that the site of inhibition by **1** is downstream of that by ordinary inhibitors [22]. Taken together, Δlac-acetogenins were revealed to be a new type of inhibitors acting at the terminal electron transfer step of Complex I.

18.3 SAR of Δlac-Acetogenins

We synthesized a series of Δlac-acetogenins in which the two alkyl side chains attached to the C_2-symmetric bis-THF portion were systematically modified, and examined their inhibitory effect on bovine heart mitochondrial Complex I. The structure-activity studies revealed that large and symmetrical hydrophobicity of both alkyl side chains is crucial for exhibiting potent inhibitory effect [20–21, 23]. It is likely that Δlac-acetogenins bind to the hydrophobic membrane arm of Complex I and the balance in hydrophobicity of the two chains decide the precise location of the hydrophilic bis-THF ring moiety at or close to the membrane interface. It is also revealed that expansion of the width of the side chain is remarkably unfavorable for the inhibitory action [23]. This is probably because the tails directly interact with the hydrophobic domain of Complex I rather than merely partitioning into the lipid membrane phase, whereupon the enzyme recognizes the molecular shape of the side chains in a strict sense. Moreover, the stereochemistry around the hydroxylated bis-THF moiety significantly affected the inhibitory potency. The *R*-configuration at all chiral centers, as shown in Figure 1, was best for the inhibition [unpublished data].

18.4 Conclusion

We revealed that Δlac-acetogenins, a new class of inhibitors of bovine mitochondrial Complex I, act differently from ordinary inhibitors and that the site of inhibition by Δlac-acetogenins is downstream of that by ordinary inhibitors. The unique inhibitory action of hydrophobic Δlac-acetogenins may be closely associated with the dynamic function of the membrane domain of Complex I.

18.5 References

1 J. E. Walker, *Q. Rev. Biophys.*, **1992**, *25*, 253–324.
2 J. Carroll, I. M. Fearnley, R. J. Shannon, J. Hirst, J. E. Walker, *Mol. Cell. Proteomics*, **2003**, *2*, 117–126.
3 L. A. Sazanov, P. Hinchliffe, *Science*, **2006**, *311*, 1430–1436.
4 T. Yagi, A. Matsuno-Yagi, *Biochemistry*, **2003**, *42*, 2266–2274.
5 J. Hirst, *Biochem. Soc. Trans.*, **2005**, *33*, 525–529.
6 T. Friedrich, P. Van Heek, H. Leif, T. Ohnishi, E. Forche, B. Kunze, R. Jansen, W. Trowitzsch-Kienast, *et al.*, *Eur. J. Biochem.*, **1994**, *219*, 691–698.
7 M. Degli Esposti, *Biochim. Biophys. Acta*, **1998**, *1364*, 222–235.
8 H. Miyoshi, *Biochim. Biophys. Acta*, **1998**, *1364*, 236–244.
9 E. A. Kean, M. Gutman, T. P. Singer, *J. Biol. Chem.*, **1971**, *246*, 2346–2353.
10 A. Majander, M. Finel, M. Wikström, *J. Biol. Chem.*, **1994**, *269*, 21037–12042.
11 J. G. Okun, P. Lümmen, U. Brandt, *J. Biol. Chem.*, **1999**, *274*, 2625–2630.
12 X. Gong, T. Xie, L. Yu, M. Hesterberg, D. Scheide, T. Friedrich, C.-A. Yu, *J. Biol. Chem.*, **2003**, *278*, 25731–25737.
13 T. Yano, W. R. Dunham, T. Ohnishi, *Biochemistry*, **2005**, *44*, 1744–1754.
14 P. M. Ahlers, K. Zwicker, S. Kerscher, U. Brandt, *J. Biol. Chem.*, **2000**, *275*, 23577–23582.
15 N. Kashani-Poor, K. Zwicker, S. Kerscher, U. Brandt, *J. Biol. Chem.*, **2001**, *276*, 24082–24087.
16 I. Prieur, J. Lunardi, A. Dupuis, *Biochim. Biophys. Acta*, **2001**, *1504*, 173–178.
17 F. Schuler, T. Yano, S. D. Bernardo, T. Yagi, V. Yankovskaya, T. P. Singer, J. E. Casida, *Proc. Natl. Acad. Sci. U.S.A.*, **1999**, *96*, 4149–4153.
18 E. Nakamaru-Ogiso, K. Sakamoto, A. Matsuno-Yagi, H. Miyoshi, T. Yagi, *Biochemistry*, **2003**, *42*, 746–754.
19 T. Ino, T. Nishioka, H. Miyoshi, *Biochim. Biophys. Acta*, **2003**, *1605*, 15–20.
20 T. Hamada, N. Ichimaru, M. Abe, D. Fujita, A. Kenmochi, T. Nishioka, K. Zwicker, U. Brandt, H. Miyoshi, *Biochemistry*, **2004**, *43*, 3651–3658.
21 N. Ichimaru, M. Murai, M. Abe, T. Hamada, Y. Yamada, S. Makino, T. Nishioka, H.Makabe, *et al.*, *Biochemistry*, **2005**, *44*, 816–825.
22 M. Murai, N. Ichimaru, M. Abe, T. Nishioka, H. Miyoshi, *Biochemistry*, **2006**, *45*, 9778–9787.
23 N. Ichimaru, M. Abe, M. Murai, M. Senoh, T. Nishioka, H. Miyoshi, *Bioorg. Med. Chem. Lett.*, **2006**, *16*, 3555–3558.

Keywords

Respiratory Inhibitor, Acetogenin, Mitochondrial Complex I, Respiratory Enzymes, Structure-Activity Relationship

III
Biology, Natural Products and Biotechnology

Pesticide Chemistry. Crop Protection, Public Health, Environmental Safety
Edited by Hideo Ohkawa, Hisashi Miyagawa, and Philip W. Lee

ISBN: 978-3-527-31663-2

19
Plant Chemical Biology: Development of Small Active Molecules and Their Application to Plant Physiology, Genetics, and Pesticide Science

Tadao Asami, Nobutaka Kitahata, Takeshi Nakano

19.1
Introduction

Combining knowledge of organic chemistry and modern aspects of plant research is very useful for investigating the interaction between chemicals and enzyme(s), or chemicals and receptor(s). For example, finding suicide substrates of abscisic acid 8'-hydroxylase were a great help in identifying this enzyme [1]. The use of biotinylated abscisic acid (ABA) derivatives demonstrated that there are proteinaceous ABA perception sites on the plasma membrane of *Vicia faba* guard cells, and direct visualization and quantitative analyses of the ABA perception sites was possible [2]. Photoaffinity labeling of membrane proteins with a photoactivatable phytosulfokine (PSK) analog characterized the PSK binding proteins, which were then purified with affinity chromatography using immobilized PSK [3]. A photoactivatable brassinosteroid analog equipped with a biotin group was also very useful for identifying the brassinosteroid binding site in the brassinosteroid receptor [4]. These are straightforward biochemical approaches. There are alternative ways to determine the function of proteins and genes using chemicals.

Genetics has been a powerful tool for biologists. A classical forward genetic analysis starts with an outward physical characteristic (called a phenotype) of interest and ends with the identification of the gene or genes that are responsible for it. In classical reverse genetics, scientists start with a gene of interest and try to find what it does by looking at the phenotype when the gene is mutated. Recently "chemical genetics" has been used as a new tool for dissecting and understanding biological systems [5]. This term impresses on us the importance of biologically active small molecules in biology. In chemical genetics, small molecules are used as a switch to turn on or turn off the biological event by affecting protein functions rather than genes. In a forward chemical-genetic screen, instead of mutating genes at random, scientists generate a lot of small molecules and then systematically introduce them into living organisms to determine their effects. Small molecules that create a change in the phenotype of interest are selected

Pesticide Chemistry. Crop Protection, Public Health, Environmental Safety
Edited by Hideo Ohkawa, Hisashi Miyagawa, and Philip W. Lee

ISBN: 978-3-527-31663-2

for further study. Since these small molecules probably change phenotype by binding to proteins inside cells, thus changing the way these proteins work, there is great interest in finding the protein targets of these small molecules. In a sense, the small molecules that bind to proteins and affect their activities mimic the random mutations used in classical genetic screens. However, there are important differences. In a genetic screen, the activity of a protein is altered indirectly by mutating its gene, but in chemical genetics this change is direct and occurs in real time (when the molecule is added). Another difference between the two approaches is that the effect of the "mutation" caused by a small molecule is reversed when the small molecule is removed. In contrast, the effect of mutating a gene is, in most cases, permanent. Therefore, chemical-genetic approaches may be more useful when scientists want to study genes that are essential to an organism's survival. A small molecule can be administered to cells or organisms for a very short time to study the function of the target protein. Thus, using this strategy of chemical genetics, it is possible to identify new reagents that act like conditional mutations, either inducing or suppressing the formation of a specific phenotype of interest. Herbicidal inhibitors are compounds that inhibit processes essential for plant growth. These inhibitors can serve as excellent tools for probing plant gene functions. Zhen and Singh gave hydroxyphenylpyruvate dioxygenase (HPPD) inhibitors as a good example of tools for plant functional genomics [6]. Plant hormones are essential for normal plant growth and therefore chemicals that can regulate endogenous plant hormone levels are good targets for chemical genetics. We selected brassinosteroids as our target molecule and tried to find the molecule that can control the endogenous level of brassinosteroid in plants. Here we look back over brassinosteroid (BR) biosynthesis inhibitors developed in our laboratory as a powerful tool for functional genomics.

Seven years have passed since the discovery of the first BR biosynthesis inhibitor [7]. One of the scientific goals of working with BR biosynthesis inhibitors is to find new functions of BRs and to identify novel components involved in BR biosynthesis and signal transduction. From the point of view of specific approaches to achieve this goal, we will discuss the following topics in this review: 1) Development of BR biosynthesis inhibitors; 2) functions of BRs in plant development unveiled by BR biosynthesis inhibitors; 3) BR biosynthesis inhibitors as a useful screening tool for BR signaling mutants.

19.2
Development of BR Biosynthesis Inhibitors

Brassinosteroids (BRs) are highly oxidized steroidal plant hormones and essential for normal plant growth. BR deficient mutants display strong dwarfism with curly, dark-green leaves in the light, and a deetiolated phenotype with short hypocotyl and open cotyledons in the dark. The characterization of BR-deficient mutants by biochemical studies and molecular genetic analysis has established the biosynthetic pathway for brassinolide (BL), the biochemically most active BR.

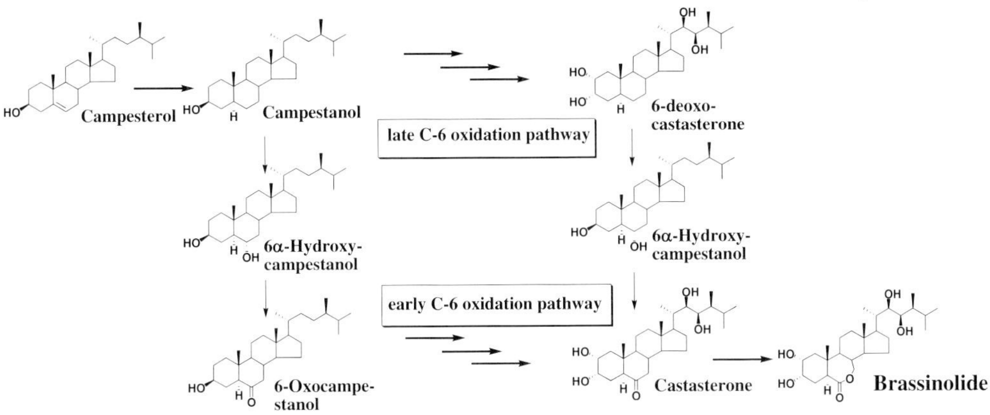

Figure 1. Brassinosteroid biosynthesis pathways.

BL is synthesized from campesterol via either early or late C-6 oxidation pathways that include cytochrome P450 monooxygeneses.

Thus, the biosynthetic pathway of BRs includes several potential active sites for cytochrome P450 inhibitors. Uniconazole, a gibberellin (GA) biosynthesis inhibitor, has been reported to inhibit BR biosynthesis, even though its main target is GA biosynthesis rather than BR biosynthesis. Various triazole compounds including uniconazole and other GA biosynthesis inhibitors have been shown to inhibit many types of cytochrome P450s. From studies of these cytochrome P450 inhibitors, the azole moiety of the inhibitors is believed to act as a ligand binding to the iron atom of the heme prosthetic group of the cytochrome P450 enzyme, forming a coordinated complex. Chemical structure other than a triazole moiety is considered to be the important factor, which results in the selective nature of the interaction. In an effort to illustrate azole-binding sites in BR biosynthesis and to identify essential structural features among azole compounds, the structure-activity relationship of uniconazole has been studied for BR biosynthesis inhibition.

19.2.1
Assay Methods for BR Biosynthesis Inhibitors

Since a good biological system for identifying BR biosynthesis inhibitors had not yet been found, we combined some biological assays. First, chemicals were assayed using a rice-stem elongation test to identify and eliminate GA biosynthesis inhibitors because rice is very sensitive to GA-deficiency and therefore a good plant for this purpose. Some of the synthesized chemicals retarded rice-stem elongation, and such retardation was reversed by treatment with GA. A second screening for BR biosynthesis inhibitors was performed to find chemicals that induce dwarfism in *Arabidopsis*, and which resemble BR biosynthesis mutants and can be rescued by the addition of BL. BL has been shown to be effective in rescuing the *Arabidopsis* BR-deficient mutants, but they cannot be rescued by other plant hormones, such as auxins and GAs. Finally, selected compounds were assayed using a cress hypocotyl elongation test. It has been demonstrated that cress is very sensitive to an internal deficiency of BRs and is therefore a useful species for evaluating BR biosynthesis inhibitors [7–8].

19.2.2 Structure-Activity Relationship Study

The presence of a *tert*-butyl group at C-2 of uniconazole and paclobutrazol could be essential for the inhibitory activity of GA biosynthesis. The chemical structure of paclobutrazol is closely related to that of uniconazole but it has no double bond. A substitution of a *tert*-butyl group of these compounds with a phenyl group caused a drastic loss of inhibition of rice stem elongation, whereas it caused strong inhibition of *Arabidopsis* and cress hypocotyl elongation [7]. This retardation was recovered by the co-application of BL but not of GA. These studies revealed that the phenyl moiety at C-2 of uniconazole and paclobutrazol is essential for the selectivity of BR biosynthesis inhibition. In addition to the substitution, an introduction of an alkyl or aryl group at C-2 of paclobutrazol caused more potent BR biosynthesis inhibition and reduced the effect on GA biosynthesis. As a result, the structural difference between paclobutrazol and brassinazole derivatives is only the existence of alkyl or aryl group and a phenyl group attached to the carbinol carbon. These groups drastically change the character of triazole derivatives from GA biosynthesis inhibitors to BR biosynthesis inhibitors.

19.2.3 Target Site(s) of BR Biosynthesis Inhibitor

To investigate the biosynthetic steps affected by brassinazole, we examined the effect of biosynthetic intermediates downstream from cathasterone on hypocotyl elongation of brassinazole-treated *Arabidopsis* [9]. The feeding experiment suggests that the target(s) of brassinazole could be the two-step conversion of 6-oxocampestanol to teasterone via cathasterone, catalyzed by DWF4 and CPD, which are *Arabidopsis* cytochrome P450s isolated as putative steroid 22- and 23-hydroxylases, respectively. In addition, we analyzed endogenous BRs in brassinazole treated and non-treated *Catharanthus roseus* cells [10]. In brassinazole-treated plant cells, the levels of campestanol and 6-oxo-campestanol levels were increased, and levels of BR intermediates with hydroxy groups on the side chains were reduced, suggesting that brassinazole treatment reduced BR levels by inhibiting the hydroxylation of the 22-position catalyzed by DWF4. Thus, DWF4 was expressed in *Escherichia coli*, and the binding affinity to brassinazole and its derivatives to the recombinant DWF4 were analyzed [11]. Among several triazole derivatives, brassinazole had both the highest binding affinity to DWF4 and the highest growth inhibitory activity. The binding affinity and activity for inhibiting hypocotyl growth were well correlated among the derivatives. On the other hand, brassinazole did not bind to the recombinant CPD proteins

(Mizutani, personal communication), which suggested that CPD was not the target site of brassinazole. In brassinazole-treated *Arabidopsis*, the CPD gene was induced within 3h, most likely because of feedback activation caused by the reduced levels of active BRs. These results indicate that brassinazole inhibits the hydroxylation of the 22-position of the side chain in BRs by direct binding to DWF4 and that DWF4 catalyzes this hydroxylation reaction. As the involvement of DWF4 protein in BR biosynthesis pathway was suggested only by comparing the phenotypes of *dwf4* mutants to that of other BR deficient mutants and feeding biosynthesis intermediates, the combination of the chemical analysis of internal BRs in brassinazole-treated plant cells and the binding assay of brassinazole to DWF4 should have been an alternative way to investigate the role of the DWF4 in BR biosynthetic pathway.

19.2.4
Searching for Novel BR Biosynthesis Inhibitors

To develop a more specific and potent BR biosynthesis inhibitor, we screened various triazole derivatives with the cress hypocotyl elongation test. Through this screening experiment, fenarimol, triadimefon, and propiconazole were selected as a likely inhibitor of BR biosynthesis. Chemical modification of fenarimol led us to the discovery of a new BR biosynthesis inhibitor DPPM4, which is specific for BR biosynthesis but not as potent as brassinazole [12]. Triadimefon shows good affinity to expressed DWF4 proteins and induces BR deficiency phenotype in plants [13]. Propiconazole is a fungicide that targets lanosterol 14α-demethylase in the ergosterol biosynthesis pathway. Propiconazole-treated cress showed dwarfism that could be rescued considerably by BL treatment. This implies that the morphological alternation of cress seedlings treated with propiconazole should be partly due to the deficiency of BL [14]. Since propiconazole showed considerable inhibitory activity in the cress hypocotyl elongation test, the synthesis of propiconazole derivatives with optimized activity and selectivity was started. Intensive study of structure-activity relationships of propiconazole led to the discovery of a more potent and specific inhibitor, Brz220 [15]. Since it contains two stereogenic carbon atoms, there are four epimeric stereoisomers of Brz220. Since the stereoisomers of azole compounds often have different biological activities, we examined the relationship between the stereochemical structure and biological activity of Brz220. The configuration of enantiomers of Brz220 was determined by a combination of asymmetric syntheses [15]. Finally Brz22012, one of the stereoisomers of Brz220, was found to be the most potent BR biosynthesis inhibitor. In inhibiting BR biosynthesis, the (*S*)-configuration of Brz220 at C-2 predicts whether a stereoisomer can bind to its receptor site on a cytochrome P450 in the BR biosynthesis pathway, as occurs with brassinazole. Further study reveals the site of action of Brz220, both *in vivo* and *in vitro*. DWF4 protein was the target site of these inhibitors including *N*-substituted hetero ring. Spironolactone has steroidal structure and was revealed to have an inhibitory activity of BR biosynthesis. The spironolactone action site was also investigated

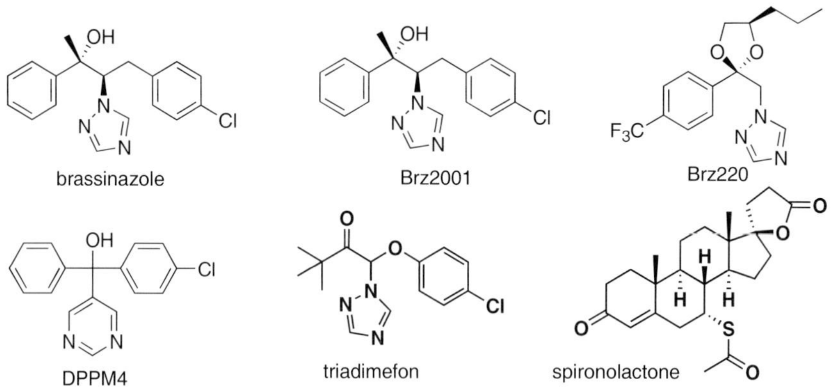

Figure 2. Chemical structures of brassinosteroid biosynthesis inhibitors.

by feeding BR biosynthesis intermediates to *Arabidopsis* grown in the dark, and the results suggested that the inhibition site of spironolactone may be the step from 4-en-3b-ol to 4-en-3-one being catalyzed by 3b-HSD. The structures of BR biosynthesis inhibitors mentioned here are listed in Figure 2.

19.3
Functions of BRs in Plant Development Unveiled by BR Biosynthesis Inhibitors

Mutant or inhibitor studies have already demonstrated quite well that BRs are essential for normal plant growth. Therefore, as a next step to understand novel functions of BR in plants, brassinazole was applied to investigate the functions of BRs in photomorphogenesis in the dark, in xylem development, in chlorella growth, in cotton fiber growth and others. To be able to use brassinazole as a tool, it is necessary to confirm in detail that various morphological and cytological changes in brassinazole-treated plants are due to inhibition of BR biosynthesis, and not to side effects of the inhibitor.

When a BR biosynthesis inhibitor, brassinazole, was applied to *Arabidopsis thaliana,* high levels of ribulose-1,5-bisphosphate carboxylase-oxygenase proteins accumulated in the plastids of the cotyledons. These results suggest that brassinazole treatment in the dark induces the initial steps of plastid differentiation, which occur prior to the development of thylakoid membranes. This is a novel presumed function of BRs [16]. Brassinazole treament also retards the development of secondary xylem in cress [17].

Brassinazole treatment suppresses BR biosynthesis in cotton ovules and inhibits fiber formation. This inhibition was rescued by brassinolide treatment. These results clearly indicate that BRs are essential for normal growth of cotton fiber [18]. Recently Bajguz and Asami reported the effect of BR biosynthesis inhibitor on *Chlorella vulgaris* cells. Treatment of cultured *C. vulgaris* cells with Brz2001 inhibited their growth in the light. This inhibition was prevented by the co-application of BR. This result suggests that the presence of endogenous BRs during the initial steps of the *C. vulgaris* cell cycle is indispensable to their normal growth in the light [19].

19.4
BR Biosynthesis Inhibitors as a Useful Screening Tool for BR Signaling Mutants

Many studies of the molecular mechanisms of plant growth have been performed using genetic methods in *Arabidopsis*. In the past decade, the identification and characterization of *Arabidopsis* BR biosynthetic mutants such as *det2* and *dwf4* has revealed the importance of BRs in plant growth regulation. These BR-deficient mutants have a pleiotropic dwarf phenotype that can be reverted to a wild-type-like phenotype by feeding with BL.

In order to analyze in detail the mechanisms of BR biosynthesis and signal transduction, we performed a screen for mutants with altered responses to Brz220 treatment in darkness in the germination stage. A screen of 140,000 *Arabidopsis* seeds that had been subjected to EMS and fast neutron mutagenesis revealed several mutants that had significantly longer hypocotyls than the wild type when grown in the dark and treated with Brz220. These plants were designated *bil* mutants (Brz-insensitive-long hypocotyl). When grown in medium containing 3-μM Brz220, wild-type plants had quite short hypocotyls, but *bil* mutants had hypocotyls as long as those of wild-type plants grown on unsupplemented medium. In parallel, *bzr1-1D* and *bes1-1D* were identified as Brz-resistant and *bri1*-suppressor mutants, respectively. Gene sequencing revealed that the *bzr1-1D* gene is the same gene as *bil1-1D*, even containing the same mutation [20]. These genes are 88% identical to *BES1*, and the *bes1* mutant has the same nucleotide substitution [21–22].

In another approach toward the understanding of BR signaling, Drs. Joanne Chory and Detlef Weigel and their colleagues have mapped quantitative trait loci (QTL) responsible for natural variations in hormone and light responses. The resulting QTL map predicted at least three strong loci that confer BR biosynthesis inhibitor insensitivity and long hypocotyls in darkness, and five weaker loci were also identified. As these strong Brz-insensitivity loci do not map near the already confirmed or potential BR biosynthesis inhibitor-insensitivity genes, a more detailed QTL analysis and more genetic screening for BR signaling mutants will be needed to clarify the mechanisms of plant growth regulation by BRs [23–24].

Recently, gene chip methods have been used to predict genes induced or reduced by BRs or brassinazole [25]. However, it is difficult to determine which genes are actually involved in BR signaling in plants from the data obtained by gene chip analyses, but reverse genetics approaches will be a great help for identifying components of BR signaling. That is, when a transgenic plant in which BRs or Brz regulating gene is overexpressed or suppressed would show insensitivity against the treatment of BRs or BR biosynthesis inhibitor, then it is possible to think that the product of such gene would be involved in BR signaling.

19.5
Usefulness of Biosynthesis Inhibitors of Biologically Active Molecules in Plant Biology

Through the development of BR biosynthesis inhibitors and their application to the study on BR function and BR biosynthesis inhibitor insensitive mutant screenings, we demonstrated that small active molecules play important roles in plant biology. In chemical genetics study, we identified not only brassinosteroid signal transduction factors such as *bil1/bzr1*, but also receptor gene *BRI1* and biosynthesis enzyme gene *DWF4*. As these genes had already been reported when we identified, we did not publish any reports about these genes, but this result strongly suggests the usefulness of biosynthesis inhibitors of biologically active small molecules. If we could find biosynthesis inhibitors of unknown or new active molecules and use them for screening mutants which are insensitive to such inhibitors, then we could identify signal transduction factors, receptors or biosynthesis enzymes of active molecules. This idea prompted us to develop new biosynthesis inhibitors.

19.6
Abscisic Acid Biosynthesis Inhibitors Targeting 9-cis-Epoxycarotenoid Dioxygenase (NCED)

As our next target, we selected biosynthesis inhibitors of carotenoid-derived molecules. Carotenoid cleavage dioxygenases (CCDs) produce various apocarotenoids that have important biological functions in animals and plants. CCDs catalyze the oxidative cleavage of double bonds at various positions in a variety of carotenoids. Several CCDs have been identified and characterized. An enzyme that cleaves β-carotene at the 15–15′ double bond produces vitamin A, which is essential for development and vision in animals. 9-cis-Epoxycarotenoid dioxygenase (NCED) is the best-characterized CCD in plants. NCED from maize, the first carotenoid cleavage enzyme identified, catalyzes the cleavage of 9-cis-epoxycarotenoid at the 11–12 double bond to produce a precursor of the plant hormone abscisic acid (ABA) [26]. CCD1 cleaves several carotenoids symmetrically at the 9–10 and 9′–10′ double bonds to yield C13-norisoprenoid compounds such as β-ionone, which plays a role in flower fragrance. Recently, it has been reported that CCD1 regulates the β-ionone content in petunia, tomato, and grape. CCD7 and CCD8 catalyze the sequential cleavage of β-carotene. As the max3/ccd7 and max4/ccd8 mutants of *Arabidopsis* show increased lateral branching, CCD7 and CCD8 appear to be involved in the biosynthesis of an unknown branch-inhibiting factor [27]. In this context we focused our attention to the development of CCD inhibitors. If we could develop an inhibitor targeting one of the CCDs, then we could utilize the information obtained through that research for the development of new CCD inhibitors because the sequence, substrate and function of CCDs are similar to each other.

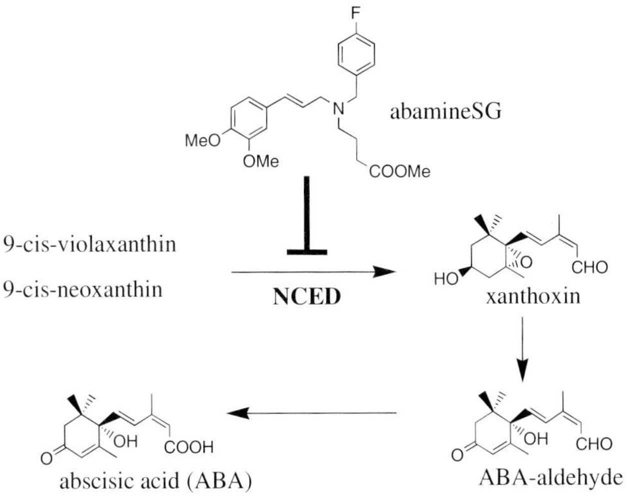

Figure 3. AbamineSG inhibits NCED in ABA biosynthesis pathway.

In view of the importance of ABA in plants, it is worthwhile to synthesize and evaluate specific ABA biosynthesis inhibitors that would be useful tools for functional studies of ABA biosynthesis and the effects of ABA in higher plants. ABA biosynthesis inhibitors provide a useful method to isolate mutants in which the genes involved in ABA signal transduction have been altered. Although carotenoid biosynthesis inhibitors such as fluridone and norflurazon have been used as ABA biosynthesis inhibitors, these compounds cause lethal damage during plant growth because carotenoids play an important role in protecting photosynthetic organisms against damage by photooxidation.

To find a lead compound for ABA biosynthesis inhibitor, firstly we tested the inhibitory activity of NDGA (nordihydroguaiaretic acid) against NCED *in vitro* because NDGA was reported to decrease ABA content in treated plants. In this test, we found that NDGA inhibited about 45% of NCED activity at 100 μM. Then we started the modification of the chemical structure of NDGA to increase its specificity as ABA biosynthesis inhibitor because NDGA has been reported to inhibit lipoxygenases and several events in cells. A number of compounds were synthesized. One of the synthesized compounds clearly inhibited NCED, ABA accumulation, and stomatal closing. This result indicated that this compound should be an ABA biosynthesis inhibitor [28–29]. On the basis of this finding, we started the structure-activity relationship study on this compound and finally found abamineSG to be the most specific and potent NCED inhibitor [30]. Treatment of plants with abamine, the first NCED inhibitor identified, inhibits ABA accumulation. Treatment of osmotically stressed plants with 100-μM abamineSG inhibited ABA accumulation by 77% as compared to the control. The expression of ABA-responsive genes and ABA catabolic genes was strongly inhibited in abamineSG-treated plants under osmotic stress. AbamineSG is a competitive inhibitor of the enzyme NCED, with a K_i of 18.5 μM.

In conclusion, we found that abamineSG should be an ABA biosynthesis inhibitor that inhibits NCED. AbamineSG will be useful in studying ABA function and the mechanism of ABA biosynthesis or catabolism in plants. More importantly, by use of chemical genetic approaches to plant biology, abamineSG should prove useful for finding mutants in genes involved in ABA signal transduction, receptors and biosynthesis.

19.7 Conclusion

Now there are at least two characterized BR biosynthesis inhibitors and one ABA biosynthesis inhibitor. They act like conditional mutations in these hormone biosyntheses. They allow the investigation of the functions of hormones in a variety of plant species. Applications of these biosynthesis inhibitors to a standard genetic screen to identify mutants that confer resistance to these biosynthesis inhibitors allow us to identify new components of the BR or ABA signal transduction pathways. This method has advantages over mutant screening using hormone deficient mutants as a background. Thus, development of chemicals which induce phenotypes of interest is now emerging as a useful and supplementary way to study biological systems of plants, enhancing classical biochemical and genetic methods.

19.8 References

1 A. Cutler, P. Rose, T. Squires, M. Loewen, A. Shaw, J. Quail, J. Krochko, S. Abrams, *Biochemistry*, **2000**, *39*, 13614–13624.

2 D. Yamazaki, S. Yoshida, T. Asami, K. Kuchitsu, *Plant J.*, **2003**, *39*, 129–139.

3 Y. Matsubayashi, M. Ogawa, A. Morita, Y. Sakagami, *Science*, **2002**, *296*, 1470–1472.

4 T. Kinoshita, A. Cano-Delgado, H. Seto, S. Hiranuma, S. Fujioka, J. Chory, *Nature*, **2005**, *433*, 167–171.

5 S. Schriber, *Bioorg. Med. Chem.*, **1998**, *6*, 1127–1152.

6 R. G. Zhen, B. K. Singh, *Weed Sci.*, **2001**, *49*, 266–272.

7 Y. K. Min, T. Asami, S. Fujioka, N. Murofushi, I. Yamaguchi, S. Yoshida, *Bioorg. Med. Chem. Lett.*, **1999**, *9*, 425–430.

8 K. Sekimata, T. Kimura, I. Kaneko, T. Nakano, K. Yoneyama, Y. Takeuchi, S. Yoshida, *Planta*, **2001**, *213*, 716–721.

9 T. Asami, S. Yoshida, *Trends Plant Sci.*, **1999**, *4*, 348–353.

10 T. Asami, Y. K. Min, N. Nagata, K. Yamagishi, S. Takatsuto, S. Fujioka, N. Murofushi, I. Yamaguchi, *et al.*, *Plant Physiol.*, **2000**, *123*, 93–99.

11 T. Asami, M. Mizutani, S. Fujioka, H. Goda, Y. K. Min, Y. Shimada, T. Nakano, S. Takatsuto, *et al.*, *J. Biol. Chem.*, **2001**, *276*, 25687–25691.

12 J. M. Wang, T. Asami, S. Yoshida, N. Murofushi, *Biosci. Biotech. Biochem.*, **2001**, *65*, 817–822.

13 T. Asami, M. Mizutani, Y. Shimada, H. Goda, N. Kitahata, K. Sekimata, S. Y. Han, S. Fujioka, *et al.*, *Biochem. J.*, **2003**, *369*, 71–76.

14 K. Sekimata, S. Y. Han, K. Yoneyama, Y. Takeuchi, S. Yoshida, T. Asami, *J. Agr. Food Chem.*, **2002**, *50*, 3486–3490.

15 K. Sekimata, J. Uzawa, S. Y. Han, K. Yoneyama, Y. Takeuchi, S. Yoshida, T. Asami, *Tetrahedron: Asymm.*, **2002**, *13*, 1875–1878.

16 N. Nagata, Y. K. Min, T. Nakano, T. Asami, S. Yoshida, *Planta*, **2000**, *211*, 781–790.

17 N. Nagata, T. Asami, S. Yoshida, *Plant Cell Physiol.*, **2001**, *42*, 1006–1011.

18 Y. Sun, M. Fokar, T. Asami, S. Yoshida, R. D. Allen, *Plant Mol. Biol.*, **2004**, *54*, 221–232.

19 A. Bajguz, T. Asami, *Planta*, **2004**, *218*, 869–877.

20 Z. Y. Wang, T. Nakano, J. Gendron, J. X. He, M. Chen, D. Vafeados, Y. L. Yang, S. Fujioka, *et al.*, J. Chory, *Dev. Cell*, **2002**, *2*, 505–513.

21 Y. H. Yin, Z. Y. Wang, S. Mora-Garcia, J. M. Li, S. Yoshida, T. Asami, J. Chory, *Cell*, **2002**, *109*, 181–191.

22 Y. H. Yin, D. Vafeados, Y. Tao, S. Yoshida, T. Asami, J. Chory, *Cell*, **2005**, *120*, 249–259.

23 J. O. Borevitz, J. N. Maloof, J. Lutes, T. Dabi, J. L. Redfern, G. T. Trainer, J. D. Werener, T. Asami, *et al.*, *Genetics*, **2002**, *160*, 683–696.

24 J. N. Maloof, J. O. Borevitz, T. Dabi, J. Lutes, R. B. Nehring, J. L. Redfern, G. T. Trainer, J. M. Wilson, *et al.*, *Nature Genet.*, **2001**, *29*, 441–446.

25 H. Goda, Y. Shimada, T. Asami, S. Fujioka, S. Yoshida, *Plant Physiol.*, **2002**, *130*, 1319–1334.

26 S. H. Schwartz, B. C. Tan, D. A. Gage, J. A. Zeevaart, D. R. McCarty, *Science*, **1997**, *276*, 1872–1875.

27 T. Bennett, T. Sieberer, B. Willett, J. Booker, C. Luschnig, O. Leyser, *Curr. Biol.*, **2006**, *16*, 553–563.

28 S. Y. Han, N. Kitahata, T. Saito, M. Kobayashi, K. Shinozaki, S. Yoshida, T. Asami, *Bioorg. Med. Chem. Lett.*, **2004**, *14*, 3033–3036.

29 S. Y. Han, N. Kitahata, K. Sekimata, T. Saito, M. Kobayashi, K. Nakashima, K. Yamaguchi-Shinozaki, K. Shinozaki, *et al.*, *Plant Physiol.*, **2004**, *135*, 574–1582.

30 N. Kitahata, S. Y. Han, N. Noji, T. Saito, M. Kobayashi, T. Nakano, K. Kuchitsu, K. Shinozaki, *et al.*, *Bioorg. Med. Chem.*, **2006**, *14*, 5555–5561.

Keywords

Chemical Genetics, Brassinosteroids, Biosynthesis, Inhibitors, Abscisic Acid, Carotenoid Cleavage Dioxygenase, Nine-cis-carotenoid Dioxygenase, Chemical Biology

20
An Overview of Biopesticides and Transgenic Crops

Takashi Yamamoto, Jack Kiser

20.1
Introduction

The natural ecosystem maintains a delicate balance between pests and predators. Pest insects can be controlled by the artificial release of predators. One example is a parasitic wasp, *Diadegma insulare*. The adult female wasp lays eggs in a *Plutella xylostella* larva and pupates inside the cocoon of the mature larva. This and other insect predators are available commercially, but the usage is limited. Protozoa and nematodes are also used in insect pest control. One example of a protozoan that effectively infects locusts and controls the population is *Nosema locustae*. A commercially available nematode insect control agent is *Steinernema carpocapsae*. This nematode parasitizes scarab larvae with a symbiotic Photorhabdus bacterium that produces insecticidal toxins.

Insects are susceptible to microbial pathogens such as bacteria, viruses, and fungi [1]. All of these insect pathogens have been utilized as biopesticides. In the 1970's and 1980's, insect specific baculoviruses were widely used to control lepidopteran pests of various crops such as cotton, soybean, and leaf vegetables. Recently, the usage has declined significantly for several reasons including the high cost of producing the virus, its slow mode of action, and short field persistence. Fungi, especially *Beauveria bassiana*, are used in insect pest control. Fungal insecticides have a broader host spectrum than some other microbial pathogens. They are often sprayed as a biopesticide in situations where *Bacillus thuringiensis* is not effective, such as for ant and beetle control.

Among these different biopesticides, bacterial biopesticides are the most intensively studied and widely used. Several insect pathogenic bacteria are known to produce proteins toxic to certain insects. *Bacillus thuringiensis* (Bt) is the most well-known bacterium for its potent insecticidal proteins. These proteins are highly specific to certain orders of insects. Insects sensitive to Bt include those of Coleoptera, Diptera, and Lepidoptera. *Bacillus sphaericus* and *Clostridium bifermentans* are known for their mosquitocidal proteins. *Paenibacillus popilliae* produces a scarab active toxin structurally similar to common insecticidal proteins

Pesticide Chemistry. Crop Protection, Public Health, Environmental Safety
Edited by Hideo Ohkawa, Hisashi Miyagawa, and Philip W. Lee

ISBN: 978-3-527-31663-2

of Bt [2]. Toxic proteins called Tc were found in *Serratia entomophila* [3] and *Photorhabdus luminescens* [4]. These insecticidal toxins are similar to the toxin of *Yersinia pestis* [5].

20.2 *Bacillus thuringiensis*

Bacillus thuringiensis (Bt) is a rod-shaped, spore-forming, Gram-positive bacterium known for its production of a wide array of insecticidal toxins. A Japanese scientist first described the bacterium as a virulent pathogen of silkworm in 1901 [6]. In his report, he suggested that this bacterium produces an insecticidal toxin. In the 1950's, Edward Steinhaus at the University of California at Berkeley promoted commercial development of a spray-on insect control agent using Bt. Since then, Bt has been utilized as the most successful biopesticide. In part due to commercial interest, numerous Bt strains were isolated and characterized. As a result, several new commercially useful strains were discovered. These commercial strains include the high potency Lepidoptera-specific subsp. *kurstaki* (Btk) HD1 strain [7], mosquito-specific subsp. *israelensis* (Bti) strain [8], and beetle-specific subsp. *tenebrionis* (Btt) strain [9].

In order to manufacture Bt spray-on insecticide formulations, the bacterial culture is grown in industrial-scale fermentation tanks as large as 400,000 liters. Bt multiplies in an ordinary bacterial culture medium until it reaches a density of 10^9 cells per mL or higher. When the bacterium exhausts the nutrients in the medium, it enters the sporulation stage. During this stage, it produces a massive amount of insecticidal protein that crystallizes in a bipyramidal shape. Sometimes, other shapes such as cuboids can be seen. When the spores mature, the cells lyse, releasing free spores and crystals into the culture medium. The spore and crystal complex is then harvested and formulated into spray-on insecticides. Bt spray-on products have been used for many decades but remain niche products because of the specific characteristics of Bt. Currently the major use is on vegetables, fruit and nut trees, and grape vines. Bt has never been widely and consistently used on large acreage row crops like cotton and soybean.

20.3 Spray-On Bt Insecticide Formulations

Bt used in spray-on formulations is, perhaps, the most successful biopesticide. Reasons for Bt's success include ease of handling and a very high specific activity against sensitive insects. For example, a Bt insecticidal protein called Cry1Ac has a LC_{50} on *Heliothis virescens* neonate larvae as low as 0.07 ppm [10]. However, most individual Bt insecticidal proteins have a narrow activity spectrum. As mentioned, Cry1Ac is very active against *H. virescens*, while another Bt toxin called Cry1Ca is not. Cry1Ca is active against *Spodoptera exigua*, but Cry1Ac is not. This is perhaps one of the reasons why Bt spray-on products have not grown as much as it was expected. When insect pests that are not sensitive to Bt must be eradicated, applicators often choose other broad spectrum chemical treatments. The native Bt has evolved to overcome this limitation by acquiring multiple genes encoding proteins having different activity spectra. For example, the commercially superior HD1 strain contains as many as seven insecticidal protein genes.

Bt is a proteinaceous insecticide that must be ingested by insects to be effective. This is a fundamental difference from most of the small molecule chemical insecticides that generally have some contact activity. Since Bt creates a strong feeding inhibition, sensitive insects that ingest a sub-lethal dose of Bt toxins often recover from intoxication and cause damage to the crop. Bt sprayed on crops in the field is easily inactivated by sunlight. An additional cause of the short field persistence of Bt is moisture. Bt is very susceptible to loss in the field from rain. Because of its size and crystalline structure, it is washed off the plants even when formulated with a good sticking agent. Bt is sensitive to alkaline pH, and high pH occurring on leaves in the field will cause inactivation. On average, the half-life of Bt in the field is 2 to 3 days. The combination of these Bt characteristics is particularly problematic. Consumption of non-lethal doses of Bt by insects results in feeding inhibition, insects thus affected can remain alive for a few days without eating during which time the Bt becomes inactivated. Once Bt is inactivated or washed off, these insects recover and start eating the crop again.

20.4
Discovery of Multiple Toxins in One Bt Strain

Due to its economical importance, numerous Bt strains have been isolated. Scientists at the Pasteur Institute in France established a classification scheme of these Bt cultures by serotyping on flagellar antigen supplemented with biochemical tests [11]. One serotype, 3a3b of subsp. *kurstaki*, has been used in commercial formulations since 1970 because of its high potency and relatively wide activity spectrum. Among strains that belong to this serotype, Krywienczyk *et al.* [12] found two kinds of proteins within one strain, a major 135 kDa protein and a minor 70 kDa protein. When isolated by column chromatography, the 70 kDa protein was found to be an alkaline protein with unique dual mosquito and lepidopteran specificities [13]. The 135 kDa protein is acidic and has lepidopteran specificity only. Indeed, those strains producing this 70 kDa protein had the mosquitocidal activity in addition to the lepidopteran activity. The 70 kDa protein was later called Cry2 and the 135 kDa protein Cry1.

Cloning of the first gene encoding the 135 kDa Cry1 protein of a kurstaki strain was made in 1981 [14]. The cloned gene was called *cry1Aa* for its crystal-forming phenotype and the protein Cry1Aa. Two additional genes, *cry1Ab* and *cry1Ac*, were cloned from similar kurstaki strains. The gene encoding the 70 kDa Cry2 protein was cloned in 1988 [15]. The sequence of the Cry2 protein gene (*cry2Aa*) revealed significant amino acid sequence homology between Cry1 and Cry2 in the first 620 amino acid residues. Genes encoding other proteins such as *cry3Aa* of Coleoptera-specific subsp. *tenebrionis*, and *cry4*, *cry10*, and *cry11* of Diptera-specific subsp. *israelensis* were cloned and sequenced. When amino acid sequences of all of these cloned genes were compared, a significant level of homology was observed. Höft and Whitely [16] proposed to classify these genes into different *cry* types using Roman numerals based on insect specificity such as Lepidoptera-specific *cryI*, Lepidoptera/Diptera-specific *cryII*, Coleoptera-specific *cryIII*, and Diptera-specific *cryIV*. Since then, additional *cry* genes have been cloned including those with no known insecticidal activity and others encoding binary toxins called Cry34 and Cry35. Now, the classification is based on only the homology of amino acid sequences, and Arabic numerals are used instead of Roman. As of June 2006, over 330 *cry* genes have been reported and classified into 50 *cry* types. A Cry protein nomenclature committee has been established [17] and the latest record can be seen at the URL [18] maintained by the committee.

20.5 Mode of Action of Bt Insecticidal Proteins

When insects ingest Bt crystals, the crystals are solubilized and activated in the insect's gut fluid. The 135 kDa Cry1-type protein is a non-active protoxin. The C-terminal half and a small portion of about 30 amino acid (aa) residues of the *N*-terminus are digested by an insect gut protease similar to trypsin. This digestion process converts the protoxin into a fully active toxin. The resulting protein of about 65 kDa is substantially resistant to any further protease digestion. Bt also produces truncated proteins such as Cry2 and Cry3. Morse *et al.* [19] have shown that the undigested full-length Cry2Aa is already active.

The 3-D structures of several Bt Cry proteins including Cry1Aa, Cry2Aa and Cry3Aa have been determined by X-ray crystallography. While these Cry proteins are only about 60% homologous in the amino acid sequence, they have remarkably similar 3-D structures consisting of three distinctive domains called Domain I, II, and III. As shown in Figure 1, Domain I (the left portion of the molecule shown in the figure) is made of seven α-helices. Domain II (right bottom) and III (right top) are composed of repeating β-strands. There are three distinctive loops in

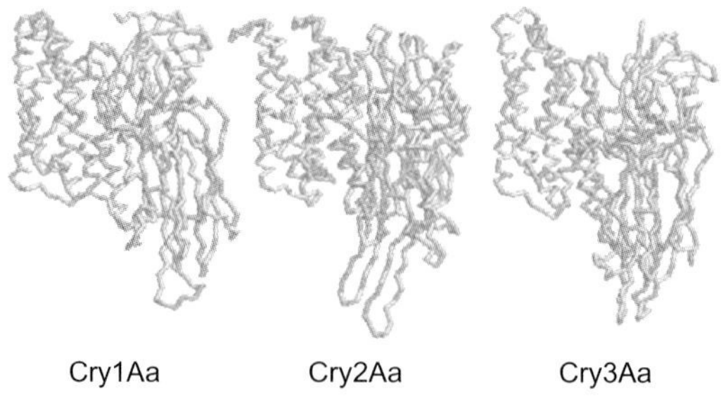

Figure 1. Backbone structure of three Bt Cry proteins.

Domain II. These loops are extruding into the solvent. The structure of Cry1Aa was determined from the protease-digested (or activated) form. The structures of the C-terminal half and the first 27 aa N-terminal residues of Cry1Aa remain unknown. The Cry2Aa structure was determined from the full length, undigested protein. The Cry2Aa structure revealed additional α-helices on its *N*-terminus.

Numerous studies (see the review by Schenpf *et al.* [20] and its citations) have indicated that the three loops in Domain II are involved in receptor binding. Bt toxin receptors found on the insect gut epithelium cells include a cadherin-like structural protein and a glycoprotein having aminopeptidase activity. After binding to the receptor, Bt toxin polymerizes and the Domain I portion of the protein penetrates into the cell membrane forming an ion channel. Once the channel forms, osmotic pressure regulation is lost by the cell resulting in cell death. As cell death occurs, the insect gut becomes porous allowing gut juice containing Bt spores to enter into the bloodstream. Eventually, Bt spores germinate in the insect body cavity and multiply.

20.6 Transgenic Bt-Crops

Many of the inherent limitations of Bt as a sprayable application, make it well suited to use as a plant-incorporated biopesticide. The active ingredient is a protein, which makes it amenable to expression in a plant through a single gene insertion. The highly specific activity of Bt proteins means they have no effect on other non-target organisms, including livestock or humans that eat plant products expressing the protein. In addition, expressing a Bt insecticidal protein in the plant where it is protected from environmental inactivation eliminates most of the limitations associated with spray-on Bt formulations. The predominance of Bt proteins as the only transgenic bioinsecticides in a decade of commercialization perhaps speaks to the rarity in nature of materials as well-suited for plant-incorporated insecticides. The first commercially significant transgenic Bt-crop was cotton [21] followed by corn [22]. Within a few years, cultivation of Bt-crops had grown substantially. According to a USDA survey for 2005 [23], 50% of the US cotton acreage are Bt-cotton and 30% of corn are Bt-corn. Cultivation of Bt-crops has reduced the use of chemical insecticides. For example, a study done in India on Bt-cotton in 2002 and 2003 showed reduction of chemical insecticides by 60 to 70% in comparison with non Bt-cotton [24]. This study found that yield of Bt-cotton not treated with insecticides was significantly higher than that of non-recombinant cotton grown with a traditional insecticide application regime. Since Bt insecticidal proteins expressed in Bt-crops are highly specific to target insects, beneficial insects, aquatic animals, birds, and other higher animals are not affected. On the other hand, most chemical insecticides are relatively non-selective. Another example is the report [25] of a study in the US summarizing results across the cotton-growing region with 59 locations compared. Insecticide use was reduced from 2.81 applications on non-Bt cotton to 1.69 applications on the most advanced Bt-cotton product.

After initial success in Bt-cotton and Bt-corn in the 1990's, the use of Bt insecticidal genes in transgenic crops has expanded. Besides *cry1Ac* and *cry1Ab* used in cotton and corn, other 135-kDa protein genes have been expressed in plants. For example, the *cry8Da* gene was cloned in turf grass lines and showed impressive control of scarab larvae in soil [26]. Cry2/3-type truncated proteins were expressed in cotton (Cry2Ab) [27] and potato (Cry3Aa) [28] to control the *Heliothis/Helicoverpa* complex in cotton and *Leptinotarsa disseminate* in potato. Some Bt cultures produce not only crystalline endotoxins but also exotoxins that are secreted into the culture medium during the vegetative growth stage. One of those exotoxins, called VIP3Aa (Vegetatively produced Insecticidal Protein) is being studied for use in transgenic crops [29].

20.7
Selection of Bt Genes for Transgenic Cotton

The first major commercial application of Bt in transgenic crops was cotton. At the time Bt cotton was developed, three genes, *cry1Aa, cry1Ab,* and *cry1Ac,* were available. Among those, *Cry1Ac* had the highest activity against the major cotton pests, *Heliothis virescens* and *Helicoverpa zea.* This is, perhaps, the major reason why this *cry1Ac* gene was selected for cotton. The Cry1Ac protein is also active against *Pectinophora gossypiella,* another important cotton pest. When amino acid sequences for these genes were compared, it suggested that the three genes were produced by swapping domains. All three proteins share a highly homologous Domain I sequence. Domain II sequences of Cry1Ab and Cry1Ac are very similar while this domain of Cry1Aa is rather unique. Cry1Aa and Cry1Ab have highly homologous Domain III sequences. Domain III of Cry1Ac has no significant homology with those of Cry1Aa and Cry1Ab. Indeed, Domain III of Cry1Ac is so unique among all known Bt Cry proteins, that only Cry1Bd has a similar Domain III [30]. It is generally understood that Domain II of all three proteins recognizes one receptor while the same domain of Cry1Ab and Cry1Ac binds to an additional receptor on *H. virescens.* This probably contributes to the higher activity of Cry1Ab and Cry1Ac against this insect than Cry1Aa. In addition, it has been shown that Domain III of Cry1Ac finds the third receptor possibly contributing even higher activity than Cry1Aa and Cry1Ab [31].

Recently, the *cry2Ab* gene was introduced to cotton. This gene was discovered along with Cry2Aa in a strain of subsp. *kurstaki.* Unlike Cry2Aa, Cry2Ab has no mosquitocidal activity but is active against the *Heliothis/Helicoverpa* complex, especially *H. zea,* which is less sensitive to the previously commercialized Cry1Ac in cotton. Since Cry2Ab recognizes a receptor that differs from those of Cry1A's, it is believed to be good for minimizing the development of insect resistance to transgenic Bt-cotton when paired with the Cry1Ac in a commercial product.

20.8
Corn Insect Pests and Bt Genes

Corn has a number of important insect pests and several Bt genes are used to control these insects. The first Bt-corn product was to control *Ostrinia nubilalis*, and the gene used in the product was *cry1Ab*. Although the Cry1Ab protein controls this insect very well, other proteins, Cry1Ac for example, appear to be as good against the same insect. However, Cry1Ab has somewhat better activity against other corn pests such as *Spodoptera frugiperda*, against which Cry1Ac has little activity. Recently, another gene, *cry1Fa*, whose protein has a better activity spectrum for corn pests than that of Cry1Ab, has become available in corn for growers [32].

The genus *Diabrotica* contains the other major corn pests. Cry3Bb has been found to be active against these pests but its activity is not high. The activity of this protein has been improved by protein engineering, and the gene of the modified Cry3Bb protein has been commercialized for *Diabrotica* control in corn [33]. Similarly, a modified *cry3Aa* has been used [34]. A new class of toxins that is active against *Diabrotica* was found [35]. These proteins, Cry34 and Cry35, that work together as a unit like many other bacterial binary toxins have been commercialized in corn.

20.9
Potential Issues of Bt-Crops

Concerns have been raised regarding the use of recombinant DNA technology to express Bt insecticidal genes in crops. These concerns are for the potential environmental impact of releasing these genes into the environment and for the potential development of resistance by insects to the genes and subsequent loss of Bt spray-on formulations to the vegetable industry and organic growers. Issues regarding the environmental release of the genes include the transfer of the genes to weedy species making the weeds more competitive and the impact of the Bt protein on non-target insect species or on the soil ecosystem by affecting residue turnover or organisms involved in decomposition of organic matter. Scientists in academia and the industry have conducted extensive studies working with the USEPA to address these issues. An assessment published by the EPA [36] has concluded that it is not likely that cross-pollination occurs between the crops that currently have been engineered to contain Bt genes and weedy relatives, that Bt is highly specific to target insects and promotes less use of wide spectrum chemical insecticides, and that Bt decomposes in the soil rapidly. Native Bt cultures exist ubiquitously in different environments without causing adverse effects. Bt spray-on formulations have been used for over three decades worldwide in agriculture, forestry, and vector control without environmental and human health issues. As it is reviewed in the following sections, insect resistance against Bt is a rare occurrence in the field. However, the EPA and the industry have developed monitoring and prevention programs.

20.10
Insect Resistance to Bt

In 1985, McGaughey [37] reported that *Plodia interpunctella* became resistant to Bt when reared for several generations in the laboratory with a sublethal dose of a Bt commercial product, presumably *Bt kurstaki* HD1. When Van Rie *et al.* [38] examined Bt-resistant *P. interpunctella*, they found that the receptor sites on BBMV for Cry1Ab were greatly reduced. The binding site for the Cry1Ca protein, however, remained in the resistant colony. In 1990, Tabashnik *et al.* [39] found a Bt resistant *Plutella xylostella* in the field where a Bt spray formulation was heavily used. A more recent example with relevance to transgenic cotton is *Helicoverpa armigera* in Australia [40].

Since 1996, a steady growth of cultivation of transgenic crops containing insecticidal Bt genes, especially Bt-cotton and Bt-corn, has occurred. Nevertheless, it appears that only limited cases of Bt-resistant insects have been found in the open field. There are several possible reasons for the rarity of resistance development. In most cases, if not all, resistance management programs have been implemented where Bt-crops are grown. No significant data is available for the frequency of resistance occurrence in the field against transgenic crops without resistance management programs. Without such data, it is difficult to conclude whether the management programs are effective in delaying resistance development.

20.11
Resistance Mechanism

As mentioned above, the Bt-resistant *Plodia interpunctella* developed by McGaughey lost its receptor to Cry1Ab. Ferré *et al.* [41] reported that a Bt-resistant *Plutella xylostella* colony showed no binding to Cry1Ab but retained the binding sites for Cry1Ba and Cry1Ca. In these two Bt-resistant insects, the binding site for Cry1Ab differed from the site for Cry1Ca and possibly Cry1Ba also. In another paper published by Ferré and Van Rie [42], they proposed a Bt receptor scheme for *P. xylostella* (Figure 2). In this scheme, Cry1Ab shares the same binding site with Cry1Ac, Cry1Fa and Cry1Ja, while Cry1Ba, Cry1Ca and Cry1Aa have their own three different binding sites. In order to overcome resistance in *P. xylostella* to spray-on formulations based on subsp. *kurstaki*, whose major components are Cry1Ab and Cry1Ac, a new formulation (EPA Reg No. 73049-40) was developed based on a subsp. *aizawai* strain producing Cry1Ca in addition to Cry1Ab.

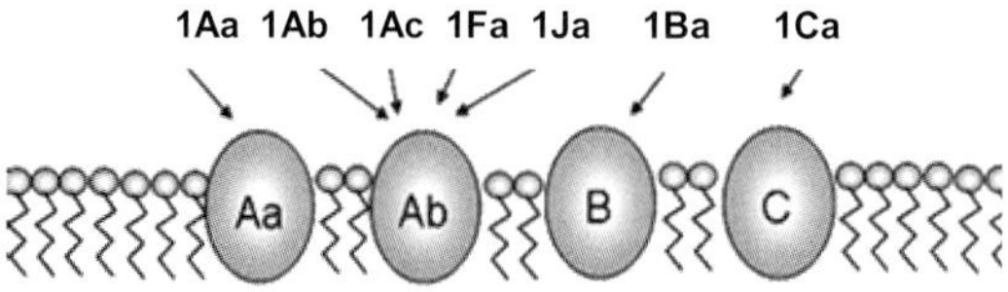

Figure 2. *Pxylostella* receptor model.

Oppert *et al.* reported another resistance mechanism [43]. A lower level of protease activity was found in midgut extracts of a Cry1Ac-resistant colony of *P. xylostella*, 198^{r}. Further study [44] showed a significant level of loss in sensitivity in the same resistant insect colony to the undigested, full-length Cry1Ab protein. The colony was still sensitive to the activated toxin that was produced *in vitro* by protease digestion. This result indicates that the 198^{r} colony has acquired resistance by blocking the activation process of the Cry1Ab protoxin.

20.12 Resistance Management Program for Bt Transgenic Crops

Due to the high effectiveness of the Bt transgenic crop technology and the fear of losing the technology to insect resistance, a combined resistance management strategy has been mandated by the US EPA as a condition of registration of these products. The strategy has several components: (i) Field monitoring to find early signs of resistance development; (ii) Use of a highly active gene. This "high-dose strategy" delivers a dose that kills 99+% of the insect population. Since few insects can survive, there is less chance of developing a resistant population. An example of a high dose is Cry1Ac in cotton for *Heliothis virescens*; (iii) Maintenance of a relatively large sensitive population by planting a non-Bt crop growing area as refuge. This maintains an insect population that is not exposed to the selection pressure. The unexposed population provides a large number of sensitive individuals to mate with any heterozygous-resistant individuals that develop. The heterozygous individuals are still controlled by a high dose of the toxin thus preventing the homozygous-resistant population from developing. This practice has been widely implemented.

Another approach believed to be effective in resistance management is combining two or more proteins having different modes of action. In this case, it is more difficult for insects to develop resistance as the insects need to react to two or more different toxins. An example is cotton containing the *cry1Ac* and *cry2Ab* genes.

Resistance management programs are still evolving and will most likely change as greater information is obtained. A critical requirement of a resistance management strategy is that it be economically feasible for the participants, the product providers and the growers to implement. There is a large effort undertaking by product providers to find additional sources of plant incorporated insecticides to extend the life of their products.

20.13 Conclusion

We believe that the use of biopesticides in transgenic crop applications will steadily increase. However, there are technical and political hurdles to be crossed. In the technical area, we need new potent genes that can be used for high-dose applications against not only one primary pest but also against one or more secondary pests. We also need new potent genes to control insects that are relatively insensitive to existing Bt proteins. No Bt insecticidal proteins significantly active against sucking insects such as aphids and plant bugs have been reported. The discovery of new potent genes with new modes of action different from the existing genes is important for delaying the development of resistance by target insects to the genes in current products.

An important aspect of transgenic crop technology is site-specific gene expression. If a Bt gene is expressed just in the tissues of the plant where insects attack, this is more desirable than global expression. Expressing a corn rootworm-active gene only in corn roots minimizes unnecessary exposure of the toxin to non-target insects. Although Bt is highly specific to target insects, this approach lends itself to metabolic efficiency and energy utilization as protein synthesis occurs only in tissues where it is required.

Another important aspect for the future of biopesticides is public acceptance. The use of Bt transgenic crop technology is growing throughout the world because of its value to agriculture and the environment, but regulatory and public relation hurdles to the commercialization of Bt products remain, particularly in Europe and Japan. Besides additional research on environmental and consumer safety issues, public education is needed. We are hoping that this presentation will make a small contribution towards this goal.

20.14
References

1 Y. Tanada, H. Kaya, *Insect Pathology*, Academic Press, San Diego, CA, USA, **1992**.

2 J. Zhang, T. C. Hodgman, L. Krieger, W. Schnetter, H. U. Schairer, *J. Bacteriol.*, **1997**, *179*, 4336–4341.

3 M. R. Hurst, T. R. Glare, T. A. Jackson, C. W. Ronson, *J. Bacteriol.*, **2000**, *182*, 5127–5138.

4 M. Blackburn, E. Golubeva, D. Bowen, R. H. ffrench-Constant, *Appl. Environ. Microbiol.*, **1998**, *64*, 3036–3041.

5 N. R. Waterfield, D. J. Bowen, J. D. Fetherston, P. D. Perry, R. H. ffrench-Constant, *Trends Microbiol.*, **2001**, *9*, 185–191.

6 S. Ishiwata, *Dainihon Sanshi Kaiho* (in Japanese), **1901**, *114*, 1–5.

7 H. T. Dulmage, *J. Invertebr. Pathol.*, **1970**, *15*, 232–239.

8 L. J. Goldberg, J. Margalit, *Mosq. News*, **1977**, *37*, 355–358.

9 V. A. Krieg, A. M. Huger, G. A. Langenbruch, W. Schnetter, *Z. Angew. Ent.*, **1983**, *96*, 500–508.

10 R. G. Luttrell, L. Wan, K. Nighten, *J. Econ. Entomol.*, **1999**, *92*, 21–32.

11 H. de Barjac, A. Bonnefoi, *C. R. Acad. Sci. (Paris) Ser. D*, **1967**, *264*, 1811–1813.

12 J. Krywienczyk, H. T. Dulmage, P. G. Fast, *J. Invertebr. Pathol.*, **1978**, *31*, 372–375.

13 T. Yamamoto, R. E. McLaughlin, *Biochem. Biophys. Res. Commun.*, **1981**, *103*, 414–421.

14 H. E. Schnepf, H. R. Whiteley, *Proc. Natl. Acad. Sci. USA*, **1981**, *78*, 2893–2897.

15 W. P. Donovan, C. C. Dankocsik, M. P. Gilbert, M. C. Gawron-Burke, R. G. Groat, B. C. Carlton, *J. Biol. Chem.*, **1988**, *263*, 561–567.

16 H. Höfte, H. R. Whiteley, *Microbiol. Rev.*, **1989**, *53*, 242–55.

17 N. Crickmore, D. R. Zeigler, J. Feitelson, E. Schnepf, J. Van Rie, D. Lereclus, J. Baum, D. H. Dean, *Microbiol. Mol. Biol. Rev.*, **1998**, *62*, 807–13.
18 http://www.lifesci.sussex.ac.uk/home/Neil_Crickmore/Bt/index.html.
19 R. J. Morse, T. Yamamoto, R. M. Stroud, *Structure*, **2001**, *9*, 409–417.
20 E. Schnepf, N. Crickmore, J. Van Rie, D. Lereclus, J. Baum, J. Feitelson, D. R. Zeigler, D. H. Dean, *Microbiol. Mol. Biol. Rev.*, **1998**, *62*, 775–806.
21 F. J. Perlak, R. W. Deaton, T. A. Armstrong, R. L. Fuchs, S. R. Sims, J. T. Greenplate, D. A. Fischhoff, *Biotechnology*, **1990**, *8*, 939–943.
22 J. J. Estruch, N. B. Carozzi, N. Desai, G. W. Warren, N. B. Duck, M. G. Koziel, *Insect Resistant Maize: Recent Advances and Utilization* (conference proceedings), **1994**, 172–174.
23 http://www.ers.usda.gov/Data/BiotechCrops.
24 R. M. Bennett, Y. Ismael, U. Kambhampati, S. Morse, *AgBioForum*, **2004**, *7*, 96–100.
25 W. Mullins, D. Pitts, in *Beltwide Cotton Conferences*, National Cotton Council, **2005**, 1822–1824.
26 S. Asano, H. Bando, M. Horita, H. Sekiguchi, T. Iizuka, 6th Pacific Rim Conference on the Biotechnology of *Bacillus thuringiensis* and Its Environmental Impact, **2005**.
27 US EPA, *Federal Register*, **2002**, *67*, 11973–11974.
28 US EPA, *Federal Register*, **1998**, *63*, 38805–38806.
29 US EPA, *Federal Register*, **2004**, *69*, 55605–55608.
30 N. Crickmore, in *Entomopathogenic Bacteria: From Laboratory to Field Application*, Kluwer Academic Publishers, The Netherlands, **2000**, 65–79.
31 J. Van Rie, S. Jansens, H. Höfte, D. Degheele, H. Van Mellaert, *Eur. J. Biochem.*, **1989**, *186*, 239–247.
32 US EPA, *Federal Register*, **2001**, *66*, 42220–42221.
33 T. Vaughn, T. Cavato, G. Brar, T. Coombe, T. DeGooyer, S. Ford, M. Groth, A. Howe, *et al.*, *Crop Science*, **2005**, *45*, 931–938.
34 US EPA, *Federal Register*, **2004**, *69*, 62678–62680.
35 R. T. Ellis, B. A. Stockhoff, L. Stamp, H. E. Schnepf, G. E. Schwab, M. Knuth, J. Russell, G. A. Cardineau, *et al.*, *Appl. Environ. Microbiol.*, **2002**, *68*, 1137–1145.
36 US EPA, Biopesticides Registration Action Document, **2001**, 127 pp.
37 W. H. McGaughey, *Science*, **1985**, *229*, 193–195.
38 J. Van Rie, S. Jansens, H. Höfte, D. Degheele, H. Van Mellaert, *Appl. Environ. Microbiol.*, **1990**, *56*, 1378–1385.
39 B. E. Tabashnik, N. L. Cushing, N. Finson, M. W. Johnson, *J. Econ. Entomol.*, **1990**, *83*, 1671–1676.
40 R. V. Gunning, H. T. Dang, Fr. C. Kemp, I. C. Nicholson, G. D. Moores, *Appl. Environ. Microbiol.*, **2005**, *71*, 2558–2563.
41 J. Ferré, M. D. Real, J. Van Rie, S. Jansens, M. Peferoen, *Proc. Natl. Acad. Sci. USA*, **1991**, *88*, 5119–5123.
42 J. Ferré, J. Van Rie, *Ann. Rev. Entomol.*, **2002**, *47*, 501–533.
43 B. Oppert, K. J. Kramer, D. Johnson, S. J. Upton, W. H. McGaughey, *Insect. Biochem. Mol. Biol.*, **1996**, *26*, 571–583.
44 S. Herrero, B. Oppert, J. Ferré, *Appl. Environ. Microbiol.*, **2001**, *67*, 1085–1089.

Keywords

Biopesticide, *Bacillus thuringiensis*, Transgenic Crop, Resistance Management

21
Essential Oil-Based Pesticides: New Insights from Old Chemistry

Murray B. Isman, Cristina M. Machial, Saber Miresmailli, Luke D. Bainard

21.1
Introduction

To defend themselves against herbivores and pathogens, plants naturally release a variety of volatiles including various alcohols, terpenes, and aromatic compounds. These volatiles can deter insects or other herbivores from feeding, can have direct toxic effects, or they may be involved in recruiting predators and parasitoids in response to feeding damage. They may also be used by the plants to attract pollinators, protect plants from disease, or they may be involved in interplant communication [1–2]. Based on these natural properties, essential oils containing these compounds have been recently touted as potential alternatives to current commercially available insecticides.

In reality, pesticides of botanical origin have been used for centuries to protect crops and stored products and to repel pests from human habitations. Among the most well known are pyrethrum, neem, rotenone, nicotine and plant essential oils, although more than 2,000 plant species have been found to possess insecticidal activity [3–4]. However, while most botanical pesticides are known solely for their insecticidal activity, plant essential oils are also known for their uses as fragrances, flavorings, condiments or spices, and many are also considered to have medicinal uses. Given this widespread use, numerous plant essential oils are already widely available and their chemistry is generally well-understood.

21.2
Essential Oil Composition

Typically consisting of highly complex mixtures of mono- (C_{10}), sesquiterpenes (C_{15}), 49 d phenols that confer the scent of the plant from which they are derived, plant essential oils are obtained through steam distillation of plant material from a relatively select group of plants [5]. As a result, most essential oils come from highly aromatic species such as those in the Apiaceae (carrot), Lamiaceae (mint),

Pesticide Chemistry. Crop Protection, Public Health, Environmental Safety
Edited by Hideo Ohkawa, Hisashi Miyagawa, and Philip W. Lee

ISBN: 978-3-527-31663-2

Table I. Examples of essential oils, their major constituents and insects or mites which are known to be affected by these oils.

Essential Oil	Major Constituents	Insects/Mites Affected	Reference
Basil *Ocimum* spp.	eugenol	*Callosobruchus maculatus* *Rhyzopertha dominica* *Sitophilus zeamais* *Sitotroga cerealella*	[7, 9]
Clove *Syzygium* spp.	eugenol	*Sitophilus zeamais* *Tribolium castaneum*	[10]
Eucalyptus *Eucalyptus* spp.	eucalyptol	*Sitophilus oryzae*	[11]
Lavender *Lavendula* spp.	linalool	*Acanthoscelides obtectus* *Cydia pomonella*	[12–13]
Mint *Mentha* spp.	menthol, pulegone	*Tetranychus urticae*	[14]
Rosemary *Rosmarinus officinalis*	1,8-cineole, camphor, α-pinene	*Sitophilus oryzae* *Tetranychus urticae*	[11, 15–16]
Thyme *Thymus, Thymbra* spp.	thymol, carvacrol	*Plutella xylostella* *Pseudaletia unipuncta*	[17]

Myrtaceae (myrtle), and Rutaceae (citrus) plant families. Table I provides examples of a few of the better known essential oils, the plants from which they are derived, and the major constituents found in each of these oils. It is important to note that the composition of these oils can vary dramatically, even within species. Factors impacting the composition include the part of the plant from which the oil is extracted (i.e., leaf tissue, fruits, stem, etc.), the phenological state of the plant, the season, the climate, the soil type, and other factors. As an example, rosemary oil collected from plants in two areas of Italy were demonstrated to vary widely in the concentrations of two major constituents, 1,8-cineole (7% to 55%) and α-pinene (11% to 30%) [6]. Such variation is not uncommon and has also been described for the oils derived from *Ocimum basilicum* [7] and *Myrtus communis* [8].

21.3 Biological Activities of Essential Oils

Essential oils demonstrate a wide range of bioactivities from direct toxicity to insects, microorganisms and plants, to oviposition and feeding deterrence as well as repellence and attraction. How these effects are mediated is still being

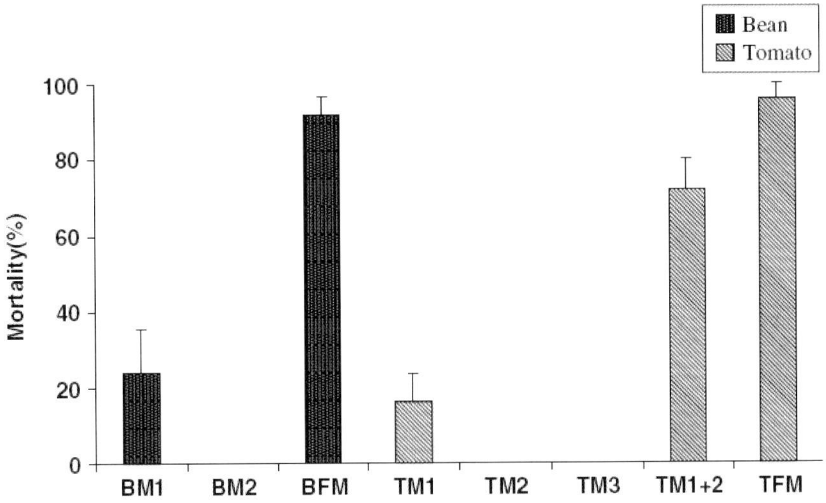

Figure 1. Mortality caused by selected blends of active and inactive constituents of rosemary oil to *Tetranychus urticae* when applied at levels equivalent to those found in the 100% lethal concentration of the pure oil (LC_{100} = 20 mL $litre^{-1}$ for *T. urticae* on beans and 40 mL $litre^{-1}$ on tomato). Error bars represent the standard error of the mean of five replicates. **BM1** ('actives') = α-pinene + 1,8-cineole + α-terpineol + bornyl acetate + *p*-cymene; **BM2** ('inactives') = β-pinene + borneol + camphor + camphene + D-limonene; **BFM** = full mixture of all constituents; **TM1** ('very active') = α-pinene + 1,8-cineole + α-terpineol + bornyl acetate; **TM2** ('moderately active') = β-pinene + *p*-cymene + borneol; **TM3** ('inactive') = camphor + camphene + D-limonene; **TM1 + 2** = TM1 + TM2; **TFM** = full mixture of all constituents. Adapted from [16].

elucidated; however, there is a growing set of results which point to membrane disruption (plants, microbes, and possibly insects) and effects on the nervous system of insects. And while individual constituents can mediate some of the effects, it is evident that complete essential oils are more effective than individual constituents or even a combination of constituents (see Figure 1).

21.3.1
Insecticidal/Deterrent Effects

Numerous studies have assessed the ability of plant essential oils and their constituents to protect plants and/or their crops from insect pests with much of this research focusing on controlling stored product pests using essential oils as fumigants and repellents. Essential oils from a wide range of plants have been tested against the rice weevil (*S. oryzae*), the maize weevil (*S. zeamais*), the red-flour beetle (*T. castaneum*), the bean weevil (*A. obtectus*), and other stored product pests (see Table I for some examples). In particular, nutmeg oil has been determined to significantly impact both *S. zeamais* and *T. castaneum* and demonstrates both repellent and fumigant properties (concentration dependent) [18]. Other essential oils with bioactivity against stored product pests include oils of basil, citrus peel oil, eucalyptus oil, oils from various mint species, lavender oil, and rosemary oil, although not all essential oils are active against all the insect pests [12, 19–20]. There is also growing research documenting the effects of these oils as contact toxicants, antifeedant compounds and repellents against a range of other insects.

Three human pests, the human louse *Pediculus humanus*, the tick *Ixodes ricinus*, and the yellow fever mosquito *Aedes aegypti* have been the target of several studies assessing the toxic and repellent effects of essential oils. In one study, the monoterpenoid (+)-terpinen-4-ol was found to be the most toxic against adult lice, followed by pulegone, while nerolidol and thymol were the most active against *P. humanus* eggs [21]. Other research with the essential oil from the carnation *Dianthus caryophyllum* found that a 10% solution in ethanol was more repellent than a 10% solution of DEET in ethanol against *I. ricinus* at 4 hours post application (100% vs. 83%) and only slightly less repellent at 8 hours (92% vs. 95%). The same oil also demonstrated 95% repellency against *A. aegypti*, although repellency dropped to 79% after 8 hours. In contrast to DEET though, the carnation oil is not known to irritate skin or eyes, etc., suggesting that carnation essential oils might be suitable alternatives to DEET [22].

In research conducted with the two-spotted spider mite *Tetranychus urticae*, rosemary oil has been demonstrated to have contact toxicity while the oils of caraway seed, citronella java, lemon eucalyptus, pennyroyal, and peppermint all exhibit fumigant activity [14–15]. Perhaps more important, however, is that commercial formulations tested against both *T. urticae* and the predatory mite *Phytoseiulus persimilis* demonstrated high levels of toxicity to *T. urticae* but not *P. persimilis*, suggesting that these commercial formulations may work well in conjunction with an integrated pest management program (see Figure 2).

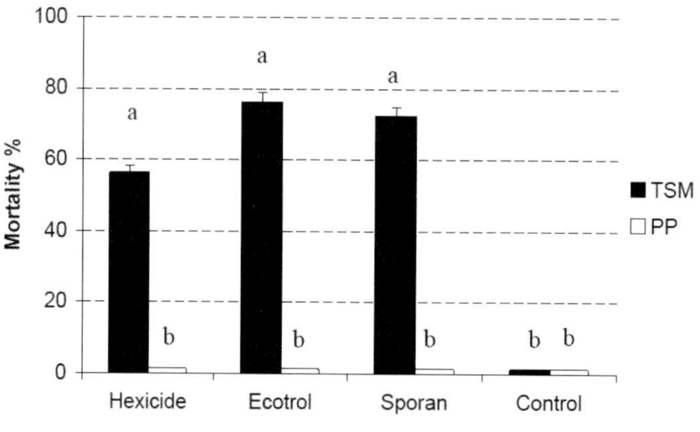

Figure 2. Efficacy (% mortality) of three commercial rosemary oil-based pesticides directly sprayed on *P. persimilis* (PP) and *T. urticae* (TSM) on tomato plants. TSM = two-spotted spider mite, PP = *P. persimilis*. Bars representing means (± SE), *n* = 5 replicates with 5 adult mites per replicate. Bars marked with the same letter do not differ significantly, Tukey. Adapted from [15].

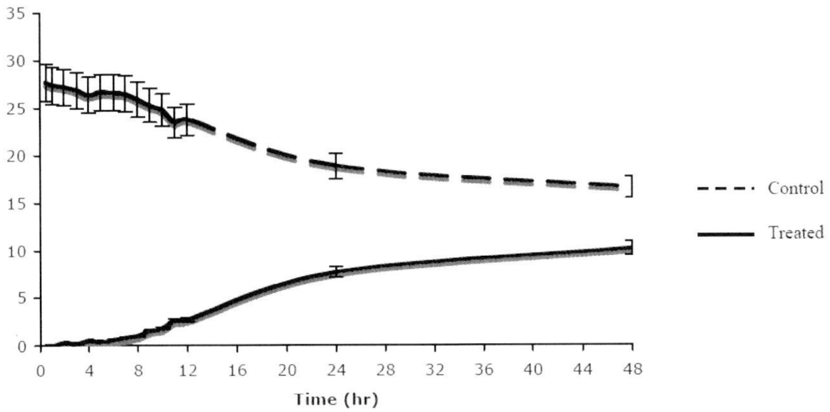

Figure 3. Number of two-spotted spider mites staying on leaf discs when given a choice between leaf discs treated with rosemary oil at 1% or untreated discs. Error bars representing standard error, $n = 10$ replicates with 30 adult spider mites per replicate. Adapted from [15].

Additional research from our laboratory shows that several essential oils possess antifeedant and oviposition deterrent effects. In particular, thymol, the major constituent of thyme oil, has been shown to be a deterrent to the lepidopteran pest species *Plutella xylostella* and *Pseudaletia unipuncta*. It should be noted that larvae experienced with thymol showed reduced deterrence, suggesting that repeated application of feeding deterrent chemicals in the field could limit their effectiveness [17]. Also, as Figure 3 shows, while some essential oils may possess deterrent properties, this deterrence is reduced over time, possibly due to habituation, the volatilization of the essential oils, or a combination of both.

While the mode of action of essential oils is still relatively unknown, new research is providing insights. As previously mentioned, essential oils are likely neurotoxic to insects and mites, and research using individual constituents seems to suggest this. Thus far, evidence has been provided suggesting that some constituents such as eugenol or thymol may work by blocking octopamine (a neurotransmitter in arthropods) receptors and/or by potentially working through the tyramine receptor cascades [23–24]. Physical effects such as membrane disruption or blockage of the tracheal systems may also be involved; however, conclusive evidence is still lacking.

21.3.2 Herbicidal Activity

Some essential oils are not only insecticidal but also possess strong phytotoxic effects. In many cases, this would be considered a serious drawback to the use of these essential oils for insect pest control; however, this also opens the door to the use of some oils as herbicides. Although few studies have addressed this herbicidal activity, work completed by Tworkoski [25] demonstrated that the

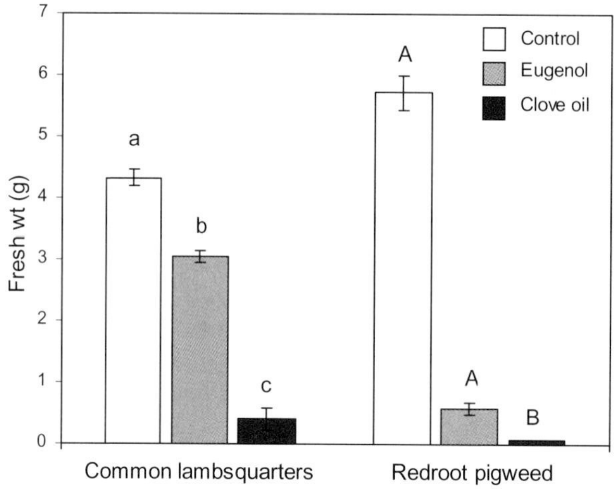

Figure 4. Effect of clove oil (2.5%) and eugenol (1.5%) on the fresh weight of common lambsquarters and redroot pigweed seedlings 14 days after treatment. Values are means ± SE of two experiments with six replicates per experiment. Bars with different letters for common lambsquarters (lower case) and for redroot pigweed (upper case) indicate significant differences at P = 0.05. Adapted from [26].

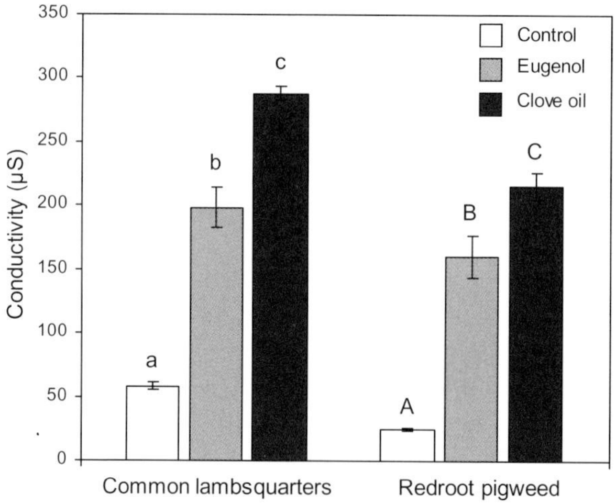

Figure 5. Effect of clove oil (2.5%) and eugenol (1.5%) on leakage of electrolytes from leaf discs of common lambsquarters and redroot pigweed. Values are means ± SE of two experiments with six replicates per experiment. Bars with different letters for common lambsquarters (lower case) and for redroot pigweed (upper case) indicate significant differences at P = 0.05. Adapted from [26].

essential oils of red thyme (*Thymus vulgaris*), summer savory (*Satureja hortensis*), cinnamon (*Cinnamomum zeylanicum*) and clove (*Syzyium aromaticum*) are highly phytotoxic. Further analysis of the major constituents of cinnamon oil found that the herbicidal activity was due to eugenol, which is also the major constituent in clove oil.

The herbicidal activity of clove oil and eugenol was further studied in our laboratory using broccoli, common lambsquarters, and redroot pigweed seedlings in an attempt to determine the role of leaf epicuticular wax in susceptibility of these plants to damage. Seedling growth was significantly inhibited by both clove oil and, to a lesser extent, eugenol, while those plants with more epicuticular wax (e.g., broccoli or common lambsquarters) showed reduced electrolyte leakage (an indicator of cell membrane damage) (see Figures 4 and 5) [26].

Besides the direct use as herbicides, one specific use of phytotoxic essential oils could be for chemical thinning of fruit trees. In initial trials with a commercially available clove oil-based herbicide (Matran, produced by EcoSMART Technologies Inc), apple blossom thinning effects were observed; however, extensive leaf and fruit russeting was also observed, with effects dependent on concentration and the apple cultivar tested [27]. While the initial results are promising, further studies will be required before such herbicides can be used as chemical thinning agents.

21.3.3 Antimicrobial Activity

The use of plant-derived compounds to treat infectious diseases or to protect crops dates back several centuries and essential oils are no exception [4, 28]. The essential oils from Ceylon cinnamon, rosemary, thyme, and willow have all been described as possessing activity against a wide range of microbes (e.g., bacteria, fungi, and viruses), and have been suggested to work specifically though membrane disruption [28]. In one recent study assessing the effects of three essential oils against 13 bacterial strains and 6 fungal species, oregano (*Origanum vulgare*) essential oil was effective in reducing bacterial growth by up to 60% using a 20% solution, and was even effective in reducing bacterial growth in penicillin-resistant strains by up to 50% [29]. In another study assessing the ability of clove oil to protect chicken frankfurters from *Listeria monocytogenes*, both 1% and 2% solutions of clove oil were effective in inhibiting growth, with the only major concern being the effect of the clove oil on flavor [30]. The essential oils of oregano and thyme were also found to be active against *Phytophthora infestans*, the fungus causing late blight in tomato and potato crops world wide, while the essential oils from rosemary, lavender, fennel, and laurel also all showed reduced bioactivity [31].

21.4
Challenges and Future Opportunities

Regardless of the growing number of scientific studies that have been published demonstrating the pesticidal and repellent properties of essential oils, very few products have been commercialized. There are likely several reasons for this including availability of sufficient quantities of plant material to produce the pesticides, the standardization and refinement of pesticide products, and regulatory approval and patents (protection of technology) [5]. While some essential oils are currently available in large quantities (for aromatherapy, food flavoring or other uses), essential oils from other plants, particularly rare plants, may be difficult to obtain in sufficient quantities. Accordingly, commercial essential oil-based pesticides may be restricted to those which are readily available or easily cultivated [32]. In addition, as was stated above, the composition of essential oils within species can vary drastically requiring some sort of standard to be established to ensure that active constituents are present at minimum levels and that efficacy is maintained. However, the largest challenge that must be overcome is the regulatory challenge. While some essential oils are EPA-exempt in the United States due to their extensive use as food additives, etc., not all essential oils are included, and at this point, only the United States and Mexico recognize any exemptions, limiting their use to these two countries at this point. Still, should these challenges be overcome, essential oils could be of particular use, especially for high value crops. Because these pesticides are naturally derived, they can also be used in organic production, providing new pest control options for these growers. Finally, owing to their low persistence and relative safety to mammals, essential oil-based pesticides or repellents may become an alternative to more toxic chemicals.

21.5
Conclusion

While public pressure is increasing, the switch to environmentally safe pesticide alternatives such as essential oil-based pesticides, commercial success of essential oil-based products is likely to be dependent on the ability to overcome the challenges suggested above. Continuing to conduct research to improve the effectiveness of essential oil-based pesticides and to establish the best method of application will also help bring these pesticides forward as useful controls against insects, mites, weeds, and microbes. However, for a commercial essential oil pesticide to thrive, it will be necessary to ensure that appropriate markets are identified. Due to their rapid volatilization, essential oil-based pesticides are unlikely to have use in field crops; however, this property is conducive to using them to control urban pests or for high value crops where a sufficiently high premium is placed on human and environmental safety. Ultimately, as current conventional pesticides continue to lose their registration, natural pesticides will play a larger role including those based on plant essential oils.

21.6 References

1 P. W. Paré, J. H. Tumlinson, *Plant Physiol.*, **1999**, *121*, 325–331.
2 E. Pichersky, J. Gershenzon, *Curr. Opin. Plant Biol.*, **2002**, *5*, 237–243.
3 M. B. Isman, *Annu. Rev. Entomol.*, **2006**, *51*, 45–66.
4 B. J. R. Philogène, C. Regnault-Roger, C. Vincent in *Biopesticides of Plant Origin*, B. J. R. Philogène, C. Regnault-Roger, C. Vincent (Eds.), Lavoisier, Paris, France, **2005**.
5 M. B. Isman in *Biopesticides of Plant Origin*, B. J. R. Philogène, C. Regnault-Roger, C. Vincent (Eds.), Lavoisier, Paris, France, **2005**.
6 G. Flamini, P. L. Cioni, I. Morelli, M. Macchia, L. Ceccarini, *J. Agric. Food Chem.*, **2002**, *50*, 3512–3517.
7 M. J. Pascual-Villalobos, M. C. Ballesta-Acosta, *Biochem. Syst. Ecol.*, **2003**, *31*, 673–679.
8 G. Flamini, P. L. Cioni, I. Morelli, S. Maccioni, R. Baldini, *Food Chem.*, **2004**, *85*, 599–604.
9 A. J. Bekele, D. Obeng-Ofori, A. Hassanali, Int. *J. Pest Manag.*, **1996**, *42*, 129–142.
10 S. H. Ho, L. P. L. Cheng, K. Y. Sim, H. T. W. Tan, *Postharvest Biol. Technol.*, **1994**, *4*, 179–183.
11 B. H. Lee, W. S. Choi, S. E. Lee, B. S. Park, *Crop Prot.*, **2001**, *20*, 317–320.
12 D. P. Papachristos, D. C. Stamopoulos, *J. Stored Prod. Res.*, **2002**, *38*, 117–128.
13 P. J. Landolt, R. W. Hofstetter, L. L. Biddick, *Environ. Entomol.*, **1997**, *28*, 954–960.
14 W. I. Choi, S. G. Lee, H. M. Park, Y. J. Ahn, *J. Econ. Entomol.*, **2004**, *97*, 553–558.
15 S. Miresmailli, M. B. Isman, *J. Econ. Entomol.*, in press.
16 S. Miresmailli, R. Bradbury, M. B. Isman, *Pest Manag. Sci.*, **2006**, *62*, 366–371.
17 Y. Akhtar, M. B. Isman, *Entomol. Exp. Appl.*, **2004**, *111*, 201–208.
18 Y. Huang, J. M. W. L. Tan, R. M. Kini, S. H. Ho, *J. Stored Prod. Res.*, **1997**, *34*, 11–17.
19 D. Obeng-Ofori, C. Reichmuth, *Int. J. Pest Manag.*, **1997**, *43*, 89–94.
20 K. N. Don-Pedro, *Pestic. Sci.*, **1996**, *46*, 79–84.
21 C. M. Priestley, I. F. Burgess, E. M. Williamson, *Fitoterapia*, **2006**, *77*, 303–309.
22 H. Tunón, W. Thorsell, A. Mikiver, I. Malander, *Fitoterapia*, **2006**, *77*, 257–261.
23 E. E. Enan, *Arch. Insect Biochem. Physiol.*, **2005**, *56*, 161–171.
24 E. E. Enan, *Insect Biochem. Mol. Biol.*, **2005**, *35*, 309–321.
25 T. Tworkoski, *Weed Sci.*, **2002**, *50*, 425–431.
26 L. D. Bainard, M. B. Isman, M. K. Upadhyaya, *Weed Sci.*, 2006, 54, 833–837.
27 C. M. Machial, M. B. Isman, unpublished results.
28 M. M. Cowan, *Clin. Microbiol. Rev.*, **1999**, *12*, 564–582.
29 B. Bozin, N. Mimica-Dukic, N. Simin, G. Anackov, *J. Agric. Food Chem.*, **2006**, *54*, 1822–1828.
30 N. Mytle, G. L. Anderson, M. P. Doyle, M. A. Smith, *Food Control*, **2006**, *17*, 102–107.
31 E. M. Soylu, S. Soylu, S. Kurt, *Mycopathol.*, **2006**, *161*, 119–128.
32 J. D. McChesney in *Bioregulators for Crop Protection and Pest Control*, P. A. Hedin (Ed.), ACS Symp. Ser., Washington, DC, **1994**.

Keywords

Essential Oils, Natural Pesticides, Botanicals

22
Eco-Chemical Control of the Potato Cyst Nematode by a Hatching Stimulator from Solanaceae Plants

Akio Fukuzawa

22.1
Introduction

The potato cyst nematodes (PCN), *Grobodera rostochiensis* and *G. pallida*, are two of the most important pests of potato, causing a significant yield loss in Europe, Russia, America, and Japan. They are characterized by the adult females which assume a saccate shape enclosing hundreds of eggs, each one of which develops to contain an infective, dormant second-stage juvenile (J2). On the death of the female, her cuticle hardens and forms a protective cyst around the eggs. The J2 of *G. rostochiensis* hatch from the eggs and emerge from the cysts in response to diffusates from host plant roots of the Solanaceae plant.

The eggs hatch in spring under suitable temperature and moisture conditions and additional host plant diffusates. Without host plant diffusate, J2s do not hatch in the soil for 20 years. Therefore, the substances that stimulate the hatching are considered to be effectively utilized for controlling J2s in the farm field, since the hatched J2s in the absence of host plant are to be starved. In view of this, we have searched for hatching stimulators produced by the host plants of cyst nematodes, aiming at utilizing them for potential available nematicides.

22.2
Classification of Cyst Nematodes and Research History to Elucidate the Naturally Occurring Hatching Stimulants

It should be noted that most important agricultural plants are parasitized by cyst nematodes that hatch in response to the root diffusates. However, the nematodes generally have very narrow and limited host ranges, suggesting the involvement of a specific host recognition process by the nematodes via the chemical components of host plants.

The secretion of hatching stimulants for PCN from the potato root was first observed by Baunacke in 1922 [1]. Todd's group attempted the isolation of hatching

Pesticide Chemistry. Crop Protection, Public Health, Environmental Safety
Edited by Hideo Ohkawa, Hisashi Miyagawa, and Philip W. Lee

ISBN: 978-3-527-31663-2

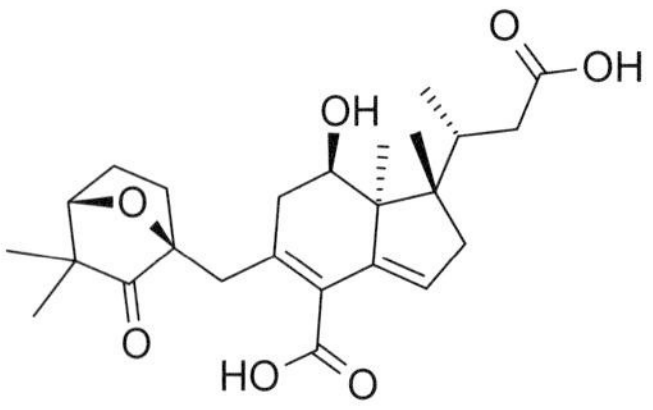

Glycinoeclepin A
(activity 10^{-11} g/mL)

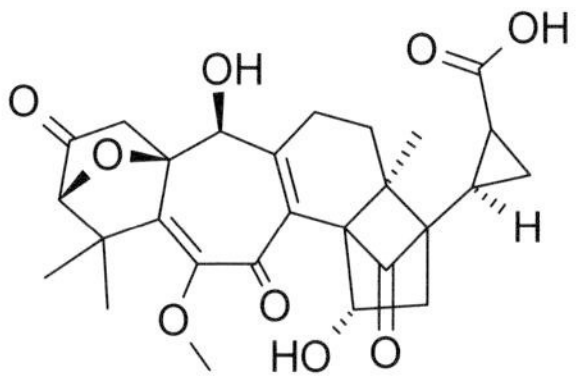

Solanoeclepin A
(activity 10^{-12} g/mL)

Figure 1. Chemical structures of hatching stimulants toward cyst nematodes.

substances from tomato root diffusate (TRD) obtained from 150,000 tomato plants [2]. They isolated "eclepic acids" showing activity at a concentration of 10^{-7}–10^{-8} g/mL following 11 years of research. Many scientists, including Widdowson, Clarke, and Perry of Rothamsted Research continued this work for 30 years, but failed to chemically characterize the stimulants [3]. Similarly, Tsutsumi and Sakurai demonstrated that the root diffusates from the host plants such as soybean, kidney bean, and azuki bean had specific hatching activity for J2 of the soybean cyst nematode (SCN), *Heterodera glycines* [4].

After an intensive effort to isolate the hatching factors from host plant roots or their diffusates, Masamune's group first established the structure of hatching stimulant to SCN, glycinoeclepin A, in 1982, which was active at a concentration of as low as 10^{-11} g/mL (Figure 1) [5–8]. Subsequently, solanoeclepin A was reported as the stimulant to PCN by Mulder *et al.* in 1992 [9]. This compound was isolated from the hydroponic culture medium of tomato and active at 10^{-12} g/mL in an aqueous solution. In spite of their ultra-high activity, neither of these compounds has been commercialized as nematicides because of their limited solubility in water. Besides, recent research reveals that the reality of chemical hatching control is not that simple.

22.3
Involvement of Multiple Factors in Hatching Stimulation for Cyst Nematodes

Devine *et al.* reported that the dose-response relationship of potato root diffusate (PRD) in terms of hatching stimulation was somewhat unstable: in some cases, the dose-response relationship did not appear in a typical sigmoidal manner, but gave multiple relative maxima [10]. We also observed a similar phenomenon regarding the freeze-dried water-soluble fraction of PRD, as shown in Figure 2, where small activity peaked in the lower concentration range around 10^{-10} g/mL after decreasing from high activity at 10^{-5} g/mL. However, the addition of fresh PRD to the freeze-dried one significantly enhanced the activity, indicating the presence of some synergistic factors in the fresh PRD. Such factors were likely to be highly unstable, resulting in relatively less reproducible assay results.

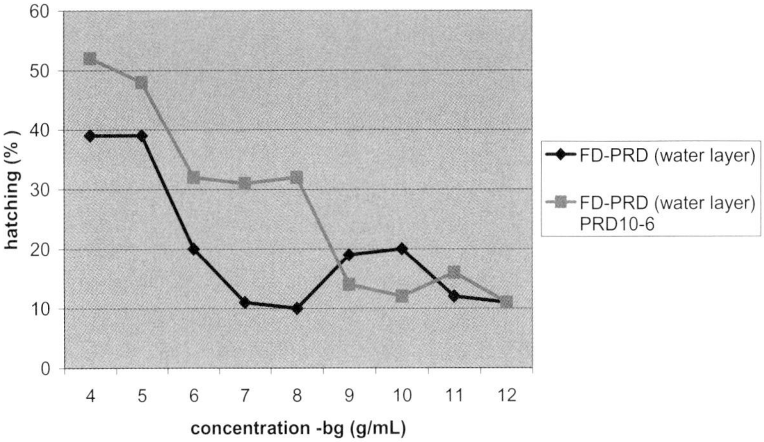

Figure 2. Hatching stimulating activity of the freeze-dried potato root diffusate (FD-PRD) and the effect of addition of freshly prepared PRD at the concentration of 10^{-6} g/mL.

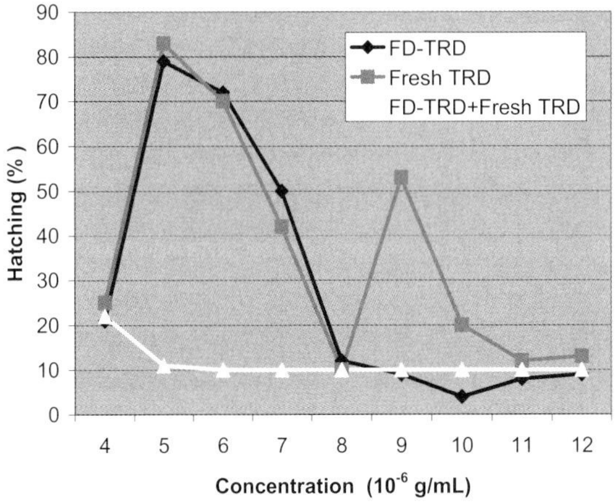

Figure 3. Hatching stimulating activity of the freeze-dried (A) and freshly prepared (B) tomato root diffusate (TRD), and the effect of addition of a highly volatile fraction of TRD (C).

In the case of tomato root diffusate (TRD), the freeze-dried material also has a relatively low activity, with a total hatch of J2 of only 28% at 10^{-4} g/mL (Figure 3, line A). By contrast, when the TRD was freshly prepared by eluting with 2-propanol, the polystyrene gel in which active principles in tomato hydroponic culture medium were adsorbed, followed by concentration under reduced pressure, a high hatching activity was observed at 10^{-6} g/mL (line B). This difference was considered to be due to the presence of synergistic factor(s) in the latter TRD sample, which had been lost in the former during the freeze-drying. Furthermore, when the highly volatile fraction of the hydroponic culture media (see below) was combined with the latter (i.e., freshly prepared) TRD sample, an additional enhancement was observed, showing hatching activity of higher than 50% even at 10^{-9} g/mL (line C). Thus, the presence of another synergist(s) was suggested in the highly volatile fraction. The dose-response curves B and C in Figure 3 look far from sigmoidal, probably as a consequence of the combination of high susceptibility of hatching stimulant to the synergistic effects and the presence of putative hatching inhibitor(s).

22.4
Isolation of Hatching Stimulators and Stimulation Synergists in TRD

The tomato root diffusate obtained from a hydroponic culture medium was concentrated with a rotary evaporator to give solid residue and concentrated aqueous extract. A vacuum line was connected to two traps in series: trap I (–45 °C) and trap II (–196 °C). The water condensed from the rotary evaporator in trap I gave a pair of synergists, which were referred to as IA and IB (SyIA and SyIB). The highly volatile synergist was obtained in the trap, and was referred to as synergist II (SyII).

On the other hand, the residue after the concentration by rotary evaporator was separated on normal phase gel adsorbents HP-20, Sephadex LH-20, and Toyopearl HW-40, to give 3 mg of a hatching stimulator. This material was active at 10^{-4} g/mL. The activity of this stimulant was increased more than 100-fold by SyIA and SyIB. SyII had a striking synergistic activity, enhancing the hatching activity *ca.* 100,000-fold.

Based on NMR spectra, the hatching stimulator is tentatively concluded to be a triterpene glycoside, whereas SyIA and IB are both presumed to diesters, and SyII to be a monoester. Detailed structure analyses are in progress.

22.5
Application of TRD to Decrease PCN Density in Soil

To examine the effect of the hatching stimulators on the density of PCN, hatching stimulators in tomato hydroponic medium were applied to a field naturally infested with PCN. Ten-fold diluted medium was applied to an area of 1 m^2 at a rate of 18 L once a month from May to August. The numbers of the remaining eggs and J2 in 20 g of dry soil were counted, and the total numbers were deemed as the PCN density. Consequently, the initial density of 3,750 in May decreased to 8.8 (0.2%) in September (mean of 18 replications). The experiments were repeated for 3 years and the similar data were obtained every year. These results show that this type of eco-chemical control of PCN by causing suicide hatch of J2 is effective and promising in the future. Chemical characterization of hatching stimulants as well as synergists is awaited to develop a practically useful controlling agent.

22.6 References

1 W. E. Baunacke, *Arb. Biol. Bund Anst. Land und Forstw.*, **1922**, *11*, 185–288.
2 C. T. Calam, H. Raistrick, A. R. Todd, *Biochem. J.*, **1949**, *45*, 513–519.
3 F. G. W. Jones, *Rep. Rothamsted Exp. Stn., Part 1*, **1972**, 154–173.
4 M. Tsutsumi, K. Sakurai, *Appl. Zool.*, **1966**, *10*, 129–137.
5 T. Masamune, M. Anetai, M. Takasugi, N. Katsui, *Nature*, **1982**, *297*, 495–496.
6 T. Masamune, M. Anetai, A. Fukuzawa, M. Takasugi, H. Matsue, K. Kobayashi, S. Ueno, N. Katsui, *Bull. Chem. Soc. Jpn.*, **1987**, *60*, 981–999.
7 T. Masamune, A. Fukuzawa, A. Furusaki, M. Ikura, H. Matsue, T. Kaneko, A. Abiko, N. Sakamoto *et al.*, *Bull. Chem. Soc. Jpn.*, **1987**, *60*, 1001–1014.
8 A. Fukuzawa, A. Furusaki, M. Ikura, T. Masamune, *J. Chem. Soc., Chem. Commun.*, **1985**, 222–224.
9 J. G. Mulder, P. Diepenhorst, P. Plieger, I. E. M. Bruggemann-Rotgans, *PCT Int. Appl.*, WO 93/02,083, **1992**.
10 K. J. Devine, J. Byrne, N. Maher, P. W. Jones, *Ann. Appl. Biol.*, **1996**, *129*, 323–334.

Keywords

Hatching Stimulator, Hatching Synergist, Potato Cyst Nematode, *Globodera rostochiensis*, *Heterodera glycines*, Tomato Root Diffusate

23
Vector Competence of Japanese Mosquitoes for Dengue and West Nile Viruses

Yuki Eshita, Tomohiko Takasaki, Ikuo Takashima, Narumon Komalamisra, Hiroshi Ushijima, Ichiro Kurane

23.1
Introduction

Dengue virus (DEV) and West Nile virus (WNV) belong to the genus Flavivirus (family Flaviviridae). Dengue virus replicates in humans as the primary vertebrate host and *Aedes* (*Stegomyia*) *aegypti* and *Ae.* (*Stg.*) *albopictus* mosquitoes as the vectors. In contrast, WNV replicates in birds as the reservoir host, while incidental infection occurs in humans and horses. Although dengue is not endemic in Japan, nor is there a stable population of *Ae. aegypti*, epidemics broke out in a port town of Nagasaki in 1942. About 200,000 typical clinical cases were recorded. Several *Ae.* (*Stg.*) mosquito species, including *Ae. albopictus*, are widely distributed in Japan. Furthermore, imported cases of the patients are increasing. On the other hand, an imported case of a West Nile fever patient was reported in 2005. Possible vector information of WNV is lacking in Japan. For these reasons, we investigated the possible vector competence of Japanese mosquitoes to DEV and WNV.

23.2
Possible Vector of Japanese Mosquitoes Against Dengue Virus

Dengue epidemics broke out in a port town of Nagasaki, Japan in 1942. It soon spread over other cities, recurring every summer until 1944. About 200,000 typical cases were recorded [1]. Nowadays, there are no dengue epidemics, nor *Ae. aegypti* mosquitoes in Japan. *Aedes albopictus* known as a secondary mosquito vector of dengue fever in Southeast Asian countries, have been however widely distributed, and also other *Stegomyia* mosquito species in Japan. Furthermore, imported cases of dengue patients incline to increase. For these reasons, we planned to research the experimentally vector competence of Japanese common mosquito species against the viruses as one of emerging and reemerging viruses [2–4].

Pesticide Chemistry. Crop Protection, Public Health, Environmental Safety
Edited by Hideo Ohkawa, Hisashi Miyagawa, and Philip W. Lee

ISBN: 978-3-527-31663-2

23.2.1
Susceptibility of Orally Infected Japanese Mosquitoes

Fourteen mosquito species found commonly in Japan were studied on their susceptibility and transmission to the virus. Dengue type 2 virus (New Guinea C) was used for the susceptibility experiment. The stock virus was prepared as a 10% suspension of suckling mice brains (SMic) diluted with 2% fatal bovine serum (FBS). The stock viruses of mouse-passaged numbers 70 and 71 (mouse-adapted viruses) were used for the experiments. The virus titer was $10^{6.6}$–$10^{6.7}$ SMicLD$_{50}$/0.02 mL. Equal parts of the stock virus, heparinized rabbit blood, and 6% sucrose as a VBS solution (virus titer: $10^{5.3}$ SMicLD$_{50}$/0.02 mL) were orally administrated to female mosquitoes by absorbent cotton-feeding procedures. Mosquitoes were maintained at 20 and 30 °C for 0 to 30 days to determine growth curves in case of a positive virus titer. Ten virus-exposed females per pool were homogenized by grinding with 1 mL of 2% FBS, and centrifuged at 2,500 rpm for 15 min. Supernatant was reserved as undiluted (10^0) inoculums. Suckling mice were inoculated intracerebrally with 0.02 mL of serial dilutions, ranging from 10^0 to 10^{-4}, of the supernatant. Fifty percent of lethal dose (LD$_{50}$) per 0.02 mL of inoculums was determined two weeks later by the method of Reed and Muench. Virus titer (number of LD$_{50}$ doses per 1 mL of suspension derived from 10 female mosquitoes) was determined by multiplying this value by 50.

Mosquitoes were allowed to feed the virus solution (VBS), and suspensions of 10-pooled mosquitoes were inoculated intracerebrally into suckling mice immediately after the ingestion and after 20 days at 30 °C.

Positive assays in this observation were done with suckling mice; each inoculated intracerebrally with 0.02 mL of ground suspension of 10 pooled virus-exposed mosquito females. The virus was demonstrated from all immediately after the virus ingestion, but positive results were obtained after 20 days at 30 °C, only in the following 4 species, *i.e., Ae. (Stg.) albopictus, Ae. (Stg.) flavopictus, Ae. (Stg.) riversi,* and *Ochlerotatus dorsalis.* On the contrary, the other 11 species, *i.e., Ae. (Aedimorphus) vexans nipponii, Och. (Finlaya) japonicus, Och. (Fin.) togoi, Armigeres (Armigeres) subalbatus, Anopheles (Anopheles) sinensis, Culex (Culex) orientalis, Cx. (Cx.) pipiens molestus, Cx. (Cx.) pipiens pallens, Cx. (Cx.) quinquefasciatus, Cx. (Cx.) tritaeniorhynchus,* and *Tripteroides (Tripteroides) bambusa* took almost the same titer of the viruses; they, however, were all negative after 20 days at 30 °C because of no replication of the viruses in these mosquito species.

In a subsequent experiment, the above-mentioned four susceptible mosquito species were infected orally with the virus and were kept up to 30 days at 30 °C. Dengue viruses were replicated in all 4 mosquito species after the eclipse phase of the virus. When mosquitoes took the higher titer viruses, the higher titer virus replications were observed in the mosquitoes. The susceptibility of these 4 species against the viruses seemed to be the same level. The virus was demonstrated after varying at different temperature levels. The virus replication ratio at 20 °C shows an extremely lower replication of viruses than those at 30 °C (Figure 1).

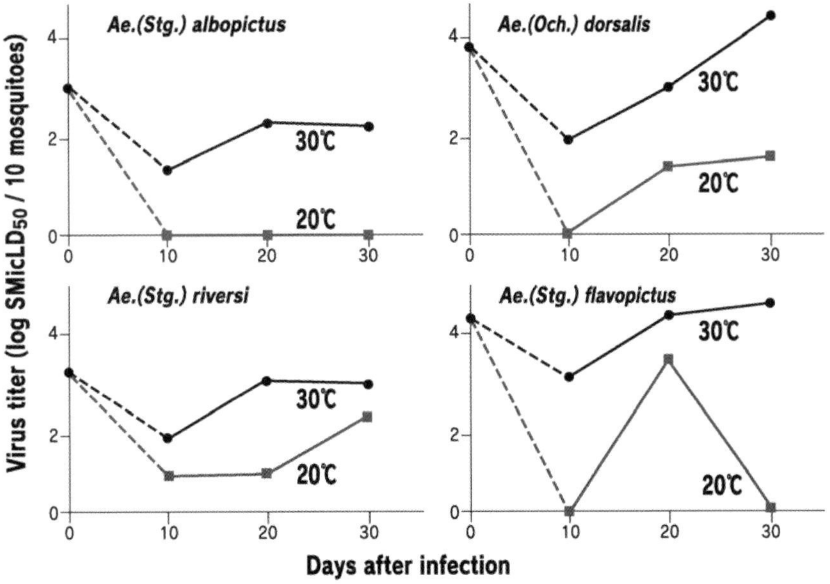

Figure 1. Dengue-2 virus titer in 10-pooled different Japanese mosquito species in 10-day intervals after oral infection at 20 and 30 °C, respectively.

23.2.2 Transmission of Mouse-Adapted or Non-mouse Passaged Dengue Viruses by the Japanese Mosquito Species

Four susceptible species, *i.e., Ae. albopictus, Ae. flavopictus, Ae. riversi,* and Thai *Ae. aegypti* as a positive control vector, were inoculated intrathoracically with a mouse-adapted strain of DEN-2 virus. The virus in the salivary glands of these mosquitoes was demonstrated by intracerebral inoculation of the suspensions of each salivary grand into suckling mice. And also transmission of the virus by mosquitoes was also achieved by feeding on suckling mice. On the contrary, orally ingested viruses in the susceptible mosquito species including *Och. dorsalis* were demonstrated both from the salivary glands and the tissue remnants of *Ae. riversi* but only from the tissue remnants of *Ae. albopictus, Ae. flavopictus,* except for *Och. dorsalis* and *Ae. aegypti.* The viral transmission however was not demonstrated in all 5 mosquito species in this experiment by using mouse-adapted dengue-2 viruses. On the other hand, when non-biting *Toxorhynchites* mosquito-passaged (non-mouse adapted) dengue-2 virus was used for oral infection of *Aedes* mosquitoes, and the mosquitoes were kept at 30 °C for 23 days, viral antigen was demonstrated in high ratios in the salivary glands of 4 *Aedes* mosquito species by an indirect immunofluorescent antibody (IFA) test after being infected by oral ingestion of the virus as well as by virus intrathoracic inoculation (data not shown). Incidentally, the titer of *Toxorhynchites* mosquito-passaged (non-mouse adapted) dengue-2 virus was 10^3–10^4 Mosquito Infectious Doses (MID_{50})/0.00017 mL, and the virus was not detected by suckling mice inoculation. Since dengue viruses passaged through 70 to 72 generations of suckling mice brain may not replicate even in certain susceptible mosquitoes, observation data obtained from *Toxorhynchites* mosquito-passaged (non-mouse adapted) dengue virus may be a more natural situation.

Since *Toxorhynchites* mosquito-passaged (non-mouse adapted) dengue-2 virus was not easily replicated in suckling mice, special apparatus for collection of mosquito saliva was developed. Wings, dorsal thorax, and abdomen of live female mosquitoes were arrested on strong sticky tape under releasing CO_2 gas, their proboscises without labia were inserted into glass capillary tubes (1-mm diameter) containing 0.5% soft agar in phosphate buffered saline (PBS) with 40% FBS and 0.01M adenosine triphosphate (ATP). Mosquito saliva that was released into the glass capillary tube was inoculated intrathoracically into live *Tx. amboinensis* mosquitoes as a laboratory host for the virus replication. And then they were kept at 30 °C for 7 days. The viral antigen was detected with different ratios by the IFA test from stamp samples of the head squash of *Tx. amboinensis* mosquitoes. Ring-forms like positive fluorescent images were found in the cell cytoplasm near and around the cell nuclei. *Culex tritaeniorhynchus* mosquitoes were not able to replicate orally ingested dengue viruses in their body; however, they could replicate the virus inoculated intrathoracically into their thorax. And also, even though oral infection of mosquitoes was permitted to replicate the viruses in their body, no dengue virus was detected in salivary glands of the positive body after extrinsic incubation periods. These findings also supported that there were two barriers for viral replication in mosquitoes, one is midgut and the other is salivary gland barriers. The reason why most of the *Stegomyia* mosquito species lack the midgut barrier is uncertain. And also thee transmission ability of dengue-susceptible *Och. dorsalis* is presently unknown.

23.2.3
Further Analysis for Vector Mosquitoes to Dengue Viruses in Japan

Serious precautions must be taken against the possible DF/DHF, since the number of dengue patients inclined to increase even in non-endemic areas of temperate regions. In Japan, 8 *Ae. (Stg.)* mosquito species at least, from Southwestern to Northeastern districts, *Ae. flavopictus miyarai, Ae. flavopictus downsi, Ae. daitensis, Ae. riversi, Ae. albopictus, Ae. wadai, Ae. flavopictus* and *Ae. galloisi*, including the putative vector of dengue viruses, are distributed in Japan (Figure 2). Out of 8 mosquito species, 4 were reported here as a proven vector of the viruses. The other 4 species are needed to survey their vector competency. And also it is keenly needed to establish vector control strategies against putative epidemics of mosquito-born diseases in Japan.

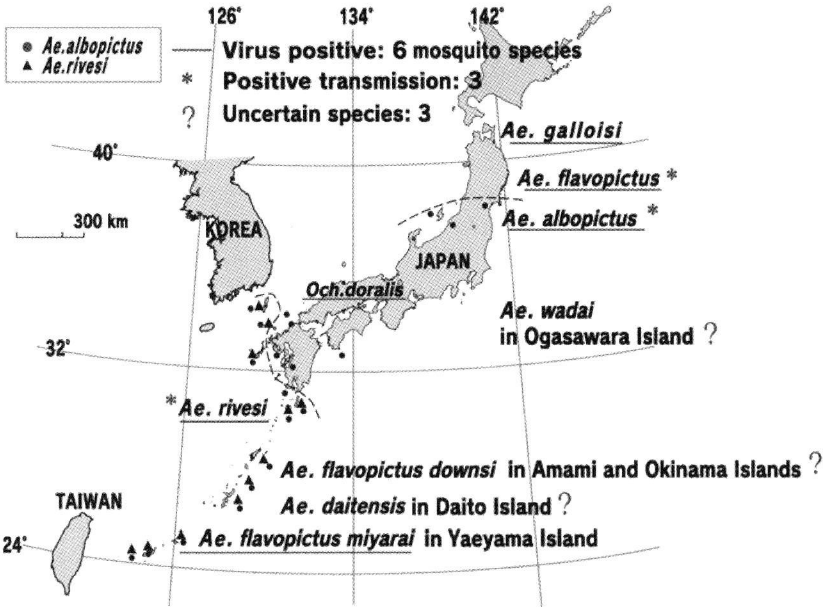

Figure 2. Distribution of nine possible dengue vectors in Japan.

23.3
Possible Vector of Japanese Mosquitoes Against West Nile Virus

Susceptibility and transmission ability of Japanese mosquitoes, *Cx. p. pallens, Cx. p. molestus, Cx. inatomii, Ae. albopictus*, and *Och. japonicus* against WNV were analyzed. The Ugandan strain of WNV was used for oral and intrathoracically-infection of the mosquito species, except *Cx. inatomii* and *Och. japonicus* with the New York strain of WNV. They were maintained for 14 days at 28 °C after the infection. And then they were kept at –80 °C until further investigation. Total RNA were extracted individually and RT-PCR was performed to examine the presence of WNV genome in mosquito parts, abdomen, thorax, legs, and head. Infection and transmission procedures were followed by the methods of Eshita *et al.* [5] and Hayashi *et al.* [6].

23.3.1
Susceptibility of Japanese Mosquitoes Against West Nile Virus

All the five mosquito species were susceptible to WNV (Table I). The West Nile virus genome was detected in *orally-infected Cx. p. pallens.* These infection rates were 43%, 12%, and 75%, at 20, 25, and 28 °C, respectively. One of four orally infected mosquitoes showed strong signals of WNV genome in the thorax and the legs. The head of the same positive mosquito was negative. As the RT-PCR

Table I. Possible Japanese Vector Mosquitoes Related to West Nile Virus Transmission.

Japanese mosquito species	Experimental infection	Experimental transmission
Culex pipiens pallens	yes	yes
Culex pipiens molestus	yes	NT
Culex inatomii	yes	yes
Aedes albopictus	yes	yes
Ochlerotatus japonicus	yes	NT

NT: not tested

reaction may be inhibited by unknown factors in mosquitoes, the purification step of the total RNA derived from individual mosquitos may be important.

We also examined WNV susceptibility of *Cx. p. molestus, Ae. albopictus* following the same procedure of *Cx. p. pallens*. They were maintained for 14 days at 15, 20, 25, and 28 °C. The viral genome was detected in both mosquito species, and also detected at 15 °C. *Culex inatomii* (Figure 3) and *Och. japonicus* were also susceptible to the New York strain of WNV with high infection rates.

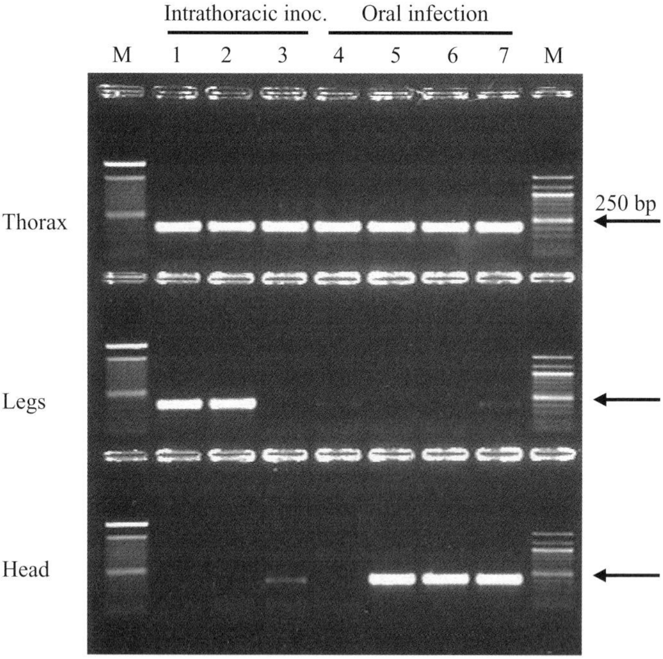

Figure 3. Detection of WNV genome from thorax, legs, and head of *Culex inatomii* by intrathoracic and oral infection, respectively (M: 100bp ladder marker, Primer2: WNNY514V-E, WNNY904-E).

23.3.2
Transmission of Japanese Mosquitoes Against West Nile Virus

Intrathoracically inoculated *Cx. p. pallens* mosquitoes were prepared for the WNV transmission experiment. They seemed to be infected theoretically with approximately 100% WNV. Anesthetized BALB/c mouse was given as feed blood for the mosquitoes 14-day post infection. Some mice were recognized with symptoms 3 to 10 days post-infection. In one of three mice, West Nile viral genome was detected in mouse blood. Similar positive data was obtained on *Cx. inatomii, Ae. albopictus* and *Och. japonicus* (Table I).

Kitaoka [7] investigated the susceptibility of field-collected three Japanese mosquito species to WNV. The orally infected mosquitoes after the incubation period were given a chance to feed on mouse blood. Some mice showed symptoms after bitings of *Cx. p. pallens.* The *Armigeres subalbatus,* however, was negative. *Culex tritaeniorhynchus* did not provide clear data because of mixed samples with *Cx. p. pallens.* It was suggested that *Cx. p. pipiens* distributed in Japan was designated as a possible vector of WNV by Kitaoka [7]. Although most mosquito species are able to transmit the virus, *Culex pipiens* complex mosquito species are recognized as the main vector. The susceptibility to WNV of the European mosquitoes, *Cx. p. molestus, Cx. p. pipiens,* and *Cx. quinquefasciatus,* was reported [8–11].

As for the dengue vector mosquito, *Aedes (Stegomyia)* subgenus mosquitoes are the main vector [12]. However, almost all mosquito species show a high susceptibility to WNV. More than 40 species of American mosquitoes were discriminated as natural vector. One of the factors that this virus distributed rapidly in the U.S.A. may be because of the numerousness of vector mosquito species as well as the change of bird migration. Transovarial transmission of WNV by mosquitoes in the natural world is unknown [13]. The reason why most of the mosquito species except the *Aedes* (*Stegomyia*) species are able to preserve their midgut barrier to DEN, while most of the species lack a midgut or other barrier to WNV, is uncertain. Midgut barriers for viral replication and dissemination may explain why only a limited number of mosquito species are able to transmit the viruses despite evidence of infection.

23.4
Conclusion

Seventeen mosquito species were analyzed on their vector competence to DEV. Positive results were obtained by oral infection in the following six species: *Ae. albopictus, Ae.* (*Stg.*) *flavopictus flavopictus, Ae.* (*Stg.*) *f. miyarai, Ae.* (*Stg.*) *riversi, Ae.* (*Stg.*) *galloisi,* and *Och. dorsalis.* The virus also was detected in the saliva of all species except *Ae. f. miyarai, Ae. galloisi,* and *Och. dorsalis.* Eleven other species, *Ae. vexans nipponii, Och. japonicus, Och. togoi, Armigeres subalbatus, Anopheles sinensis, Cx. orientalis, Cx. p. molestus, Cx. p. pallens, Cx. quinquefasciatus, Cx. tritaeniorhynchus* and *Tripteroides bambusa,* were negative.

Culex p. pallens, *Cx. p. molestus*, and *Cx. quinquefasciatus* may become epidemiologically important, if WNV is introduced in Japan. Five mosquito species, *Cx. p. pallens*, *Cx. p. molestus*, *Cx. inatomii*, *Ae. albopictus*, and *Och. japonicus*, were discriminated as possible vectors of WNV.

A reduction of the vector population through elimination of larval breeding sites and use of larvicides may be one of keys for the prevention of dengue fever and West Nile fever outbreaks. An increase in the number of WNV-infected patients on the west coast of the USA caused the first domestically imported case in Japan in 2005. Establishment of national countermeasures, such as strengthening of quarantine organizations and vector surveillance around the international air and seaports, is needed in the absence of an effective vaccine. Extensive source-reduction countermeasures of mosquito larvae by local government and individuals must be developed at a community level.

23.5 Acknowledgments

This work was supported in part by grants from the Ministry of Health, Labor, and Welfare of Japan (Grant numbers H12-Shinko-32, H15-Shinko-17 and H18-Shinko-9), the Ministry of the Environment, Japan (Consignment Research by Global Environment Research Coordination System in FY2004, FY2005), the Ministry of Education, Culture, Sports, Science, and Technology of Japan (Grant numbers 14206036 and 18580310), and the Core University Exchange Program of the Japan Society for the Promotion of Science, coordinated by the University of Tokyo and Mahidol University. We also acknowledge Raweewan Srisawat, Department of Medical Entomology, Faculty of Tropical Medicine, Mahidol University, for obtaining partial data on the dengue experiment.

23.6 References

1 S. Hotta, *Med. Entomol. Zool.*, **1998**, *49*, 267–274.
2 Y. Eshita, T. Kurihara, T. Ogata, A. Oya, *Jpn. J. Sanit. Zool.*, **1982a**, *33*, 61–64.
3 Y. Eshita, T. Kurihara, T. Ogata, A. Oya, *Jpn. J. Sanit. Zool.*, **1982b**, *33*, 65–70.
4 Y. Eshita, *Teikyo Medical Journal*, **1982**, *5*, 17–27.
5 Y. Eshita, T. Takasaki, K. Yamada, I. Kurane, in *Anthology VI: Arthropod Borne Diseases*, J. Y. Richmond (Ed.), American Biological Safety Association, Illinois, **2003**, 63–71.
6 A. Hayashi, K. Kamakura, K. Taga, H. Mori, S. Imura, Y. Eshita, Y. Uchida, *Jpn. J. of Inf. Dis.*, **2003**, *77*, 822–829.
7 M. Kitaoka, *Jpn. Med. J.* (Natl. Inst. Health Jpn.), **1950**, *3*, 77–81.
8 Z. Hubalek, *Viral Immunol.*, **2000**, *13*, 415–26.
9 A. S. Tahori, V. V. Sterk, N. Goldblum, *Am. J. Trop. Med. Hyg.*, **1955**, *4*, 1015–1027.
10 K. M. Pavri, K. R. Singh, *Indian. J. Med. Res.*, **1965**, *53*, 501–505.

11 D. L. Vanlandingham, B. S. Schneider, K. Klingler, J. Fair, D. Beasley, J. Huang, P. Hamilton, S. Higgs, *Am. J. Trop. Med. Hyg.*, **2004**, *71*, 120–123.

12 F. Rodhain, L. Rozen, in *Dengue and Dengue Hemorrhagic Fever*, D. J. Gubler, G. Kuno (Eds.), CABI International, New York, **1997**, 45–60.

13 L. B. Goddard, A. E. Roth, W. K. Reisen, T. W. Scott, *J. Med. Entomol.*, **2003**, *40*, 743–746.

Keywords

Dengue Virus, West Nile Virus, Japanese Mosquito, Vector Control, Susceptibility, Transmission

24
Life Science Applications of Fukui Functions

Michael E. Beck, Michael Schindler

24.1
Introduction

Understanding the metabolic fate and behavior of a drug or agro-chemical is of crucial importance for the development of such chemicals, as metabolism in conjunction with absorption, distribution, and excretion determines bio-availability, effect, and adverse reactions as well as toxicology of a drug or agro-chemical. Many computational approaches exist for understanding metabolism and/or toxicology a posteriori as well as prospectively. These typically use statistical modeling based on (often topological) molecular descriptors [1–4]. Cytochrome P450 enzymes play an important role in metabolism, as most oxidative metabolic reactions are mediated by these enzymes (see [5] and [6] for reviews on experimental and theoretical work). With the availability of X-ray structures of cytochrome P450 enzymes, 3D-QSAR models have been developed (for example [7–10]). Another approach is ligand based, predicting the preferred sites of hydrogen abstraction by quantum chemistry (for example [10–12]).

The most straightforward, but also most demanding, approach certainly is to model the entire process of cytochrome P450 enzyme mediated hydroxylation quantum chemically [6, 13–15]. In the most recent such calculation [14], the hydroxylation process has been modeled by coupled quantum/molecular mechanics (QM/MM). These calculations strongly support the "two-state reactivity" model (TSR), in which the putative oxidizing species in the active site of cytochrome P450 enzymes [5–6, 14–15], called "Compound I" (Cpd I), is viewed as a radical species of either doublet of quartet spin coupling. The reaction takes place in a two-step so-called rebound mechanism. In the first step, hydrogen is abstracted from the substrate. The activation barrier for this step is slightly lower in the doublet than in the quartet state. In a second step, a hydroxy group is added to the substrate. This step proceeds essentially without a barrier in the doublet but requires some activation in the quartet.

The first and only application of DFT-derived reactivity descriptors to a biological system the authors are aware of is [16], where the hard and soft acids and bases

Pesticide Chemistry. Crop Protection, Public Health, Environmental Safety
Edited by Hideo Ohkawa, Hisashi Miyagawa, and Philip W. Lee

ISBN: 978-3-527-31663-2

(HSAB) principle [17] was applied to inorganic model systems for arsenate reductase and low molecular weight phosphatase.

In a previous study [18], Fukui functions [19] were calculated for 18 drugs and 11 agrochemicals, and the relation of these functions to experimentally observed metabolism was discussed. This contribution briefly reviews the theory underlying the approach, and discusses its strengths and weaknesses along with examples for biotic as well as abiotic reactions.

24.2 Theoretical Background

24.2.1 Some Results from Conceptual DFT [19–23]

A number of concepts for general chemical reactivity exist, such as the principle of hard and soft acids and bases [17, 24], electronegativity [25–27] and frontier orbital theory [28–30].

These concepts gained a solid theoretical foundation in density functional theory, namely through the work of Yang and Parr [19–20]. Besides its formal beauty, this theory exhibits another nice property: The necessary calculations are rather easy to perform.

Within density functional theory, the chemical potential μ and the hardness η become partial derivatives of the system's energy E expressed as a functional of an external potential $V(\underline{r})$, i.e., the nuclear conformation, and a function of the number of electrons N:

$$\mu = \left(\frac{\partial E}{\partial N} \right)_{V(\underline{r})} \tag{1}$$

$$2\,\eta = \left(\frac{\partial^2 E}{\partial N^2} \right)_{V(\underline{r})} \tag{2}$$

The respective functional derivative with respect to $V(\underline{r})$ yields the electron density $\rho(\underline{r})$.

$$\rho(\underline{r}) = \left(\frac{\delta D}{\delta\, V(\underline{r})} \right)_N \tag{3}$$

It can be shown that the hardness η becomes maximal for the density $\rho(\underline{r})$, which at a given external potential $V(\underline{r})$ minimizes the energy E. The hardness is a global parameter related to chemical reactivity, as it does not depend on $\underline{r}$.

In a finite difference approach, the hardness can be approximated as

$$\eta \approx \frac{IP - EA}{2} \tag{4}$$

where

$$IP = E\,[N-1, V(\underline{r})] - E\,[N, V(\underline{r})] \approx \mu^- \tag{5}$$

$$EA = E\,[N, V(\underline{r})] - E\,[N+1, V(\underline{r})] \approx \mu^+ \tag{6}$$

are the vertical ionization potential and the vertical electron affinity, respectively.

Mulliken's electronegativity is defined as

$$\chi = \frac{IP + EA}{2} \approx -\mu^0 \tag{7}$$

As indicated in Equations 5 to 7, IP, EA and χ may be viewed as finite difference approximations to the left, right, and mean derivatives of the chemical potential, $\mu = \left(\frac{\partial E}{\partial N}\right)_{V(\underline{r})}$.

Local reactivity parameters, the Fukui functions, are obtained by mixed derivatives of the energy with respect to N and $V(\underline{r})$.

$$f^{\pm}(\underline{r}) = \left(\frac{\delta^2 E}{\partial N\, \delta V(\underline{r})}\right)^{\pm} = \left(\frac{\partial \rho(\underline{r})}{\partial N}\right)^{\pm}_{V(\underline{r})} \tag{8}$$

Again the right derivative differs from the left derivative, as indicated by the ± sign. The maxima of f^+ indicate regions in the molecule, which prefer attack by a nucleophile, while f^- exhibits maxima at sites susceptible to an attack by an electrophile. In other words, f^+ indicates where increase of electron density is energetically favorable, while f^- is maximal where decrease of electron density is preferred. Practically, the Fukui functions are calculated by finite differences, e.g.:

$$\begin{aligned} f^+(\underline{r}) &\approx \rho\,(N+1, \underline{r}) - \rho\,(N, \underline{r}) \\ f^-(\underline{r}) &\approx \rho\,(N, \underline{r}) - \rho\,(N-1, \underline{r}) \end{aligned} \tag{9}$$

with all densities evaluated at a fixed external potential $V(\underline{r})$. The mean derivative, defined by $f^0(\underline{r}) = ½\,(f^+(\underline{r}) + f^-(\underline{r}))$, is not further discussed here. Its maxima can be interpreted as sites of preferred attack by radicals.

An equation, which supports understanding of the interplay between global and local reactivity, is

$$d\mu = 2\,\eta\, dN + \int d\underline{r}\; f(\underline{r})\, dV(N, \underline{r})\,, \tag{10}$$

which connects the change in chemical potential for the transition from one groundstate to another to the change in the number of electrons and the change in the external potential via the hardness and the Fukui function, respectively.

It should be noted that according to the definition adopted here the Fukui functions restrict the description of reactivity to its electronic aspects. A more complete picture is gained by inclusion of nuclear displacement, see for example [31–32] and literature cited in these papers.

24.2.2
Why Fukui Functions May be Related to Sites of Metabolism

As roughly sketched in the introduction, metabolic hydroxylation mediated by cytochrome P450 is attributed to a ferro-oxyl species, called "Compound I" (Cpd I). From present knowledge, Cpd I may be seen as an "electrophilic oxidant" [6]. Thus the f^- function (calculated for the substrates) should help identify those positions in a molecule, which are susceptible to an attack by Cpd I. Conversely, the f^+ function, evaluated for Cpd I, should show where Cpd I prefers to be attacked by the substrate. Atomic HOMO coefficients from semiempirical calculations on agrochemicals have already been quite successfully correlated to oxidative metabolic pathways [33–34]. This procedure is essentially equivalent to calculating the f^- function in the frozen orbital approximation, in which f^- reduces to the HOMO density. In the examples section, a nice demonstration of the breakdown of this approximation will be given.

For reductive attack, calculation of the f^+ function for a substrate can often yield interesting insight.

The most severe drawback of this kind of approach is obvious: The Fukui function describes reactivity against an isotropic, abstract "reactivity bath". If a drug is positioned in a specific orientation within the active site of a cytochrome, the oxidation will not take place at the same site of the ligand as it would in an isotropic situation (e.g., in solution, with a small, sterically less demanding reaction partner). This is a common drawback of any "ligand based" approach. Combinations of docking approaches and estimations of reactivity like MetaSite [35] circumvent this problem and (at least in principle) allow for discrimination of different classes of metabolic enzymes. However, this also means that generality is lost, which is more severe in agro-applications than in pharmaceuticals: In agro sciences, the exact metabolizing enzyme is often unknown, or at least not precisely known. Moreover, reductive metabolic reactions play a more prominent role for agrochemicals when compared to pharmaceuticals.

24.3
Methods

Details on how to perform the calculations are given in [18]. Here, it shall be sufficient to roughly describe the procedure: Conformational sampling is performed by an in-house Monte-Carlo/simulated annealing algorithm [36], based on the MMFF94s [37–38] forcefield as implemented in Sybyl [39]. Typically, the Fukui functions of different conformers do not exhibit different features (conformers with significant intra-molecular interactions are of course an exception to this rule). Thus, results for only one conformation will be reported in the following.

The selected conformations are subjected to full geometry optimization at RI-DFT level of theory [40–42], using the Becke-Perdew combination of functionals [43–44] and Ahlrichs's TZVP basis sets [45]. Solvent effects are estimated using COSMO [46] with dielectric constants of 4.0, simulating the interior of an enzyme. See [16, 47–49] for a discussion of dielectric effects on reactivity measures. All calculations are performed using the Turbomole package of programs [50].

At the optimized geometries thus obtained, single point unrestricted RI-DFT calculations are performed, using the same functionals and basis sets as before: From the resulting densities, Fukui functions are evaluated numerically on a regular grid of 0.5 Å resolution and visualized using Molcad [51].

For the calculations on Cpd I, the procedure had to be modified. According to the TSR model, two relevant spin states exist of Cpd I, a doublet (^{2}A) and a quartet (^{4}A). This leads to two f^- and two f^+ Fukui functions, namely [18]:

$$
\begin{aligned}
f_M^{+\uparrow}(\underline{r}) &\approx \rho^{M+1}(N+1,\underline{r}) - \rho^{M}(N,\underline{r}) \\
f_M^{+\downarrow}(\underline{r}) &\approx \rho^{M-1}(N+1,\underline{r}) - \rho^{M}(N,\underline{r}) \\
f_M^{-\uparrow}(\underline{r}) &\approx \rho^{M}(N,\underline{r}) - \rho^{M+1}(N-1,\underline{r}) \\
f_M^{-\downarrow}(\underline{r}) &\approx \rho^{M}(N,\underline{r}) - \rho^{M-1}(N-1,\underline{r})
\end{aligned}
\tag{11}
$$

where the up and down arrows symbolize incrementation and decrementation of the multiplicity by one. $M = 2, 4$ for the ^{2}A and ^{4}A states of Cpd I.

As calculation of these functions would be far too demanding for a full Cpd I system, a model system was constructed, see Figure 4. For the justification of this model system and details on the basis sets and treatment of relativistic effects, please refer to references [13, 18, 52–53].

24.4
Example Applications

24.4.1
Parathion and Chlorpyrifos: Fukui Functions Related to Biotic Degradation

Chlorpyrifos' and parathion's observed *in vitro* and *in vivo* metabolism in mammals is dominated by oxidative attack at the P=S double bond [54]. It is assumed that the phosphooxathiiran species sketched in Figures 1 and 2 are created first, which are subsequently hydrolyzed to the observed metabolites: These are the respective oxons, diethylphosphate, diethylphosphothioate, and finally p-nitrophenol and 3,5,6-trichloro-2-pyridinol in the cases of parathion and chlorpyrifos, respectively.

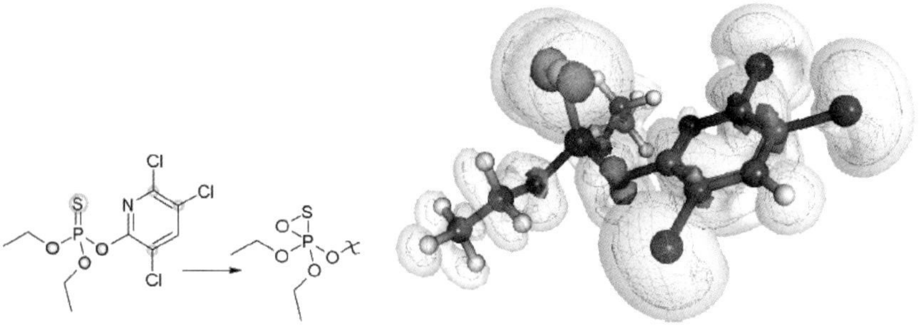

Figure 1. Fukui function f^- for attack by an electrophile of chlorpyrifos. Contour levels: 0.005 (dark, opaque), 0.001 (light, net), 0.0005 (light, transparent). In the reaction scheme shown on the left-hand side, maxima of the Fukui function are indicated by gray circles.

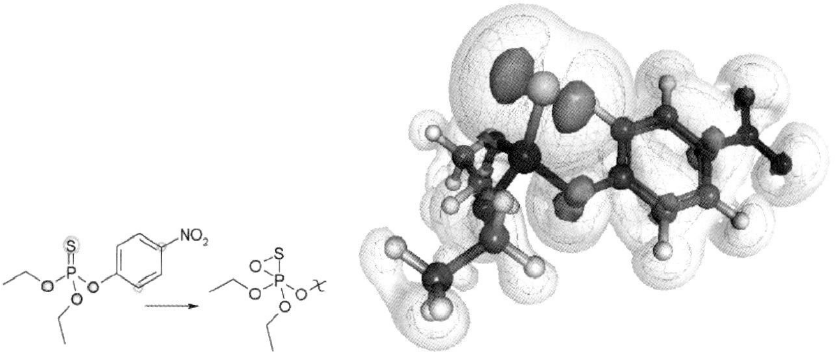

Figure 2. Fukui function $f^-(\underline{r})$ of parathion for attack by an electrophile. See Figure 1 for explanation.

Figures 1 and 2 show the Fukui functions for electrophilic attack of both compounds. The most prominent maxima f^- can be easily correlated to the observed metabolic reactions. The phenyl ring of chlorpyrifos shows three side maxima of the Fukui function. The three chlorine substituents occupy exactly these positions, thereby blocking possibly hydroxylation.

The example shows that Fukui functions do not only predict likely sites of hydroxylation, like, e.g., quantum chemical estimates of H-abstractions energies [11–12, 35]. On the other hand, interpretation of Fukui functions in the context of metabolism is not straightforward and requires a lot of expertise and experience, as will become more apparent in the following.

Interpretation of Fukui functions is easiest, if these functions are "localized", i.e., show only a few maxima, or clearly focus on certain parts of a molecule. Then, one can use chemical and biological knowledge to arrive at hypotheses for metabolic reactions. In cases where the Fukui functions are evenly smeared across the entire molecule, no reasonable predictions can be made.

Another weakness of the Fukui functions is that they frequently "underestimate" hydroxylations at terminal aliphatic sidechains. Physically, this can be easily understood. On the other hand, semiempirically derived H-abstraction energies seem not to suffer from this problem [11–12]. This is in itself a surprising finding, as homolytic dissociation energies from semiempirical Hamiltonians – and from single-determinant closed shell approaches in general – are known to be of very poor quality. Personally, we believe that it is a strength of the Fukui function approach that it does not rely on dissociation energies, as we do not consider even DFT to be a sufficient level of theory for such a purpose [20, 55–56].

24.4.2
Selective Thionation of Emodepsid

Emodepside is the active ingredient of Profender® in combination with praziquantel, a unique systemic agent for the control of gastrointestinal nematodes in cats. The reason we discuss it here is that it provides a nice example why Fukui functions provide more information than just the frontier orbitals.

Emodepsid is a cycloocta-depsipeptide exhibiting eight carbonyl moieties arranged in a makrocycle composed of alternating amide and ester functions. Interestingly, it is possible to selectively thionate just one of these using Belleau's reagent [57]. Figure 3 shows the HOMO, the HOMO-1, and the Fukui-function for attack by an electrophile. The frontier-orbitals do not at all hint at any carbonyl-group to be activated. The Fukui function f^+, however, shows that the carbonyls are amenable to electrophilic attack. It also shows a clear preference of the carbonyl in the only amide with cis configuration. It is exactly this carbonyl group which can be selectively thionated by Belleau's reagent.

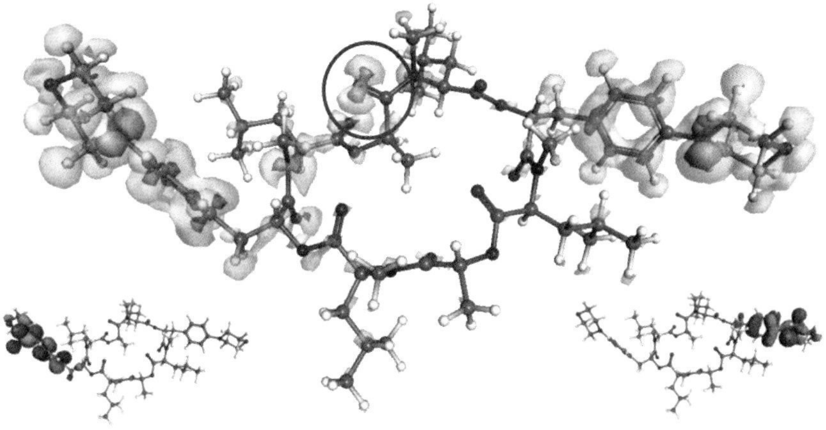

Figure 3. Fukui function $f^-(r)$ of emodepsid for attack by an electrophile. The amide function, which can be selectively thionated is indicated by a circle. The contour levels are chosen as in Figures 1 and 2. The smaller inlays to the left and the right show the HOMO and the HOMO-1, respectively.

A more detailed theoretical analysis of the mechanism for attack of carbonyls by phosphoylids in the framework of conceptual DFT is in preparation [58].

This example shows that the frozen orbital approximation, which is crucial for classical frontier orbital theory, can lead to severe artifacts and misleading interpretations. Fukui functions, on the other hand, capture electronic relaxation resulting from changes in the number of electrons.

24.4.3
Fukui Functions Reveal the Nature of the Reactive Species in Cytocrome P450 Enzymes

As mentioned in the introduction, it is believed that oxidation of substrates by cytochrome P450 enzymes is mediated by a ferro-oxo species called Cpd I [5–6]. Details of the so-called rebound mechanism could be rationalized by quantum chemistry only recently [13–14]. Based on all available data, it seems reasonable to assume that the route via a doublet state of Cpd I is more plausible than a reaction via a quartet state. Figure 4 shows Fukui functions calculated for a model system.

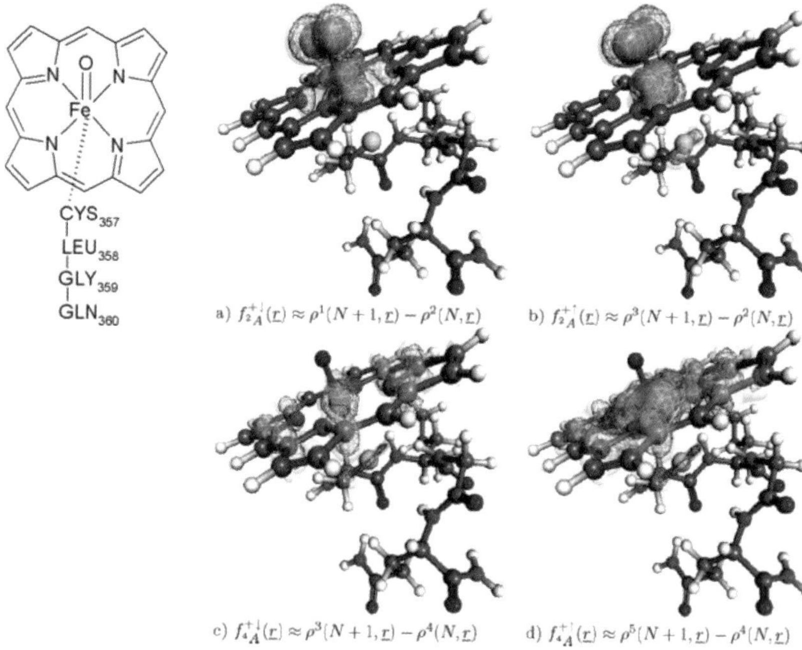

Figure 4. Fukui functions $f^+(\underline{r})$ for attack by a nucleophile, calculated for the model system of Cpd I sketched on the left-hand side. Panel a) shows the multiplicity lowering function $f^{+\downarrow}_{2A}$ arising from doublet coupling. b) shows the respective multiplicity raising function $f^{+\uparrow}_{2A}$. c) and d) show the same functions evaluated for quartet coupling, $f^{+\downarrow}_{4A}$ and $f^{+\uparrow}_{4A}$ respectively. Nomenclature follows Equation 11, contour levels as in Figures 2–3.

The four amino acids of the model system are required to arrive at a realistic description of the electron density around the cystein sulfur atom [18, 59]. Panels a) and b) of Figure 4 show Fukui functions for attack by a nucleophile, as they arise from Equation 11 with doublet spin coupling. Panels c) and d) show the respective functions assuming a quartet state. Reactivity is nicely localized around the Fe-O moiety for the doublet derived functions only, while the quartet's reactivity is located in the plane of the tetrapyrrol-system. The Fukui function provides a further argument in favor of a reaction route via a doublet state of Cpd I.

24.5 Conclusion and Outlook

Fukui functions support the understanding of biotic and abiotic reactions and degradation processes for small molecules. In conjunction with (bio-)chemical expertise and experience, Fukui functions can also be used to predict reactive behavior. In contrast to other quantum chemical approaches to reactivity, the calculation of Fukui functions does not directly depend on energies for open shell systems, which require a higher level of theory to be of acceptable quality.

The fundamental weakness of this ligand-based approach, the disregard of any enzyme specific interactions, may also be viewed as a strength: Metabolism is not restricted to oxidation mediated by cytochrome, nor is it restricted to electrophilic attack. Fukui functions (f^- as well as f^+) may thus help to understand *in vivo* metabolic reactions of unknown enzymatic origin.

Fukui calculations for Cpd I are in line with experimental and theoretical evidence. Development of a combined docking/local reactivity approach could be a less reliable, but significantly cheaper, alternative to QM/MM studies on full reaction paths.

New, exciting developments can be expected from conceptual DFT, for example the extension of the theory to "excited state Fukui functions", which may be of interest for an abiotic degradation process of particular interest for agro-chemistry: Photodegradation.

24.6 Acknowledgments

The authors want to thank Thorsten Bürger and Svend Matthiessen for their assistance in performing the calculations.

24.7 References

1 N. Greene, Computer Systems for the Prediction of Toxicity: An Update, *Adv. Drug. Deliv. Rev.*, **2002**, *54*, 417–431.

2 J. Langowski, A. Long, Computer Systems for the Prediction of Xenobiotic Metabolism, *Adv. Drug. Deliv. Rev.*, **2002**, *54*, 407–415.

3 J. Hodgson, ADMET-Turning Chemicals into Drugs, *Nature Biotech.*, **2001**, *19*, 722–726.

4 D. A. Winkler, F. Burden, Toxicity Modelling Using Bayesian Neural Nets and Automatic Relevance Detection. In *EuroQSAR 2002: Designing Drugs and Crop Protectants: Processes, Problems and Solutions*, M. Ford, D. Livingstone, J. Dearden, H. van de Waterbeemd (Eds.), Blackwell Publ. Ltd., Malden, MA, USA, **2003**, 251–254.

5 M. Sono, M. P. Roach, E. D. Coulter, J. H. Dawson, Heme-Containing Oxygenases, *Chem. Rev.*, **1996**, *96*, 2841–2887.

6 S. Shaik, S. Cohen, S. P. de Visser, P. K. Sharma, D. Kumar, S. Kozuch, F. Ogliaro, D. Danovich, The "Rebound Controversy": An Overview and Theoretical Modeling of the Rebound Step in C-H Hydroxylation by Cytochrome P450, *Eur. J. Inorg. Chem.*, **2004**, *2*, 207–226.

7 G. Cruciani, M. Pastor, S. Clementi, GRIND (Grid Independent Descriptors) in 3D Structure-Metabolism Relationships. In *Rational Approaches to Drug-Design, Proceedings of the European Symposium on Quantitative Structure-Activity Relationships, 2000*, H. D. Hoeltje, W. Sippl (Eds.), Prous Science, Barcelona, Spain, **2001**, 251–260.

8 L. Afzelius, I. Zamora, M. Ridderstrom, T. B. Andersson, A. Karlen, C. M. Masimirembwa, Competitive CYP2C9 Inhibitors: Enzyme Inhibition Studies, Protein Homology Modeling, and Three-Dimensional Quantitative Structure-Activity Relationship Analysis, *Mol. Pharmacol.*, **2001**, *59*, 909–919.

9 M. J. de Groot, S. Ekins, Pharmacophore Modeling of Cytochromes P450, *Adv. Drug. Deliv. Rev.*, **2002**, *54*, 367–383.

10 I. Zamora, L. Afzelius, G. Cruciani, Predicting Drug Metabolism: A Site of Metabolism Prediction Tool Applied to the Cytochrome P450 2C9, *J. Med. Chem.*, **2003**, *46*, 2313–2324.

11 L. Higgins, K. R. Korzekwa, S. Rao, M. Shou, J. P. Jones, An Assessment of the Reaction Energetics for Cytochrome P450-Mediated Reactions, *Arch. Biochem. Biophys.*, **2001**, *385*, 220–230.

12 S. B. Singh, L. Q. Shen, M. J. Walker, R. P. Sheridan, A Model for Predicting Likely Sites of CYP3A4-Mediated Metabolism on Drug-Like Molecules, *J. Med. Chem.*, **2003**, *46*, 1330–1336.

13 J. C. Schöneboom, H. Lin, N. Reuter, W. Thiel, S. Cohen, F. Ogliaro, S. Shaik, The Elusive Oxidant Species of Cytochrome P450: Characterization by QM/MM Calculations, *J. Am. Chem. Soc.*, **2002**, *124*, 8142–8151.

14 J. C. Schöneboom, S. Cohen, H. Lin, S. Shaik, W. Thiel, Quantum Mechanical/Molecular Mechanical Investigation of the Mechanism of C-H Hydroxylation of Camphor by Cytochrome P450cam: Theory Supports a Twostate Rebound Mechanism, *J. Am. Chem. Soc.*, **2004**, *126*, 4017–4034.

15 J. T. Groves, Y. Watanabe, Reactive Iron Porphyrin Derivatives Related to the Catalytic Cycles of Cytochrome P-450 and Peroxidase, Studies of the Mechanism of Oxygen Activation, *J. Am. Chem. Soc.*, **1988**, *110*, 8443–8452.

16 G. Roos, S. Loverix, F. De Proft, L. Wyns, P. Geerlings, A Computational and Conceptual DFT Study of the Reactivity of Anionic Compounds: Implications for Enzymatic Catalysis, *J. Phys. Chem. A*, **2003**, *107*, 6828–6836.

17 R. G. Pearson, Hard and Soft Acids and Bases, *J. Am. Chem. Soc.*, **1963**, *85*, 3533–3539.

18 M. E. Beck, Do Fukui Function Maxima Relate to Sites of Metabolism? A Critical Case Study, *J. Chem. Inf. Model.*, **2005**, *45*, 273–282.

19 R. G. Parr, W. Yang, Density Functional Approach to the Frontier-Electron

Theory of Chemical Reactivity, *J. Am. Chem. Soc.*, **1984**, *106*, 4049–4050.
20 R. G. Parr, W. Yang, Density Functional Theory of Atoms and Molecules, Clarendon Press, **1989**.
21 W. Yang, R. G. Parr, R. J. Pucci, Electron Density, Kohn-Sham Frontier Orbitals, and Fukui-Functions, *J. Chem. Phys.*, **1984**, *81*, 2862–2863.
22 P. W. Ayers, R. G. Parr, Variational Principles for Describing Chemical Reactions: The Fukui Function and Chemical Hardness Revisited, *J. Am. Chem. Soc.*, **2000**, *122*, 2010–2018.
23 P. W. Ayers, W. Yang, Perspective on "Density Functional Approach to the Frontier-Electron Theory of Chemical Reactivity", *Theor. Chem. Acc.*, **2000**, *103*, 353–360.
24 R. G. Pearson, R. G. Parr, Absolute Hardness: Companion Parameter to Absolute Electronegativity, *J. Am. Chem. Soc.*, **1983**, *105*, 7512–7516.
25 R. S. Mulliken, New Electroaffinity Scale; Together with Data on Valence States and on Valance Ionization Potentials and Electron Affinities, *J. Chem. Phys.*, **1934**, *2*, 782–789.
26 L. Pauling, *Nature of the Chemical Bond*, Cornell, **1960**.
27 R. T. Sanderson, Electronegativities in Inorganic Chemistry, *J. Chem. Educ.*, **1952**, *29*, 539–544.
28 R. Hoffmann, R. G. Woodward, Conservation of Orbital Symmetry, *Acc. Chem. Res.*, **1968**, *1*, 17–22.
29 R. B. Woodward, R. Hoffmann, Die Erhaltung der Orbitalsymmetrie, *Angew. Chem.*, **1969**, *21*, 797–869.
30 K. Fukui, Role of Frontier Orbitals in Chemical Reactions, *Science*, **1982**, *218*, 747–754.
31 M. Torrent-Sucarrat, M. Duran, M. Sola, Global Hardness Evaluation Using Simplified Models for the Hardness Kernel, *J. Phys. Chem. A*, **2002**, *106*, 4632–4638.
32 M. Torent–Sucarrat, J. M. Luis, M. Duran, A. Toro-Labbe, M. Sola, Relations Among Several Nuclear and Electronic Density Functional Reactivity Indices, *J. Chem. Phys.*, **2003**, *119*, 9393–9400.
33 T. Katagi, Application of Molecular Orbital Calculations to the Estimation of Environmental and Metabolic Fates of Pesticides. In *Rational Approaches to Structure, Activity and Ecotoxicology of Agrochemicals*, W. Fujita, T. Draber (Eds.), CRC Press, Boca Raton, Florida, USA, **1992**, 543–564.
34 D. L. Reid, C. J. Calvitt, M. T. Zell, K. G. Miller, C. A. Kingsmill, Early Prediction of Pharmaceutical Oxidation by Computational Chemistry and Forced Degradation, *Pharm. Res.*, **2004**, *21*, 1708–1717.
35 G. Cruciani, E. Carosati, B. De Boeck, K. Ethirajulu, C. Mackie, T. Howe, R. Vianello, MetaSite: Understanding Metabolism in Human Cytochromes from the Perspective of the Chemist, *J. Med. Chem.*, **2005**, *48*, 6970–6979.
36 T. Bürger, MOCCA: A Conformational Analysis Tool Combining Monte Carlo and Simulated Annealing Techniques., unpublished results.
37 T. A. Halgren, Merck Molecular Force Field, I–IV, *J. Comput. Chem.*, **1996**, *17*, 491–641.
38 T. A. Halgren, Merck Molecular Force Field, VI–VII, *J. Comput. Chem.*, **1999**, *20*, 720–748.
39 Sybyl Version 7.2, distributed by Tripos Inc., St. Louis, MO, USA, **2005**.
40 K. Eichkorn, O. Treutler, H. Oehm, M. Haeser, R. Ahlrichs, Auxiliary Basis Sets to Approximate Coulomb Potentials, *Chem. Phys. Lett.*, **1995**, *240*, 283–290.
41 K. Eichkorn, O. Treutler, H. Oehm, M. Haeser, R. Ahlrichs, Auxiliary Basis Sets to Approximate Coulomb Potentials, *Chem. Phys. Lett.*, **1995**, *242*, 652–660.
42 K. Eichkorn, F. Weigend, O. Treutler, R. Ahlrichs, Auxiliary Basis Sets for Main Row Atoms and Their Use to Approximate Coulomb Potentials, *Theo. Chem. Acc.*, **1997**, *97*, 119–124.
43 J. P. Perdew, Density Functional Approximation for the Correlation Energy of the Inhomogenous Electron Gas, *Phys. Rev. B*, **1986**, *33*, 8822–8824.
44 A. D. Becke, Density Functional Exchange–Energy Approximation with Correct Asymptotic Behaviour, *Phys. Rev. A*, **1988**, *38*, 3098–3100.
45 K. Jankowski, R. Becherer, P. Scharf, R. Ahlrichs, The Impact of Higher

Polarization Basis Functions on Molecular Ab Initio Results, *J. Chem. Phys.*, **1985**, *82*, 1413–1419.
46 A. Klamt, G. Schürmann, COSMO: A New Approach to Dielectric Sreening in Solvents with Explicit Expressions for the Screening Energy and Its Gradient, *J. Chem. Soc., Perkin Transact.*, **1993**, *2*, 799–805.
47 P. Li, Y. Bu, H. Ai, Theoretical Determinations of Ionization Potential and Electron Affinity of Glycinamide Using DFT, *J. Phys. Chem. A*, **2004**, *108*, 1200–1207.
48 B. Safi, R. Balawender, P. Geerlings, Solvent Effects on Electronegativity, Hardness, Condensed Fukui Functions, and Softness, in a Large Series of Diatomic and Small Polyatomic Molecules: Use of the EFP Model, *J. Phys. Chem.*, **2001**, *105*, 11102–11109.
49 R. Balawender, B. Safi, P. Geerlings, Solvent Effects on the Global and Atomic DFT-Based Reactivity Descriptors Using the Effective Fragment Potential Model. Solvation of Ammonia, *J. Phys. Chem. A*, **2001**, *105*, 6703–6710.
50 R. Ahlrichs, M. Bär, H.-P. Baron, R. Bauernschmitt, S. Böcker, M. Ehrig, K. Eichkorn, S. Elliott, *et al.*, Turbomole Version 5.6., Universität Karlsruhe, Germany, **2002**.
51 M. Keil, T. Exner, J. Brickmann, Molcad II (Mk) v1.4.16., Distributed by Tripos Inc., St. Louis, MO, USA, **2004**.
52 U. Wahlgren, The Effective Core Potential Method. In *Lecture Notes in Quantum Chemistry: European Summer School in Quantum Chemistry*, B. J. Roos (Ed.), Springer Verlag, Berlin, Germany, **1992**, 413–421.
53 M. Dolg, U. Wedig, H. Stoll, H. Preuss, Energy–Adjusted Ab Initio Speudopotentials for the First Row Transition Elements, *J. Chem. Phys.*, **1987**, *86*, 866–872.
54 A. A. Kousba, L. G. Sultatos, T. S. Poet, C. Timchalk, Comparison of Chlorpyrifos-Oxon and Paraoxon Acetylcholinesterase Inhibition Dynamics: Potential Role of a Peripheral Binding Site, *Toxicol. Sci.*, **2004**, *80*, 239–248.
55 A. Szabo, N. S. Ostlund, *Modern Quantum Chemistry: Introduction to Advanced Electronic Structure Theory*, McGraw-Hill, NY, USA, **1982**.
56 R. McWeeny, *Methods of Molecular Quantum Mechanics*, Academic Press, **1978**.
57 P. Jeschke, A. Harder, W. Etzel, W Gau, G. Thielking, G. Bonse, K. Iinuma, Synthesis and Anthelmintic Activity of Thioamide Analogues of Cyclic Octadepsipeptides Such As PF1022A, *Pest. Manag. Sci.*, **2001**, *57*, 1000–1006.
58 M. E. Beck, M. Schindler, P. Jeschke, in preparation.
59 F. Ogliaro, S. Cohen, M. Filatov, N. Harris, S. Shaik, The Highvalent Compound of Cytochrome P450: The Nature of the Fe-S Bond and the Role of the Thiolate Ligand as an Internal Electron Donor, *Angew. Chem., Intern. Edition*, **2000**, *39*, 3851–3855.

Keywords

Fukui Function, Local Reactivity, Cytochrome, Compound I, Metabolism, Conceptual Density Functional Theory, Emodepside, Chlorpyrifos, Parathion

IV
Formulation and Application Technology

Pesticide Chemistry. Crop Protection, Public Health, Environmental Safety
Edited by Hideo Ohkawa, Hisashi Miyagawa, and Philip W. Lee

ISBN: 978-3-527-31663-2

25
Homogeneous Blends of Pesticide Granules

William L. Geigle, Luann M. Pugh

25.1
Introduction

In order to tailor products for specific crop protection market segments, agrochemical companies often formulate two or more active ingredients into a single product to effectively target multiple pests. A formulation containing multiple active ingredients will deliver a set ratio of active ingredients in the spray application. Each formulation can typically require 6–9 months for formulation development time followed by 1–2 years of regulatory field tests and extensive environmental and toxicological test data to support product registration. A formulation containing the same active ingredients but in a different ratio will also require a complete development program. This can be cost and time intensive, since several ratios can be desired to meet the local needs of specific market segments.

Mixing separately formulated granular products provides an alternative to formulating individual products with set ratios of active ingredients. However, mixtures of these granular products can segregate and they require packaging in containers sized for a single application (unit area pack) so that variability within the package is not important. Furthermore, the applicator loses the ability to dose the container in subunits and the package size may not be convenient for the application.

To eliminate segregation in the mixture, DuPont has pioneered homogeneous blends of dry granules where the geometry of each mixture component is carefully controlled. With this technology, two or more granular products can be blended over a wide range of ratios to tailor active ingredient combinations for specific market segments. By controlling the granule size, the granules give a homogeneous mixture in the container and will remain homogeneous during transportation. These products can be used in the same manner as conventional products containing multiple active ingredients. They can be subsampled by the applicator with the guarantee that each subsample will contain the same composition as the bulk material.

Pesticide Chemistry. Crop Protection, Public Health, Environmental Safety
Edited by Hideo Ohkawa, Hisashi Miyagawa, and Philip W. Lee

ISBN: 978-3-527-31663-2

25.2
Granule Blend Products

25.2.1
General Theory of Segregation

Mixtures of solid particles can segregate during handling and lead to inhomogeneity of the bulk material. The major properties that can lead to segregation of particles include differences in particle size, density, and shape. While each of these can contribute to overall segregation, it is widely recognized that the particle size is by far the most important [1–6]. Particle median diameter ratios as low as 1.1 have been shown to cause segregation [7], with the larger the ratio the greater the tendency for segregation [2, 8–9]. Surprisingly, the contribution of differences in density is comparatively unimportant [1–2], even at density ratios up to 6 [10]. In fact, it has been demonstrated that when mixtures of different particle sizes are vibrated, the bigger particles can always be made to rise, regardless of their density [9].

Bulk blending of granular components is a process well known in the fertilizer business where it is common to blend various proportions of straight fertilizers to yield appropriate N,P,K grades for specific applications. A standard rule of thumb for fertilizer blending is to maintain a component median diameter difference of less than 5% [11]. Consideration of the component size distribution or spread is also important to consider when comparing properties of materials intended for blending [12].

25.2.2
Sampling of Granule Blends

Every mixture of solid particles, if scrutinized closely enough, will show areas of segregation – that is, the composition will vary from point to point [13]. Because mixtures are composed of particles with different compositions, it is important to consider the sample size needed to represent the comosition. In an ideal mixture of equal amounts of two particles, the arrangement will not be ordered such that each particle is directly next to a different type (as in Figure 1), but will occur as a random mixture where variations between spot samples of a known size will be found (as in Figure 2) [14].

A common way to express the randomness (or the degree of mixing) of a particle blend is by measuring the statistical variation (standard deviation) in the composition of several separate samples. For a completely random mixture it has been shown that:

$$\text{Standard Deviation} = [(P)\,(1 - P)\,/\,n]^{0.5}$$

where P is the overall proportion of one component and n is the number of particles in each sample [14].

Figure 1. Ordered mixture (not random).

Figure 2. Random mixture.

Using the above equation, the theoretical sample size needed to represent the bulk composition of a blend with a given standard deviation can be calculated. Data in the table below are normalized to the relative standard deviation based on the component concentration (P) to allow for easy comparison (relative standard deviation = standard deviation / $P \times 100$). The number of particles is also expressed as grams. It can be seen in the table that the lower the %component in the blend or the greater the confidence in the composition (lower the acceptable relative standard deviation), the greater the number of particles are required in the sample.

The required sample size for a granule blend is important to consider relative to the intended use practices for the product. The sample size required to represent the blend composition must be less than the amount an applicator would be expected to measure for use. In other words, a blend containing 25% of one component and requiring a 20–30 g sample to accurately represent the blend composition would not be a good candidate for a product that would be used in back pack applications where the user would need to sample a 1-g aliquot. However, an application in a ground sprayer at a product rate of 1 ounce per acre

Table I. Relationship between blend composition and sample size for different standard deviations.

Acceptable Relative Standard Deviation (RSD)	1.5		2.0	
% Component in blend	# particles	*grams	# particles	*grams
5	*84444*	211	*47500*	119
10	*40000*	100	*22500*	56
25	*13333*	33	*7500*	19
50	*4444*	11	*2500*	6.3

* Assumes approximately 400 particles are in 1 gram.

could easily be accommodated. This "feasibility" of a blend product needs to be considered for any commercial offering.

25.2.3 Manufacture of Granule Blends

Homogeneous blends of pesticide granules are manufactured by mixing extruded formulated granules. The components for the granule blends are prepared by an extrusion process such that the diameter is fixed and the granule length distribution is modified by a sifting operation (References [15–16]). Since the components are matched in their size, segregation does not occur.

Blends are mixed assuming the nominal concentration of each component and can be prepared in either continuous blending equipment or in a batch process. In a continuous process, careful calibration of the dispensers is critical to maintaining the correct ratio of components during the process. In batch mixers, a good knowledge of the geometry, performance and rotation speed is needed to set the appropriate operating conditions. In general, higher ratio compositions will take longer to adequately mix, and the ability to take appropriate samples is critical to determine the blending endpoint.

25.2.4 Regulatory Requirements for Granule Blends

The regulatory requirements for granule blends are similar to the requirements of the component products, including physical tests and chemical stability. However, each partner is itself a formulation with a registered assay tolerance. Because of the real-life variability associated with the component assay as well as mixing and sampling variability, the Food and Agriculture Organization (FAO), US EPA, and other regulatory agencies have established guidelines for setting upper and lower limits for formulations that are composed of formulations. These guidelines are an extension of the standard assay tolerance specification for basic formulations which are based on a sliding scale such that the lower the percent assay in the product, the more variability is allowed. Guidelines for assay specifications for basic formulations according to the FAO and US EPA conventions are shown in the table below.

To address the additional variability in a granule blend introduced by each blend component having an associated assay variability, FAO recently accepted a procedure for calculating the tolerances for active ingredient contents in proucts that are mixtures of formulated products. The limit for the active ingredient content in each component formulation is expanded by applying a corresponding tolerance to the content of the fomulation in the mixture [17]. This provides a simple empirical approach to calculate expanded tolerances and reflect limits achievable with good practice in manufacturing. An example of the calculation for a blend product containing 75% of a 60% AI component and 25% of a 50% AI component is shown in the table below.

Table II. Tolerances for active ingredient content.

Nominal Concentration = *N*	Range (as % of Nominal)	
	US EPA	FAO
$N \leq 1\%$	± 10%	
$N \leq 2.5\%$		± 25% (for WG)
$1\% \leq N \leq 20\%$	± 5%	
$2.5\% \leq N \leq 10\%$		± 10%
$10\% \leq N \leq 25\%$		± 6%
$25\% \leq N \leq 50\%$		± 5%
$N \geq 20\%$	± 3%	
$N \geq 50\%$		± 25 g/kg

Table III. Calculation of tolerance for a granule blend.

	%AI in Component (A)	%Component in Blend (B)	Calculated tolerance for AI in Blend Product (A × B)
Maximum	62.5	77.5	48.4
Component A – *Nominal*	60 ± 25 g/kg	75 ± 25 g/kg	45.0
Minimum	57.5	72.5	41.7
Maximum	52.5	26.5	13.9
Component B – *Nominal*	50 ± 5%	25 ± 6%	12.5
Minimum	47.5	23.5	11.2

25.2.5 Measurement of Homogeneity

Measurement of granule blend homogeneity has been accomplished by analyzing multiple subsamples to verify that all portions meet the required assay specifications. In a typical test, at least 10 subsamples are analyzed. Tests using granule components that differ in size distribution will show a higher variability in composition. In the figure below, both compositions average 50% of the measured component, but the variability is quite different for the two mixtures. In the homogeneous blend, the relative standard deviation of the samples is 1.3%, while in the poor quality blend the relative standard deviation is > 30%. Obviously only the blend with the low sample to sample variability could be accurately subsampled as a homogeneous product.

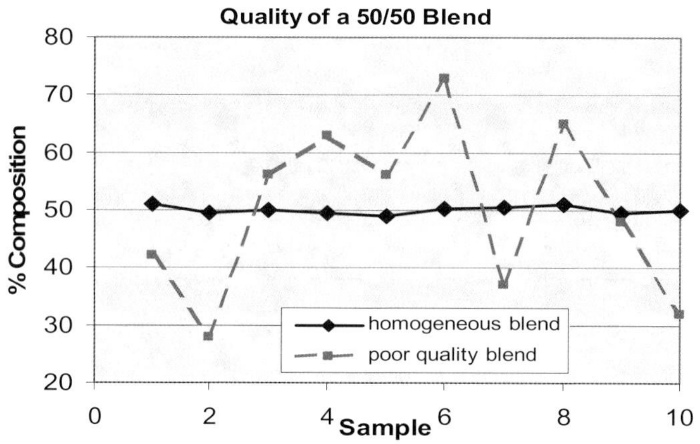

Figure 3. Compare the quality of a homogeneous blend and a poor quality blend.

25.2.6
Advantages of Granule Blends

Granule blends offer convenience to grower such that custom combinations of different blend partners can be offered to meet local or regional needs. The blend product remains homogeneous during shipment, storage, and measurement. Using a blend product requires the farmer to purchase, inventory, and measure a single product rather than separately purchase, measure, and inventory multiple products. The homogeneous granule blend product also eliminates the need to unit pack a mixture product which eliminates dose flexibility by the applicator.

25.3
Conclusion

Homogeneous granule blends are mixtures of formulated granular products with carefully controlled geometry such that the resulting blend will not segregate. The blends can be prepared in either a continuous or a batch operation. Sample size considerations are important to verify the sample amount is large enough to adequately represent the bulk composition, and also be consistent with applicator use practices. Once mixed, the blends will remain homogeneous during shipping and handling and provide an opportunity for agrochemical suppliers to tailor offerings to meet specific customer or geographic needs.

25.4 References

1 J. C. Williams, *Fuel Soc. J.*, **1963**, *14*, 29–34.
2 H. Campbell, W. C. Bauer, *Chem. Eng.*, **1966**, *73*, 19, 179–185.
3 R. L. Brown, *J. Inst. Fuel*, **1939**, *13*, 15–19.
4 G. Hoffmeister, S. C. Watkins, J. Silverberg, *J. Agr. Food Chem.*, **1964**, *12*, 64–69.
5 J. C. Williams, *Powder Technology*, **1976**, *15*, 245–251.
6 G. A. da Silva, *Braz. J. Chem. Eng.*, **1977**, *14*, 3, 259–268.
7 I. Bridle, M. S. A. Bradley, A. R. Reed, H. Abou-Chakra, U. Tuzun, I Hayati, M. Phillips, *International Fertiliser Society*, **2005**, *547*, p1–27.
8 J. C. Williams, M. I. Kahn, *Chem. Eng.*, **1973**, *19*, 269.
9 R. Julien, P. Meakin, A. Pavlovitch, *Phy. Rev. Lett.*, **1992**, *69*, 640–643.
10 A. P. Campbell, J. Bridgwater, *Trans. Inst. Chem. Eng.*, **1973**, *51*, 72.
11 M. S. A. Bradley, R. J. Farnish, *International Fertiliser Society*, **2005**, *554*, 1–15.
12 O. Miserque, E. Pirard, *Chemometrics and Intelligent Laboratory Systems*, **2004**, *74*, 215–224.
13 P. V. Danckwerts, *Research*, **1953**, *6*, 355–361.
14 P. M. C. Lacey, *Trans. Inst. Chem. Eng.*, **1943**, *21*, 52.
15 US Patent No. 6,022,552.
16 US Patent No. 6,270,025.
17 *Manual on Development and Use of FAO and WHO Specifications for Pesticides*, March 2006 Revision of First Edition (available only on the internet, http://whqlibdoc.who.int/publications/2006/9251048576_eng_update_2006.pdf)

Keywords

Homogeneous, Granule, Blend, Mixture, Pesticide, Non-segregating

26
Sprayable Biopesticide Formulations

Prem Warrior, Bala N. Devisetty

26.1
Introduction

Agriculture in the 21st century has seen significant advances that may be best defined as ground-breaking. With the introduction of genetic engineering tools, crop production has seen remarkable evolution in the use of toxins and enzymes of natural origin either to avoid or resist a pest or to modify the plant's genetic machinery to enhance the nutritional value or provide protection from potent herbicides. The first commercial attempt at the introduction of an insecticidal toxin from *Bacillus thuringiensis* into corn plants was carried out in the 1980s; this was soon followed by the introduction of herbicide-tolerant crop cultivars. The growth in sale of GM seeds (genetically modified) has continued to increase steadily since 1995, while that of conventional seeds has remained generally steady in spite of an initial decline [1].

Use of a biological pesticide typically involves the use of natural living systems [2] or non-living systems of biological origin to either protect plants from pests or diseases or to enhance crop yields. Biological pesticides, in general, are dependent on the successful establishment and maintenance of a threshold population of suppressive organisms on the crop plants, the soil or more generically the matrix, below which their efficacy is impaired or insufficient. Most of our current knowledge base on biological pesticides results from diligent studies on very few biologically active microorganisms such as *Bacillus thuringiensis, Bacillus subtilis, Trichoderma harzianum*, and *Metarrhizium anisopliae* to name a few and are now being applied to the newer active ingredients. Devisetty *et al.* [3], provided an in-depth review of the various approaches to biopesticide formulations with particular reference to Bt-based products. Burges [4] published a treatise on the formulation of microbial pesticides and has succeeded in highlighting the importance of this area of research for commercializing biological products. This document reviews some of the key aspects of commercial development of biopesticides in the 21st century.

Pesticide Chemistry. Crop Protection, Public Health, Environmental Safety
Edited by Hideo Ohkawa, Hisashi Miyagawa, and Philip W. Lee

ISBN: 978-3-527-31663-2

26.2
The Biopesticide Market

A recent report estimates the North American market for biopesticides at 109 million USD in 2005 [5] at the grower level and it has been projected to grow to 260 million USD by 2015. The largest market for biopesticides was estimated to be in the United States (89 million USD) with *Bacillus thuringiensis*-based (Bt) products taking more than 60% of total market share. The fastest growing segment in the North American market was reported to be the public health segment covering mosquito and black fly control products based on *B. thuringiensis* subsp. *israelensis* and *B. sphaericus*.

In general, the use of biopesticide products have been on high value crop segments (fruit and vegetable crops) and not as much on the high acre/lower value field crops such as corn, soybean, and wheat. The other markets for biological pesticides include nursery crops, and other container-grown ornamental species. Among the non-agronomic market segments, forestry (for control of lepidopteran insect pests) and public health use targeted at control of disease-vectoring mosquitoes such *Aedes* spp. *Anopheles* spp., and *Culex* spp. are the most important. Currently in the United States, there are more than 700 biopesticide products based on more than 195 active ingredients registered by the US EPA [6].

The key drivers for the biopesticide industry are the increasing demand for food safety and quality. There is clearly an increasing demand for higher quality produce and the more discerning consumer is seeking produce with specific attributes. There is increasing awareness and sensitivity about the environmental impact of pesticide use. These have led to a higher degree of interest in residue management as well as integrated pest management (IPM) utilizing biopesticides. One of the most influential, driving forces in the growth of the biopesticide industry as a whole is the favorable regulatory environment for environmentally compatible biorational products in the United States that has promoted the development, registration, and commercialization of such pesticides in this country. In addition, this trend is particularly indicative of several other countries such as Canada, Australia, and Mexico. Harmonization efforts on a global basis have enabled the faster adoption of biopesticides in several countries including those in Europe where a multi-national, cooperative initiative – REBECA (Regulation of Biological Control Agents) – is attempting to streamline the registration and commercialization of biopesticides.

26.3
Biological Pesticides

Biopesticides include microbial living systems primarily based on bacteria, fungi and viruses. They may also include macro-organisms such as entomopathogenic nematodes, insect predators, and parasites. Biological pesticides may also include plant-derived metabolites as well as insect pheromones and most interestingly microbial metabolites. In its broadest definition, biopesticides may be defined as pest or disease limiting agents of biological origin. The scope of this review is primarily related to the biological pesticides of microbial origin, even though references may be made to other categories to illustrate specific examples. Several candidates for biopesticides were discussed in an earlier review [7–8]. In the general field of microbial pesticides, bacteria have attracted enormous attention as potential agents for biological control since they are easier and economical to produce and stabilize, are considered aggressive colonizers of the rhizosphere or phyllosphere, and inherently possess a rapid generation time. They are also known to affect life cycles of different plant pathogens or pests by diverse mechanisms including the production of extracellular metabolites and intracellular proteinaceous toxins. In general, spore-forming bacteria (e.g., *Bacillus* spp.) survive to a greater extent even in harsh environments, compared to the non-spore forming bacteria. As noted earlier, among the *Bacillus* spp., the ones that have attracted the most attention and studied are the *B. thuringiensis* and *B. subtilis*.

The major fungal species used in biological control are Trichoderma harzianum and Gliocladium spp. Major root and foliar fungal pathogens such as Rhizoctonia, Pythium, Botrytis, and the powdery mildew fungus [9–11] are known to be controlled by these fungi. Control of multiple plant parasitic nematodes has been achieved by the use of fermentation extracts from Myrothecium verrucaria, a hyphomycetous fungus, commercialized under the trade name DiTera® [12].

Among the viruses used in the biological control of insect pests, the baculoviruses have generated the greatest interest as potential biocontrol agents against pests of Lepidoptera, Hymenoptera, and Coleoptera insect families. Baculoviruses, though very specific in their host range and effective in specific cases, have only been used on a limited basis in the US.

26.4 Factors Affecting Biopesticide Use

Several factors affect the activity and performance of biopesticide formulations. Being generally of natural origin and often being live, these unique pesticidal products need special attention compared to the synthetic active ingredients with which they generally "compete" in agrichemical markets. Several factors alter the behavior or activity of biopesticidal compositions.

26.4.1 The Living System

Screening of ecological niches for signs of antagonism such as a disease or pathogenesis in the pest eventually leading to suppression of the pathogen or pest population have been the primary source for the discovery of new biopesticide products. Thus, the phylloplane (surface of the leaf) and rhizosphere (microenvironment around the root surface) have been the focus of attention in the search for the most promising biocontrol organism. In many cases, targeted screening of the organisms collected from a specific ecological niche (e.g., pest cadavers) is the primary process; in this step, the objective is to select the most active or virulent biological agent(s) that warrants further investigation. The selected strain is subjected to further biological screening for purity of isolates and for the absence of undesirable metabolites such as exotoxins. The final selection of the strain for product development is a very important step and needs to be reconfirmed and the isolates purified and preserved.

26.4.2 The Production System

The most critical component toward commercialization is in the identification of an economic and efficient production or manufacturing system. In the case of biological products, the objective is to maintain the intrinsic advantages of the microbe while increasing the number of viable propagules through an economically acceptable manufacturing process. The most studied of these manufacturing processes relate to the *in vitro* submerged fermentation of bacterial or fungal bioinsecticides in specific media, particularly as in the case of Bt manufacturing. Other examples of *in vitro* production include the production of biological fungicides based on *Trichoderma harzianum*, *Bacillus subtilis* and the mycoherbicide DeVine® (*Phytophthora palmivora*) and the recent efforts by multiple laboratories in the production of the bacterial endoparasitic of nematodes, *Pasteuria penetrans*.

Even though many biological pesticides can be produced on solid, semi-solid or liquid media, submerged fermentation in deep tanks is still the most economical method of choice for large-scale manufacturers. Standardized procedures for quality control for fermentation processes have now been defined in most industrial laboratories. Development of suitable, commercially viable defined media is the key step in the fermentation development process [3]. While the media components and fermentation conditions may vary widely, maintenance of sterility until the final stage is essential for optimal product performance. The final product is a function of the specific media used, fermentation growth conditions, and post-fermentation or down-stream processing. Once desired activity is produced in the fermentor, it is harvested, concentrated (precipitation, centrifugation, ultrafilteration, evaporation), and spray-dried to obtain technical powder concentrate. Various methods of drying such as lyophilization, fluidized bed and spray drying are typically used. During these processes, special attention must be given to the physical and biological properties of the spray-dried concentrate since the process variables affect the ability to develop an efficacious formulation.

Whatever method of manufacturing is used, the importance of process control and quality through every step cannot be overemphasized. The specific steps include (a) monitoring and controlling the production process from stock culture stage through inoculum preparation, (b) seed, pilot plant, and production fermentation runs, (c) recovery procedures, and (d) the final formulation of the product. It is important for the manufacturer to demonstrate sufficient process control throughout the process to ensure acceptable product quality and minimize potential contaminants. Batch control procedures including potency comparisons, microbial purity, and absence of unintentional contaminants, need to be applied to each manufacturing batch.

26.4.3 Biological Activity

The proof of activity of a biological pesticide is typically evaluated by a standardized bioassay except in the case of microbial metabolites where the major active ingredient(s) may be measured by analytical methods. Biological activity measurements, besides serving as a parameter for quality control, are an essential tool in the product development and optimization process. It is important to define the assay procedure in order to compare production batches and experimental formulations. These assays are typically used for product release or may be designed to assess specific aspects of product activity such as mobility in soils, colonization on leaf surface, etc.

26.4.4 Stabilization

The manufacturing process for biological pesticides aims at preserving the integrity of the organism or the consistency of the product throughout the production process. The addition of selected ingredients to preserve the active ingredient(s) during the post-fermentation and/or processing steps may thus be important. The choice of stabilizers used depends on the final formulation, i.e., aqueous suspension, non-aqueous suspension, wettable powder, dry flowable, etc. [4]. It may also be important to ensure stability of the formulation in tank mixes; critical factors include water quality (pH, hardness, temperature, etc.), and the specific attributes of the mixes used. Use of surfactants, UV-protectants, humectants, and other diluents might be necessary to maximize the optimal application conditions and to ensure stability of the active components on the foliage for extended biological efficacy.

26.4.5 Quality Control

Successful commercialization of a biological product necessitates the integration of various production steps discussed earlier with quality control measures to ensure product consistency, stability, and efficacy. This is the most important factor that can differentiate a successful product from an ineffective one. One of the important attributes that dictate the quality of the final product is the quality of the raw materials used in fermentation and formulation itself.

All specifications should be accompanied by an internationally acceptable method of determination of physical, chemical, and biological properties, cited in ASTM (American Society of Testing Materials) Standards and CIPAC (Collaborative International Pesticides Analytical Council) Methods. Needless to further emphasize, the testing protocols should, whenever applicable, comply with GLP (Good Laboratory Practices) and GMP (Good Manufacturing Practices). Product specifications should typically include strain identity, information on metabolite production, if any, spore or viable propagule count, biological stability data, and physical properties of formulation (particle size, viscosity, bulk density, color, odor, wetting time, dispersion/suspension properties).

Packaging used for finished products can significantly affect the quality of the final product at the distribution and warehouse location. The final package should meet all IATA (International Air Transport Association), DOT (Department of Transportation), UFC (Uniform Freight Classification), uniform packaging codes and transportation requirements for both dry and liquid biopesticides.

26.4.6 Delivery

Microbial pesticides rely on their ability to either infect the pest resulting in an infection or disease or in their ability to inhibit the pest's growth by production of active metabolites. Hence it becomes essential to deliver the pesticidal organism in the proximity of the pest itself. Since the viability of the organism may be vital, the zone of desired activity needs to be considered during the development phase of a microbial pesticide.

The microenvironment on the phylloplane or in the rhizosphere is of critical relevance when the survival of the live organism is at stake. In the case of foliar application, presence of leaf exudates, other microbial colonies, the leaf surface (waxy *vs.* smooth), orientation of the leaf, and the crop species itself plays a decisive role in defining the type of formulation required. Additionally in the case of foliar-applied pesticides, as in the case of several biological insecticides, fungicides and herbicides, rainfastness, ability to resist photodegradation, etc., are critical components to be considered. When Bt was applied to conifer foliage, oil-based formulations provided greater rainfastness than the aqueous formulations [13]. In the case of soil-applied products, however, as in the case of nematicides and soil insecticides, the factors that affect activity include soil type, soil pH, CEC (cation exchange capacity), organic matter content, soil moisture, and other microbial factors around the root zone of the plant.

26.5 Role of Formulations in Sprayable Biopesticides

The key objective of a formulation is to deliver an optimal dose of the agent at the optimal site and time and is essentially similar to chemical pesticides. However, in the production of biological formulations, many additional obstacles may have to be overcome, mainly due to the fact that the active ingredient itself may be living. Unlike a chemical active ingredient, the development of a stable, active biological formulation begins from the selection of the production process and can even be an integrated process. Unfortunately, these aspects have received very little attention and only limited research has been carried out (and much less published since some of these may be considered "trade secrets") on biomass, fermentation, and delivery of microbes for control of pathogens and pests.

There are significant differences in the performance characteristics and desired attributes of a soil-applied *vs.* foliar-applied biological product. The performance of biological pest control agents applied to the soil may depend largely on physical and chemical characteristics of soil, pH, moisture, temperature regiments, as well as their ability to compete with the native microflora. Similarly, environmental factors such as temperature, free moisture, protection against UV irradiation and desiccation can influence biological control in the phyllosphere [14]. The above-ground parts are often directly exposed to harsh climatic conditions that could be hostile for microorganisms and may be significantly different from the rhizosphere which is generally considered more conducive for the survival of microbes, especially with regard to its structure, ecology, and nutrient status [15]. Also, the compatibility of the biocontrol agent with existing cultural practices and chemical control methods is an important criterion for successful use in formulation [16].

Biological formulations applied to seeds greatly help deliver the microbial agents to the spermosphere of plants, where, in general, extremely conducive environments prevail. Significant advances in seed treatment technology have been made in the past few years, and the approach is considered an attractive means for introducing biological control agents into the soil-plant environment. Much of the work on seed treatment has been carried out on *Bacillus subtilis* strains as in the case of the commercial product Kodiak®, used for control of *Rhizoctonia* and *Fusarium* in cotton and peanut. Several other gram-negative bacterial antagonists such as *Pseudomonas fluorescens* have also been evaluated for potential commercial use; however, in many instances stability issues have impeded large scale use of many of these strains.

Liquid formulations are applied to foliar parts of the plants for control of insect pests. Ultra low volume (ULV) aerial applications using high potency formulations such as emulsifiable suspensions (ES) or aqueous suspension concentrates (SC) have been developed for forestry applications. Development of high potency, cost-effective, formulations with good suspension properties, and good stability have contributed substantially to the successful global adoption of biological Bt-based formulations in forestry. The recent introduction of the dry flowable (DF) or wettable granule (WG) formulations for Bt strains is a significant step in this direction.

The formulated biological product may also be directly applied to plant roots in the form of a root-dip, spray, drip or flood application for control of soil pests and diseases. This is primarily applicable for control of fungal diseases such as

Fusarium, *Pythium*, or *Rhizoctonia*, delivery of insect-parasitic nematodes and for management of plant parasitic nematode populations. In all the above instances, the objective of the formulation is to stabilize and preserve the active ingredient and also to distribute it evenly in the rhizosphere. Formulation ingredients targeted to protect the viable propagules such as spores may also be required. Uniform lateral and horizontal distribution in the soil matrix is the most important goal when the target is a soil pest such as a nematode. In the case of the soil-applied nematicide DiTera, an emulsifiable suspension has been used successfully to deliver this killed-microbial product for commercial nematicide treatments on grapes [17].

26.6 Future Outlook and Needs

Biological pesticides will remain an integral part of global agriculture. The continuing need for safer, environmentally compatible pest and disease control agents will drive the need for newer, cost-effective active ingredients as well as commercial products. On the technical side, there will be a continued need to discover and develop newer products with newer modes of action. Newer technologies in formulations to preserve biological agents as well as their by-products specifically targeted to delivery at the right place at the right time will have to be developed. Continued research on technologies to help preserve the specific active ingredients (live propagules as well as metabolites) will help extend as well as enhance the activity of biopesticides is needed.

Even though newer chemistries and GM plants will continue to be developed, there will be a continuing demand to develop newer alternatives using living systems as well as products derived from living systems. On the commercial side, there is a continued need to identify specific grower needs where the biological pesticides can be uniquely effective. Recognition of the specific grower needs and integration of the biological pest control opportunity into the existing grower practice is essential. This necessitates a good understanding of the customer value equation in terms of the input costs as well as the return on investment and to provide the grower with a solution that makes economical sense. This may also involve education of the grower, the applicator, the distributor, or in other words, education across the entire value chain. Finally, it is essential to manage the expectations on a biological pesticide; it is important to recognize that biopesticides may not work in every situation and may not be ideal for every pest/disease. Biopesticides do work well when used properly. They are here to stay and will remain an integral part of crop production as either stand-alone pesticides or as "life-extenders" of chemical active ingredients and/or genetically modified crop plants.

26.7
References

1 Phillips McDougall, *The Global Crop Protection Markets – Industry Prospects*, June **2006**.
2 P. Warrior, *Pest Management Sci.*, **2000**, *56*, 681–687.
3 B. N. Devisetty, Y. Wang, P. Sudershan, B. L. Kirkpatrick, R. J. Cibulsky, D. Birkhold in *Pesticide Formulations and Application Systems*, J. D. Nalewaja, G. R. Goss, R. S. Tann (Eds.), **1998**, *18*, ASTM STP 1347, 242–272.
4 H. D. Burges (Ed.), *Formulation of Microbial Pesticides*, Kluwer Academic Publishers, **1998**.
5 R. Quinlan, A. Gill, in *North America: The Biopesticides Market*, CPL consultants, Wallingford, Oxfordshire, UK, **2006**, 135.
6 US EPA website, http://www.epa.gov/pesticides/biopesticides/whatarebiopesticides.htm.
7 P. Warrior, K. Konduru, P. Vasudevan, in *Biological Control of Crop Diseases*, S. S. Gnanamanickam (Ed.), Marcel Dekker, Inc., NY, **2002**.
8 B. N. Devisetty, Production and Formulation Aspects of *Bacillus thuringiensis*. In *Proc. 2nd Canberra Meeting on Bacillus thuringiensis*, R. J. Akhurst (Ed.), CPN, Australia, **1994**, 95–102.
9 J. S. Cory, D. H. L. Bishop, in *Methods in Molecular Biology*, Vol. 39: Baculovirus expression protocols, C. D. Richardson (Ed.), Totowa, NJ, Humana, **1995**, 277–294.
10 R. R. Belanger, M. Benyagoub, *Can. J. Plant Pathol.*, **1997**, *19*, 310–314.
11 Z. K. Punja, *Can. J. Plant Path.*, **1997**, *19*, 315–323.
12 Y. Elad, I. Chet, Practical Approaches for Biocontrol Implementation. In *Novel Approaches to Integrated Pest Management*, R. Reuveni (Ed.), Lewis Publishers, CRC Press, Boca Raton, FL, **1995**, 323–328.
13 A. Sundaram, J. W. Leung, B. N. Devisetty, Rain-fastness of Bacillus thuringiensis Deposits on Confer Foliage. In *Pesticide Formulations and Application Systems: 13th Volume*, ASTM STP 1183, P. D. Berger, B. N. Devisetty, F. R. Hall (Eds.), American Society of Testing and Materials, Philadelphia, **1993**.
14 C. D. Boyette, P. C. Quimby Jr., A. J. Caesar, J. L. Birdsall, W. J. Connick, Jr., D. J. Daigle, M. A. Jackson, G. H. Eagley, *et al.*, *Weed Technol.*, **1996**, *10*, 637–644.
15 J. H. Andrews, Biological Control in the Phyllosphere, *Annu. Rev. Phytopathol.*, **1992**, *30*, 603–635.
16 B. J. Jacobson, P. A. Backman, *Plant Dis.*, **1993**, *77*, 311–315.
17 P. Warrior, L. Rehberger, M. Beach, P. A. Grau, G. W. Kirfman, J. M. Conley, *Pestic. Sci.*, **1999**, *55*, 376–379.

Keywords

Biological Pesticides, *Bacillus thuringiensis*, Integrated Pest Management, Myrothecium, Formulations, *Bacillus subtilis*, Biopesticides, Larvicides, Biological Control

V
Mode of Action and IPM

Pesticide Chemistry. Crop Protection, Public Health, Environmental Safety
Edited by Hideo Ohkawa, Hisashi Miyagawa, and Philip W. Lee

ISBN: 978-3-527-31663-2

27
Molecular Basis of Selectivity of Neonicotinoids

Kazuhiko Matsuda

27.1
Introduction

Insect nervous systems are targets of many commercial insecticides showing high selective toxicity. Neonicotinoids acting on insect nicotinic acetylcholine receptors (insect nAChRs) provide examples of such selectivity. Binding assays and voltage-clamp electrophysiology have revealed that the selective toxicity of neonicotinoids stem, at least in part, from a degree of selectivity to their targets, insect nicotinic acetylcholine receptors. The author has studied the mechanism for the receptor-based selectivity of neonicotinoids in collaboration with Professor David Sattelle's team in Oxford. Semi-empirical molecular orbital calculations of neonicotinoid features, combined with two-electrode voltage-clamp electrophysiology and homology modeling of nAChRs with imidacloprid bound, have contributed to our understanding of the mechanism. In this chapter, the selectivity of neonicotinoids is considered in terms of the structural features of neonicotinoids and the agonist binding loops of insect nAChRs involved in the neonicotinoid-nAChR interactions.

27.2
Interactions with Basic Residues Induce a Positive Charge in Neonicotinoids which Mimics the Quaternary Ammonium of Acetylcholine

The lead compound, which initiated interest in the neonicotinoids, is a nitromethylene heterocycle, nithiazine (Figure 1) [1]. Nithiazine, with its unique nitromethylene moiety, showed low mammalian toxicity, but its insecticidal potency and field stability were inferior to commercial organophosphates and pyrethroids. Thus, nithiazine was not of commercial use for pest control. However, the introduction of 6-choloronicotinyl and 2-nitroimino-imidazolidine moieties led to the development of the first commercial neonicotinoid imidacloprid (Figure 1) [2]. The key features of imidacloprid are its high insecticidal efficacy for plant-

Pesticide Chemistry. Crop Protection, Public Health, Environmental Safety
Edited by Hideo Ohkawa, Hisashi Miyagawa, and Philip W. Lee

ISBN: 978-3-527-31663-2

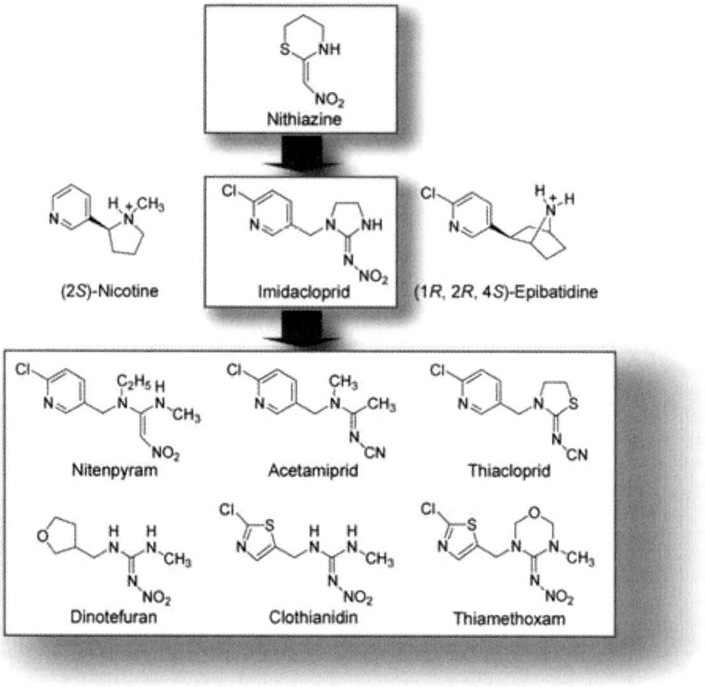

Figure 1. Natural nicotinic ligands nicotine and epibatidine and the development of neonicotinoids.

sucking pests and good systemic activity in plants. Also, imidacloprid was less prone to photo-decomposition than nithiazine and related nitromethylenes, which offered the stability required for field applications. The successful introduction of imidacloprid into the market ignited the development of the 2nd and 3rd generations of neonicotinoids (Figure 1).

The structural features of neonicotinoids are the nitro and cyano groups, which are conjugated with heteroatom(s) via C=C or C=N bonds to form "toxophores" with high affinity binding to insect nAChRs. Whereas protonated nicotine, epibatidine and acetylcholine possess a positively charged nitrogen (Figure 1), imidacloprid has no such nitrogen but instead negatively charged oxygens. A simple explanation for the selectivity of neonicotinoids is that the negatively charged oxygen(s) form hydrogen bonds and the bond donors therefore determine the binding potency. However, the negativity of the nitro oxygens is not the only factor contributing to the binding affinity and selectivity of neonicotinoids.

It has been proposed that the quaternary ammonium of acetylcholine (ACh) makes contact directly with the tryptophan residue in loop B of the nAChR via cation-π interactions [3–4]. This has been extended to the neonicotinoid-nAChR interactions, where the imidazolidine ring makes contact with the tryptophan residue [5]. This theory, however, has a problem because the negative nitrogens of the imidazolidine ring do not favor interactions with the π-electrons of the

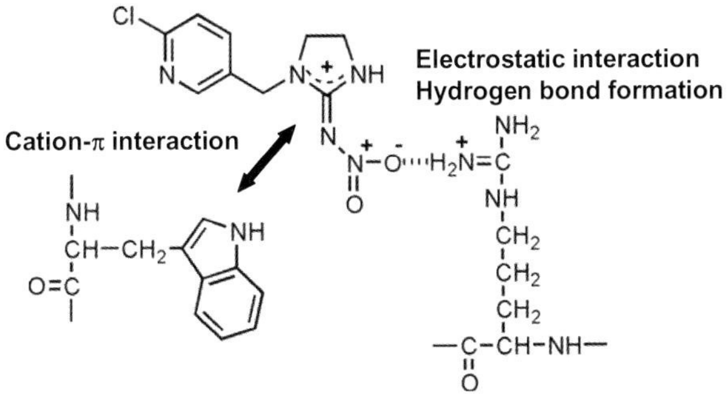

Figure 2. Schematic representation of imidacloprid-insect nAChR interactions.

tryptophan ring. Furthermore, the tryptophan residue is unable to account for the selective interactions with neonicotinoids because it is present not only in insect, but also in vertebrate nAChR agonist binding sites. Calculation of the electrostatic potentials of the imidazolidine moiety of imidacloprid before and after interactions with ammonium showed that the two nitrogens of the imidazolidine are made positive by the electrostatic and hydrogen bond formation of the nitro oxygens with ammonium [6]. The positively charged nitrogens then mimic the quaternary ammonium of ACh, involved in the cation-π interactions with the loop B aromatic residue [7]. It is conceivable that the interactions of the nitro or cyano groups with basic residues, which are present only in insect nAChRs, play a key role in the selective neonicotinoid-insect nAChR interactions (Figure 2). Accordingly, the most important problem to be solved in elucidating the mechanism for the selectivity of neonicotinoids is to identify such basic residues present somewhere in insect nAChRs.

27.3
Exploring Structural Features of nAChRs Contributing to the Selectivity of Neonicotinoids Employing the α7 nAChR

The well-studied nicotinic acetylcholine receptors of vertebrates usually consist of two α and three non-α subunits, but α9 forms a heteromer with α10 with a stoichiometry of two α9 and three α10 subunits when expressed in *Xenopus* oocytes [8]. Also, α7, α8, and α9 subunits can form functional homomers in the oocyte expression system [9–11]. The cation-permeable ion channels of nAChRs open in response to binding of ACh. Most neonicotinoids are agonists of native [12] and recombinant [13–14] nAChRs, thereby probably sharing their binding site, at least in part, with ACh. The agonist binding site consists of 6 loops A–F, of which loops A–C are located on α subunits whereas loops D–F are present on the non-α subunits [15]. However, α7–α9 subunits forming homomers possess all the 6 loops. Although it is a vertebrate subunit, α7, unlike some other vertebrate subunits, is sensitive to neonicotinoids and a great deal is known of the physiology and pharmacology of the native and recombinant α7 receptor. Unfortunately, at this time, there is no robust functional recombinant receptor made only of insect nAChR subunits.

The loop we first investigated to identify the possible site of interactions with neonicotinoids was loop F, because it was earlier referred to as a negative subsite involved in the electrostatic interactions with quaternary ammonium of ACh [16]. Site-directed mutagenesis together with two-electrode voltage-lamp electrophysiology were employed to investigate the effects of mutations in loop F of the homomer-forming α7 nAChR expressed in *Xenopus* oocytes. It was found that the G189D and G189E mutations markedly reduced the responses to imidacloprid and nitenpyram of the α7 nAChR, whereas G189N and G189Q scarcely influence the responses, suggesting that the effects on the neonicotinoid response of the mutations to acidic residues is due to changes of the electrostatic features in loop F rather than its steric features. By contrast, the effects of the G189D and G189E mutations on the response to desnitro-imidacloprid (DN-IMI), which lacks the nitro group, were minimal [17]. The result suggested that the marked reduction of the response to neonicotinoids by the mutations in loop F resulted from electrostatic repulsion between the acidic residues and the negative nitro group oxygens of neonicotinoids tested.

Even though these findings were interesting, basic residues were not found in loop F of insect nAChRs. A breakthrough which helped resolve this situation came with the publication of the crystal structure of the acetylcholine binding protein (AChBP) from a pond snail *Lymnaea stagnalis* [18]. The AChBP, which is secreted from glia cells to control the ACh concentration in synapses shows homology to the agonist-binding domain of the α7 nAChR and forms a water-soluble homopentamer [19]. Like the α7 nAChRs, the AChBP possesses five

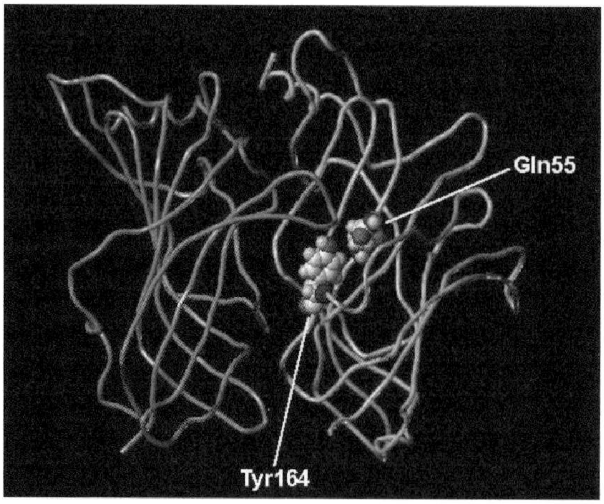

Figure 3. The crystal structure (PDB file: IUX2) of the acetylcholine-binding protein from *Lymnaea stagnalis*. Tyr164 in loop F and Gln55 in loop D are shown by space-fill models.

agonist binding pockets, each composed of 6 loops A–F and located at the subunit interfaces. In the crystal structure of AChBP, Tyr164 corresponding to Gly189 of the α7 nAChR faces the agonist binding site. Exploring amino acids in the vicinity of Tyr164 led to a finding of Gln55 in loop D [20]. It was postulated that if the nitro group of neonicotinoids interacts electrostatically with acidic residues added to loop F in the α7 nAChR, then similar electrostatic interactions are likely to be observed when Gln79, corresponding to Gln55 in AChBP, is mutated to acidic residues. Based on this hypothesis, the effects of mutations of Gln79 to not only acidic, but also basic residues in loop D on the response to neonicotinoids of the α7 nAChR were investigated. The Q79E mutation markedly reduced the α7 current amplitude recorded in responses to neonicotinoids, whereas the Q79K and Q79R mutations enhanced responses. The Q79E mutation enhanced the response to DN-IMI, which lacks the nitro group but instead possesses a positive charge, whereas the Q79K and Q79R had the opposite effect [20]. The maximum current amplitude of the neonicotinoid-induced response was scarcely influenced by prolongation of the exposure time, excluding the contribution of an open channel blocking action to the responses. These results suggest that the striking changes in neonicotinoid sensitivity of the α7 nAChR following the site-directed mutagenesis in loop D result from the direct interactions of the nitro group oxygens with the acidic and basic residues added to loop D.

27.4 Homology Modeling of nAChRs has Assisted in the Identification of Key Amino Acid Residues Involved in the Selective Interactions with Neonicotinoids of Heteromeric Nicotinic Acetylcholine Receptors

The loop D mutations in the α7 nAChR affected the maximum responses to neonicotinoids more strongly than the affinity. In addition, the striking effects of the site-directed mutagenesis studies on loop D might be limited to the case for the α7 nAChR expressed in *Xenopus* oocytes. Nevertheless, insect nAChR non-α subunits possess basic residue clusters in loop D, whereas such clusters are not seen in vertebrate non-α and α7 subunits (Table I). If the basic residues are those we have explored, then introduction of such residues to loop D might be expected to enhance the neonicotinoid sensitivity of heteromeric vertebrate nAChRs. This hypothesis was examined employing the chicken α4β2 and *Drosophila* Dα2/chicken β2 hybrid nAChRs expressed in *Xenopus* oocytes [21].

Imidacloprid was ineffective on the α4β2 nAChR, consistent with our earlier reports [13–14]. This low imidacloprid sensitivity of the α4β2 nAChR was, however, not enhanced significantly by mutations of the β2 subunit Thr77 corresponding to the Gln79 in loop D of the α7 subunit to the amino acid residues observed at insect non-α subunits. To investigate this, homology models of the α4β2 nAChR with imidacloprid bound were constructed based on the crystal structures of the AChBPs with nicotine [22] and epibatidine [23] bound (see Reference [21] for methods).

Table I. Amino acid sequences in loop D of vertebrate and insect nicotinic acetylcholine receptors.

Subunits		**Amino acid number of chicken β2 subunit**[a]									
		73	74	75	76	77	78	79	80	81	82
Vertebrates	Chicken α7	N	I	W	L	Q	M	Y	W	T	D
	Chicken β2	N	V	W	L	T	Q	E	W	E	D
	Chicken β4	N	V	W	L	N	Q	E	W	I	D
	Human β2	N	V	W	L	T	Q	E	W	E	D
	Human β4	N	V	W	L	K	Q	E	W	T	D
	Human δ	N	V	W	I	E	H	G	W	T	D
	Human ε	S	V	W	I	G	I	D	W	Q	D
	Rat β1	K	V	Y	L	D	L	E	W	T	D
	Rat β2	N	V	W	L	T	Q	E	W	E	D
	Rat β4	S	I	W	L	K	Q	E	W	T	D
Insects	Fruit fly Dβ1	N	V	W	L	R	L	V	W	Y	D
	Fruit fly Dβ2	N	L	W	V	K	Q	R	W	F	D
	Fruit fly Dβ3	H	C	W	L	N	L	R	W	R	D
	Locust migratoria β	N	V	W	L	R	L	V	W	N	D
	Myzus persicae β1	N	V	W	L	R	L	V	W	R	D
	Heliothis virescens β1	N	V	W	L	R	L	V	W	M	D

[a] Residue numbering is from the start methionine.

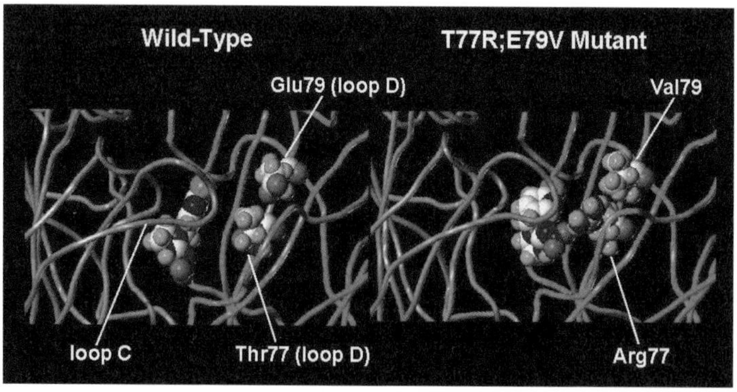

Figure 4. Homology model of the wild-type and T77R;R79V double mutant of the chicken α4β2 nAChR agonist binding site with imidacloprid bound. Reproduced from reference [21] with permission of American Society for Pharmacology and Experimental Therapeutics.

The homology models for the wild-type and mutant α4β2 nAChRs with imidacloprid bound (Figure 4) indicated that the Thr77 of the β2 subunit is likely to make contact with one of the nitro group oxygens of imidacloprid. The model also showed that the weak effects of the Thr77 mutations to basic residues on the imidacloprid sensitivity of the α4β2 nAChRs might be due to an electrostatic interference by Glu79 with the interactions between the nitro group and the amino acids added to the position 77. Thus, combined mutations of the Thr77 and Glu79 to amino acid residues seen in the insect non-α subunits (Table I) were generated on the grounds that they might enhance the neonicotinoid sensitivity of the α4β2 nAChR. The T77R;E79V double mutation was found to enhance markedly imidacloprid sensitivity in terms of the shift of the concentration-response curve to the left. In addition, this structural change in loop D of the β2 subunit increased the maximum normalized response to imidacloprid (Figures 5 and 6). The effect of the T77K;E79R mutations on neonicotinoid sensitivity was weaker than that of the T77R;E79V mutant recombination. This result is probably due to alteration of the agonist binding site conformation, counteracting the enhancement of imidacloprid affinity by the basic residues, a view supported by the observed change in the EC_{50} value for ACh.

The Dα2β2 hybrid nAChR was much more sensitive to imidacloprid than the α4β2 nAChR, resembling earlier observations [13–14]. When tested using this hybrid nAChR, even a single amino acid mutation of Thr77 to basic residues significantly shifted the imidacloprid concentration-response curve to the left. Further enhancement of neonicotinoid sensitivity was observed not only for the T77R;R79V mutation, but also for the T77N;E79R and T77K;E79R mutations [21]. The different effects of mutations in loop D between the α4β2 and Dα2β2

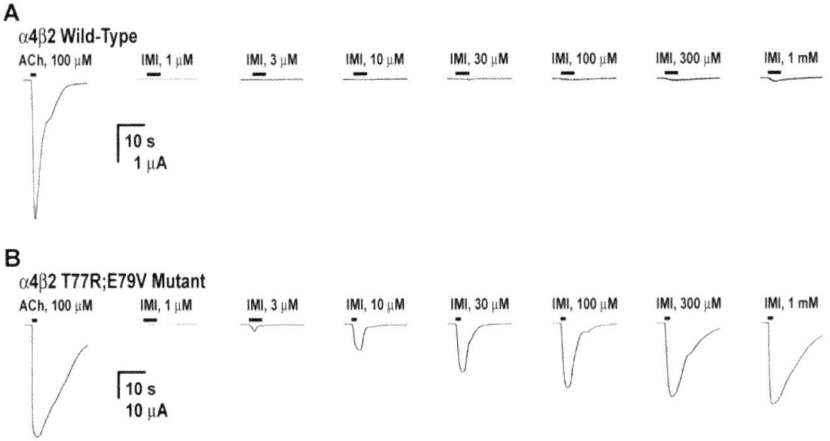

Figure 5. Inward currents induced by bath-applied acetylcholine (ACh) and imidacloprid of the wild-type (A) α4β2 nicotinic acetylcholine receptor and the T77R;E79V double mutant (B) expressed in *Xenopus laevis* oocytes. Reproduced from reference [21] with permission of American Society for Pharmacology and Experimental Therapeutics.

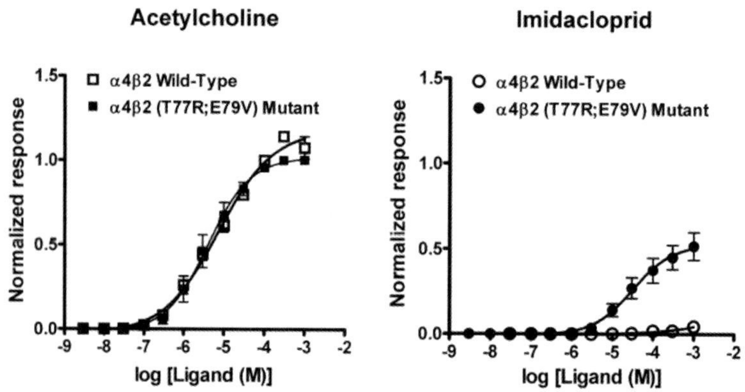

Figure 6. Concentration-response curves of acetylcholine and imidacloprid obtained for wild-type and T77R;E79V mutant of the α4β2 nicotinic acetylcholine receptor expressed in *Xenopus laevis* oocytes.
Each point plotted represents mean ± standard error of the mean.
Reproduced from reference [21] with permission of American Society for Pharmacology and Experimental Therapeutics.

nAChRs reflect differential binding modes of imidacloprid. The homology model of the wild-type Dα2β2 hybrid nAChR with imidacloprid bound revealed that the nitro group is located closer to Thr77 than in the α4β2 nAChR, enabling stronger interactions in the hybrid nAChRs [21].

Compared to the marked effects of the double mutations in loop D of the β2 subunit on the neonicotinoid sensitivity of the α4β2 and Dα2β2 nAChRs, the effects of the mutations on the concentration-response curves of ACh were minimal (Figure 6). These findings suggest that loop D of the insect nAChR plays a role in recognizing the structural features of neonicotinoids.

27.5 Conclusion

We have studied the selectivity of neonicotinoids using two-electrode voltage-clamp electrophysiology, site-directed mutagenesis and homology models of nAChRs with imidacloprid bound. The results obtained by studies on site-directed mutants and nAChR models suggest that basic residues in loop D of non-α subunits contribute to the selectivity of neonicotinoid-nAChR interactions. Since the mutations in loop D barely affect the ACh sensitivity of the nAChR, such mutations in pest species could lead to neonicotinoid-resistance. It has been found recently that a Y151S mutation in loop B results in a reduction in the neonicotinoid sensitivity of sucking pests [24]. It will be of interest to explore whether other resistant species possess mutations in loop D of the nAChRs. We have found that the region upstream of loop B and an amino acid in loop C in the *Drosophila* Dα2 subunit contribute to the high neonicotinoid sensitivity of the Dα2β2 hybrid nAChRs [25–26]. The homology models should also prove useful for exploration of structural features in the region upstream of loop B involved in the selectivity of neonicotinoids.

27.6 Acknowledgments

The author thanks Professor David B. Sattelle of The University of Oxford, UK, for helpful discussion on the ms and Dr. Miki Akamatsu of Kyoto University for her kind help in the preparation of Figure 3.

27.7 References

1 S. B. Soloway, A. C. Henry, W. D. Kollmeyer, W. M. Padgett, J. E. Powell, S. A. Roman, C. H. Tieman, R. A. Corey, *et al.*, in *Advances in Pesticide Science Part 2*, H. Geissbühler, G. T. Brooks, P. C. Kearney (Eds.), Pergamon, Oxford, **1979**, pp. 206–217.

2 S. Kagabu, *Rev. Toxicol.*, **1997**, *1*, 75–129.

3 W. Zhong, J. P. Gallivan, Y. Zhang, L. Li, H. A. Lester, D. A. Dougherty, *Proc. Natl. Acad. Sci. U.S.A.*, **1998**, *95*, 12088–12093.

4 H. A. Lester, M. I. Dibas, D. S. Dahan, J. F. Leite, D. A. Dougherty, *Trends Neurosci.*, **2004**, *27*, 329–336.

5 M. Tomizawa, N. Zhang, K. A. Durkin, M. M. Olmstead, J. E. Casida, *Biochemistry*, **2003**, *42*, 7819–7827.

6 K. Matsuda, S. D. Buckingham, D. Kleier, J. J. Rauh, M. Grauso, D. B. Sattelle, *Trends Pharmacol. Sci.*, **2001**, *22*, 573–580.

7 K. Matsuda, M. Shimomura, M. Ihara, M. Akamatsu, D. B. Sattelle, *Biosci. Biotechnol. Biochem.*, **2005**, *69*, 1442–1452.

8 P. V. Plazas, E. Katz, M. E. Gomez-Casati, C. Bouzat, A. B. Elgoyhen, *J. Neurosci.*, **2005**, *25*, 10905–10912.

9 S. Couturier, D. Bertrand, J.-M. Matter, M.-C. Hernandez, S. Bertrand, N. Millar, S. Valera, T. Barkas, *et al.*, *Neuron*, **1990**, *5*, 847–856.

10 V. Gerzanich, R. Anand, J. Lindstrom, *Mol. Pharmacol.*, **1994**, *45*, 212–220.

11 A. B. Elgoyhen, D. S. Johnson, J. Boulter, D. E. Vetter, S. Heinemann, *Cell*, **1994**, *79*, 705–715.

12 M. Ihara, L. A. Brown, C. Ishida, H. Okuda, D. B. Sattelle, K. Matsuda, *J. Pestic. Sci.*, **2006**, *31*, 35–40.

13 K. Matsuda, S. D. Buckingham, J. C. Freeman, M. D. Squire, H. A. Baylis, D. B. Sattelle, *Br. J. Pharmacol.*, **1998**, *123*, 518–524.

14 M. Ihara, K. Matsuda, M. Otake, M. Kuwamura, M. Shimomura, K. Komai, M. Akamatsu, V. Raymond, *et al.*, *Neuropharmacology*, **2003**, *45*, 133–144.

15 P.-J. Corringer, N. Le Novère, J.-P. Changeux, *Annu. Rev. Pharmacol. Toxicol.*, **2000**, *40*, 431–458.

16 A. Karlin, M. H. Akabas, *Neuron*, **1995**, *15*, 1231–1244.

17 K. Matsuda, M. Shimomura, Y. Kondo, M. Ihara, K. Hashigami, N. Yoshida, V. Raymond, N. P. Mongan, *et al.*, *Br. J. Pharmacol.*, **2000**, *130*, 981–986.

18 K. Brejc, W. J. van Dijk, R. V. Klaassen, M. Schuurmans, J. van der Oost, A. B. Smit, T. K. Sixma, *Nature*, **2001**, *411*, 269–276.

19 A. B. Smit, N. I. Syed, D. Schaap, J. van Minnen, J. Klumperman, K. S. Kits, H. Lodder, R. C. van der Schors, *et al.*, *Nature*, **2001**, *411*, 261–268.

20 M. Shimomura, H. Okuda, K. Matsuda, K. Komai, M. Akamatsu, D. B. Sattelle, *Br. J. Pharmacol.*, **2002**, *137*, 162–169.

21 M. Shimomura, M. Yokota, M. Ihara, M. Akamatsu, D. B. Sattelle, K. Matsuda, *Mol. Pharmacol.*, **2006**, *70*, 1255–1263.

22 P. H. N. Celie, S. E. van Rossum-Fikkert, W. J. van Dijk, K. Brejc, A. B. Smit, T. K. Sixma, *Neuron*, **2004**, *41*, 907–914.

23 S. B. Hansen, G. Sulzenbacher, T. Huxford, P. Marchot, P. Taylor, Y. Bourne, *EMBO J.*, **2005**, *24*, 3635–3646.

24 Z. Liu, M. S. Williamson, S. J. Lansdell, I. Denholm, Z. Han, N. S. Millar, *Proc. Natl. Acad. Sci. USA*, **2005**, *102*, 8420–8425.

25 M. Shimomura, M. Yokota, K. Matsuda, D. B. Sattelle, K. Komai, *Neurosci. Lett.*, **2004**, *363*, 195–198.

26 M. Shimomura, H. Satoh, M. Yokota, M. Ihara, K. Matsuda, D. B. Sattelle, *Neurosci. Lett.*, **2005**, *385*, 168–172.

Keywords

Acetylcholine Binding Protein, Chicken α4 Subunit, Chicken β2 Subunit, *Drosophila melanogaster* Dα2 Subunit, Homology Modeling, Loop D, Neonicotinoid, Nicotinic Acetylcholine Receptor, Two-Electrode Voltage-Clamp

28
Target-Site Resistance to Neonicotinoid Insecticides in the Brown Planthopper *Nilaparvata lugens*

Zewen Liu, Martin S. Williamson, Stuart J. Lansdell, Zhaojun Han, Ian Denholm, Neil S. Millar

28.1
Introduction

Neonicotinoid insecticides (Figure 1) are potent selective agonists of insect nicotinic acetylcholine receptors (nAChRs) and are used extensively for both crop protection and animal health applications. The use of neonicotinoids has grown considerably in recent years and annual worldwide sales are currently estimated to be approximately one billion US dollars [1–2]. Since the introduction of the first neonicotinoid insecticide (imidacloprid) in 1991, resistance has been slow to develop, but is now established in some insect field populations and is a major worldwide threat to the effective control of insect pests [3]. For several major insecticide classes (including organophosphates, carbamates, and pyrethroids), both target-site modifications and enhanced detoxification have been identified as being important resistance mechanisms. However, in most cases where mechanisms of resistance to neonicotinoid insecticides have been resolved, resistance has been attributed to enhanced oxidative detoxification of neonicotinoids by overexpression of mono-oxygenase enzymes [3].

28.2
Identification of Target-Site Resistance in *Nilaparvata lugens*

In 2003, studies at Nanjing Agricultural University, China identified resistance to imidacloprid in a population of the brown planthopper, *Nilaparvata lugens*, a major rice pest in many parts of Asia. A field population of *N. lugens* was isolated which, after several generations of laboratory selection, exhibited strong resistance to imidacloprid [4]. After 25 generations of laboratory selection, the resistance ratio to imidacloprid of this population of *N. lugens* was 73-fold [4] and, after further laboratory selection, the resistance ratio after 35 generations had increased to 250-fold [5–6]. The lack of cross-resistance to other insecticide

Pesticide Chemistry. Crop Protection, Public Health, Environmental Safety
Edited by Hideo Ohkawa, Hisashi Miyagawa, and Philip W. Lee

ISBN: 978-3-527-31663-2

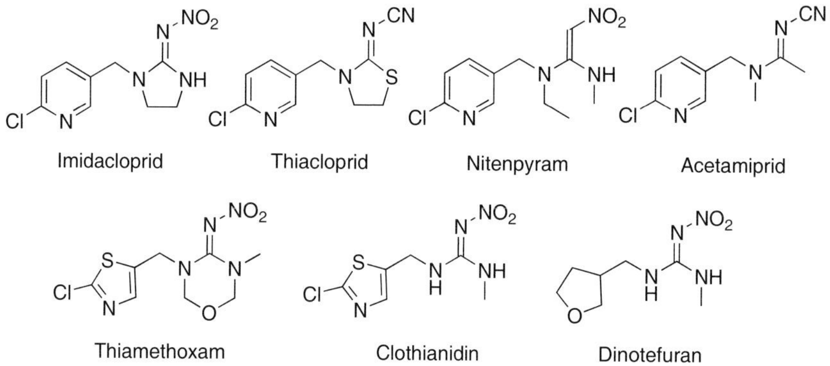

Figure 1. Chemical structures of neonicotinoid insecticides.

classes, and the lack of marked synergism by inhibitors of detoxifying enzymes, suggested that resistance might be due to a target-site mutation rather than due to enhanced detoxification [4]. Radioligand binding studies with [^{3}H]imidacloprid with susceptible and resistant *N. lugens* provided strong evidence that resistance might be due to a change in the insecticide target site (the nAChR) [5].

28.3
Characterization of a nAChR (Y151S) Mutation in *N. lugens*

Molecular cloning of nAChR subunits from *N. lugens* led to the identification of a single point mutation in two nAChR α subunits (Nlα1 and Nlα3) at a position close to the predicted agonist binding site [5]. In both nAChR subunits, a tyrosine at position 151 was altered to a serine (Y151S). By use of allele-specific PCR, it was found that 100% of individuals in the susceptible population were homozygous wildtype. After 25 generations of laboratory selection (when the resistance ration was 73-fold), 16% of individuals were homozygous for the Nlα1^{Y151S} mutation and 84% were heterozygous [5]. In contrast, after 35 generations of selection (when the resistance ratio was 250-fold), 100% of individuals were homozygous for the mutation. In order to investigate the influence of the Y151S mutation in more detail, cloned wildtype (Nlα1) and mutant (Nlα1^{Y151S}) nAChR subunits were expressed in cultured *Drosophila* S2 cells. Because of practical difficulties associated with the heterologous expression of recombinant nAChRs, wildtype and mutant *N. lugans* Nlα1 subunits were co-expressed with mammalian (rat) β2 subunit [7]. Radioligand binding studies with recombinant nAChRs provided direct evidence that the Y151S mutation was responsible for the loss of specific [^{3}H]imidacloprid binding [5]. To examine the influence of the Y151S mutation upon the functional properties of nAChRs, further studies were performed with recombinant nAChRs expressed in *Xenopus* oocytes, using two-electrode voltage clamp recording [7]. The maximal current observed in response to application of acetylcholine in oocytes expressing Nlα1^{Y151S} + β2 nAChRs was not significantly different from that observed in oocytes expressing Nlα1 + β2 nAChRs [7]. In contrast, oocytes expressing Nlα1^{Y151S} + β2 nAChRs gave significantly smaller maximal whole-cell currents in response to imidacloprid (13 ± 2.8%) compared with responses to imidacloprid in oocytes expressing Nlα1 + β2 nAChRs [7]. Additional experiments were performed to examine other commercially available neonicotinoid insecticides (acetamiprid, clothianidin, dinotefuran, nitenpyram, thiacloprid, and thiamethoxam). Two-electrode voltage-clamp recording was used to examine the influence of the Y151S mutation on agonist potency of these neonicotinoid compounds. The Y151S mutation resulted in significantly lower maximal current for all neonicotinoid compounds examined (acetamiprid 15 ± 1.1%, clothianidin 21 ± 1.1%, dinotefuran 81 ± 3.8%, nitenpyram 21 ± 1.9%, thiacloprid 22 ± 1.1%, and thiamethoxam 20 ± 1.3%) [7].

28.4 Discussion

Despite evidence that the potency of all neonicotinoid compounds is reduced by the Y151S mutation, this effect is less pronounced for the tertrahydrofuryl compound dinotefuran. In contrast to dinotefuran, all of the other neonicotinoid insecticides examined contain a chlorinated heterocyclic (chloropyridyl or chlorothiazolyl) group (Figure 1). The importance of the heterocyclic group in determining the extent to which the Y151S mutation influenced agonist potency is illustrated clearly by comparisons of two neonicotinoids, clothianidin and dinotefuran, which are otherwise chemically identical (Figure 1). As has been discussed previously [5, 7], the Y151 residue is highly conserved in both insect and mammalian nAChR subunits. Consequently, despite the profound effect of the Y151S mutation upon the binding and agonist potency of neonicotinoids, it cannot be responsible for the remarkable selectivity of neonicotinoid compounds for insect nAChRs. It has been proposed that the selectivity of neonicotinoids for insect nAChRs is due to interactions between an electronegative pharmacophore in these compounds with a cationic subsite in insect nAChRs [8]. An atomic resolution three-dimensional structure has been determined for a soluble pentameric acetylcholine binding protein (AChBP) from the mollusc *Lymnaea stagnalis* [9] which displays significant sequence similarity to the nAChR ligand-binding domain. Comparison of nAChR and AChBP amino acid sequences reveals that Y151 in nAChR α subunits is at a position analogous to a histidine residue (H145) in the AChBP [5, 7, 9]. X-ray diffraction studies of the *Lymnaea* AChBP reveals that H145 lies in close proximity to the agonist binding site of the AChBP [9], but the side chain of H145 points away from the agonist binding pocket. Thus, despite the likely close proximity of Y151 to the nAChR agonist binding site, it may not be involved directly in the binding of neonicotinoids. Rather, it is possible that the Y151S mutation might induce a conformational change within the nAChR binding site which results in a substantial effect on neonicotinoid binding but only a relatively minor effect on the binding of acetylcholine.

28.5 Conclusion

A point mutation has been identified that is responsible for conferring target-site resistance to neonicotinoid insecticides. Despite evidence that this mutation can have a substantial detrimental effect on biological fitness [6], studies with recombinant nAChRs indicate that the Y151S mutation has little, if any, significant effect on the potency of the endogenous agonist acetylcholine. Although the precise subunit composition of native *N. lugens* nAChRs is not known, our findings would suggest that insects containing the Y151S mutation may retain functional nAChRs, despite this mutation having a significant effect on neonicotinoid binding. Two issues which remain to be answered are 1) the importance of Y151S mutations within nAChR subunits other than Nlα1 and 2) the comparative potency of Y151S in conferring resistance in heterozygous and homozygous form. Both of these issues may need to be considered when attempting to assess the potential significance of these findings upon insecticide use in the field. As yet, there has been no work to establish the prevalence of the Y151S mutation in field populations of *N. lugens*; however, this is being investigated in conjunction with ongoing surveys of neonicotinoid resistance in several countries. An important next step in understanding the practical significance of the mutation is to relate data reported here with the phenotypic expression of resistance in laboratory bioassays and under field treatment regimes.

28.6 References

1 M. Tomizawa, J. E. Casida, *Ann. Rev. Pharmacol. Toxicol.*, **2005**, *45*, 247–268.
2 K. Matsuda, S. D. Buckingham, D. Kleier, J. J. Rauh, M. Grauso, D. B. Sattelle, *Trends Pharmacol. Sci.*, **2001**, *22*, 573–580.
3 R. Nauen, I. Denholm, *Arch. Insect Biochem. Physiol.*, **2005**, *58*, 200–215.
4 L. Zewen, H. Zhaojun, W. Yinchang, Z. Lingchun, Z. Hongwei, L. Chengjun, *Pest Manag. Sci.*, **2003**, *59*, 1355–1359.
5 Z. Liu, M. S. Williamson, S. J. Lansdell, I. Denholm, Z. Han, N. S. Millar, *Proc. Natl. Acad. Sci.*, **2005**, *102*, 8420–8425.
6 Z. Liu, Z. Han, *Pest Manag. Sci.*, **2006**, *62*, 279–282.
7 Z. Liu, M. S. Williamson, S. J. Lansdell, I. Denholm, Z. Han, N. S. Millar, *J. Neurochem.*, **2006**, *99*, 1273–1278.
8 M. Tomizawa, N. Zhang, K. A. Durkin, M. M. Olmstead, J. E. Casida, *Biochem.*, **2003**, *42*, 7819–7827.
9 K. Brejc, W. J. van Dijk, R. V. Klaassen, M. Schuurmans, J. van der Oost, A. B. Smit, T. K. Sixma, *Nature*, **2001**, *411*, 269–276.

Keywords

Neonicotinoid, Insecticide, Resistance, Nicotinic Acetylcholine Receptor, *Nilaparvata lugens*

29
QoI Fungicides: Resistance Mechanisms and Its Practical Importance

Karl-Heinz Kuck

29.1
Introduction

Due to its broad spectrum of activity and its high fungicidal efficacy, QoI fungicides have rapidly gained high market importance since the launch of the first representatives in 1996. Accordingly, only 10 years later, QoIs are considered nowadays to be the second important fungicide group behind the DMI fungicides [1].

The first report by Timm Anke and coworkers in 1977 on antifungal metabolites of the Basidiomycete *Strobilurus tenacellus* [2] opened a new field of fungicide research based on natural product lead structures. The first group named 'strobilurins' for the expanding fungicide class was later changed to 'Quinone outsite Inhibitors' (QoIs) which refers to the target site at the cytochrome (cyt) b protein in Complex III of fungal respiration [2] and which covers more precisely the manifold of structures than the original strobilurin designation.

Although the toxophore of the strobilurins and of related molecules such as oudemansins had been well-characterized by a (E)-β-methoxyacrylate substructure, a wide range of toxophore variations has been developed for agricultural use. Besides the original methoxyacrylates, QoI fungicides showing as toxophore moieties oximino-acetates, oximino-acetamides, methoxy-carbamates, dihydro-dioxaxines, imidazolinones, and oxazolidine-diones are now available on the market level. In spite of this chemical variability, binding in the target in Complex III is similar in all chemical subclasses of QoI fungicides which results in a general cross-resistance as far as target mutations are concerned.

29.2
Resistance Risk Assessments Before Market Introduction

As the molecular target site of QoI fungicides was known quite early, this fungicide group is one of the rare examples which allowed extensive resistance risk assessments well before and during the market launch of the first representatives.

Pesticide Chemistry. Crop Protection, Public Health, Environmental Safety
Edited by Hideo Ohkawa, Hisashi Miyagawa, and Philip W. Lee

ISBN: 978-3-527-31663-2

Mutagenesis studies with *Saccharomyces cerevisiae* revealed several target mutations within two relatively short interhelical regions of the cytochrome b gene associated with a reduction in sensitivity to QoI fungicides many of which result in impaired respiration and, hence, reduced fitness [3–5].

Strobilurins are natural antibiotics synthesized by the Basidiomycete genera *Strobilurus, Mycena,* and *Oudemansiella,* small agarics which grow on decaying wood. Accordingly the strobilurin producers have the task to protect themselves against the fungicidal activity of these metabolites. To understand the basis of this natural resistance, Kraiczy *et al.* [6], analyzed the relevant cytochrome b sequences in three species belonging to the genera *Strobilurus* and *Mycena.* Five substitutions of amino acids within the QoI binding site causing reduced QoI sensitivity have been detected, among them the G143A substitution. However, it was difficult from the data of Kraiczy to draw conclusions on the resulting practical resistance risk of individual substitutions.

Besides target site mutations, a second resistance mechanism has been identified in a number of phytopathogens. This involves the activity of an alternative oxidase (AOX) which serves as a bypass of Complex III and IV in respiration while reducing oxygen to water. Under *in vitro* conditions, this bypass effectively overcomes the inhibition of Complex III by QoI fungicides. In the presence of a host plant, however, alternative respiration appears to have limited impact on the level of disease control achieved with QoI fungicides. A model implying that expression of AOX is suppressed during pathogenesis by the presence of constitutive plant antioxidants, such as flavones, tried to explain these findings [7–9]. This model was however questioned by Avila-Adame and Köller who showed distinct but not complete protection of AOX against azoxystrobin in studies with *Magnaporthe grisea* [10].

29.3
Resistance Mechanisms of QoI Fungicides in Field Isolates

From 1998 onward, reports on decreased control of pathogens with QoI fungicides under field conditions or studies with less sensitive isolates isolated from field have become available [11]. In the following the different resistance mechanisms and their respective practical relevance will be discussed.

29.3.1
Mutation Upstream Complex I

A new but still unknown resistance mechanism was described by Steinfeld *et al.*, after studies on the resistance mechanism of apple scab isolates originating from a Swiss trial site [12]. The isolates were characterized by high resistance factors (> 1000) and showed no obvious fitness penalty under *in vitro* conditions. Measurements of oxygen and NADH consumption as well as ATP levels suggested a resistance mechanism based on compensation of energy deficiency following QoI treatment upstream of NADH dehydrogenase in the respiratory chain. Steinfeld *et al.* postulated therefore a monogenic mutation upstream of complex I in fungal respiration. As isolates of this type were never detected elsewhere the practical importance of the described mechanism remained limited.

29.3.2
Metabolization by Fungal Esterases

In the late 1990s reports on decreased efficacy of kresoxim-methyl against apple scab (*Venturia inaequalis*) became known from Germany and Belgium. In a short report from Jabs *et al.* [13], an external fungal esterase produced *in planta* by the pathogen was shown to hydrolyze the ester moiety in the toxophore of kresoxim-methyl. In the case of other strobilurins with the same toxophores, such as trifloxystrobin, the affinity of the enzyme was found later to be significantly lower. Accordingly, efficacy decreases via metabolization of the active ingredient are confined to kresoxim-methyl and apple scab and did not reach general importance for QoI fungicides.

29.3.3
Target Mutation G143A

From 1998 onward, reduced control with QoI fungicides under field conditions became obvious with several pathogens. The first pathogens showing resistance issues with QoIs were all classical high-risk pathogens well known for their genetic flexibility from earlier resistance cases with other fungicidal modes of action. Substantiated reports on the resistance mechanism of these pathogens were published from 2000 on. It could be shown that all these pathogens (*Blumeria graminis* f.sp. *tritici, Mycosphaerella fijiensis, Sphaerotheca fuliginea, Pseudoperonospora cubensis,* and *Plasmopara viticola)* were commonly characterized by a single amino acid exchange at the position 143 of the cyt b protein [14–17]. The substitution is based on a single nucleotide polymorphism in the cyt b gene from GGT (coding for glycine) to GCT (coding for alanine). As already shown earlier with the strobilurin-producer *Mycena galopoda,* this amino acid exchange causes a significant loss in fungicide sensitivity and no or little fitness penalties. Resistance factors vary – according to the pathogen and the test system between 100 and 1000. As a result of the monogenic G143A substitution a disruptive selection process is easily found in sensitivity monitorings. From Table I, it becomes clear that the G143A mutation has meanwhile been detected in 17 plant pathogens covering not only high risk pathogens but also more and more medium- and low-risk pathogens.

29.3.4
Target Mutation F129L

Although early studies on the resistance risk of QoIs had detected a multitude of mutations all lowering the sensitivity to QoI fungicides, up to now, besides the G143A mutation, only one further mutation could be proved in field isolates. At

Codon 129 of the cyt b gene characterized in wild type strains by the base triplet TTT, mutations to CTA, CTT or CTG could be detected in 6 pathogens [18–19]. At the amino acid level, this mutation causes the incorporation of leucine instead of phenylalanine into cytochrome b. In regard to the corresponding decrease in sensitivity, the consequences of the F129L exchange are distinctly less dramatic than with the G143A mutation. Typically, resistance factors of 5 to 15 (and only in rare cases resistance factors of up to 50) could be determined with several pathogens [19]. Accordingly, at full-dose rates, a decrease in field efficacy of QoI fungicides can hardly be detected. Only at low-dose rates of a solo product does the F129L mutation show clear effects.

29.4
Practical Importance of Individual Resistance Mechanisms to QoIs

Early (pre-launch) risk assessments were essentially not able to predict correctly the high resistance risk of QoI fungicides. This reflects a general problem with mutagenesis studies delivering mostly only limited information on the impact of a mutation on fitness and competitiveness.

A current balance shown in Table I reveals that – ten years after the market introduction of the first commercial QoI fungicides – the G143A mutation has meanwhile been detected in 17 pathogens. Whereas in the beginning all pathogens with this mutation belonged to the well-known, high-risk pathogens in the field of resistance development, now also pathogens known to bear a medium- or low-resistance risk are associated with QoI resistance. Due to the low resistance factors resulting in a lower selection advantage, the F129L mutation is found less often than the G143A mutation. Three pathogens have been identified to carry only this mutation and in three others both target site mutations, G143A and F129L, have been detected in parallel.

Due to the high selection advantage of the G143A mutation in a fungal population, all other resistance mechanisms have been revealed to be of secondary importance for different reasons. So, for example, lack of regional spread of a mutation (Section 3.1), lack of cross-resistance to other QoI fungicides (Section 3.2), or low-resistance factors (Section 3.4) may have a negative impact on the spread of a mutation within the fungal population.

Besides the intrinsic properties of each resistance mechanism, the economic impact of QoI resistance on the QoI market is also governed by the availability of alternative fungicidal modes of action for disease control and resistance management in each crop. For example, the early and rapid development of resistance of cereal powdery mildew (*Blumeria graminis*) in European wheat and barley production from 1998 on had a remarkably low impact on QoI consumption. The broad disease spectrum controlled by QoIs and the availability of several other modes of action for the control of powdery mildews in cereals such as DMIs, amines, cyprodinil, quinoxyfen, and metrafenone resulted in a nearly unchanged use frequency of QoIs in European cereal production.

Table I. Emergence of target mutations to QoIs.
Adapted from FRAC QoI Working Group (www.frac.info).

Pathogen	Host	Region	Target mutation		Detected*
			G143A	F129L	
Blumeria graminis f.sp. *tritici*	wheat	EU	x		1998
Mycosphaerella fijiensis	banana	America, Africa, SE Asia	x		1998
Pseudoperonospora cubensis	cucurbits	EU, Asia	x		1998
Venturia inaequalis	apple	EU, Chile	x		1998
Sphaerotheca fuliginea	cucurbits	EU, Asia	x		1998
Blumeria graminis f.sp. *hordei*	barley	EU	x		1999
Plasmopara viticola	grape	EU	x	x	2000
Alternaria solani	potato	USA		x	2002
Didimella bryoniae	cucurbits	USA	x		2002
Mycosphaerella graminicola	wheat	EU	x		2002
Pyricularia grisea	turf grass	USA	x	x	2002
Pythium aphanidermatum	turf grass	USA		x	2002
Alternaria alternata	pistachio	USA	x		2003
Alternaria tenuissima	pistachio	USA	x		2003
Alternaria arborescens	pistachio	USA	x		2003
Alternaria mali	apple	USA	x		2003
Colletotrichum graminicola	turf grass	USA	x		2003
Corynespora cassiicola	cucumber	Japan	x		2004
Glomerella cingulata	strawberry	Japan	x		2004
Mycovellosiella nattrassii	eggplant	Japan	x		2004
Pyrenophora teres	barley	EU		x	2004
Pyrenophora tritici-repentis	wheat	Sweden	x	x	2004
Alternaria alternata	European pear	Japan	x		2006
Botrytis cinerea	tomato, gentian	Japan	x		2006

* First detection of field problems, literature references at FRAC website.

However, in 2002, when G143A-based resistance was detected in *Mycosphaerella graminicola* (anamorph *Septoria tritici*), only DMIs and chlorothalonil-based products were available to substitute QoIs. Accordingly, because of the high economic importance of *Septoria tritici* in European wheat production, QoI applications dropped sharply after the rapid spread of this resistance in the years 2003 and 2004.

29.5 Resistance Management

Resistance management for QoI fungicides is coordinated in the FRAC QoI Working Group. This industry group yearly publishes a resistance survey and adapted recommendations for QoI resistance management (www.frac.info).

The basic tools of QoI resistance management are:
- limitation of the number of applications per season,
- use of QoIs only in mixture with non-cross resistant fungicide partners,
- avoidance of curative/eradicative uses.

Specific recommendations for important uses are published yearly.

29.6 Perspectives

The detection of the G143A mutation in more and more plant pathogens raises the question whether in the near future most – if not all – important target pathogen populations will be able to acquire this mutation.

Fortunately, already within the currently known examples, important differences in the dynamics of resistance spread and concurrently in the practical consequences can be found. Worst-case examples, represented by pathogens such as cereal powdery mildew or *Septoria tritici*, are characterized by a rapid spread of resistance within the fungal population over continent-wide distances but are obviously rare. In other cases such as with the apple scab pathogen (*Venturia inaequalis*), problems remain localized to certain regions or even individual orchards for many years and give a good chance for consequent resistance management.

Moreover, Grasso *et al.* [20] published recently an interesting observation which gives occasion for some optimism. Analyzing the sequence of the cytochrome b gene around the 129 and the 143 position, these researchers remarked that in several rust fungi an intron was inserted just beside the 143 codon.

According to these authors
- The G143A mutation will significantly affect the splicing process from pre-mRNA to mature mRNA if a type I intron is present after codon 143 in the cyt b gene.

- A nucleotide substitution in codon 143 would prevent splicing of the intron, leading to deficient cytochrome b which is lethal.
- QoI resistance based on G143A is not likely to evolve in pathogens carrying an intron directly after this codon.

Consequently, if the postulates of Grasso *et al.* should be further substantiated, there is the chance that the important group of rust fungi and some other pathogens such as *Alternaria solani* will not be affected by G143A resistance (Table II).

On the other hand, it has to be considered that other – still unknown – resistance mechanisms to QoIs could possibly gain importance in the future. The examples given in Sections 3.1 and 3.2 illustrate this possibility. As in monitoring studies, quite regularly individual resistant isolates are identified in which neither the G143A mutation nor other known resistance mechanisms can be detected. Some efforts have to be made to elucidate the resistance mechanism of such isolates.

Table II. Comparison of the Cyt b gene structure in the region of the amino acid residues 120–170 in different plant pathogens. From Grasso *et al.*, 2006.

Pathogen	129 [1]	143 [1]	Introns
Puccinia. spp.[2]	F	G	G143 (1474–1734 bp)
Phakopsora pachyrhizi	F	G	G143 (1337 bp)
Uromyces appendiculatus	F	G	G143 (1458 bp)
Hemileia vastatrix	F	G	Y132 (1396 bp), G143 (1657 bp)
Alternaria solani	L	G	A126 (1140 bp, G143 (2157 bp) V146 (1740 bp, F164 (1292 bp)
Alternaria alternata	F	A	–
Blumeria graminis	F	A	–
Magnaporthe grisea	L	A	–
Mycosphaerella graminicola	F	A	–
Mycosphaerella fijiensis	F	A	L169 (1064 bp)
Plasmopara viticola	L	A	–
Venturia inaequalis	F	A	P135 (360 bp), F169 (1623 bp)
Saccharomyces cerevisiae	F	G	G143 (1404 bp), F169 (1623 bp)

1 amino acid coded by codons 129 or 143, respectively, in sensitive or resistant isolates (bold),

2 following *Puccinia* species had been included: P. *coranata* f sp *avenae*, P. *graminis* f sp *tritici*; P. *hordei*, P. *horiana*, *recondita* s sp *tritici*, P. *recondita* f sp *secalis*, P. *sorghi*, P. *striiformis* f sp *tritici*.

29.7 Conclusion

Despite multiple efforts to assess the practical resistance risk of QoI fungicides in the years before and around market launch, the initial classification of QoIs as 'medium' risk fungicides strongly underestimated the resistance risk of this fungicide group. The rapid upcoming of severe resistance problems shortly after market entrance involved a re-classification of QoIs to 'high-risk' fungicides.

Among the resistance mechanisms found in fungal populations after market launch, the G143A mutation is by far the most important one for practical performance. All other resistance mechanisms are of low relevance for the overall performance of QoIs under practical conditions. This includes metabolism of the active ingredient which is relevant only for one pathogen/a.i. combination and the F129L mutation which can be readily managed due to relatively small resistance factors.

Accordingly, as long as no other resistance mechanisms are detected, resistance management concepts should be focused to prevent or to delay the development of the G143A mutation in fungal populations.

29.8 References

1 K. H. Kuck, U. Gisi, *Modern Crop Protection Compounds*, W. Krämer & U. Schirmer (Eds.), Wiley, January **2007**.

2 T. Anke, F. Oberwinkler, W. Steglich, G. Schramm, *J. Antibiot.*, **1977**, *30*, 806–810.

3 A. M. Colson, *J. Bioenerg. Biomem.*, **1993**, *25*, 211–220.

4 J. P. Di Rago, S. Herrmann-Le Denmat, F. Paques, J. Risler, P. Netter, P. P. Slonimski, *J. Mol. Biol.*, **1995**, *248*, 804–811.

5 G. Brasseur, A. Sami Saribas, F. Daldal, *Biochim. Biophys. Acta*, 1996, 1275, 61–69.

6 P. Kraiczy, U. Haase, S. Gencic, S. Flindt, T. Anke, U. Brandt, G. von Jagow, *Eur. J. Biochem.*, **1996**, *235*, 54–63.

7 D. Zheng, G. Olaya, W. Köller, *Curr. Genet.*, **2000**, *38*, 148–155.

8 A. Mizutani, N. Miki, H. Yukioka, H. Tamura, M. Masuko, *Phytopathology*, **1996**, 295–300.

9 B. N. Ziogas, B. C. Baldwin, J. E. Young, *Pestic. Sci.*, **1997**, 28–34.

10 C. Avila-Adame, W. Köller, *MPMI*, **2002**, *15*, 493–500.

11 K. H. Kuck, A. Mehl, *Pflanzenschutz-Nachrichten Bayer*, **2003**, *56*, 313–325.

12 U. Steinfeld, H. Sierotzki, S. Parisi, U. Gisi, Modern Fungicides and Antifungal Compounds III, H. W. Dehne, U. Gisi, K. H. Kuck, P. E. Russell, H. Lyr (Eds.), *Agroconcept Bonn*, **2002**, 167–176?

13 T. Jabs, K. Cronshaw, A. Freund, *Phytomedizin*, **2001**, *2*, 15.

14 H. Sierotzki, H. Wullschleger, U. Gisi, *Pest. Biochem. Physiol.*, **2000**, *68*, 107–112.

15 H. Sierotzki, S. Parisi, U. Steinfeld, I. Tenzer, S. Poirey, U. Gisi, *Pest. Management. Sci.*, *56*, 833–841.

16 S. Heaney, A. A. Hall, S. A. Davis, G. Olaya, *Proc. Brighton Crop Prot. Conf.*, **2000**, 755–762.

17 H. Ishii, B. A. Fraaije, T. Sugiyama, K. Noguchi, K. Nishimura, T. Takeda, T. Amano, D. W. Hollomon, *Phytopathology*, **2001**, *91*, 1166–1171.

18 J. S. Pasche, L. M. Piche, N. C. Gudmestad, *Plant Dis.*, **2005**, *89*, 269–278.

19 FRAC QoI Working Group, www.frac.info.

20 V. Grasso, S. Palermo, H. Sierotzki, A. Garibaldi, U. Gisi, *Pest. Management Sci.*, **2006**, *62*, 465–472.

Keywords

QuI Fungicides, Strobilurins, Resistance Mechanism

30
Chemical Genetic Approaches to Uncover New Sites of Pesticide Action

Terence A. Walsh

30.1
Introduction

The wealth of genomic information from an increasing number of organisms of interest to pesticide discovery has great potential for the identification of new sites of action. However, approaches for mining and exploiting this information that rely solely on genetic or molecular biological techniques often do not provide sufficient confidence that a potential site of interest can be effectively modulated by chemical intervention. For example, the effect of a genetic knockout of a gene may be far more profound than can realistically be achieved by the effect of an applied chemical inhibitor. Conversely, genetic redundancy may underestimate the potential effect of an inhibitor that is able to interact with two or more members of a target encoded by a gene family. Genetic techniques can be designed to circumvent some of these issues but often add more difficulty and complexity to the execution and interpretation of these types of experiments. In addition, the discovery or design of new chemistry that interacts with a novel site of interest predicted from genetic evidence requires significant investment of resources and a high level of risk that is becoming increasingly difficult to undertake in the pesticide discovery arena. Consequently there is a great need for shortcuts in this discovery process that can take advantage of genomic information (and information from other -omic technologies) while simultaneously providing insights into chemistry that can effectively interact with new sites of action. A chemical genetic approach can offer such a methodology.

30.2
The Chemical Genetic Approach

Chemical genetics can be defined as the use of small molecules to mimic the effect of genetic mutations in a biological system of interest [1]. Thus a chemical that produces a specific phenotype in a treated organism or cell can be investigated

Pesticide Chemistry. Crop Protection, Public Health, Environmental Safety
Edited by Hideo Ohkawa, Hisashi Miyagawa, and Philip W. Lee

ISBN: 978-3-527-31663-2

in many of the same ways as a genetic mutant. Chemical genetics is gaining increasing interest for biological investigations [2–4] as it can have several convenient advantages over conventional genetic methods. Compounds can be applied and removed at specific times and locations to rapidly produce their effects and their effects are readily titratable in a dose response. Also a compound may be able to affect several members of a gene family thus avoiding difficulties created by genetic redundancy. The use of chemistry to interrupt or modulate key biological processes is familiar territory for pesticide chemists and biologists as compounds of interest invariably produce profound, often lethal, "phenotypes". Thus the principles of forward genetic screening for distinct and desired phenotypes can be readily used to organize a chemical genetic approach for pesticide discovery. This can provide a robust and reasonably rapid connection between genomic information and chemistries of interest and the advantages of both chemical screening and genomic resources can be combined to maximum advantage.

Classical screening of chemistry for lethality on the pests of interest (Figure 1A) has been highly productive in generating leads but relies on sources of novel chemistry for screening that are becoming increasingly difficult to sustain. Also the biochemical targets of effective chemistries are generally unknown at this early lead generation stage so that target site information cannot be exploited for lead elaboration early in the Discovery process.

Genomic screening (Figure 1B) in model organisms can be used to screen for phenotypes of interest that may allow potential novel target sites to be identified. However, no chemical starting points are available at this point. Validation that a target has the potential to be chemically modulated can be difficult to achieve and may require considerable high-risk resource investments in high-throughput screening or structure-based design. The additional barrier of translating *in vitro* to *in vivo* activity is also a considerable practical hurdle.

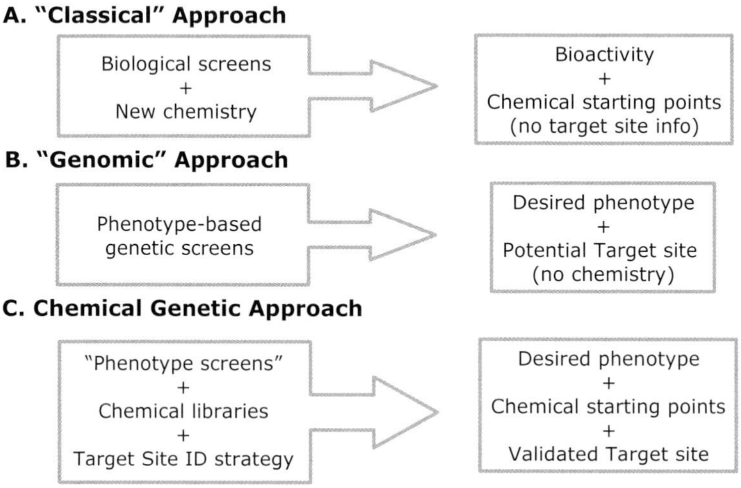

Figure 1. Comparison of three screening approaches and their deliverables.

A chemical genetic approach (Figure 1C) combines the use of an organized chemical library with phenotype screens and a robust target identification method to produce novel targets of interest coupled with interacting chemistry. This approach requires more upstream tools than the other approaches but the reward can be high-quality, information-rich leads acting at defined novel target sites. Dow AgroSciences partnered with Exelixis (South San Francisco, CA) and Exelixis Plant Sciences (Portland, OR) to implement such a multidisciplinary approach to find novel herbicide sites of action.

30.3 Components of a Chemical Genetic Process

The three components of a chemical genetic process to uncover novel sites of pesticide action, chemical libraries, phenotype screens and target site identification, can be organized and deployed in a variety of different ways and with varying degrees of complexity, depending on project objectives.

30.3.1 Chemical Libraries

Chemical libraries for screening should be of high sample quality to minimize the time and effort spent in follow-up of poorly characterized or impure samples. They should also be diverse to maximize the potential for novel activity. We routinely annotated compounds with information on known pharmacophores, chemistries with known modes of action and undesirable chemistries (reactive, unstable, etc.) within our library selections so that these compounds could be rapidly removed (either virtually or physically) from a selected set. Further annotations with physicochemical properties, lead-like properties, synthetic accessibility, etc., allowed for both structure and property-based data-mining. For academic programs, developing appropriate chemical libraries can be costly and difficult. However, agricultural chemical companies generally have access to large libraries of compounds enriched for bioactivity that can act as an initial source for library development. For our initial exploration, we selected a set of compounds that had displayed some degree of bioactivity in primary and secondary herbicide screens. No potency criteria were applied at this preliminary stage.

30.3.2 Phenotype Screens

Data on lethality and spectrum of activity on key pests are routinely gathered for all compounds tested within agricultural chemistry discovery programs. In a chemical genetics program, more specific and targeted screening information is required to facilitate novel target site discovery. Activity on genetically tractable model organisms is essential. For herbicides, this can include *E. coli*, green and blue-green algae, *Arabidopsis*, etc., to enable target site identification using genomic tools. Biological tests with low uptake and translocation barriers such as hydroponic systems that allow direct application of compounds to the tissue of interest are also of great utility. These may reveal compounds that may be otherwise overlooked in conventional whole organism screens. Compounds that produce specific desired phenotypes can be sought using additional screens that score specific physiological effects such as bleaching, epinasty, senescence, root elongation, etc. At this stage, a robust and desired phenotypic effect due to the compound is of principal importance rather than potency on whole organisms. More sophisticated phenotype screens can be readily envisaged using metabolomic profiles or appropriate reporter genes.

30.3.3
Target Site Identification

Target site identification is the rate-limiting step in the chemical genetic process [5–6]. Many strategies have been described with varying degrees of applicability. These can be protein-based (e.g., affinity methods using the active compound as "bait") or gene-based (e.g., mutational methods) [7–9]. At present, no methodologies are universally applicable or have a guaranteed high probability of success. We elected to use a genetic approach by selecting resistant mutants in genetically tractable model organisms that had available genomic information. This unbiased screening method has a proven track record of success. It can be readily implemented based on the original phenotype screen and requires no upfront biochemical knowledge or investigation. We also chose to introduce mutations by ethylmethanesulfonate (EMS) mutagenesis to generate point mutations. This necessitates more downstream time and effort in identifying the sites of mutation but greatly increases the chances of recovering mutations that reveal useful and actionable target site information.

30.4
Three Examples of Chemical Genetic Target Identification

The combination of in-depth chemical annotation of a bioactive library with data from a variety of phenotypic screens creates an information-rich environment for selection of compounds of interest. Selection criteria can be adjusted depending on the goals of the project. Compounds that have uniformly favorable scores (e.g., strong activity on key pests with good synthetic accessibility) are likely to have already been identified and synthetically investigated. However, a chemical genetic approach allows for the identification of compounds that may not necessarily display all of the desirable attributes of a high-quality lead but are nevertheless indicative of a phenotype of specific interest. Alternatively, chemistry criteria can be used to select compound classes that are amenable to synthetic follow-up, for example, by parallel or combinatorial synthesis methods. The following three examples were selected using different chemical and biological criteria and all resulted in target identification. They include a complex natural product with a novel chlorotic phenotype, a simple aryltriazoleacetate with an interesting bleaching phenotype, and a novel picolinate exhibiting a potent auxinic phenotype on *Arabidopsis*.

30.4.1
NP-1, a Complex Natural Product

NP-1 is a complex natural product that had poor synthetic accessibility and physical properties and so was unappealing as an initial synthetic lead. However, it had attractive biological properties including activity on grass and broadleaf weeds. Of compelling interest was the novel chlorotic phenotype. The compound was active on the blue-green alga *Synechocystis* making it amenable to rapid genetic techniques. A screen of EMS-mutagenized *Synechocystis* recovered two NP-1-resistant strains. Genomic DNA from one of the resistant strains was transposon-tagged and transformed into wild type *Synechocystis* via homologous recombination. Five NP-1-resistant colonies were recovered and the genomic location of the insertions determined by sequencing from the transposon tag. These defined a 5.9-kb region surrounding three genes (Figure 2A). One gene encoded an enzyme in primary metabolism and was a likely candidate for the NP-1 target site. Sequencing of this gene from the two original NP-1-resistant mutants revealed that they both contained different single base pair mutations that introduced changes in the encoded polypeptide sequence.

The *Synechocystis* enzyme was then expressed in *E. coli* allowing the enzyme to be readily assayed. The wild type recombinant enzyme was potently inhibited by NP-1 whereas enzymes containing the mutations were 30-fold resistant to inhibition by NP-1. Thus the target of NP-1 in *Synechocystis* was rapidly established (Figure 2B).

It was then necessary to validate that NP-1 targeted the same enzyme in higher plants. A metabolomic analysis showed that metabolites in the target pathway that

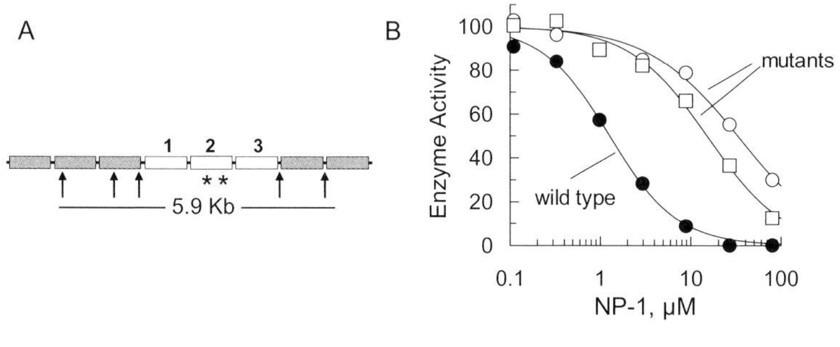

Figure 2. A. Genomic mapping of transposon-tagged insertions conferring NP-1-resistance to wild type *Synechocystis*. Arrows show the insertion sites, asterisks show the mutation sites determined by sequencing.
B. Effect of NP-1 on Gene 2 wild type and mutant enzymes.

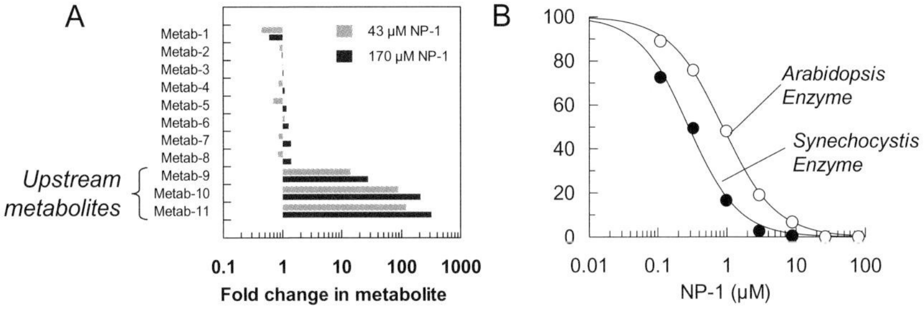

Figure 3. A. Effect of NP-1 on metabolites of *Arabidopsis* in candidate enzyme pathway. B. Effect of NP-1 on target enzymes from *Synechocystis* and *Arabidopsis*.

are upstream of the candidate enzyme were significantly elevated in NP-1-treated plants (Figure 3A), indicating that the putative target was indeed inhibited *in planta*. The plant enzyme was then expressed in *E. coli* and shown to be potently inhibited *in vitro* by NP-1 (Figure 3B). These data confirmed that the target site identified in *Synechocystis* was also the target in higher plants. The advantage of the chemical genetic method was further exploited as crystal structures of the target enzyme were available. This allowed the site of interaction of the natural product ligand with the target enzyme to be predicted by the location of the two resistance mutations in the homologous enzyme crystal structure.

30.4.2
ATA-7, a Bleaching Phenotype

In contrast to NP-1, ATA7 (Figure 4A) is a compound that presented excellent synthetic accessibility but had significantly less whole plant activity. However, phenotypic assays had identified three interesting characteristics of the compound. It elicited distinctive bleaching on new growth (Figure 4B); it had excellent activity on the genetic model plant *Arabidopsis* and the phenotype could be suppressed by the addition of the adenine in hydroponic assays.

The compound had no effect on the microbial model organisms tested so an in-depth chemical genetic screen for resistant *Arabidopsis* mutants was employed. Seven resistant mutants exhibiting 5- to 120-fold resistance to ATA7 relative to wild type were obtained from a screen of 420,000 EMS-mutagenized M2 seedlings

Figure 4. A. Structure of aryltriazole acetates (R1 = alkyl; R2 = aryl). B. Effect of ATA7 on new growth.

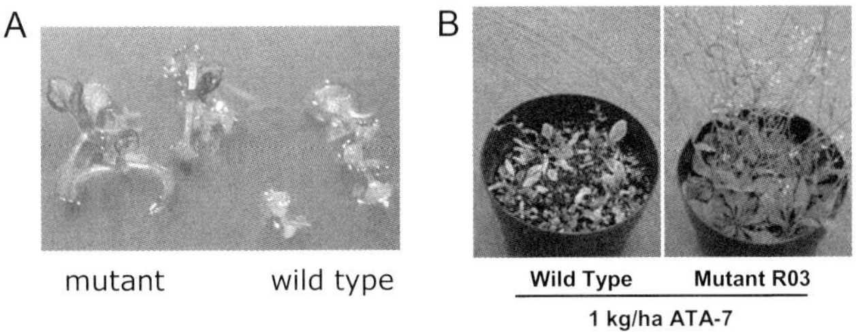

Figure 5. A. Mutant and wild type *Arabidopsis* seedlings growing on media containing ATA7. B. Effect of postemergent treatment of ATA7 on wild type and mutant *Arabidopsis* plants.

(Figure 5). Genetic testing showed that the mutants were dominant and appeared to be allelic.

The site of one of the mutations conferring ATA7 resistance was mapped to a genomic interval containing over 150 genes on chromosome 4. Inspection of gene annotations within the interval indicated that only one gene was involved in purine biosynthesis, At4g34740 encoding glutamine phosphoribosylpyrophosphate amidotransferase (GPRAT), the first enzyme in the purine biosynthetic pathway. At4g34740 is one of three functional genes encoding GPRATs in the *Arabidopsis* genome [10].

Sequencing of the GPRAT2 gene from the mutants showed that they all contained mutations introducing changes in the polypeptide sequence. AtGPRAT2 was functionally expressed in *E. coli* and ATA7 was found to be a potent time-dependent inhibitor of the enzyme activity. In contrast, a mutant enzyme with an R264K mutation found in the highly ATA7-resistant plants was not inhibited by the compound.

This example demonstrates an important advantage of the chemical genetic approach to target site identification. An insertional knockout mutant of GPRAT2 *atd2* shows the same bleached phenotype as treatment with ATA7 [11]. Bleaching of *atd2* can be prevented by addition of adenine to the media, thus treatment with ATA7 phenocopies the *atd2* genetic mutant. *atd2* plants can be readily identified in a genetic phenotype screen to suggest that GPRAT2 may be a candidate target site [12]. However, considerable additional work is needed to find chemistry that may effectively inhibit the enzyme. In contrast, a chemical genetic approach identifies the same phenotype allowing the target site to be found and associated with an *in vivo* active chemical starting point.

30.4.3
DAS534, a Picolinate Auxin

The final example of our chemical genetic approach to dissect herbicide action involves auxin biology. A subset of our bioactive compound library that exhibited auxinic herbicidal symptoms was tested for inhibition of root growth of *Arabidopsis* seedlings. The most potent of these compounds was a novel picolinate auxin DAS534. Previous mutational studies with 2,4-D have elucidated an auxin signaling pathway involving targeted ubiquitin-mediated proteolysis of auxin gene repressors [13–14]. However, it is presently unclear if all herbicidal auxins operate via the same mechanism. DAS534 is structurally differentiated from 2,4-D and is five-fold more potent in the *Arabidopsis* seedling root inhibition assay, therefore an *Arabidopsis* resistance screen was undertaken to find DAS534-resistant mutants. A screen of 780,000 EMS-mutagenized M2 seedlings recovered 33 robust DAS534-resistant mutants. A secondary screen was used to identify seven mutants that were *not* cross-resistant to 2,4-D.

Interestingly, all seven mutants had strong (> 30-fold) cross-resistance to the commercial picolinate auxin picloram (Figure 6). Further testing showed that the mutants were also cross-resistant to other picolinate auxins such as clopyralid and analogs of DAS534. However, they were not cross-resistant to auxins that do not contain a picolinate core such as dicamba, fluroxypyr analogs, and naphthylacetic acid. They also lacked cross-resistance to the native hormone IAA. Thus the chemical genetic approach uncovered a novel auxin response phenotype.

Genetic analysis of these lines showed that they consisted of two complementation groups. A mutant allele from one of the lines was positionally mapped to an interval of 47 genes on chromosome 5. Inspection of the genes within this interval highlighted a previously uncharacterized homolog of the F-box LRR protein, TIR1. TIR1 has recently been shown to be an auxin receptor for IAA and 2,4-D [15–16]. Sequencing of At5g49980 revealed that all four resistant lines in the complementation group contained single mutations leading to amino acid residue changes in the encoded polypeptide. Transformation of one of the resistant mutant lines with a wild type copy of At5g49980 restored normal DAS534 sensitivity. This established that picolinate auxin-selective resistance was due to mutations in At5g49980 [17].

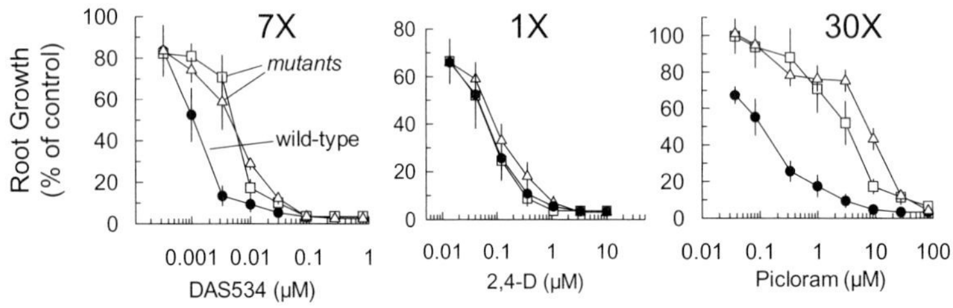

Figure 6. Seedling root growth response of DAS534-resistant *Arabidopsis* mutants to DAS534, 2,4-D, and picloram.

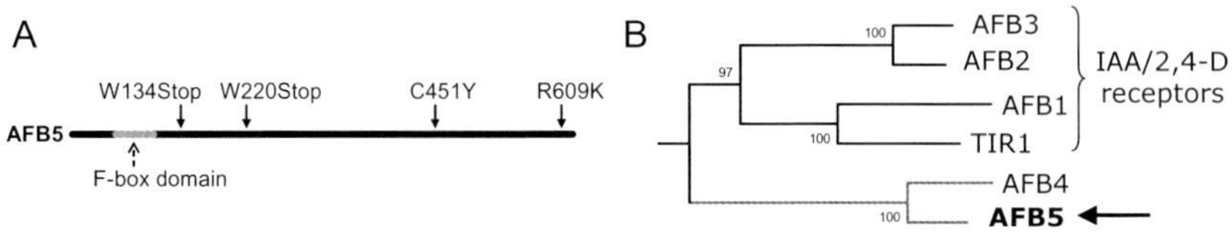

Figure 7. A. Mutation sites conferring picolinate auxin resistance in AFB5. B. Phylogenetic tree of TIR1 auxin receptor gene family in *Arabidopsis*.

There are three characterized homologs of the 2,4-D and IAA receptor TIR1 in the *Arabidopsis* genome, called Auxin response F-Box protein (AFB)1, AFB2, and AFB3 that appear to be functionally redundant [15, 18]. Two additional uncharacterized members of this gene family, AFB4 (AT4g24390) and AFB5 (At5g49980), can be identified in the genome by sequence homology. Our chemical genetic study indicates that AFB5 is involved in the response to picolinate auxins but not to 2,4-D and other auxins. Thus, our chemical genetic approach uncovered a novel member of the auxin receptor gene family that is not revealed by screening with 2,4-D.

30.5 Key Learnings

The chemical genetic approach we used was successful in uncovering new herbicidal target sites coupled with bioactive ligands. Our strategy involved assembling a chemical library biased for biological activity and identifying compounds with interesting novel phenotypes from it. Target site identification was enabled by in-depth unbiased screens for resistant mutants using genetically accessible model organisms. Our success rate in identifying useful mutants was improved by careful selection of phenotypes of interest with chemistry that had robust activity on the genetic models (*Synechocystis* and *Arabidopsis* in the examples described here). Identification of the precise sites of EMS-induced point mutations in *Arabidopsis* requires time-consuming positional mapping; therefore, it was important to establish that the mutants were likely to be informative. This was accomplished by detailed characterization of mutants prior to mapping. The response of the mutants to the lead chemistry was carefully assessed, as well as the response to any available analogs. They were also compared to similar known mutants wherever possible. The use of deep mutagenesis screens enabled recovery of several mutant alleles at each locus. These proved to be useful in confirming gene identifications as well as giving additional information about the site of interaction with the ligands.

30.6 Conclusion

A chemical genetic approach offers an opportunity to circumvent some of the issues associated with validating and exploiting pesticidal target sites identified via genetic knockouts or down-regulation. Our examples show that a high success rate in identifying target site/bioactive ligand pairs can be achieved. However, not all target sites can be necessarily identified using the resistance screening method we employed. Also our process emphasized quality over throughput. Nevertheless, even a modest success rate can deliver quality target site and chemical information that can be immediately exploited by synthetic chemists.

30.7 Acknowledgments

The following scientists made significant contributions to this work: Teresa Bauer, Mendy Foster, Nick Irvine, David McCaskill, Ann Owens Merlo, Jon Mitchell, Roben Neal, and Paul Schmitzer (Dow AgroSciences, Indianapolis, IN); John Davies, Karen Wolff, and Wendy Matsumura (Exelixis Plant Sciences, Portland, OR); Glenn Hicks, Mary Honma, and Cathy Hironaka (Exelixis, South San Francisco, CA).

30.8 References

1 B. R. Stockwell, *Nature Reviews Genetics*, **2000**, *1*, 116–125.
2 T. Asami, T. Nakano, H. Nakashita, K. Sekimata, Y. Shimada, S. Yoshida, *J. Plant Growth Regul.*, **2003**, *22*, 336–349.
3 H. E. Blackwell, Y. Zhao, *Plant Physiology*, **2003**, *133*, 448–455.
4 T. U. Mayer, *Trends in Cell Biology*, **2003**, *13*, 270–277.
5 L. Burdine, T. Kodadek, *Chemistry & Biology*, **2004**, *11*, 593–597.
6 R. S. Lokey, *Current Opinion in Chemical Biology*, **2003**, *7*, 91–96.
7 K. Szardenings, B. Li, L. Ma, M. Wu, *Drug Discovery Today: Technologies*, **2004**, *1*, 9–16.
8 X. S. Zheng, T. F. Chan, H. H. Zhou, *Chemistry & Biology*, **2004**, *11*, 609–618.
9 C. P. Hart, *Drug Discovery Today*, **2005**, *10*, 513–519.
10 R. Boldt, R. Zrenner, *Physiologia Plantarum*, **2003**, *117*, 297–304.
11 E. van der Graaff, P. Hooykaas, W. Lein, J. Lerchl, G. Kunze, U. Sonnewald, R. Boldt, *Frontiers in Bioscience*, **2004**, *9*, 1803–1816.
12 J. Lerchl, T. Ehrhardt, U. Sonnewald, R. Boldt (Inventors), **1999**, PCT Appl. 19949000.
13 G. Parry, M. Estelle, *Current Opinion in Cell Biology*, **2006**, *18*, 152–156.
14 M. Quint, W. M. Gray, *Current Opinion in Plant Biology*, **2006**, *9*, 448–453.
15 N. Dharmasiri, S. Dharmasiri, M. Estelle, *Nature*, **2005**, *435*, 441–445.
16 S. Kepinski, O. Leyser, *Nature*, **2005**, *435*, 446–451.
17 T. A. Walsh, R. Neal, A. Owens Merlo, M. Honma, G. R. Hicks, K. Wolff, W. Matsumura, J. P. Davies, *Plant Physiology*, **2006**, *142*, in press.
18 N. Dharmasiri, S. Dharmasiri, D. Weijers, E. Lechner, M. Yamada, L. Hobbie, J. S. Ehrismann, G. Juergens, *et al.*, *Developmental Cell*, **2005**, *9*, 109–119.

Keywords

Chemical Genetics, Target Site Identification, Resistant Mutants, Herbicide Mode of Action, Auxins, *Arabidopsis*

31
The History of Complex II Inhibitors and the Discovery of Penthiopyrad

Yuji Yanase, Yukihiro Yoshikawa, Junro Kishi, Hiroyuki Katsuta

31.1
Introduction

MTF-753 (penthiopyrad), (*RS*)-*N*-[2-(1,3-dimethylbutyl)thiophen-3-yl]-1-methyl-3-trifluoromethyl-1*H*-pyrazole-4-carboxamide is a novel fungicide that belongs to the carboxanilide family. Early carboxanilide fungicides such as carboxin have activity against basidiomycetes such as rust and rhizoctonia diseases but have limited activity on other pathogens. However, penthiopyrad shows remarkable activity against not only these diseases but covers also pathogens belonging to the ascomycetes such as gray mold, powdery mildew, and apple scab. Here, we describe the history of the carboxanilide fungicide class starting from carboxin and leading to the discovery of penthiopyrad. We also describe its biological properties and discuss mode-of-action and resistance studies.

The first generation compound, carboxin, was developed about 40 years ago and has been used as an important seed treatment fungicide. Mepronil and flutolanil, developed in the eighties, are also used to control some diseases caused by basidiomycetes such as rhizoctonia. Then furametpyr and thifluzamide were developed in the late nineties, and they showed higher activity but their spectrum was not broadened.

Benzamide (BC723), discovered by Mitsubishi Chemical Corporation, only had activity against gray mold (*Botrytis cinerea*) [1]. However, we paid special attention to the fact that some early *N*-phenyl benzamide lead compounds had moderate, but broad-spectrum activity against pathogenic fungi. After an extensive research effort, we found a highly active novel carboxanilide derivative that contained two heteroaromatic rings and found that branched alkyl substitution on the heteroaromatic ring on the amino part of the carboxanilide expanded its antifungal spectrum. With further research, we finally discovered penthiopyrad that is a unique carboxanilide fungicide candidate containing both a pyrazole and thiophene ring.

Interestingly, there was literature precedent suggesting that *ortho*-substituted carboxanilides could show broader spectrum activity. For example, Edgington

Pesticide Chemistry. Crop Protection, Public Health, Environmental Safety
Edited by Hideo Ohkawa, Hisashi Miyagawa, and Philip W. Lee

ISBN: 978-3-527-31663-2

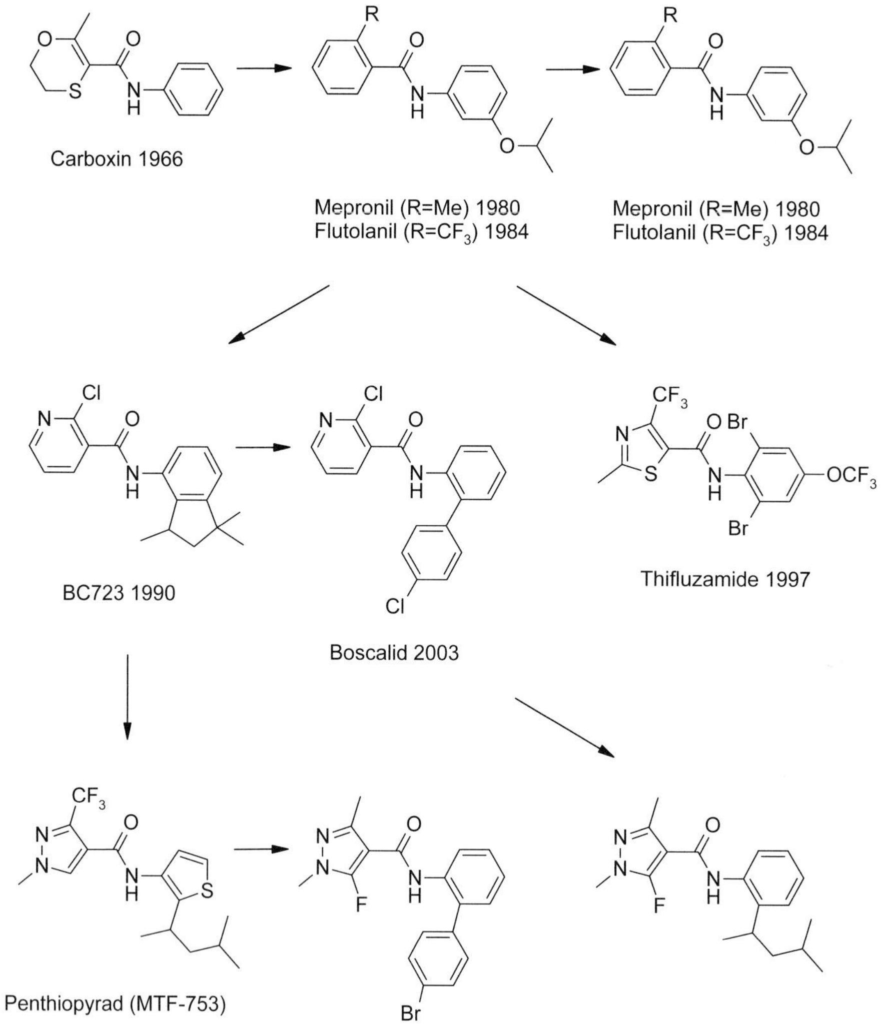

Figure 1. History of carboxanilide compounds.

reported that some oxathiin compounds showed activity against not only basidiomycetes but also *Alternaria solani* and *Botrytis* sp. [2]. For example, F427, an ortho-biphenyl carboxanilide, has activity not only on basidiomycetes but also on deuteromycetes (Table 1).

Table 1. Activity of oxathiin carboxanilides against fungi.

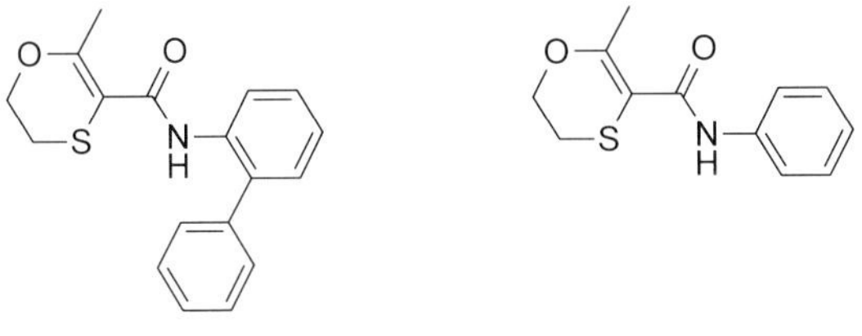

F427 D735 (carboxin)

Fungi		EC_{50} (10^{-6} M)	
		F427	D735
Deuteromycetes			
Phialosporae	*Aspergillus niger*	< 5	> 50
	Colletotrichum lagenaium	> 50	21
Porosporae	*Alternaria solani*	< 20	> 50
Blastosporae	*Botrytis* sp.	< 5	54
Phycomycetes	*Mucor* sp.	22	> 50
Basidiomycetes	*Ryzoctonia solani*	< 5	< 5

F427: (5×10^{-6} M = 1.6 ppm), D735: (5×10^{-6} M = 1.1 ppm)

31.2 Discovery of Penthiopyrad (MTF-753)

To develop a new carboxanilide fungicide, our goals and objectives were as follows. The target compound would have the same level of activity against fungal strains resistant to other chemistry classes and possesses broad-spectrum antifungal activity. In our study of carboxanilide derivatives with different *ortho*-substituent groups, we felt that BC723 possessed enough broad spectrum activity that we chose it as a lead compound. On the carboxylic acid side of the molecule, we synthesized and evaluated many heterocyclic ring modifications. On the amide side, we changed the indane ring to hetero rings substituted with various alkyl chains as *ortho*-substituents. We evaluated the activities of those compounds against rice blast, kidney bean gray mold, cucumber powdery mildew, and wheat brown rust, and we especially focused on gray mold and powdery mildew.

Table 2 summarizes the antifungal activity of alkyl-thiophene derivatives against kidney bean gray mold, cucumber powdery mildew, rice blast, and wheat brown rust with pot test. All compounds showed high activity against brown rust. But against gray mold and powdery mildew, the activity is strongly influenced by the number of carbon and type of alkyl side chain branching. Penthiopyrad had optimum activity with excellent control of many diseases, not only basidiomycetes but also ascomycetes and deuteromycetes.

Table 2. Antifungal activity of alkyl-thiophene derivatives.

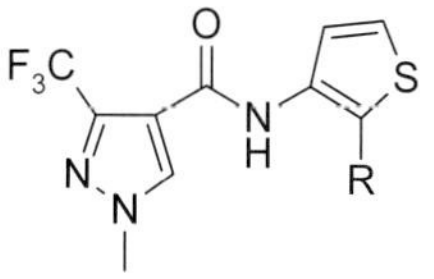

Compound No.	R	Kidney Bean Gray Mold	Cucumber Powdery Mildew	Rice Blast	Wheat Brown Rust
		MIC (ppm)	MIC (ppm)	MIC (ppm)	MIC (ppm)
8		125	210	> 250	22
7	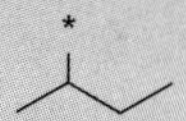	75	148	> 250	39
9	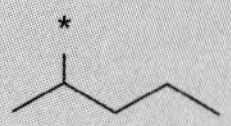	15	23	> 250	5
penthiopyrad	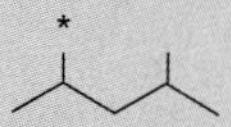	12	8	41	5
10	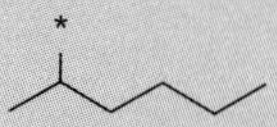	118	96	224	22
11	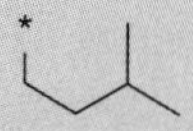	< 50	68	N.T.	< 6
12	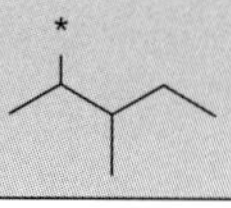	17	13	> 250	36

31.3
Biological Attributes

Penthiopyrad shows excellent broad antifungal activity on a wide range of crops. It is effective in controlling a range of pathogens including: gray mold, powdery mildew, tomato leaf mold, cucumber corynespora leaf spot, rust, southern blight, apple and pear scab, apple blossom blight, peach and cherry brown rot. On cereals and turf, penthiopyrad also has high activity against wheat Septoria diseases, rust, brown patch, dollar spot, fairly rings, anthracnose, and snow mold.

Furthermore, penthiopyrad has a mode of action different from that of many commercial fungicides used on these kinds of diseases. No cross resistance has been observed to benzimidazole, dicarboximide, anilinopyrimidine, DMI, and strobilurin. Also there is very low risk of phytotoxicity to crops.

31.3.1 Target Site of Penthiopyrad

The main target sites of penthiopyrad in the life cycle of *Botrytis cinerea* are spore germination and sporulation. Penthiopyrad also inhibits mycelium elongation, but we recommend to it be used preventively in order to obtain optimum results. Figure 2 shows the inhibitory activity of penthiopyrad on sporulation of *Botrytis cinerea*. Each fungicide was treated on the surface of a mycelial colony on potato dextrose agar (PDA) media, then sporulation was accelerated on each colony under a Blacklight Blue (BLB) lamp for seven days. The number of spores was then counted.

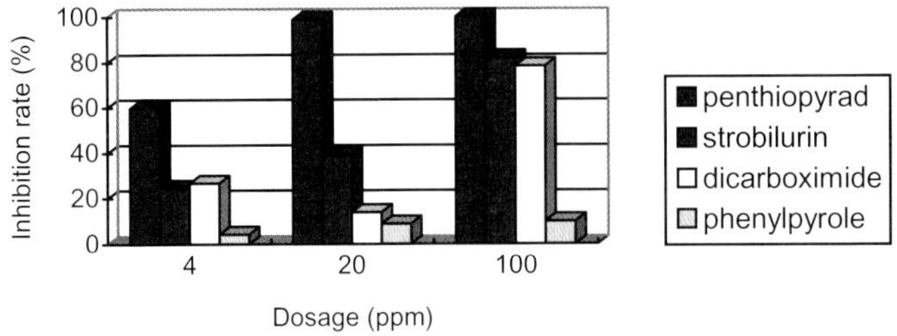

Figure 2. Inhibitory activity of penthiopyrad on sporulation of *Botrytis cinerea*.

31.3.2 Mode of Action

Penthiopyrad interrupts electron transport in the mitochondrial respiratory chain. As a result of this effect, the fungus cannot produce vital energy in the form of ATP.

Mitochondria were extracted from mycelia of *Rhizoctonia solani*, *Botrytis cinerea*, and *Fusarium oxysporum*. Succinate-ubiquinone oxidoreductase activity was assayed spectrophotometrically following the method of Miyoshi [3] and the results are listed in Table 3 expressed as I_{50} (50% inhibition).

Table 3. Mitochondrial Complex II inhibition activity of penthiopyrad as indicated by I_{50} (nM).

	Rhizoctonia solani	*Botrytis cinerea*	*Fusarium oxysporum*
Penthiopyrad	50	14	4 ~ 8
Flutolanil	372	> 8,000	800 ~ 2,000
Boscalid	N.T.	40	N.T.

N.T.: not tested

Penthiopyrad showed high inhibitory activity against the enzyme Complex II derived from not only *Rhizoctonia solani* but also *Botrytis cinerea* and *Fusarium oxysporum*.

31.3.3
Effect on Resistant Strains of Other Fungicides

Although there is no resistance problem of DMI on apple scab in Japan, it is already a serious problem in Europe and the U.S.A. Figure 3 shows the result of a field trial of apple scab in France in 2002. The details are as follows.

Crop:	Apple (Variety: Golden delicious)
Location:	Nimes, France
Scale:	3 trees/plot, 4 replicates
Spray:	March 27, April 10, 22, May 6, 24, 31, June 14
Assessment:	May 10, 30, June 11, 26

Alliage, kresoxim-methyl, showed a remarkable effect, but % control of Anvil, hexaconazole, is at most 40%.

Figure 4 shows the result of a field trial in Italy in 2004. The details are as follows:

Crop:	Apple (Variety: Royal Gala)
Location:	Boschi di Baricella, Italy
Scale:	20 trees/plot, 4 replicates
Spray:	April 2, 10, 15, 22, May 3, 14, 26, June 7, 17, 28
Assessment:	May 3, 10, June 3, 7, July 6

The effect of Flint was gradually decreasing during this trial presumably because QoI resistant strains were present in the orchard.

From these trials, it became clear that penthiopyrad shows no cross-resistance to DMI fungicides or strobilurin fungicides.

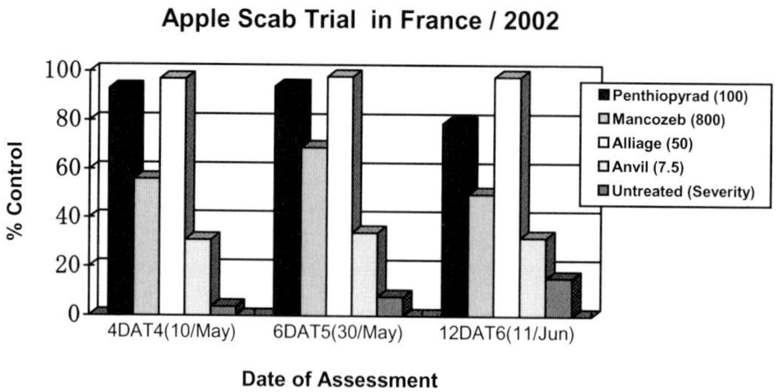

Figure 3. Field trial of apple scab in France in 2002.

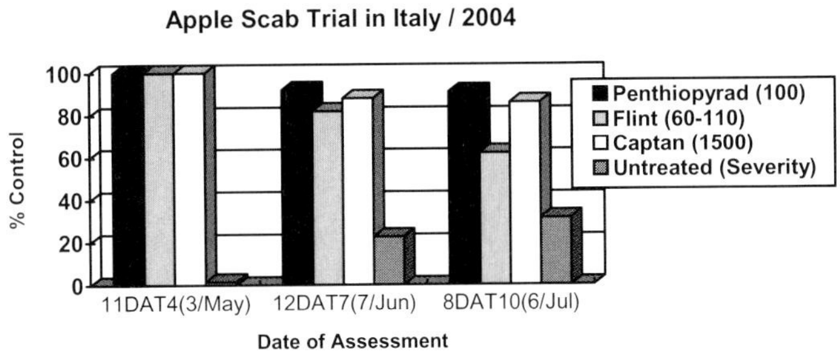

Figure 4. Field trial on apple scab in Italy in 2004.

31.3.4
The Risk of Occurrence of Resistance to Penthiopyrad

There have been reports on the risk of resistant pathogen strains developing to Complex II inhibitors. Mutants resistant to carboxin were found to contain a single amino-acid substitution in the third cysteine-rich domain of the Ip protein. These mutations resulted in the conversion of a highly conserved His residue, located in a region of the protein associated with the [3Fe-4S] high-potential, non-heme iron sulfur-cluster (S3), to either Tyr or Leu [4].

And there is a comment that resistance management is required if used in risky pathogens on Fungicide Resistance Action Committee (FRAC) web site (www.frac.info).

On the other hand, G. A. White, *et al.*, showed the existence of negative correlation between carboxin and other carboxin analogs [5]. Diethofencarb and benzimidazole are the most famous example of a negatively correlated cross-resistance relationship, but those structures are very different. Table 4 shows a 50% inhibitory concentration of succinate-2,6-dichlorophenol reductase activity

Table 4. Negatively correlated cross-resistance in carboxanilide.

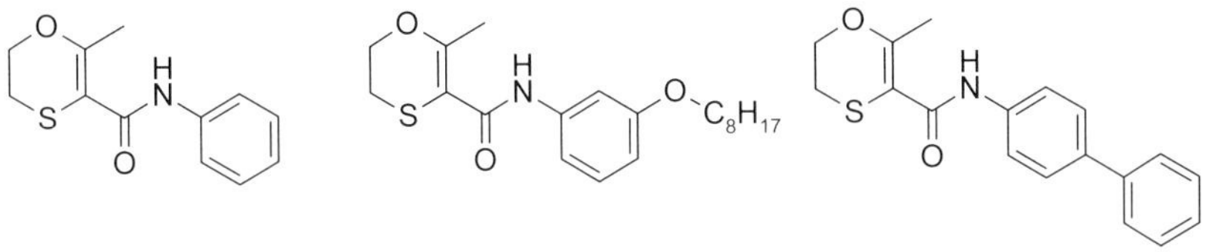

I VII XXII

Compound No.	Wild-type No. 14826		Mutant No. 724	
	I_{50} (μM)	Relative sensitivity	I_{50} (μM)	Resistance level
I	0.36	1.0	8.4	23.3
VII	0.008	45.0	0.12	15.0
XXII	4.8	0.075	0.44	0.092

in mitochondrial preparations from wild-type and carboxin-resistant mutants of *U. maydis*. Compounds I and VII are stronger against wild-type strain than mutant No. 724 strain. On the other hand, Compound XXII is stronger against mutant No. 724 strain than wild-type strain.

31.4 Conclusion

Carboxin is the one of the oldest members of the carboxanilide fungicide family and has been used as an important seed treatment fungicide for about 40 years. A new and more active fungicide, penthiopyrad, with the same mode of action as carboxin, has been discovered where its antifungal spectrum is much broader.

Although the carboxanilide family is one of the oldest groups of fungicides, the discovery of penthiopyrad will open a new avenue for the future research and development of novel compounds from this family of fungicides. The biological properties and favorable attributes of penthiopyrad as a commercal candidate are as follows:

Broad spectrum antifungal activity
Inhibition of electron transport in the mitochondrial respiratory chain
No cross-resistance to other fungicide classes
A good tool for resistance management of other fungicides
Resistance management will be required on risky pathogens

31.5 Acknowledgments

We thank Dr. Miyoshi of the Department of Applied Life Science, Kyoto University, for the study of the mode of action of penthiopyrad, and Dr. Ishii of National Institute of Agro-Environmental Sciences for work on the resistant fungi.

31.6 References

1 M. Oda, N. Sasaki, T. Sasaki, N. Nonaka, K. Yamagishi, H. Tomita, *J. Pestic. Sci.*, **1992**, *17*, 91.

2 L. V. Edgington, G. L. Barron, Fungitoxic Spectrum of Oxathiin Compound, *Phytopathological Notes*, **1967**, 1256–1257.

3 A. Yamashita, H. Miyoshi, T. Hatano, H. Iwamura, Direct Interaction Between Mitochondrial Succinate-Ubiquinone and Ubiquinol-Cytochrome C Oxidoreductase Probed by Sensitivity to Quinone-Related Inhibitors, II, *J. Biochem. (Tokyo)*, **1996**, *120*, 377–384.

4 W. Skinner, A. Bailey, A. Renwick, J. Keon, S. Gurr, J. Hargreaves, A Single Amino-Acid Substitution in the Iron-Sulphur Protein Subunit of

Succinate Dehydrogenase Determines Resistance to Carboxin in *Mycosphaerella graminicola* (*Septoria tritici*), *Curr. Genet.*, **1998**, *34*, 393–398.
5 G. A. White, G. D. Thorn, S. G. Georgopoulos, Oxathiin Carboxamides Highly Active Against Carboxin-Resistant Succinic Dehydrogenase Complexes from Carboxin-Selected Mutants of *Ustilago maydis* and *Aspergillus nidulans*, *Pestic. Biochem. Physiol.*, **1978**, *9*, 165.

Keywords

Complex II Inhibitor, Penthiopyrad, Carboxanilide, MTF-753

32
The Costs of DDT Resistance in *Drosophila* and Implications for Resistance Management Strategies

Caroline McCart and Richard ffrench-Constant

32.1
Introduction

To prevent further rapid development of resistance to new and currently effective insecticides in important crop pests, it is necessary to develop effective control strategies. Models of the evolution of resistance to xenobiotics and pathogens often share the central assumption that resistance carries a fitness cost. This assumption is based on studies of evolutionary biology, population genetics, and physiology. Evolutionary biology dictates that any resistance must involve a large modification of the previous phenotype and that a large phenotypic modification is therefore deleterious within the ancestral environment. Population genetics predicts that although resistance confers an advantage in the presence of the selective force, resistance genes are assumed not to approach fixation in natural populations and the observed frequency is assumed to be the net result of the selective advantage in the presence of selection and its cost in the absence of selection. Finally, our knowledge of the molecular basis of resistance-associated mutations has reinforced the concept that mutations carry a cost as we can derive hypotheses based on the altered function of the associated enzymes or receptors.

32.2
Global Spread of DDT Resistance

DDT (dichlorodiphenyltrichloroethane) was widely used in the 1940s and 1950s, in particular to control the vectors of typhus and malaria. Use of DDT declined as resistance occurred and, in the 1970s and 1980s, use of DDT was banned due to the persistance of the compound in the environment and concerns about toxicity to bird populations [1]. DDT use is still advocated for malaria vector control due to the high levels of mortality caused by malaria; however, use is strictly governed by the protocols of the Stockholm Convention on Persistent Organic Pollutants (POPs) and use for disease vector control is negligible [2].

Pesticide Chemistry. Crop Protection, Public Health, Environmental Safety
Edited by Hideo Ohkawa, Hisashi Miyagawa, and Philip W. Lee

ISBN: 978-3-527-31663-2

DDT-R is a dominant gene that confers resistance to DDT and cross-resistance to a range of other insecticides [3–5]. DDT resistance has been shown to be associated with the overtranscription of the Cytochrome P450 gene *Cyp6g1* [3]. Cytochrome P450 genes encode a large family of enzymes which are important in the metabolism of a range of compounds in insects including hormones and xenobiotics [6]. Microarray data have shown that in the DDT-resistant, field-isolated strain, Hikone-R *Cyp6g1* is the only known P450 to be overexpressed relative to the susceptible field-isolated strain Canton-S [5]. In addition, when *Cyp6g1* is overexpressed in transgenic flies carrying a copy of *Cyp6g1* driven by the Gal4/UAS expression system, it is shown to be necessary and sufficient for P450 DDT resistance [4]. The overexpression of *Cyp6g1* has been shown to correlate with the presence of a 491bp insertion within the 5′ region of the gene [4]. This insertion has sequence homology to the terminal repeat of a transposable *Accord* element. A strong reduction in variability at flanking microsattelite loci [7] and an absence of variability in the first intron of *Cyp6g1* [4] strongly suggests selection in this area.

DDT-R in *Drosophila* is a useful model system for a number of reasons. DDT was one of the earliest and most widespread pesticides ever used. In addition, individual flies can be readily genotyped for the presence of the resistance-associated mutation using a simple Polymerase Chain Reaction (PCR) based diagnostic that exploits the insertion of the *Accord* transposable element. This allows us to identify all three genotypes; *DDT-S*/*DDT-S* (DDT sensitive), *DDT-R*/*DDT-S* and *DDT-R*/*DDT-R* (or SS, RS and RR) in individual flies. Recently, in an outstanding example of parallel evolution, insertion of a different transposable element in the 5′ end of the *Cyp6g1* homolog in *D. simulans* was also shown to be associated with insecticide resistance [8]. *DDT-R* is therefore a widespread, representative and current mechanism of insecticide resistance.

A detailed population survey of the *Accord* insertion showed that in most non-African populations the frequency of the *Accord* was 85–100% [7]. In Africa, the highest frequency of the *Accord* was found in West and North African populations at a frequency of 70–90% compared with a frequency of 32–55% in East African populations. This is despite the ban on DDT use in most countries outside of Africa. There are a number of possible reasons why the *Accord* element is still so prevalent. First, *Cyp6g1* shows broad cross resistance to organophosphate and carbamate insecticides, which may be selecting for *Cyp6g1* overexpression. Second, *Cyp6g1* is capable of metabolizing a number of xenobiotics and endogenous compounds. This may give DDT-resistant strains an advantage over the susceptibles in the field even in the absence of DDT. Finally, low levels of migration in *Drosophila* and no measurable costs to overexpression of *Cyp6g1* would be expected to result in no loss of the *Accord* in populations after DDT use is removed.

32.3
Lack of Fitness Cost

Studies of the potential costs associated with xenobiotic resistance in the absence of the selective agent can suffer from several confounding experimental factors. First, fitness costs associated with strains in which resistance has been repeatedly selected for in the laboratory are unlikely to represent fitness costs associated with resistance mechanisms found in the field. Second, the resistant and susceptible strains compared are also often genetically unrelated and any observed costs may therefore be independent of the resistance trait itself. Third, when insects are used they are often not checked for the presence of microbial pathogens, such as *Wolbachia*, which can influence the outcome of crosses between infected and uninfected strains.

There have been a number of studies that have attempted to determine whether insecticide resistance confers any fitness costs in the absence of insecticide [9–10]. However there has been no general consensus as to whether insecticide resistance is always costly in the absence of selection [11]. We use a resistant strain of *Drosophila* that has been back-crossed repeatedly for five generations to the susceptible laboratory strain, Canton-S. This was to reduce the effect of differing genetic backgrounds. Following five generations of back-crossing, we have replaced ~98% of the genome of Hikone-R with the susceptible Canton-S background. A population homozygous for the *Accord* insertion in the Canton-S was established using the *Accord* PCR diagnostic. Both strains were cured of *Wolbachia* to prevent alteration of life-history characteristics by the intercellular bacterium [12].

Traits commonly used to quantify fitness, fecundity, viability of eggs, larvae and pupae, lifespan and developmental rate were recorded for homozygous resistant (*RR*) and homozygous susceptible (*SS*) flies. We also compared heterozygous flies where resistance was inherited from the female (*RS*) and where resistance was inherited from the male (*SR*) [13].

The results showed a significantly higher egg and larval viability in *RS* genotypes, where resistance was inherited from the female, than *SR* genotypes where resistance is inherited from the male (Figure 1). The pupal viability and fecundity of the offspring showed a higher fitness where resistance was inherited from either parent. These results were not temperature-dependent.

In a trend similar to that seen for viability, the larval and pupal development of the *RS* genotype is accelerated in comparison with the *SR* genotype at 20 °C. This advantage is, however, obscured when development is faster at 25 °C.

A comparison of mRNA in early embryos (3 hours old) and late embryos (15 hours old) shows that in both *RR* and *SS* embryos there is mRNA detectable in early embryos indicating the transfer of *Cyp6g1* mRNA from the female parent to the offspring. In *RR* embryos, there is approximately 10 fold more transcript present compared with *SS* embryos. This transfer of mRNA may have positive impact on the embryo and therefore be resulting in the higher embryo viability seen in the life history study. Thus, although the female-linked fitness advantage could formally be associated with a closely linked gene, the simplest explanation is that the advantage is associated with *DDT-R* itself.

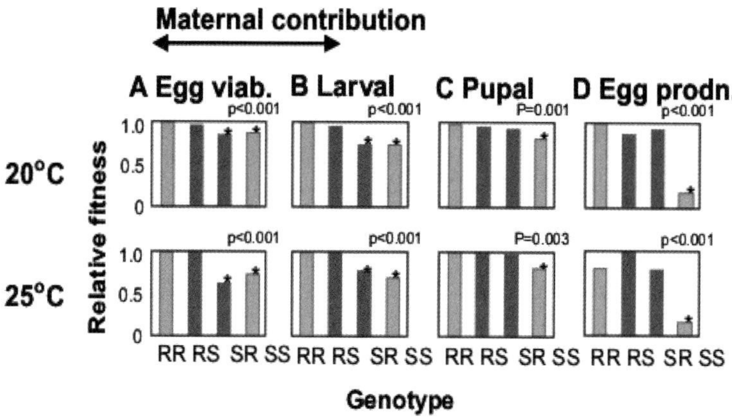

Figure 1. Life history analysis of the different *DDT-R* genotypes. The fitness of each genotype is plotted relative to the most fit which is given a value of 1.0. An asterisk indicates significant differences at the 5% level (ANOVA with Tukey post-hoc pairwise comparisons) and individual P values are given above each histogram.

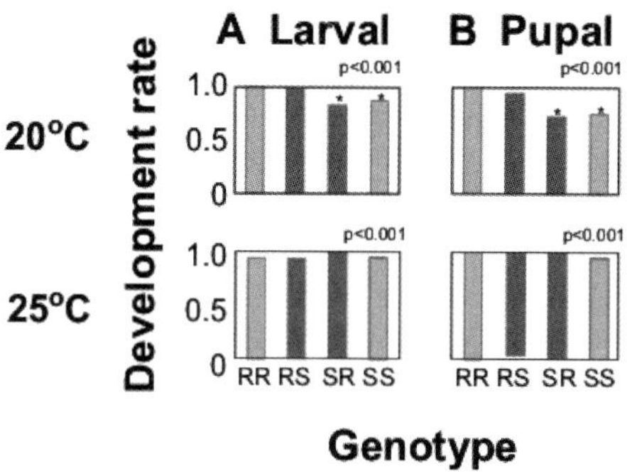

Figure 2. Differences in the development rate of larvae and pupae of the *DDT-R* genotypes.

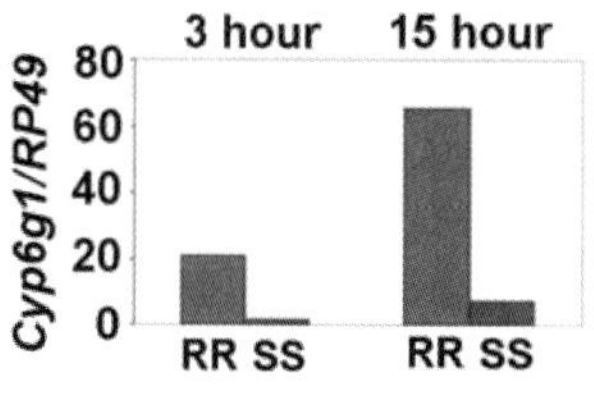

Figure 3. Quantitation of mRNA levels in 3-hour (pre-transcription) and 15-hour embryos.

32.4 Single Genes in the Field and Many in the Laboratory

Our finding that a single P450 gene is over-expressed in all field-collected strains of *D. melanogaster* stands in stark contrast to a review of the literature where many genes have been implicated in resistance [14–16]. We believe that the central reason for this is the difference between selection in the field and selection in the laboratory. Specifically, single genes are selected for in the field but one of a number of different alternative genes can be selected for by chronic selection with insecticides in the laboratory. We specifically tested this hypothesis by taking a field-derived resistant strain, only over-expressing *Cyp6g1*, and then subjecting it to a number of different selection regimes in the laboratory [4]. When we continued to select the field strain with DDT in the laboratory, a second P450 gene *Cyp12d1* was also over-produced, as noted by others [2]. Moreover, when we excluded both of these genes by genetic recombination, further selection led to over-production of a third P450 gene, *Cyp6a8*. This demonstrates that further selection in the laboratory will select for other P450 genes besides *Cyp6g1* and suggests, as other authors have, that many different P450s have the capability to confer DDT resistance. However, when field strains are examined that have not been pressured with DDT in the laboratory, only *Cyp6g1* is found to be over-transcribed. This suggests that mutation in the field is a rate-limiting step in the appearance of DDT resistance in *D. melanogaster* and is consistent with the observation that a single allele of *Cyp6g1* carrying a single *Accord*-transposable element causes DDT resistance globally [5]. In other words, despite the fact that other P450 genes are capable of causing DDT resistance, *Cyp6g1* is the only one in recent evolutionary history that has been targeted by an appropriate mutation to cause over-expression, and also apparently an increase in female linked fitness [6].

32.5 Implications for Resistance Management

The key to managing resistance is to reduce selection pressure. High pesticide doses rapidly select for resistance. The Insecticide Resistance Action Committee (IRAC) recommends a number of resistance management guidelines to keep pesticides for crop pests and vectors working effectively and keep costs down [17]. The number of treatments and the concentration of treatments may be varied in practices considered to be good-resistant management [18]. In theory, selection for resistance can be reduced by using low-frequency doses of insecticide to increase survivorship of susceptibles.

One of the most commonly used resistance management strategies involves rotating or mixing products from different classes based on modes of action and, where there are multiple applications per year, alternate products of different classes. This assumes a reduction in the frequency of the resistant genotype in the absence of the insecticide. If the resistant phenotypes show increased fitness, however, it can be assumed that the frequency of resistant genotypes would increase under these conditions. Furthermore, the strategy assumes no cross-resistance to alternative compounds. *DDT-R* shows cross-resistance to a wide variety of compounds [4] due to the broad spectrum of compounds that can be metabolized by *Cyp6g1*.

Alternative strategies that do not assume a reduced fitness of the resistant phenotype are important in trying to reduce the development of resistance in the field. The IRAC recommends that for crops, options for minimizing insecticide use include selecting early maturing or insect-resistant varieties and managing the crop for 'earliness'. In addition to chemical treatments, efficient cultural and biological control practices in pest control programs can be used, in particular the careful selection of crop protection tools not only for cost and effectiveness but also for the ability to maintain beneficial insects [17].

It is important that the pest or vector populations are monitored for resistance and the effectiveness of the control pesticide is monitored. In the event of a control failure that can be linked with resistance, it is important that the pest is not re-sprayed with an insecticide from the same class. Our data suggest a difference in the development rate of the susceptible and resistant strains at higher temperatures. This difference may be exploited in an attempt to control and prevent the spread of resistance in the population.

32.6 Conclusion

There are few examples of thorough studies looking at the fitness costs associated with insecticide resistance. Modeling predicts resistance to be costly but makes assumptions which, in the case of *DDT-R* in *Drosophila*, are not accurate. Resistance genes are assumed not to approach fixation in natural populations and the observed frequency is assumed to be the net result of the selective advantage in the presence of selection and its cost in the absence of selection. *DDT-R* in *Drosophila*, however, is a single mutation which has spread globally and is now found to be fixed in populations outside of East Africa.

This analysis shows that *RS* flies significantly outperform their *SR* counterparts in counts of both egg and larval viability; however, this advantage disappears during the pupal stage. The same results were observed at the two different temperatures, 20 and 25 °C. This disappearance of the *RS* advantage in the pupal stage is similar to effects seen in other *Drosophila* traits that are conferred via a maternal contribution to the eggs and developing larvae. Similarly, in a study of both larval and pupal development rates, although no significant differences in rates of development were observed at the higher temperature, dropping the temperature to 20 °C again revealed faster development in *RS* rather than *SR* genotypes. Therefore, both the viability and rate of development of larvae and pupae is improved when resistance is inherited via the female. *Cyp6g1* is detected in embryos before the onset of transcription indicating transmission from the female parent to the offspring.

Although the precise molecular mechanism whereby over-expression of CYP6G1 enzyme confers a maternally derived advantage remains to be elucidated, these results are both striking and highly significant for the population genetics of insecticide resistance. First they demonstrate that, contrary to conventional population genetic theory, insecticide resistance can show a benefit in the absence of selection, rather than a cost. This, together with the continued use of other insecticides to which *DDT-R* confers cross-resistance, may help explain the fact that globally *DDT-R* is approaching fixation in non-African *D. melanogaster* populations, long after the withdrawal of widespread DDT use. These findings suggest that a re-examination of our models for managing resistance to drugs and pesticides is necessary. Most pesticide-resistance strategies rely on a reduction in the frequency of the resistant phenotype due to the assumed fitness cost associated with resistance in the absence of the pesticide. This approach may be flawed in the case of some resistance alleles as our data indicate that not all resistance is costly. The implications for pesticide management are that strategies that do not assume a pesticide cost should be developed. Beyond the study of insecticide resistance, these results suggest that other xenobiotic-resistance mechanisms, or indeed other adaptive traits, may not incur significant fitness costs in the absence of the selective agent.

32.7 References

1 WHO (2005). Frequently asked questions on DDT use for disease vector control, WHO/HTM/RBM/2004.54.

2 UNEP (2001). Stockholm Convention on Persistent Organic Pollutants (POPs), UNEP/Chemicals/2001/3.50.

3 P. Daborn, S. Boundy, J. Yen, B. Pittendrigh, R. ffrench-Constant, *Mol. Genetic Genomics*, **2001**, *266*, 556–563.

4 P. Daborn, J. Yen, M. Bogwitz, G. Le Goff, E. Feil, S. Jeffers, N. Tijet, T. Perry, *et al.*, *Science*, **2002**, *297*, 2253–2256.

5 G. Le Goff, S. Boundy, P. J. Daborn, J. L. Yen, L. Sofer, R. Lind, C. Sabourault, L. Madi-Ravazzi, R. H. ffrench-Constant, *Insect Biochem Mol. Biol.*, **2003**, *33*, 701–708.

6 R. Feyereisen, *Annu. Rev. Entomol.*, **1999**, *44*, 507–533.

7 F. Catania, M. O. Kauer, P. J. Daborn, J. L. Yen, R. H. ffrench-Constant, C. Schlotterer, *Mol. Ecol.*, **2004**, *13*, 2491–2504.

8 T. A. Schlenke, D. J. Begun, *Proc. Natl. Acad. Sci. USA*, **2004**, *101*, 1626–1631.

9 D. Bourguet, T. Guillemaud, C. Chevillon, M. Raymond, *Evolution Int. J. Org. Evolution*, **2004**, *58*, 128–135.

10 B. R. Levin, V. Perrot, N. Walker, *Genetics*, **2000**, *154*, 985–997.

11 C. Coustau, C. Chevillon, R. H. ffrench-Constant, *Trends in Ecology and Evolution*, **2000**, *15*, 378–383.

12 A. A. Hoffmann, M. Hercus, H. Dagher, *Genetics*, **1998**, *148*, 221–231.

13 C. McCart, A. Buckling, R. ffrench-Constant, *Current Biology*, **2005**, *15*, 587–589.

14 A. Brandt, M. Scharf, J. Pedra, G. Holmes, A. Dean, M. Kreitman, B. Pittendrigh, Differential Expression and Induction of Two *Drosophila* Cytochrome P450 Genes near the Rst(2)DDT Locus, *Insect Mol. Biol.*, **2002**, *11*, 337–341.

15 J.-B. Berge, R. Feyereisen, M. Amichot, *Cytochrome P450 Monooxygenases and Insecticide Resistance in Insects*, Philosophical Transactions of the Royal Society of London, Series B, **1998**, *353*, 1701–1705.

16 R. A. Festucci-Buselli, A. Carvalho-Dias, M. de Oliviera-Andrade, C. Caixeta-Nunes, H. Li, W. Muir, M. Scharf, B. Pittendrigh, Expression of Cyp6g1 and Cyp12d1 in DDT Resistant and Susceptible Strains of *Drosophila melanogaster, Insect Mol. Biol.*, **2005**, *14*, 69–77.

17 IRAC, **2006**. http://www.irac-online.org

18 J. Mckenzie *in* Ecological and Evolutionary Aspects of Insecticide Resistance, R. G. Landes (Ed.), Academic Press, Austin, Texas, **1996**.

Keywords

DDT Resistance, *Drosophila*, *Accord*, Cytochrome P450, Fitness Cost, Life History Study, *Cyp6g1*

VI
Human Health and Food Safety

Pesticide Chemistry. Crop Protection, Public Health, Environmental Safety
Edited by Hideo Ohkawa, Hisashi Miyagawa, and Philip W. Lee

ISBN: 978-3-527-31663-2

33
New Dimensions of Food Safety and Food Quality Research

James N. Seiber

33.1
Introduction

The focus in food toxicology and food safety research has shifted, subtly but noticeably. In the last few decades of the 20th century, pesticide residues, persistent organic pollutants (POPs), and other industrial chemicals, including solvents and by-products, and heavy metals and other inorganics were of major concern. Microbial toxins, particularly aflatoxins and other mycotoxins, mutagens formed during cooking (e.g., polyaromatic hydrocarbons (PAHs) and charcoal-broiled meat), food allergens, food additives and, of course, foodborne diseases, such as hepatitis, were also emphasized as new scientific findings or episodes occurred [1].

These are still important today. Indeed, the finding of a carcinogen, acrylamide, in fried potato products, and other foods [2] continues to generate significant attention, as do safety issues associated with perchlorate, industrial contaminants such as mercury and other contaminants that arise from time to time.

However, much more attention is now devoted to pathogenic microorganisms in food, *E. coli* O157:H7 *Salmonella* spp., *Listeria monocytogenes, Campylobacter* spp. [3], and to genetically engineered foods, which are perceived to afford a risk although the mounting evidence does not support this [4]. And in the past 3–5 years, obesity-enhancing substances in foods, including those involved in coronary disease and diabetes, are receiving much more attention [5].

Awareness of healthful constituents in foods that can exert a positive influence on health, by reducing risk from cancer, heart disease, arthritis, Alzheimer's disease, and others has also increased [6]. These include antioxidants, soluble fibers, trace elements, and anti-microbials. Much of the available information is anecdotal or from the non-peer reviewed literature, and awaits scientific research to catch up with observations from advocates of various types and sources of foods. Also, dietary supplements are expanding in the market place as the source of nutraceuticals in place of natural sources in foods, stimulating questions regarding their bioavailability and relative benefits.

Pesticide Chemistry. Crop Protection, Public Health, Environmental Safety
Edited by Hideo Ohkawa, Hisashi Miyagawa, and Philip W. Lee

ISBN: 978-3-527-31663-2

Much of the research in the USDA's Agricultural Research Service (ARS), as well as that of other federal and state agencies, universities, and the food industry, is aimed at developing improved methods for foodborne disease detection and prevention, and supporting and optimizing the health benefits of naturally occurring food constituents. In ARS's Western Regional Research Center, in Albany, CA, one of four regional centers in ARS's national network, roughly half of the research is devoted to food safety and optimizing health benefits of foods. The Center also has a significant program in controlling orchard pests and invasive weeds using means that eliminate or minimize the use of synthetic chemical pesticides, also benefiting food safety, and much of the Center's research is supported by strong programs in plant and microbial molecular biology and natural products chemistry that further underpin food safety and quality programs.

The Center is located on the west coast, in the Pacific West Area, one of seven geographical zones of USDA-ARS. This is a prolific and diverse producing area, including primarily fruit, vegetable, small grains, and vineyard production in the west coast states, animal agriculture throughout the region, irrigated cropland in the desert southwest, tropical fruits and vegetables in Hawaii, and aquaculture and fisheries in Idaho and Alaska. Much of the work done at WRRC, as with other ARS locations, is conducted in collaboration with universities or in conjunction with technology transfer to industrial partners.

Foodborne illness is very much in the news. In the produce area alone, multiple outbreaks (two or more cases, same strain) and sporadic cases (individual illness) are reported each year due to pathogenic microorganisms [7]. While seafood consumption generates the largest number of outbreaks nationally, produce is second, and ahead of poultry, beef, or eggs. In 2005, produce generated the largest number of cases of food poisoning among all food categories. The increase in produce-associated disease may be due to increased consumption of uncooked vegetables, in salads, sandwiches, ethnic foods and others, and to the increase in fresh-cut processing of fruits and vegetables. Sprouts, cilantro, lettuce, tomatoes, strawberries, cantaloupes, and almonds are among the produce types involved in disease incidents. Vegetable production is located occasionally in the vicinity of animal operations (dairy, grazing, feedlots), a factor that may contribute to the occurrence of enteric pathogens in agricultural produce and sometimes in processed food.

33.2
New Analytical Methods for Identification and Source Tracking

Source-tracking is one of the rapidly growing areas of food safety research and outbreak investigation, increasingly involving high-resolution genomic and/or proteomic molecular fingerprinting methods. A relatively new method is based on MALDI Time-of-Flight mass spectrometry [8]. This is a useful and now routine method for analyzing proteins and other higher molecular weight biopolymers.

MALDI-TOF MS generates fingerprints of foodborne microorganisms based upon their protein content which is unique and diagnostic. The bacterial genus *Campylobacter*, for example, includes > 10 species, and subspecies, and strains that have food disease potential, but the species and strains vary significantly in their virulence. *Campylobacter jejuni*, the major cause of bacterial foodborne illness in US and other parts of the world, is a pathogen of particular concern, frequently encountered in uncooked poultry, raw or underpasteurized milk, and occasionally in fruits and vegetable. Two subspecies of *C. jejuni* subspecies *jejuni* and doylei, show related, but still diagnostic, mass spectral profiles and both subspecies are distinguishable from other species of *Campylobacter* [9]. The individual peaks in these spectra represent intact protein biomarkers, many of which have been identified or at least tentatively identified by mass spec-based proteomics tools [10]. This is a good example of a context in which chemists and biologists work together to refine and apply a technique useful to both.

ARS has invested heavily in DNA sequencing the genomes of pathogenic microorganisms. The DNA sequences of *Campylobacter jejuni* [11], *C. lari, Listeria monocytogenes, Salmonella enteritidis*, and others have been determined. Sequence information provides the fundamental information for developing highly specific assays. Genome sequences that vary between strains can be identified, facilitating development of microarrays of genes or oligonucleotides used as probes of test strains to see if any of these same genes are present or absent. Sequence-based methods minimize ambiguity and are conducive to development of databases that can be interrogated with new sequence data and for determining strain relatedness (phylogeny).

The application of genotyping to outbreak events is illustrative. There were two outbreaks associated with *Salmonella enteritidis* and raw almonds occurring a few years apart. One outbreak occurred in 2000–2001 involving greater than 50 cases of people consuming raw (unblanched and unroasted) almonds. A second similar outbreak in 2004 involved greater than 40 cases including one death [12]. The almonds in both cases were harvested in California's San Joaquin Valley, but not necessarily from the same grower and/or area. The two strains appeared different by conventional methods, i.e., pulse field electrophoresis and phage typing, indicating that these were likely unrelated events. But use of *Salmonella* microarray gene-indexing based on greater than 3000 gene features revealed a close relatedness between the two outbreak strains suggesting a common or related source. This was quite helpful during follow-up investigations aimed at correcting the problem by recognizing potential contributing factors that occurred during production.

A recent ongoing investigation involves *E. coli* O157:H7 contaminating lettuce from the Salinas Valley in California [13]. This lettuce and other leafy vegetables from the Valley are distributed and sold nationwide, so there is much concern by the growers as well as state and federal officials, and consumers, over multiple outbreaks. A sampling of water and soil throughout the vegetable-growing areas of the central valley was conducted, and strains isolated from samples were subjected to gene-typing analysis. The location of samples yielding positives appeared to be involved with storm events and heavy water flow, a suggested possible source of the *E. coli* O157:H7 strains involved in the outbreaks. This is still under study.

33.3
Methods for Reducing Aflatoxins in Foods

Another food safety issue in which natural product chemistry and molecular biology come into play involves aflatoxins, a family of naturally occurring furanocoumarins produced by *Aspergillus flavus* and related molds, in tree nuts, corn, cottonseed, and peanuts [14]. Oily products of these crops are among items which sometimes also show contamination with aflatoxins, near, at, and above the U.S. FDA Action Level. Some jurisdictions, such as the European Union, have adopted even stricter residue standards than U.S. FDA presenting a trade obstacle for the U.S. and other producing nations. A useful observation is that, among tree nuts, aflatoxin B_1 occurrence and levels declined in the order almond, pistachio, and walnut. One hypothesis, that walnut contains a natural product(s) or other material that is a barrier for the fungus, and/or for the fungal biosynthesis leading to aflatoxins, has been confirmed by USDA-ARS scientists [15]. The phytochemicals in question are phenols and polyphenols (hydrolyzable tannins), antioxidant chemicals naturally present in the nut seed coat, such as ellagic and gallic acids. A variety of other phenolics, such as vanillic or caffeic acids, also can reduce or completely stop aflatoxin biosynthesis in exposed *Aspergillus* colonies. These findings provide targets for plant breeders who can now select for antioxidant-rich nut varieties for commercial nurseries. And using *Aspergillus*-based microarrays, the specific genes responsible for aflatoxin biosynthesis in the fungus were identified, and used to provide additional clues regarding other chemical and physical factors that might be manipulated to shut down biosynthesis of aflatoxins.

Another way of preventing aflatoxin in nuts is to control the insect pests that feed on nuts, like the navel orangeworm, a pest of almonds, or on other hosts like corn (corn earworm and corn rootworm), cottonseed (cotton bollworm and pink bollworm), and peanuts (southwest corn borer). Pests like these bore tunnels that allow *Aspergillus* to invade and infect. One USDA-ARS research path focuses on discovery of host plant volatiles that function as kairomone attractants of the codling moth, a pest of walnut [16]. The volatiles from pear fruit yielded ethyl (2*E*,4*Z*)-2,4-decadienoate, the "Pear Ester", which is highly attractive to adult codling moths and, unlike the codling moth pheromone which is used by the female to call male moths, the pear ester attracts both males and females in roughly equal numbers. This tool can support several strategies to monitor and control codling moths. In one strategy, small amounts of pear ester added to insecticide solutions before spraying reduced worm-caused damage and *Aspergillus* invasion access to as little as $1/10^{th}$ than in orchards sprayed with insecticide alone. In another strategy, the pear ester was used as a lure in pesticide-laced traps, resulting in an effective attractive approach to controlling codling moth populations.

All of these avenues – use of kairomone attractant, and spray adjuvant, phenolic antioxidants, and other factors that reduce aflatoxin biosynthesis – have the same potential result, to reduce aflatoxin levels in commercial nuts so as to preserve traditional markets and keep the products safe as well as healthy for consumers, and do so without, or with minimal use of synthetic insecticide and fungicides.

33.4 Molecular Biology and Food Safety

Molecular biology can be used in other ways to improve food safety. For example, glycoalkaloids, potent inhibitors of cholinesterase, can be reduced in concentration in potatoes by blocking key biosynthetic routes, using antisense RNA transgenes [17]. The pathway proceeds from cholesterol through solanidine, an alkaloid aglycone, through a series of glycosyl transferase reactions producing the much more toxic glycoalkaloids α-chaconine and α-solanine. The antisense technology can selectively block various stages of glycosylation, altering the chaconine:solanine ratio, or reducing the amounts of both relative to non-transformed potatoes. This is an example of the use of metabolomics to characterize changes in metabolites, based on genetic engineering in this case.

The food supply chain includes a number of steps, from production, through harvesting, through packing and/or processing, through transport and wholesaler/retailer distribution ultimately to consumers where food safety issues can surface, requiring constant vigilance. Mad Cow disease is contracted by animals upon consuming infected material during feeding operations. The disease may only be detected several steps beyond this, at the slaughterhouse for example, and then only after a fairly cumbersome post mortem analysis and bioassay of brain material. Better detection methods are clearly needed in this case, preferably involving a relatively sensitive blood test. Developments in antibody-base immunoassay, or mass spectrometry, afford such opportunities, but are yet to be proven in practice.

Acrylamide is formed from high-temperature cooking (such as frying; other methods of cooking form acrylamide also) of food, that contains a mix of carbohydrates and proteins [2]. The finding of relatively high levels of acrylamide in foods like potato chips and French-fried potatoes was somewhat accidental – certainly not part of a designed monitoring program. This example underscores the need for constant vigilance of the safety of the food supply.

There is also a need for new ways to remove, or at least reduce levels of microbial and/or chemical contaminants. Infrared and microwave heating of some foods, such as almonds, can "pasteurize" them so that *Salmonella*, and potentially other pathogens, are inactivated. Chemical sanitizers, such as bleaching agents, ozone, and mild acid or base are effective for some produce items as well [18]. The effectiveness of routine processing steps, such as cooking, blanching, and extruding, in destroying and denaturing contaminants is often not known for many contaminants which occur in various food matrices.

Genetically modified foods are a reality already in some parts of the world. In the U.S., 90% of soybeans are genetically transformed to be herbicide-resistant; 80% of cotton is transformed for resistance to herbicides and, through incorporation of *Bacillus thuriengensis*, to various insect pests; and 50% of field corn is similarly modified for both herbicide and insect resistance [4]. None of the crops are used directly for human food, although food use is made of corn and cottonseed oil from genetically modified corn and cotton and the bulk of the transformed soybeans and corn, and cottonseed meal, are fed to animals which enter the human food supply. End-user and consumer nonacceptance of genetically modified foods continues in many quarters.

Research is under way to reduce or eliminate any real or perceived risks due to genetically modified food, so that public and political acceptance can proceed [19]. One avenue involves developing new molecular tools for precise integration and expression of transgenes, rather than doing so using the relatively less precise tools available for transformations to date. Site-specific or tissue-specific transformation, for example, could confine transformation to a non-edible part of the plant, without altering the composition of the edible part [20]. As more precise transgenic modification tools are developed, plants can be modified in such a way so that consumers can accept, and benefit from the modifications. An example of a consumer-beneficial target might be reduction or elimination of wheat seed proteins that cause Celiac disease, or Baker's asthma. Thus, another approach to reducing the perceived risks of food biotechnology would entail use of only all-native, or intragenic, within-plant based DNA modifications, which do not involve introduction of "foreign" DNA, e.g., from animals or microorganisms into plants [21]. When consumers are queried, the acceptance of GM increases substantially for intragenically modified foods.

33.5
Healthy Food Constituents

Research into food safety, both microbial and chemical food safety, has been accompanied by a parallel effort to identify, test, and optimize healthy constituents of foods. The interest in chemicals in foods extends beyond traditional areas – vitamins, essential minerals, etc. – to secondary chemicals sometimes termed phytonutrients or neutraceuticals which have positive health benefits including prevention or alleviation of diabetes, heart disease, Alzheimer's disease, cancer, arthritis, and many other diseases. Food producers use this information as a marketing tool. Compounds of interest include phenols/polyphenols, flavonoids, carotenoids, anthocyanins, and several other classes that function as antioxidants, i.e., reduce oxidative damage to cells that cause aging, inflammation, and a variety of other symptoms [22]. One research approach is to enhance the level of beneficial phytonutrients in foods that lack them completely, as exemplified by "golden" rice (genetically modified to produce carotenoids) as a means to combat vitamin A deficiency in whole populations in less developed areas of the world.

Increasing consumption of fruits, vegetables, nuts, and whole grains is a goal of USDA and many other 'eat healthy' programs. USDA-ARS research projects that address this goal include: developing new technologies for production of 100% fruit health bars [23] that are shelf stable and convenience/snack ready; developing cast films from fruit and vegetable processing streams that can be used to preserve and/or enhance quality of preserved food products; and developing better ways to preserve fruits and vegetables in fresh, unprocessed or minimally processed forms using processing and/or chemical preservatives (e.g., Apple Dippers® recently introduced in McDonald's restaurants) [24]. Maintaining product safety and health in these formats requires research into such things as pathogen behaviors in fresh cut, packaged product, and potential use of naturally occurring preservatives in the films ('anti-microbial films').

33.6 Conclusion

Advances in food safety and food quality research will increasingly involve interdisciplinary science. Basic sciences, such as chemistry, biochemistry, and molecular biology are required along with more applied sciences (food technology, fermentation science) and with health sciences including nutrition, pharmacology and clinical sciences. Programs which can combine multidisciplinary approaches through teamwork, within the research organization or by forming partnerships with external cooperators, are particularly well set up to be successful. Communicating the underlying science to the public continues to be a challenge and one that may require input from professionals skilled in science as well as public policy and communication.

33.7 References

1 W. Helferich, C. K. Winter (Eds.), *Food Toxicology*, CRC Press, Boca Raton, FL, **2001**.

2 E. Tareke, P. Rydberg, P. Karlsson, S. Eriksson, M. Toernqvist, *J. Agric. Food Chem.*, **2002**, *50*, 4998–5006.

3 J. James (Ed.), *Microbial Hazard Identification in Fresh Fruits and Vegetables*, John Wiley and Sons, Inc., Hoboken, NJ, **2006**, 312 pp.

4 J. Fernandez-Cornejo, M. Caswell, *The First Decade of Genetically Engineered Crops in the United States*, Electronic Report Economic Research Service, www.ers.USDA.gov, **2006**, 30 pp.

5 J. O. Hill, H. R. Wyatt, G. W. Reed, J. C. Peters, *Science*, **2003**, *299*, 853–855.

6 A. Eaglesham, C. Carlson, R. W. F. Hardy, Integrating Agriculture, Medicine and Food For Future Health, National Agricultural Biotechnology Council Report 14 on Foods for Health, *National Agriculture and Biotechnology Council*, Ithaca, NY, **2002**, 340 pp.

7 Center for Science in the Public Interest, Outbreak Alert: Closing the Gap in Our Federal Food Safety Net, *Center for Sciences in the Public Interest*, Washington, D.C., Nov. 2005.

8 C. L. Wilkins, J. O. Lay, Jr., *Identification of Microorganisms by Mass Spectrometry*, John Wiley and Sons, Inc., Hoboken, N.J., **2006**.

9 R. Mandrell, L. A. Harden, A. Bates, W. G. Miller, W. F. Hadden, C. K. Fagerquist, *Applied Environ. Microbiol.*, **2005**, *71*, 6292–6307.

10 C. Fagerquist, W. Miller, L. Harden, A. Bates, W. Vensel, G. Wong, R. Mandrell, *Anal. Chem.*, **2005**, *77*, 4897–4907.

11 D. Fouts, E. Mongodin, R. Mandrell, W. Miller, D. Rasho, J. Ravel, L. Brinkac, R. Dehoy, *et al.*, *PLoS Biology*, **2005**, *3*, 72–85.

12 Center for Disease Control, Outbreak of *Salmonella* Serotype Enteritidis Infections Associated with Raw Almonds – United States and Canada, 2003–2004, *Morbidity and Mortality Weekly Report*, CDC, Atlanta, GA, June 11, 2004.

13 Food Navigator, FDA Targets Lettuce industry with *E. coli* guidance, Food USA Navigator.com, Nov. 9, 2005.

14 B. Campbell, R. Molyneux, T. Schatzki, *Food Sci. Technol.*, **2005**, *151*, 483–515.

15 J. Kim, N. Mahoney, K. Chan, R. Molyneux, B. Campbell, *Applied Microbiol. Biotechnol.*, **2006**, *70*, 735–739.

16 D. Light, A. Knight, C. Henrick, D. Rajapasha, B. Lingren, J. Dickens, K. Reynolds, R. Buttery, *et al.*, *Naturwissenschaften*, **2001**, *88*, 333–338.

17 K. McCue, L. Shepherd, D. Rockhold, P. Allen, H. Davies, W. Belknap, *Plant Science*, **2005**, *168*, 267–273.

18 M. Isabel, M. Selma. In *Microbial Hazard Identification in Fresh Fruits and Vegetables*, J. James (Ed.), Boca Raton, Florida, **2006**.

19 A. E. Blechl. In *Agricultural Biotechnology*, ACS Symposium Series 866, **2004**, 53–65.

20 V. Srivastara, D. W. Ow, *Trends in Biotechnol.*, **2004**, *22*, 627–629.

21 C. M. Rommens, J. M. Numara, J. Ye, H. Yan, C. Richael, L. Zhang, R. Perry, K. Swords, *Plant Physiol.*, **2004**, *135*, 421–431.

22 F. Shahidi, C. T. Ho, *Compounds in Foods and Natural Health Products*, ACS Symposium Series 909, **2005**, 308 pp.

23 T. McHugh, C. Huxsoll, *Lehensmittel-Wissenschaft Technologie*, **1999**, *32*, 513–520.

24 C. Chen, T. Trezza, D. Wong, W. Camirand, A. Pavlath, *PCT Int. Appl.*, U.S. Patent 5,939,117, August 17, 1999.

Keywords

Food Safety, Food Quality, Pathogenic Microorganisms, Agricultural Research Service (ARS), MALDI Time-of-Flight, Mass Spectrometry, Aflatoxins, Molecular Biology, Healthy Food Constituents

34
Impact of Pesticide Residues on the Global Trade of Food and Feed in Developing and Developed Countries

Jerry J. Baron, Robert E. Holm, Daniel L. Kunkel, Hong Chen

34.1
Introduction

Commercial growers of fruit, vegetables, herbs, spices, nursery crops, landscape plants, flowers, forest trees, interior plants, and other specialty horticultural crops face many obstacles including availability of adequate land, water, and work force to grow and harvest the crops. Destructive pests (insects, mites, nematodes, plant diseases, weeds, etc.) further challenge the successful production of these high-value crops. Crop protection products (including fungicides, insecticides, miticides, herbicides, plant growth regulators, and biopesticides) are often necessary tools in the integrated "war" against destructive pests. However, most crop protection products are developed for large markets on the major crops such as maize (*Zea mays*), soybean (*Glycine max.*), cotton (*Gossypium* sp.), rice (*Oryza sativa)*, and small grains where the cost of discovery, development, registration and production can be offset by significant sales of the product. This is necessary because the cost of bringing a new chemical crop protection product to the market is very high, yet this leaves horticultural crops with few pest control product options. Horticultural crops are low acreage, high-value crops. In 2004, specialty crops in the United States accounted for 2.9% of the harvested cropland acreage and for 40% of the cropland value of production [1]. Despite their high value and nutritional importance, horticultural crops are less attractive to chemical manufacturers because of their relatively smaller production scales. This is referred to as "The Minor Use Problem", though there is nothing minor about either the scope or the seriousness of the problem.

Further complicating the minor use problem is increasing global trade of specialty crops and the associated problem of chemical residues on imported and exported foods. Global trade of specialty food crops is increasing. In fiscal year 2005, the United States imported about 25.8 billion USD of specialty crops and exported about 14.5 billion USD [1]. Horticulture crops which were once deemed a minor part of the farm economy are now growing in stature. For example, in the United States, the cash receipts for specialty crops are collectively 52.2 billion

Pesticide Chemistry. Crop Protection, Public Health, Environmental Safety
Edited by Hideo Ohkawa, Hisashi Miyagawa, and Philip W. Lee

ISBN: 978-3-527-31663-2

USD which is more than the combined value of five major commodity crops. The same trend is occurring in other countries. We believe that this increased economic importance in global trade of specialty crops is being fueled by health recommendations to eat more fruits and vegetables as well as population shifts and immigration. New residents to the United States from other parts of the Americas as well as from Asia desire the specialty or ethnic crops they remember from "home".

Global trade of specialty crops itself is not a problem. The problem is with the unharmonized use of crop protection products in different countries. In most countries, a use of a crop protection product must be registered with a competent regulatory authority prior to its use by farmers. Furthermore, there must also be an associated Maximum Residue Limits (MRLs) established to support the commodity entering the channels of trade. If a specialty crop is grown and sold for consumption in that country, the grower has a clear indication on what can be legally used. However, if the crop is being grown for export markets, the grower must know where the commodity will be exported to prior to making any chemical applications. If the country importing the produce does not have MRLs for that particular chemical or if the chemical residue level is higher than the established MRL, the crop is considered adulterated. For example, if a grower of California oranges is growing the crop for domestic consumption, the grower could use any chemical registered in the U.S. for use on the crop. If the oranges are destined for Japan, the grower would only be able to use chemicals that are not only registered on oranges in California but also have an associated MRL in Japan. If the grower was unsure where the oranges were going to be sold, the situation is several orders of magnitude more complicated.

The trade conflicts caused by pesticide residues often place greater impact on the developing countries, as they have less technical and financial support to establish specialty crop programs. Therefore, they have less chance to register newer and safer chemicals and obtaining CODEX MRLs as in many developed countries. When the growers from these developing countries export their commodities into developed countries, they could face rejection if residues of older chemicals are found in the shipment, or their ethnic or minor crops are not included in the group of crops listed under the MRLs, which could be established based mainly on crops that are more popular in the developed countries. Another important negative impact of using older chemicals in the developing countries is the food safety concerns, and such cases are often highlighted on various news reports.

34.2 Potential Solutions

The minor use problem and the unharmonized MRL issue poise some real significant challenges for those associated with specialty food crops. The world's specialty crop growers, consumers, and regulators in the countries and regions have all experienced the difficulties caused by discrepancies in residue data requirements and MRL regulations and have all suffered the consequences of trade barriers and food safety problems. There are several opportunities being sought to solve or at least minimize the impact of these issues and enhance global trade of specialty crops.

34.2.1
The IR-4 Model and Other Minor Use Programs

In the United States, growers have had the benefit of the IR-4 Project to assist them with these minor use problems. IR-4 was established in 1963 as a partnership research program with the United States Department of Agriculture, the state agricultural research universities, the private sector, the crop protection industry, specialty crop growers, and the food processing industry. IR-4's main objective is to develop the appropriate residue exposure data on specialty food crops to support the registration through the US Environmental Protection Agency (EPA) for uses of crop protection products which are not being developed by the crop protection industry due to economic reasons. IR-4's efforts also support registrations on minor uses of chemicals on large major acreage crops as well as the development of efficacy and/or crop safety data to support the registration of crop protection products on non-food "ornamental" crops. Since the IR-4 Project's inception, its submissions have supported over 10,000 clearances of crop protection products on food crops and over 10,000 clearances on ornamental crops. Over 80% of the U.S. clearances on food crops in the past 5 years were the result of the IR-4 tolerance petitions with a major focus on newer and reduced risk pest management technology.

Inspired by the success of the IR-4 Project in the United States, other countries have looked into ways in developing systems similar in scope and nature. In Canada, the Ministry of Agriculture and Agri-Food Canada (AAFC) established a substantially similar program to IR-4 called the Pest Management Centre (PMC) in 2003. In fact, prior to this formal program, the AAFC and the Canadian Horticultural Council (CHC) had collaborated with IR-4 on several joint residue programs starting in 1996. This collaboration resulted in over 90 joint field research trials between 1996 and 2002 based on mutual US and Canada grower priorities. Concurrently, a joint Health Canada Pest Management Regulatory Agency (PMRA) and US EPA workshare of an IR-4 petition submission resulted in the first North American Free Trade Agreement (NAFTA) approval in 2002. The number of joint research programs within the US and Canada significantly increased after the establishment of the Pest Management Centre by the Canadian government in 2003. Since 2003, there have been 47 joint IR-4/PMC projects (42 for residue and 5 for efficacy) involving 154 residue trials and 190 efficacy trials conducted. The PMRA/EPA partnership and workshare between US's IR-4 and Canada's PMC has expanded with the support of the NAFTA Technical Working Group on Pesticides, hence four petitions were jointly reviewed in 2004 with two approvals as NAFTA registration's anticipated in 2006.

There are also minor use data development programs similar to the IR-4 Project in Europe, Asia, South America, and Australia/New Zealand. In Europe, significant minor use data development programs are working on local needs in Belgium, France, Germany, The Netherlands, United Kingdom, and possibly other countries. Other developed countries with known programs include Australia and Japan. Minor use data programs have been established in the developing countries such as Colombia and Mexico.

34.2.2 Tools for Harmonization

As noted previously, the rate at which products are being labeled in one country compared to others causes a number of complications. Although one country may have access to newer products that have lower risk characteristics, it may preclude some growers from using them if the produce is going to be shipped to countries that do not have MRLs established for these new products. In the end, the growers would likely resort to using the older, riskier products. Therefore, if a product could be registered globally, rather than segmented country by country, there would be no clear advantage for one country over another and the newer safer products could be integrated more rapidly into production systems providing even greater protection for the applicators, consumers, and the environment. Jim Jones, Director of EPA's Office of Pesticide Programs, noted in his opening remarks at the Public Meeting of the December 2004 North American Free Trade Agreement's (NAFTA) Technical Working Group on Pesticides (TWG) in Merida, Mexico, "we need to find a way to harness the global regulatory resources to work smarter in registering pest control products". A number of regulatory issues have been brought up as major focuses within the NAFTA countries as well as internationally by the CODEX, the EU, Australia, and other countries. These include: work sharing, harmonized guidelines and templates, crop zones, data requirements, risk assessment, and crop grouping.

34.2.2.1 Crop Grouping

The idea of crop grouping has been around for over 40 years and was initially discussed in the first edition of Food and Feed Crops in the United States published in 1971. Basically, crops that are botanically or taxonomically related or culturally similar are grouped together with a few representative crops selected for GLP research purposes. This allows for tolerances or MRLs to be established on crop groups based on residue data from the representative crops. The second edition of Food and Feed Crops in the United States [3] documents the currently established 508 crops in 19 U.S. Crop Groups. However, many more specialty crops that are not included in the current crop groupings scheme are being grown in or imported into the U.S. based on demand from an expanding and diverse ethnic population. The USDA and IR-4 hosted an International Crop Grouping Symposium in Washington, D.C., in 2002 to obtain stakeholder input into this increasingly important issue. The over 125 symposium attendees came up with over 500 additional specialty crops that could be added to current crop groups and proposed doubling the number of crop groups while expanding the representative crops and subgroups. This led to the formation of the IR-4/EPA Crop Grouping Working Group in 2004 co-chaired by the EPA and IR-4. Plans were formed by this Working Group to submit new crop grouping petitions to the EPA for review and approval by their Chemistry Scientific Advisory Council (ChemSAC) prior to final rule making in the Federal Register. IR-4 was aware that these activities would have NAFTA and global regulatory implications, so it formed the International Crop Grouping Consulting Committee (ICGCC) to include crop, agrichemical, and regulatory experts and authorities from around the world. In over two years, the ICGCC has grown from about 60 member from NAFTA countries to over 180 members representing over 40 countries, and the first ICGCC meeting was held prior to the 2005 IR-4 annual Food Use Workshop to review the progress and further explore international cooperation opportunities. To date, the ICGCC has submitted new crop grouping petitions to ChemSAC for bulb vegetables, berries/small fruits, edible fungi, fruiting vegetables, oil seeds, citrus fruits, and pome fruits, and is currently preparing petitions for stone fruits and herbs/spices. ChemSAC has approved the bulb vegetables, berries/small fruits and edible fungi crop group petitions and established a new crop group 21 for the edible fungi. Petitions for tropical/subtropical fruits (edible and inedible peel), leafy vegetables (except Brassica), root and tuber vegetables and stalk/stem vegetables are planned for the next crop group submissions. The leveraging power of crop groupings can be observed in the bulb vegetables approved as the first crop group reviewed by ChemSAC. The representative crops green onion and bulb onion initially covered only 7 minor crops. The new crop grouping expands this number to 25.

Currently, there are no representative commodities in the CODEX system which makes it less practical to use for global specialty crop residue harmonization. The CODEX system was developed to provide a complete listing of food and feed commodities and to track commodities in trade. And in fact, the CODEX system and the U.S. system were both originated from the work of Dr. Roe Duggan of the USDA. As the U.S. system later developed the concept of representative crops, the CODEX system stayed in its original structure. Therefore, even if a commodity had a MRL in one country, that MRL could not be extended to CODEX if the commodity is not currently listed in the crop grouping tables. The purposes of using representative crops are not only to facilitate MRL establishment but more importantly to protect specialty crop production by extrapolating MRLs from representative crops to "minor" crops in the same group, The EU Crop List of Regulation also uses representative crops. In order to better harmonize the world crop grouping/classification systems and minimize trade conflicts, the CODEX system needs to expand its utility to allow for representative commodity MRLs to cover a broader range of other crops, especially those specialty crops where it would be too costly to generate data for registration.

The internationalization of the crop grouping concept was addressed by the CODEX Committee on Pesticide Residues (CCPR) at their 38th Meeting in Brazil in April 2006. This meeting led to CCPR proposing an extended revision proposal of the CODEX Classification of Foods and Animal Feeds. The CODEX Alimentarius Commission at its 29th meeting in July 2006 approved the CCPR proposal. The authors hope that this will lead to more global harmonization of tolerances and MRLs through the use of representative crops to set the MRLs.

34.2.2.2 **Work Sharing**

Over the past several years, regulators from Canada, Mexico, and the United States have moved to make work sharing a standard way of doing business. As a result, the governments have developed processes for sharing resources regarding the review of pesticide residue data, and have implemented efforts to streamline registration procedures, as well as eliminated a number of repetitive regulatory requirements across borders. Many of these new processes have been successful due to the support from registrants and other stakeholders and the openness of all three parties to working together, because they are compelled by a growing North American outlook for free trade in food products, and the desire of maintaining a high level of health and environmental protection. By allowing for quick, coordinated efforts to make decisions on pesticides and minimize trade barriers, while ensuring the sound and sustainable management of new and older products, the cooperation among Canada, Mexico, and the United States has not only improved working relationships, but also facilitated the free flow of trade in crop protection products and agricultural goods across borders.

Guidelines and report formats have also been harmonized to make them more consistent and easier to share among reviewers not only within a given agency, but also when sharing reviews across the borders. The NAFTA Technical Workgroup on Pesticides continues to refine the North American Crop Zones and data requirements both for domestic requirements and for studies that will be conducted on a NAFTA basis seeking registration in all three countries. Much effort has also been put into harmonizing risk assessments and a method has been developed to statistically determine MRLs based on field data (MRL calculator).

Canada and the United States now have a process in place where data for specialty crop stakeholder needs are generated by Canada's AAFC Pest Management Centre and the IR-4 Program. These data are submitted to the respective regulatory agencies simultaneously. The EPA and PMRA have the framework in place to make assignments as to which agency will conduct the review for a given submission. Once those reviews are completed, the reports are peer reviewed by the companion country and the registration is approved in both countries at approximately the same time with harmonized tolerances/MRLs. The review and approval process is expected to take as little as eight months for these joint review minor use requests.

Based on success in North America and building from past workshops and discussions, the OECD sponsored a workshop in early 2005 to advance work sharing on an international basis. The workshop examined national reviews to identify specific barriers to work sharing, to develop recommendations to eliminate or reduce such barriers, and to promote work savings. The workshop also intended to increase the experience and confidence of government evaluators and registrants in using dossiers and monographs and to identify to what extent the current procedures and processes in countries can be improved to facilitate work sharing.

Some of the main points resulting from the workshop were to have the production of common data requirements and guidelines, and to standardize MRLs on a global basis. It was pointed out that problems had resulted from the use of different methodologies in different countries and from differences in hazard and risk assessments. Finally, it was noted that there was a need for adoption of common data review formats by the various national governments (common templates) and harmonized residue guidelines. Thus, the development of harmonized guidelines and terminology by the OECD is critical to the advancement of work sharing on the international level. There continues to be substantial progress in harmonization of guidelines and report formats within OECD member countries.

34.2.2.3 Rationalized Global Data Requirements

A significant effort has been made into determining the feasibility of developing global zones for generating field residue data to determine pesticide levels in agricultural crops. After a long review process, the US implemented crop field trial zones in the mid-1990's detailing the production zones for various crops and data requirements (number of field trials) for individual crops across these zones. NAFTA zone maps extending through all three countries, based on agronomic geographic regions that overlap from one country to the other, were approved by the NAFTA TWG on Pesticides in September of 2001. As a result, requirements for residue studies on a NAFTA basis are significantly reduced compared to conducting studies in individual countries. In order to further promote NAFTA registrations, the NAFTA TWG has also approved further data reductions when studies are conducted on a NAFTA basis in order to promote NAFTA registrations. For a number of crops, the NAFTA requirements for residue studies are essentially equal to the maximum number of trials required by an individual country.

The OECD conducted a comprehensive review of residue data to determine if global zones could be considered to facilitate international cooperation. After this review, the team could not discern zones because of high variability in residues from comparable trials. In many cases, the data reviewed showed just as much variability within a zone as compared to across zones. The data also showed that the pre-harvest climate may not have had as strong of an influence on the residue levels as would have been expected. This indicated that the zoning effect may not be a major factor in determining residues. Unfortunately, neither final zone recommendations nor data requirements for international registrations could be suggested based on the review.

The EU is also considering a more flexible approach to pesticide registration using geographical zones rather than the country zones. The current proposal considers three zones across all of the member states; these include a north, a central, and a south region for residue studies. The scheme in the EU is to also consider having zonal evaluations that could be shared by other member states within the same regions. There are a number of pilot projects evaluating these new zonal and review schemes to see if they are feasible.

Considering that zones are established for all of North America extending from the tropical southern states of Mexico throughout the United States, including the island of Hawaii and into the Northern most regions of Canada, it could be speculated that essentially all of the regions of the world would be represented by these 21 zones used by the NAFTA. As well, zones exist for Europe and the OECD zoning project indicated that zones may not be a major factor in residue variability. The establishment of international zones could greatly facilitate the development of residue data for both major and minor crops, as well as prevent duplication of trials in various countries, thereby reducing the overall cost for industry to develop data globally. This approach could also greatly reduce the time spent on data review for regulatory agencies and ultimately bring the growers pest control solutions more efficiently. With all of these factors in mind, global zoning certainly is an area that needs further discussion to make it possible for industry to pursue global registrations as they set out to register new products.

The authors would like to see the current OECD global review of new products such as the one currently under way for Rynaxypyr™, a new DuPont Crop Protection insecticide, extended to residue data. The Rynaxypyr™ basic registration package is currently under joint review in the U.S., Canada, EU (Ireland), Australia, and New Zealand with the goal of a global new active ingredient registration. However, individual crop tolerance will be determined on a country-by-country basis based on residue data generated in the countries of interest under Good Agricultural Practices (GAPs). IR-4 is attempting to form coalitions with registrants beginning to commercialize new products to determine their interest in a global registration that would include residue data. Several countries such as the UK, Germany, Australia, Canada, and others have expressed interest in this approach. The challenge will be for international regulatory bodies to agree upon proposed global residue zones and common GAPs to conduct the trials. The opportunities to harmonize global MRLs in order to reduce trade barriers and promote international specialty crop trade appear to adequately justify the challenge.

34.3 References

1 J. E. Noel, The U.S. Specialty Crop Industry: Significance and Sustainability. http://cissc.calpoly.edu/farmbill/USSpecCropIndSigandSust.pdf

2 N. Brooks, E. Carter, Outlook for U.S. Agricultural Trade, Electronic Outlook Report from the *Economic Research Service* and *Foreign Agricultural Service*, http://usda.mannlib.cornell.edu/usda/ers/AES//2000s/2005/AES-11-28-2005_revision.pdf

3 G. M. Markle, J. J. Baron, B. A. Schneider, *Food and Feed Crops of the United States*, 2nd Edition, revised, Meister Publishing Co., Willoughby, Ohio, **1998**.

Keywords

Minor Use Problem, Global Trade, Speciality Crops, Ethnic Crops, Maximum Residues Limits, Trade Barriers, Food Safety, IR-4 Model, Crop Grouping, MRL Harmonization, Good Agricultural Practices, GAP

35
Pesticide Residue Assessment and MRL Setting in China

Yibing He, Wencheng Song

35.1
Introduction

In 1978, China began to supervise pesticide use. In 1986, pesticide quality control began. Pesticide management regulation was publicized and enforced in 1997. Now, China is one of the countries with the largest pesticide production and application. China ranks first in pesticide production and second in pesticide usage. There are over 2,000 manufacturers, 20,000 registered products, and 600 active ingredients in China. China produces about 400,000 tons of technical grade active ingredients every year and uses it on over 200 million hectares. The pesticides in China are used mainly on rice, cotton, vegetables, and fruits, and are also exported to over 120 countries and regions.

Since the 1980s, China has paid special attention to pesticide management; some administrative measures have been taken and many regulations as well as national standards have been issued to control pesticide residue in/on raw agricultural commodities, feed and food items.

35.2
Regulations and National Standards for Pesticide Residue Management in China

35.2.1
Key Components of Pesticide Management Regulation of China for Pesticide Residue

In 1997, pesticide management regulation was promulgated and enforced; corresponding rules such as Provisions for the Implementation of Pesticide Management Regulation were subsequently issued. Currently, China is paying special attention to the regulation of persistent pesticides and highly toxic pesticides, canceling the use of highly toxic pesticides on vegetables, fruits, tea, Chinese medicinal herbs, etc. The supervision of pesticide application and

Pesticide Chemistry. Crop Protection, Public Health, Environmental Safety
Edited by Hideo Ohkawa, Hisashi Miyagawa, and Philip W. Lee

ISBN: 978-3-527-31663-2

pesticide residue was also emphasized. Crop, food, and feed which contain pesticide residue over MRLs are not allowed to be sold in China.

35.2.2
Key Components of National Standards for Pesticide Residue Management in China

In setting national standards for pesticide residue management in China, the data requirements for pesticide registration are most important. Listed below are detailed data requirements for pesticide residue to support pesticide registration. A series of technological criteria were established:

1) Application of pesticides in China should follow Guidelines of Safety Application of Pesticides. There are over 50 laboratories that can conduct pesticide field trials certified by ICAMA. The good agriculture practices (GAP) for the safe and effective use of the product must be established. Over 1,000 field trials with over 300 active ingredients in more than 30 crops have been conducted in China. On the basis of these field trials, a series of GAP have been drawn up and new revisional MRLs have been proposed. The main contents in this GAP are common name, crops, application rate and method, maximum number of application, PHI (pre-harvest interval), and recommended MRLs.

2) There are 478 MRLs for 136 pesticides in over 30 varieties of agricultural products issued in 2005. We also have some other standards, such as Residues Experiment Guideline of Pesticide Registration, Test Guideline for Safety Assessment of Chemical Pesticide in Environment, etc.

35.3
Summary of Date Requirements of Pesticide Registration

The applicant should submit a dossier including efficacy, identification of active ingredient, physico-chemical properties, analytical method, toxicity and metabolism, residue, environmental fate, toxicology, and label information. To process a new pesticide registration in China, there are 3 stages: (1) field efficacy, (2) temporary registration, and (3) full registration. The data required for residue assessment in each stage are as follows. In the field trial stage, the applicant should submit acute toxicity data, residue data, and registration status in other countries, so that ICAMA can assess the product safety under the specific label/use pattern and environmental conditions in China. If the product exhibits low acute toxicity and shows no chronic toxicological effects, supervised residue trials in China can be conducted after the temporary registration is granted and the field residue data are submitted during the full registration process. For products that are classified as toxic compounds, residue data are required to support the temporary registration consideration.

Residue trials should follow the Residues Experiment Guideline of Pesticide Registration; the number of residue field trials of each compound should be held in at least 2 locations conducted over for 2 years.

35.4 Residue Data Requirements

Final residue trial reports should include key information for field trials, analysis method, and experimental results.

35.4.1 The Protocol of Field Trials

The protocol of field trials should include the following information:

1) Field application rate (or concentration) of pesticide, application method, timing, and equipment.

2) Plot size of the field (or number of crops), treatment intervals and number of applications, and PHI.

3) pH value of soil in the experiment area, soil nature, organic matter content, climate conditions, and cultivation system.

4) Sampling, locations, sample handling, and storage conditions.

35.4.2 Residue Analysis Method

The detailed residue analysis method must be reported including the sample extraction procedure and clean-up. The method of detection must be simple and feasible, and the instrumentation and its operating conditions are to be acceptable for residue laboratories. Secondly, the recovery, RSD (relative standard deviation), and sensitivity (limit of detection and minimum concentration of detection) of the analytical method should meet the requirements of the Residues Experiment Guideline of Pesticide Registration.

35.4.3 Experimental Results

The following information should be reported in the final report:

1) The relationship between residue in crops and time, residue in soil cultivation layer (0–20 cm) and time, residue in water (only for the paddy field), and time

should be described in text as well as in tables and figures (i.e., degradation curve).

2) Final residues in the crops under maximum registered uses, i.e., maximum application dosage, maximum number of applications, and minimum pre-harvest interval, should be contained in the table.

3) Residue Data Considerations
   a. The analytical method should determine the residue of parent pesticide and toxic metabolites in crops, soil, and water.
   b. Metabolism in the crops and animals, the absorption, distribution, excretion, transformation, terminal metabolite and degradation product, and their toxicity data must be provided.
   c. The residue data generated in other countries and regions should be also provided.
   d. MRL accepted by CAC or established by other country authorities should be submitted.

35.5 Procedures for Establishing MRLs and Setting Up PHI in China

Necessary components for establishment of MRLs in China involve acceptable daily intake (ADI), Chinese national dietary patterns, and residue data generated at GAP. Below is the procedure for setting MRLs:

1) The highest residue level reported from residue trial studies in target crops following the critical GAP are used to estimate MRL.

2) Simultaneously, the full toxicology data is evaluated to derive the ADI. The ADIs that have been already proposed by JMPR or other registration authorities should be considered in risk assessment when there is no ADI in China.

3) Daily intake assessment
   TMDI (Theoretical Maximum Daily Intake) will be calculated as follows:

   $$\mathrm{TMDI} = \sum [\mathrm{MRL}_i \times F_i]$$
   MRL_i = Maximum Residue Limit for a given food commodity
   F_i = Corresponding national consumption of that food/person/day

   If TMDI is less than ADI, this registration will be granted, otherwise it will be rejected.

4) Recommended MRLs will be proposed to the national standard authority for approval.

35.6 Examples of MRL Setting in China

The following are cases to describe the procedure details for MRL setting in China.

All ADIs in Table I are abstracted from CAC after careful evaluation by JMPR. Table II contains detailed consumption data. The national diet in this example is derived from the Chinese national survey data conducted in 2002.

Supervised trials of 7 pesticides were conducted on 6 crops at registered doses from 1986–2003. Table III summarizes these 7 pesticides applications.

The summary of supervised trials conducted at different PHIs is presented in Table IV. The conduct of residue trials on these crops for these 7 compounds was consistent with their GAPs. The ranked residue data are presented in Table IV. Double-underlined data are highest residues from treatments and used for estimating maximum residue level. Recommendation MRLs are listed in Table V.

Table VI lists the dietary intake calculation for abamectin. The other pesticides calculations were done and listed in Table VII. All TMDIs of residues of 7 compounds from this case did not exceed individual ADI and the dietary intake is unlikely to present a public heath concern. The MRLs proposals will be submitted to the China authority for approval.

Table I. Examples of acceptable daily intake (ADI) from JMPR.

Compound	ADI (mg/kg bw/day)
abamectin	0.002
carbendazim	0.03
cyhalothrin	0.002
diflubenzuron	0.02
fipronil	0.0002
imidacloprid	0.06
iprodione	0.06

Table II. Examples of the Chinese national dietary pattern.

Commodity	Diet (g/person)
rice and rice products	240
wheat flour and products	138
other cereal grains	23
root and tuber vegetables	50
pulses	4
soybean products	12
leafy vegetables	91
other vegetables	184
vegetables in preservative	10
fruits	46
tree nuts	4
meat from poultry and mammals other than marine mammals	79
milk and milk products	26
eggs	24
fish and crustaceans	30
vegetable oil	33
poultry and mammalian fats	9
sugar and starch	4
salt	12
sauce	9
Total	1028

Table III. Example of registered or approved pesticide uses on food crops.

Compound	Crop	Locations, year	Form	Rate
abamectin	cucumber	Anhui, Hebei, 1999–2000	EC/1.8%	10.8 g ai/ha
carbendazim	citrus	Zhejiang, Guangxi, 2002–2003	WP/50%	50 g ai/hL
cyhalothrin	wheat	Beijing, Shandong, 1994–1995	EW/2.5%	7.5 g ai/ha
diflubenzuron	wheat	Beijing, Henan, 1986–1988	WP/25%	37.5 g ai/ha
fipronil	cabbage	Beijing, Jiangsu, 1996–1997	SP/5%	26.55 g ai/ha
imidacloprid	chietqua	Guangdong, Hainan, 2001–2002	EC/5%	8.9 g ai/ha
iprodione	tomato	Beijing, Zhejiang, 2000–2001	SP/50%	750 g ai/ha

Table IV. Residues from the supervised trials.

Compound	Crop	Application Number	PHI (day)	Residues in ranked order (mg/kg)
abamectin	cucumber	2	1	0.008, 0.006, 0.007, 0.006
			2	0.005, 0.004, 0.004, 0.003
			3	0.004, 0.001, 0.003, 0.001
		3	1	0.007, 0.006, 0.007, 0.006
			2	0.005, 0.004, 0.005, 0.004
			3	0.003, 0.002, 0.003, 0.002
carbendazim	citrus	3	20	1.91, 2.04, 2.62, 3.78
			30	1.20, 1.33, 1.69, 2.44
		4	20	2.40, 2.74, 4.00, 4.39
			30	1.35, 1.59, 2.90, 2.81
cyhalothrin	wheat	1	30	< 0.002(4)
			60	< 0.002(4)
		2	15	0.008, 0.01, 0.012, 0.032
			30	< 0.002(2), 0.008, < 0.002, 0.01
diflubenzuron	wheat	2	20	0.015, 0.017, 0.013, 0.068
			30	< 0.012(2), 0.017, 0.046
		3	20	0.017, 0.023, 0.017, 0.101
			30	< 0.012(2), 0.024, 0.050
fipronil	cabbage	2	3	0.009, 0.022, 0.016, < 0.005
			5	< 0.005(3), 0.01
		3	3	0.018, 0.029, 0.016, 0.013
			5	< 0.005(3), 0.01
imidacloprid	chietqua	3	3	0.04, 0.03, 0.08, 0.07
			5	0.04, 0.03, 0.06, 0.04
			7	0.02, 0.02, 0.04, 0.04
		4	3	0.04, 0.04, 0.06, 0.07
			5	0.04, 0.04, 0.06,0.06
			7	0.02, 0.02, 0.05, 0.04
iprodione	tomato	3	2	0.15, 0.18, 0.085, 0.076
			4	0.14, 0.15, 0.064, 0.059
			6	0.13, 0.14, 0.040
		4	2	0.18, 0.19, 0.11, 0.12
			4	0.17, 0.18, 0.081, 0.087
			6	0.15, 0.16, 0.057

Table V. Recommendations MRLs.

Compound	Crop	MRLs recommendation
abamectin	cucumber	0.01
carbendazim	citrus	3
cyhalothrin	wheat	0.05
diflubenzuron	wheat	0.2
fipronil	cabbage	0.05
imidacloprid	chietqua	0.1
iprodione	tomato	0.5

Table VI. TMDI of abamectin.

Commodity	Diet (g/person/d)	MRLs (mg/day)	TMDI (mg/day)	% of the ADI
Rice and rice products	240			
Wheat flour and products	138			% ADI =
Other cereal grains	23			TMDI/60/ADI
Root and tuber vegetables	50			
Pulses	4			
Soybean products	12			
Leafy vegetables	92	0.01	0.001	
Other vegetables	184	0.01	0.002	
Vegetables in preservative	10	0.01	0.0001	
Fruits	46			
Tree nuts	4			
Meat from poultry and mammals other than marine mammals	80			
Milk and milk products	26			
Total			0.0031	2.6

Table VII. Summary of TMDI of seven pesticides.

Compound	ADI [mg/kg · (bw)]	% of the ADI
abamectin	0–0.002	3%
carbendazim	0–0.03	44%
cyhalothrin	0–0.002	62%
diflubenzuron	0–0.02	31%
fipronil	0–0.0002	38%
imidacloprid	0–0.06	0.7%
iprodione	0–0.06	36%

35.7 Perspective

China adopted the JMPR procedure for risk assessment and MRL setting. The critical data such as ARfD, refined dietary pattern, should be taken into consideration in risk assessment.

China is amending pesticide registration data requirements to generate more residue data for major crops in China. Data requirements for pesticide registration have been in the process of revision for several years to meet today's global regulatory standards. ICAMA is coordinating residue studies within the various major crop-growing regions in China.

China will refine the present national diet based on further dietary surveys for accurate intake and risk assessment.

China will continue cooperation with other countries regarding MRLs setting. As the new host country of CCPR, China will work together with CAC member countries to harmonize global MRL setting.

Keywords

Pesticide Residue Assessment, MRL, China,
Guidelines of Safety Application of Pesticides, Good Agriculture Practices,
Pre-Harvest Interval, ICAMA, Residue Data Requirements, Field Trials,
Residue Analysis Method

36
Harmonization of ASEAN MRLs, the Work towards Food Safety and Trade Benefit

Nuansri Tayaputch

36.1
Introduction

Food safety is a major concern for both government and the public. This has been further emphasized by recent reports of both microbial and chemical contamination of food. Moreover, the consumers' preferences have resulted in large increases in food imports which, in some cases, do not provide traceability in terms of production processes, or details of pesticide applications pre- and post-harvest. Therefore, to secure food safety and provide information for consumers, national authorities have to supply all information concerning food quality and regulate the amount of pesticide residues that could remain without posing any risk to consumers.

36.2
Role of Codex MRLs in Regulating Food Quality

Internationally recognized food standards, i.e., Codex standards, have been established for more than three decades as standards for facilitating international trade and solving trade disputes. One of the many standards established by Codex is the maximum residue limit (MRLs) for pesticides on food commodities in international trade. Codex MRLs are used as national standards by many countries; however, some countries continue to establish their own MRLs or tolerances and impose zero tolerance to residues of pesticides on imported crops which do not have nationally/regionally agreed-upon MRLs. Therefore, the acceptance of Codex MRLs among countries is different. An example of the variation in MRLs for carbaryl in some commodities is shown in Table I.

The difference in pesticide residue limits in different countries results in trade irritations.

Pesticide Chemistry. Crop Protection, Public Health, Environmental Safety
Edited by Hideo Ohkawa, Hisashi Miyagawa, and Philip W. Lee

ISBN: 978-3-527-31663-2

Table I. Comparison of carbaryl MRLs (mg/kg) in some commodities among selected countries [1].

Country	Rice	Wheat	Apple	Grape	Strawberry
Codex	5	5	7	5	7
USA	5	3	10	10	10
Canada	2	2	10	5	7
Japan	1	–	1	1	–
Korea	1	3	0.5	0.5	0.5

36.3
Pesticide Residues in Developing Countries

A majority of ASEAN member countries are basically food exporters which rely on using agrochemicals to grow high-quality produce. The use of pesticides may result in residues on food and feed items. There were several reports on findings of pesticide residues in agricultural produce in the export countries; residue levels were found in several commodities with wide variations in amounts and types of pesticide. Detailed information revealed that pesticide residues in some commodities were found to be above Codex MRLs as illustrated in Table II.

The exceeding residues in those commodities might have been caused by many factors such as:

- Post-harvest treatment
- Seed treatment
- The differences in cultural practices and climatic conditions, etc.

Table II. Pesticide residues exceeded Codex MRLs.

Country	Commodity	Pesticide
Indonesia	Vegetables	carbendazim, carbofuran, chlorpyriphos, fenvalerate, dithiocarbamate
Malaysia	Fruits & Vegetables	chlorpyriphos, profenophos, iprodione, monocrotophos, dithiocarbamate
Thailand	Fruits & Vegetables	methamidophos, chlorpyrifos, cypermethrin, profenophos
Vietnam	Fruits & Vegetables	monocrotophos, cypermethrin, methamidophos, lamda-cyhalothrin

Source: International Seminar on Food Safety and Quarantine Inspection, 2000.

Several kinds of pesticides are used on tropical fruits and vegetables because of a wide variation of insects and diseases on these crops. The export of agricultural produce which do not comply with the regulations of the import country might face clearance rejection, which has happened many times with exports from developing countries. In general, tropical fruits and vegetables are considered to be minor crops because of the small cropping area when compared to major crops such as rice, wheat, soybean, tomato, potato, etc., which are grown all over the world. A few Codex MRLs were set up only on some fruits such as banana, papaya, and pineapple, whereas some other tropical fruits which are economically viable for developing countries, especially in Asia, such as durian, longan, litchi, starfruit, guava, mangosteen, pomelo, etc., are not available.

In general, the pesticide residues in/on minor crops with unavailable Codex limits were often treated at lowest residue values or at a limit of determination (LOD).

36.4 Issues on Minor Crops

The minor crops which are mainly tropical fruits and vegetables are considered important as being the export produce from developing countries. In order to get these produce accepted into overseas markets, their maximum residue levels must be available for reference. The Codex MRLs are required to cover all pesticides/commodities in the region. However, the work towards setting up Codex MRLs is complicated and needs expertise as well as financial support in doing local supervised residue trials and dietary risk assessments. Therefore, extrapolation from major crops to minor crops is a way to make use of already generated data for setting up new MRLs. For developing countries, it is also the way to save resources and manpower to gain MRLs for minor crops.

The basic requirement for residue extrapolation has already been described [2–4] that information such as methods of application and use pattern, formulation type, climatic conditions and crop morphology, etc., has to be comparable to that of the main crops. However, the extrapolation of data to minor crops has occurred rarely in JMPR.

36.5 The Work of ASEAN Expert Working Group on Pesticide Residues

A concern over trade barriers related to unavailability of Codex MRLs on important crops in the region and the problem of different MRLs in different countries is prevailing in the region. As a result of an initiative by the Sectoral Working Groups on Crops of the ASEAN Ministries of Agriculture and Forestry, the ASEAN Expert Working Group on Pesticide Residues was founded to carry out the task of harmonizing MRLs among ASEAN Member Countries which include Brunei,

Cambodia, Indonesia, Laos PDR, Malaysia, Myanmar, Philippines, Singapore, Thailand, and Vietnam.

The ASEAN Expert Working Group on Pesticide Residues also aims to obtain regional cooperation and the pooling of technical and financial resources to overcome the common issues of pesticide residue problems in the region. The first meeting of the Expert Working Group was held in 1996 in Malaysia with the objective of facilitating ASEAN trade in agricultural commodities while protecting consumer health.

The outcome of the first meeting was several agreements on:
- Procedures in setting ASEAN harmonized MRLs
- Principles of harmonization
- Prioritized pesticide/commodity for harmonization
- Collation of GAP information relating to the identified pesticides

36.6
Principles of Harmonization

From the first meeting in 1996, the EWG (Expert Working Group) agreed upon the following principles:

- Pesticides proposed for setting up ASEAN MRLs should have registered uses in all member countries.
- Each member country is urged to set up a national committee for establishing national MRLs.
- If Codex MRLs are available and applicable, they should be adopted for harmonization as ASEAN MRLs.
- If Codex MRLs are not acceptable, modification of MRLs should be supported with residue trial data and dietary risk assessment.
- If Codex MRLs are not available, member countries could propose MRLs to the ASEAN EWG for consideration with supporting data based on Codex procedure.

The ASEAN Expert Working Group has met annually since 1996 to work on establishment of ASEAN MRLs which started in the third meeting in 1998. At that time, 10 ASEAN MRLs for 5 pesticides were adopted for harmonization and the work continued until 2005 with the total number of harmonized ASEAN MRLs at 559 for 42 pesticides.

36.7
Several Observations Made During the Process of Harmonization from 1998 to 2005

- The 559 ASEAN harmonized MRLs were mainly adopted from Codex MRLs; it was considered that the dietary risk was acceptable.
- There were few cases where Codex MRLs were not accepted by some member countries due to risk assessment based on national diet higher than ADI.
- Extrapolation had been made where there were similar crops with Codex MRLs (Table III).

The ASEAN MRLs received from extrapolation were considered on a case-by-case basis and a consensus was required for adoption.

The revision/deletion of the reference Codex MRLs has affected harmonization of ASEAN MRLs and 185 MRLs need to be re-considered. The EWG has agreed on the following:

- If Codex MRLs have been changed, the EWG would consider amending ASEAN MRLs on a case-by-case basis.
- If Codex MRLs have been deleted on some commodities, the EWG should retain the ASEAN MRLs, except MRLs which have acute intake concerns.
- If Codex MRLs have been deleted due to lack of data supported, the EWG should delete or retain the ASEAN MRLs supported by residue trial data and dietary intake studies from respective resources.

Therefore, some deleted Codex MRLs still remain as ASEAN MRLs because they are needed as reference standards, necessary for regional or international trading.

Table III. ASEAN MRLs set up by extrapolation from similar crops.

Pesticide	Crop	ASEAN MRLs (mg/kg)	Similar crops with Codex MRLs
cypermethrin	cabbage	1	brassica vegetables
cypermethrin	crucifers	1	brassica vegetables
cypermethrin	garlic stem	0.5	leek
cypermethrin	shallot bulb	0.1	onion bulb
diazinon	garlic	0.05	onion bulb
malathion	chilli	0.5	pepper
malathion	stringbean	2	common beans
malathion	Chinese cabbage	8	cabbage
metalaxyl	maize	0.05	cereal grain
methomyl	chilli	1	pepper
methomyl	shallot bulb	0.2	onion bulb

In case that deletion was due to lack of supporting data, the member countries were requested to submit inputs on the earlier harmonized ASEAN MRLs.

36.8 Future Outlook

36.8.1 ASEAN MRLs with Quality Data Conducted at Regional Levels on Tropical Crops Should be Established as International Standards

Due to financial constraints and inadequate facilities in the region, member countries are initiating regional collaborations in the generation of residue trial data using an internationally accepted protocol. As a result, some member countries agreed to submit local residue trial data for the establishment of ASEAN MRLs, of which details follow:

Regional Residue Data Proposed for Submission in the Next EWG Meeting:

Malaysia:	cypermethrin	for mango
	monocrotophos	for oil palm
	methamidophos	for oil palm
Singapore:	profenophos	for pepper, chilli
Thailand:	cypermethrin	for mango
	chlorpyrifos	for litchi, longan
	carbosulfan	for asparagus, yard long bean
	cyhalothrin	for asparagus, mango, okra

36.8.2 Member Countries Should Have Comprehensive Knowledge on MRLs' Establishment Consistent with International Guidelines

The lack of technical expertise in MRLs establishment in ASEAN member countries is the problem in setting up MRLs. A few member countries have already initiated conducting local residue trials as well as collecting national dietary intake. Generally, among ASEAN countries, the work toward setting up MRLs, especially supervised residue trials, were undertaken only by national governments. The pesticide industry was not interested in generating residue trial data for minor crops due to economic reasons. The member countries willing to conduct the residue studies always face problems concerning inadequate technical and financial support. Therefore, there is a strong need for collaborative efforts between governments and the pesticide industry in generating data for establishment of ASEAN MRLs, as well as to fill up the gap of knowledge and technology among member countries. In this respect, CropLife Asia has contributed to ASEAN EWG in supporting several workshops covering the process of MRLs establishment. The workshops provided EWG members with in-depth knowledge and the opportunity to exchange experience with international experts on residue issues. Such collaboration could assist ASEAN member countries in establishing more MRLs and increase their competency in performing the work.

36.9 Conclusion

A demand for MRLs for many tropical crops has activated the work of ASEAN EWG on the harmonization of ASEAN MRLs. Despite the original plan to evaluate and adopt MRLs of crops with no Codex MRLs, EWG progress could be made mainly from adoption of Codex MRLs for pesticides/commodities which have been registered for use in regional countries. The work was also intensified to cover extrapolation from similar crops. Moreover, there is a strong possibility that in the very near future the ASEAN EWG might receive more regional residue data from work sharing among member countries as well as from single country residue field trials. However, the number of residue field trials proposed to submit for evaluation was very low compared to a large number of MRLs needed for many tropical crops. There was a consensus from ASEAN EWG about the need to build up capabilities of their resource personnel in the area of MRLs establishment. It was also agreed to seek assistance from industry and other concerned agencies to share the burden in conducting residue trials as well as to provide the regional countries with technical knowledge in the relevant issues. In this respect, cooperation and collaboration among governments and industry which has already begun, needs to move on until member countries have enough expertise for MRLs establishment, with international acceptance, to use them for their consumers' health protection and to facilitate trade in the regional and international markets.

36.10 References

1 B. Y. Oh, in *Proceedings of International Seminar on Food Safety and Quarantine Inspection*, October 16–21, **2000**, Suwon, Korea, 24.
2 FAO, *Submission and Evaluation of Pesticide Residues Data for the Estimation of Maximum Residue Levels in Food and Feed*, FAO Plant Production and Protection Paper, **2002**, *170*, 25–26.
3 K. Hohgardt, M. Theurig, R. Savinsky, W. Pallutt, *Registration of Plant Protection Products in Case of Minor Uses – A Concept to Facilitate the Testing of Residue Behaviour*, 9th IUPAC International Congress of Pesticide Chemistry, **1998**, Abstract 7E-005.
4 OECD, *Registration and Work Sharing*, Report of the OECD/FAO Zoning Project, OECD Series on Pesticides, Number 19, ENV/JM/MONO, **2003**, *4*.

Keywords

Harmonization, ASEAN MRLs, Food Safety, Trade Benefit, Codex MRLs, Pesticide Residues, Minor Crops, ASEAN EWG

37
Possible Models for Solutions to Unique Trade Issues Facing Developing Countries

Cecilia P. Gaston, Arpad Ambrus, and Roberto H. González

37.1
Introduction

Developing countries face a number of problems in international trade, these include problems associated with the lack of Codex Maximum Residue Limits (MRLs) or unharmonized MRLs imposed by different importing countries. These problems have been widely discussed on the international stage and are well-documented; however, the initiatives taken by developing countries to find solutions to these trade issues have not been reported in any detail.

Commodities produced and exported from developing countries are normally specialty crops or minor crops. The lack of MRLs for these crops is largely due to the lack of scientific data required to support establishment of these MRLs.

Codex MRLs are established through a lengthy process which starts with a review of technical data by the Joint FAO/WHO Meeting on Pesticide Residues (JMPR). MRLs are estimated from supervised residue trials conducted according to use patterns reflecting the national authorized uses (GAPs). Results of processing and animal transfer studies, data on the chemistry and composition of pesticides, environmental fate, metabolism in farm animals and crops, and methods of analysis for pesticide residues [1] are also taken into account during the MRL setting process. National MRLs are established using a similar approach.

Generally, residue data on major crops are provided by the pesticide manufacturers supporting the establishment of the MRLs. Due to the high cost of producing the data, manufacturers have to set priorities and, consequently, some minor crops may not be supported. In order to establish MRL for minor or specialty crops, the farmers and exporters from developing countries may need to generate the data.

Realizing the situation and the need to protect their export crops, developing countries have, over the last 15–20 years, been taking initiatives to resolve their problems in trade. Three successful examples of how specific problems were resolved are presented below.

Pesticide Chemistry. Crop Protection, Public Health, Environmental Safety
Edited by Hideo Ohkawa, Hisashi Miyagawa, and Philip W. Lee

ISBN: 978-3-527-31663-2

37.2
Possible Solution to the Lack of Analytical Facilities and Expertise on Developing Data to Support Establishment of MRLs

In the early 1990s, many of the developing countries neither had the facilities nor the know-how to generate acceptable data for establishing Codex MRLs. Realizing the need to generate local data locally for minor/specialty intended for exports, governments of these developing countries sought technical assistance from international organizations.

In 1997, the FAO and the International Atomic Energy Agency (IAEA) responded with the establishment of the FAO/IAEA Training and Reference Centre for Food and Pesticide Control (TRC), as part of the activities of the Joint FAO/IAEA Division of Nuclear Techniques in Food and Agriculture, based in Vienna, Austria.

The Centre has been given the mandate to assist Member Countries and their institutions to fulfill requirements to support the development and implementation of international standards/agreements relevant to food safety and control. The Centre provides training, quality assurance services, and technology transfer.

As part of this mandate, TRC has organized a number of training workshops aimed at supporting the generation of reliable pesticide data locally, i.e.:

- Quality assurance and quality control principles in pesticide residue analysis;
- Planning and implementation of supervised field trials to provide data for establishing maximum residue limits;
- Evaluation of safety of pesticide residues in food;
- Introduction of principles of Good Agricultural Practice (GAP) in growing tropical fruits and vegetables.

Since 1997, "hands-on" training workshops of 4–6 weeks' duration have been organized, e.g., in China, Democratic Republic of Korea, Kenya, Malaysia, and Thailand. Shorter (1–2 weeks) workshops were also held in several countries. The training workshops were complemented with fellowship training programs providing the opportunities for in-depth training in specific topics for 3 to 12 months. Over 300 scientists and government officials received training during the last 8 years.

The training programs have assisted in upgrading analytical facilities and greatly strengthened capabilities of national institutions, e.g.,

- Establishment of QA/QC systems in national pesticide analytical laboratories;
- Acceptance of locally generated residue data by JMPR and importing countries;
- Establishment of Codex MRLs in a number of tropical fruits (e.g., mango, papaya, passion fruit)

The countries that have benefited from these training programs have since provided and continue to provide data to the JMPR to support the establishment of Codex MRLs in a number of pesticides used in tropical fruits and vegetables.

37.3
A Solution to the Lack of MRLs on Spices

The international trade in spices is unique in that about 90% of the exports are from developing countries and almost all import markets are in more industrialized nations. The world trade in spices amounts to 1.5 million tons valued at about 3 billion USD with projected increases of 2% to 5% annually [2]. Of these exports, 33% come from developing countries with per capita incomes of less than 1,000 USD [3]. The leading exporters are China, India, Madagascar, Indonesia, Vietnam, Brazil, Guatemala, and Sri Lanka while the main importers are the United States, Europe, and Japan.

Numerous trade disruptions have occurred over the past years as a result of the lack of national and Codex MRLs for spices. Dried chili peppers have suffered the most detentions. In India alone, from 1999 to 2001, the reported losses in revenues ranged from 4.4 to 6.1 million USD [4]. Notifications in the EU Food Alert System often show rejection of consignments of spices from Asia due to residues of pesticides. Considering the main producers are small farm holders in developing countries that rely mainly on trade in spices for subsistence, the importance of establishing residue limits for pesticides used on spices was brought to the attention of the CCPR in 2000.

37.3.1
Rationale for an Alternative Approach to Setting MRLs for Spices

Spices are grown primarily in developing countries where production is at subsistence level agricultural operations. For example, in India, more than 3 million tons of spices are grown annually in about 105 million small holdings. Approximately 62 million of these small holdings (59%) belong to marginal farmers that own less than 1 hectare of land and 20 million (10%) consist of farms of 1 to 2 hectares of land. Only 1.5% of these holdings represent areas with more than 10 hectares [4]. Not only are spices cultivated in 1 to 2 hectare plots, but also frequently they are grown in between a variety of other crops. It is not unusual to find plots with several pepper vines, a nutmeg tree, a few banana trees, tapioca bushes, and cotton plants all growing in close proximity. The key difference between spices and other commodities considered for MRLs by the CCPR is this subsistence level of farming.

In addition to this, the CCPR was convinced that due to the large number of spice varieties available in international trade and the equally large number of pesticides that can be used on them, it was not possible to follow the conventional practice of establishing MRLs based on GAPs for each pesticide/spice combination.

After considering the above situations and the fact that intakes would not be a concern because the per capita consumption of spices is very low (representing less than 0.5% of the diet, based on the WHO regional diets), the CCPR agreed to establish MRLs for spices based on monitoring data [5].

37.3.2
Codex MRLs for Spices

The CCPR made it clear that the use of monitoring data to set MRLs would be applicable only to "spices" as defined by the Codex Classification of Foods and Animal Feeds: Spices (Group 028): "Spices are dried aromatic seeds, buds, roots, bark, pods, berries or other fruits from a variety of plants, which are used in relatively small quantities as seasoning, flavoring, or imparting aroma in foods" [5–6]. For purposes of establishing MRLs, spices were divided into subgroups, on the basis of parts of plants from which they are obtained (seeds; fruits or berries; roots or rhizomes; bark; buds; aril and flower stigma).

The JMPR in 2004 reviewed the available monitoring data submitted by governments and the International Organization of Spice Trade Associations (IOSTA) and proposed MRLs for spices using statistical methods and the following basic principles [7]:

- All monitoring data were considered; no data point was excluded as an outlier.
- Maximum residue level (MRL), high residue (HR), and median residue (STMR) for each subgroup were recommended where the monitoring data enabled the estimation of the > 95th percentile of the residue population with 95% confidence (probability) level. That required a minimum of 58–59 samples. Where the required number of samples for the subgroups was insufficient, estimates were made for the entire group 028.
- A maximum residue level was proposed at the limit of determination for pesticides in which all residues were non-detectable, even if the minimum sample requirements (59) were not met for satisfying the specified probability (> 95th percentile) and confidence (95%) limits for any of the sub-groups.
- In cases where all data were non-detectable and different LOQ values were reported for a particular pesticide by the different data sources, the maximum residue level was proposed at the highest LOQ provided for the pesticide. The median was calculated with the values corresponding to the reported LOQ levels.

37.3.3
Codex MRLs for Dried Chili Peppers

Dried chili peppers do not fall within the definition of spices. Therefore, the use of monitoring data was not applicable to dried chili peppers. Instead, Codex MRLs for dried chili peppers were set based on the existing MRLs for fresh peppers, applying an agreed default dehydration factor of 10 [7]. This dehydration factor was based on the loss of moisture after sun-drying fresh peppers.

Chronic and short-term intake calculations showed that there were no dietary exposure concerns for any of the pesticides for which MRLs for spices and dried chili peppers were proposed by CCPR. Consequently, the Codex Alimentarius Commission adopted the MRLs for spices (Table I) and for dried chili peppers (Table II) in its sessions in 2005 [8] and 2006 [9], respectively.

Table I. List of Codex MRLs in and on spices.

Pesticide	Codex MRL (mg/kg)
Acephate	0.2*
Azinphos-methyl	0.5*
Chlorpyrifos	5 (seeds), 1 (fruits or berries), 1 (roots or rhizomes)
Chlorpyrifos-methyl	1 (seeds), 0.3 (fruits or berries), 5 (roots or rhizomes)
Cypermethrin	0.1 (fruits or berries), 0.2 (roots or rhizomes)
Diazinon	5 (seeds), 0.1* (fruits or berries), 0.5 (roots or rhizomes)
Dichlorvos	0.1*
Dicofol	0.05* (seeds), 0.1 (fruits or berries), 0.1 (roots or rhizomes)
Dimethoate	5 (seeds), 0.5 (fruits or berries), 0.1* (roots or rhizomes)
Disulfoton	0.05
Endosulfan	1 (seeds), 5 (fruits or berries), 0.5 (roots or rhizomes)
Ethion	3 (seeds), 5 (fruits or berries), 0.3 (roots or rhizomes)
Fenitrothion	7 (seeds), 1 (fruits or berries), 0.1* (roots or rhizomes)
Iprodion	0.05* (seeds), 0.1(roots or rhizomes)
Malathion	2 (seeds), 1 (fruits or berries), 0.5 (roots or rhizomes)
Metalaxyl	5
Methamidophos	0.1*
Parathion	0.1* (seeds), 0.2 (fruits or berries), 0.2 (roots or rhizomes)
Parathion-methyl	5 (seeds), 5 (fruits or berries), 3 (roots or rhizomes)
Permethrin	0.05*
Phenthoate	7
Phorate	0.5 (seeds), 0.1* (fruits or berries), 0.1* (roots or rhizomes)
Phosalone	2 (seeds), 2 (fruits or berries), 3 (roots or rhizomes)
Pirimicarb	5
Pirimiphos-methyl	3 (seeds), 0.5 (fruits or berries)
Quintozene	0.1 (seeds), 0.02 (fruits or berries), 2 (roots or rhizomes)
Vinclozolin	Entire Group 028

* MRL set at Method LOQ

Table II. List of Codex MRLs in and on dried chili peppers.

Pesticide	Codex MRL (mg/kg)	Pesticide	Codex MRL (mg/kg)
Abamectin	0.2	Benalaxyl	0.5
Acephate	50	Bromide ion	200
Azinphos-methyl	10	Carbaryl	50
Carbendazim	20	Malathion	1
Chlorothalonil	70	Metalaxyl	10
Chlorpyrifos	20	Methomyl	7
Chlorpyrifos-methyl	5	Methoxyfenozide	20
Cyfluthrin	2	Permethrin	10
Cyhexatin	5	Piperonyl butoxide	20
Cypermethrin	5	Pirimicarb	20
Cyromazine	10	Procymidone	50
Diazinon	0.5	Profenofos	50
Dichlofluanid	20	Propamocarb	10
Dicofol	10	Pyrethrins	0.5
Dinocap	2	Quintozene	0.1
Dithiocarbamates	10	Spinosad	3
Ethephon	50	Tebuconazole	5
Ethoprophos	0.5	Tebufenozide	10
Fenarimol	5	Tolufluanid	20
Fenpropathrin	10	Triadimefon	1
Fenvalerate	5	Triadimenol	1
Imidacloprid	10	Vinclozolin	30

37.4 Difficulties of Complying with Unharmonized MRLs, Including 'Private' MRLs

In addition to compliance with rigid quality criteria, the exported produce have to meet the different food safety standards set by importing countries. The biggest problem faced by Chilean exporters has been the need to apply different pest management strategies for the same crop just to ensure the exports comply with the MRL in each of the importing countries. In recent years, the problem has been compounded with the implementation of 'private MRLs' from additional mandatory certification schemes by retail establishments in some of these importing countries (EuroGAP, British Retail, TESCO Nature's Choice, ChileGAP, etc.). These schemes neither follow national MRLs nor Codex standards and often require more restrictive limits than the official MRLs. Exporters complain that some of these limits do not seem to take into account efficacy studies.

Any change in plant protection strategy would result in increases in the cost of production that are ultimately shouldered by local growers. Table III illustrates an example for apples, showing the extent of disparity in MRLs among the international markets for Chile.

Using the information from Table III, if the pesticide chlorpyrifos is applied to apples according to good agricultural practices in Chile and the residues at harvest are less than the Codex MRL of 1 mg/kg, then the apples would be acceptable for the domestic market and for the markets in Japan and the US, where the MRLs are at the same level or higher. However, the same apples would not be accepted in Europe, nor would they be accepted by the retailers in the UK since their respective MRLs would be exceeded.

Table III. MRLs (mg/kg) for apples in key trading partners of Chile.

Apples	EPA	EC (harmonized)	UK[1]	Japan	Codex[2]
Acetamiprid	1	–	–	5	–
Azinphos methyl	1.5	0.5	1	2	2
Captan	25	3	2.5	5	15
Carbaryl	10	3	5	1	5
Chlorpyriphos	1.5	0.5	0.5	1	1
Diazinon	0.5	0.3	0.3	0.1	0.3
Dodine	5	1	5	5	5
Methoxyfenocide	1.5	–	2	2	2
Phosmet	10	–	0.02	10	10
Thiacloprid	0.3	–	0.5	2	–
Thiabendazol	10	5	5	3	3

1 Private sector retailers
2 MRLs adopted by Chile

37.4.1
Program to Facilitate Exports of Chilean Fruits

In light of the situation and in an effort to facilitate acceptance of their fruits in different markets worldwide, the Chilean Exporters Association with the technical support of the University of Chile developed a program which includes the publication of a bulletin, "Agenda de Pesticidas – Asociación de Exportadores de Chile", referred to herewith as "Pesticide Agenda", made available in print and online, at least every three weeks [10].

For nearly two decades, supervised trials have been conducted following the FAO guidelines, with the ultimate goal of determining the dissipation rate of pesticide residues until a target MRL for a particular pesticide/crop combination is attained. Local trials for about 25 fruit crops (berries included) have been carried out in representative growing areas, under conditions of GAPs in Chile, using approved plant protection schemes. Residue experts evaluate the dissipation pattern of a given pesticide/crop combination to determine the appropriate preharvest interval for each selected market for a particular crop. These pre-harvest intervals and supporting data are incorporated in the "Pesticide Agenda" and disseminated to growers for implementation.

Under the program, when the produce is exported, fruit lots are carefully selected in the field according to their final destination. Fruit lots designated for a particular importing country follow a pesticide regime conforming to that allowed by the importing country. The fruits are harvested at the appropriate PHI recommended in the "Pesticide Agenda" that would ensure residues are within the MRLs for pesticides acceptable in that country.

To ascertain the success of the program, exporters are encouraged to check for alerts issued by the importing country for trade violations related to MRLs. The program provides this type of information to exporters from publications such as the British PRC Quarterly Reports, the USDA Pesticide Data Program, or from internet sites of importing country regulatory offices.

37.4.2
Generating Pre-Harvest Interval Data

Supervised trials included in the "Pesticide Agenda" are not conducted to support petitions for establishing MRLs. They are conducted for the exclusive purpose of generating pre-harvest interval (PHI) data. In general, in these trials, applications are made at two dose rates, one at the GAP and the other at a higher rate. Samples are taken for residue analysis at intervals up to harvest. Results are analyzed statistically and plotted using the polynomial equation. The recommended PHI is derived from the trials at the higher rate. Recovery data, limit of detection (LOD), and limit of quantitation (LOQ) are also determined from untreated control samples.

All supervised trials are conducted by University-associated personnel, each receiving the same instructions for procedures, equipment, and data interpretation. A single national laboratory is responsible for analysis of the samples. Relevant information, such as fruit species, market area, local MRL in the importing country, is also included in the "Pesticide Agenda".

Trials are designed to cover a range of representative field conditions. Chile is a rather long and narrow country and the fruit regions extend for nearly 2,000 km, from the desert area in the north (no rain at all) to rainy areas receiving an annual average of 1.5 meters of rainfall in the southernmost part of the country where berries are produced. Since climatic conditions have an important influence on the persistence and performance of selected chemicals, representative growing areas are carefully selected as trial sites. In a period of 12 years, over 1,500 supervised trials have been conducted.

37.4.3 An Example of a Supervised Trial Model in the "Pesticide Agenda"

An example of a supervised trial model in the "Pesticide Agenda" is shown in Figure 1, using bifenthrin at the highest recommended dosage on clementines. Application method, dosage, and equipment used are indicated in the information accompanying the decline curve. Five samples of 16 fruits each were collected at 0, 3, 7, 14, and 21 days after application and processed for residue analysis. The dissipation curve was statistically calculated and plotted accordingly [11]. Corresponding values for LOD, LOQ, and percent recovery from non-treated samples are shown below. In this particular case, to reach, for example, a citrus fruit tolerance of 0.1 mg/kg for bifenthrin, the "Pesticide Agenda" would recommend a pre-harvest interval of 28 days to the growers.

As a result of these initiatives, Chile has emerged as the largest fresh fruit exporting country from the Southern hemisphere. Over 70 markets worldwide are supplied with assorted Chilean fresh produce, primarily table grapes, kiwifruit, pome fruits, and stone fruits, and to a lesser extent, lemons, clementines, and avocados.

37.5 Conclusion

During the past several years, tropical fruits and other minor crops (mango, papaya, mangosteen, longan, cherimoya, star fruit, etc.) produced and exported from different countries in the developing world have become increasingly available to markets worldwide. While seemingly impossible a few years back, there now exist MRLs for spices and dried chili peppers, which would go a long way towards facilitating trade in these commodities. All these have come about through significant initiatives taken by developing countries to resolve the problems they faced with their exports, in particular, the problems related to residues in and on their exported commodities.

The FAO/IAEA Training and Reference Centre for Food and Pesticide Control has provided countries with a mechanism to improve analytical facilities and

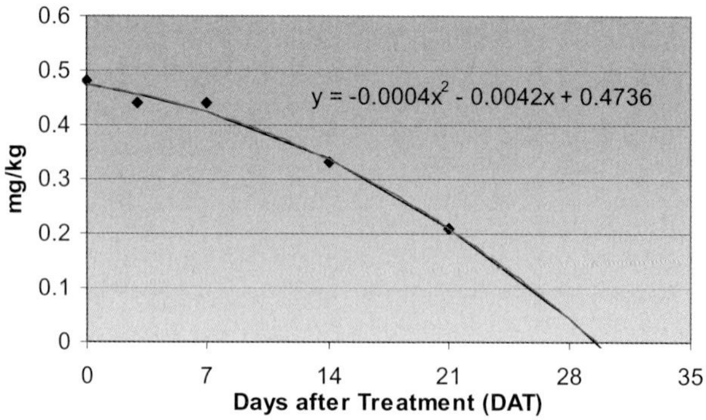

DAT	mg/kg
0	0.48
3	0.44
7	0.44
14	0.33
21	0.21

Fruit species:	Clemenules, 500 trees/hectare
Phenological stage:	Fruits 3–3.5 cm diameter
Formulation used:	Talstar 10EC (bifentrin)
Concentration used:	50 cc (comm. product)/100 L
Dosage/hectare:	250 cc (a.i.)/hectare
Date:	March 21, 2006
Application equipment:	High volume, hand gun

LOD: 0.01 mg/kg	LOQ: 0.02 mg/kg	% Recovery: 107%

Figure 1. Residue decline curve for bifenthrin in clementines (Metropolitan, summer 2006).

strengthen capabilities of national institutions to enable the generation of scientific data to support the establishment of Codex MRLs. The case of spices has illustrated that alternative approaches for setting MRLs are possible. Chile has provided a model of how cost-effective plant protection strategies can be implemented to suit the diverse requirements of each of its many trading partners.

37.6
References

1 Manual on the Submission and Evaluation of Pesticide Residues Data for the Estimation of Maximum Residue Levels in Food and Feed, FAO, **2002**.

2 World Markets in the Spice Trade (2000–2004), International Trade Centre (ITC), UNCTAD/WTO, April, **2006**.

3 Global Spice Markets Imports (1994–1998), ITC, UNCTAD/WTO, September, **2000**.

4 Spices Board, India, **2001**.

5 Alinorm 04/27/24: Report of the 36th Session of CCPR, 19–24 April 2004 [paras. 235–247].

6 CX/PR 04/13: Discussion Paper on Elaboration of MRLs on Spices, April, **2004**.

7 Pesticide Residues in Food – 2004, JMPR Report, pp. 19–24, 212–226, September, **2004**.

8 Alinorm 05/28/24: Report of the 37th Session of CCPR, April **2005**, Appendix IV, adopted by the 28th Session of the

Codex Alimentarius Commission, July **2005**.

9 Alinorm 06/29/24: Report of the 38th Session of CCPR, April **2006**, Appendix II, adopted by the 29th Session of the Codex Alimentarius Commission, July **2006**.

10 Agenda de Pesticidas – Asociación de Exportadores de Chile (www.ASOEX.cl).

11 R. H. Gonzalez, Degradación de Residuos de Plaguicidas en Huertos Frutales en Chile, Univ. de Chile, Serie: Ciencias Agronómicas N° 4, 165 p., 97 figs., **2002**.

Keywords

Possible Solutions, Trade Issues, Developing Countries, FAO/IAEA, Unharmonized MRLs, Spices, Dried Chili Peppers, Supervised Trials, Pesticide Agenda

38
Genetically Modified (GM) Food Safety

Gijs A. Kleter, Harry A. Kuiper

38.1
Introduction

Genetic modification through recombinant DNA technology, also known as genetic engineering or as modern biotechnology, allows for the transfer of genes into organisms in ways that are impossible or difficult to achieve by alternative and traditional methods, such as hybridization of naturally crossable plants. This technology therefore expands the tools available to plant breeders to impart desirable characteristics to crops.

The large-scale commercial cultivation of genetically modified (GM) crops commenced in 1996 and has since undergone a steady increase in terms of area cultivated with these crops and the number of countries where these crops are grown. For example, the total area of GM crops in 2005 amounted to 90 million hectares [1], which is comparable to 2.4 times the national area of Japan. The major GM-crop-growing countries include the USA, Argentina, Canada, Brazil, South Africa, China, and India [1], and this list is likely to expand in the coming years.

The main traits with which GM crops have been modified are herbicide resistance and insect resistance [1]. Herbicide resistance allows for the topical application of broad-spectrum herbicides, such as glyphosate and glufosinate, which would otherwise be detrimental to the crop. Insect resistance has in most cases been achieved by the introduction of insecticidal proteins that naturally occur in the soil bacterium *Bacillus thuringiensis*, which is also used as a biological pesticide in agriculture, in particular in organic agriculture.

Before GM crops are allowed into the marketplace in many countries, they have to be approved by the authorities. The regulatory procedure for approval commonly includes an assessment of the safety of the GM crop for human and animal health, and for the environment. The following sections discuss the principles of safety assessment of GM crops for food use and highlight some recent developments in this area of science.

Pesticide Chemistry. Crop Protection, Public Health, Environmental Safety
Edited by Hideo Ohkawa, Hisashi Miyagawa, and Philip W. Lee

ISBN: 978-3-527-31663-2

38.2
General Principles of GM Food Safety Assessment

Years before the first foods derived from GM crops entered the market, international organizations like the Food and Agriculture Organization (FAO), World Health Organization (WHO), Organization for Economic Cooperation and Development (OECD), and International Life Sciences Institute (ILSI) convened meetings with scientists and issued reports in order to build international scientific consensus on how the safety of these foods could be assessed [2]. These international consensus building activities were consolidated by the publication in 2003 of the FAO/WHO Codex Alimentarius guidelines for the safety assessment of foods consisting of plants or micro-organisms derived through modern biotechnology [3].

The internationally harmonized approach is based on the principle of comparative safety assessment, also known as "substantial equivalence" [4]. This entails the comparison of the genetically modified organism (GMO) with a conventional counterpart that has a history of safe use. For example, a GM maize crop that has been genetically modified with a gene encoding an insecticidal protein from *B. thuringiensis* can be compared with conventional maize, which is widely grown in many countries. Ideally, the conventional maize should be as similar as possible in genetic background to the GM maize. The rationale for this comparison is based on the insight that foods are complex mixtures that can contain both beneficially and adversely acting substances. The balance between these substances in conventional crops is considered to be positive based on extensive experience in breeding and food processing for human consumption, through which any health-damaging characteristics have been averted.

Based on the differences identified in the comparison, the safety assessment can then further focus on these differences. The initial comparison thus serves as the starting point for the safety assessment and not as the end-point. Because the traits that have been introduced into GM crops and the nature of the crops themselves can vary widely, there are no standard protocols for which specific tests should be done. Instead, the choice for the tests that have to be done further for the assessment is made on a case-by-case basis.

The following items are commonly addressed during the safety assessment of GM crops for food use (see also Figure 1):

- General data on DNA donor and recipient organisms;
- Molecular characterization of the introduced DNA;
- Comparison of the GMO with a conventional counterpart;
- Potential toxicity of introduced foreign proteins;
- Potential toxicity of introduced foreign proteins of the whole food;
- Potential allergenicity of the introduced foreign proteins;
- Potential allergenicity of the introduced foreign proteins of the whole food;
- Potential horizontal gene transfer;
- Nutritional characteristics; and
- Potential unintended effects of the genetic modification.

These issues will be discussed in more detail in the following sections.

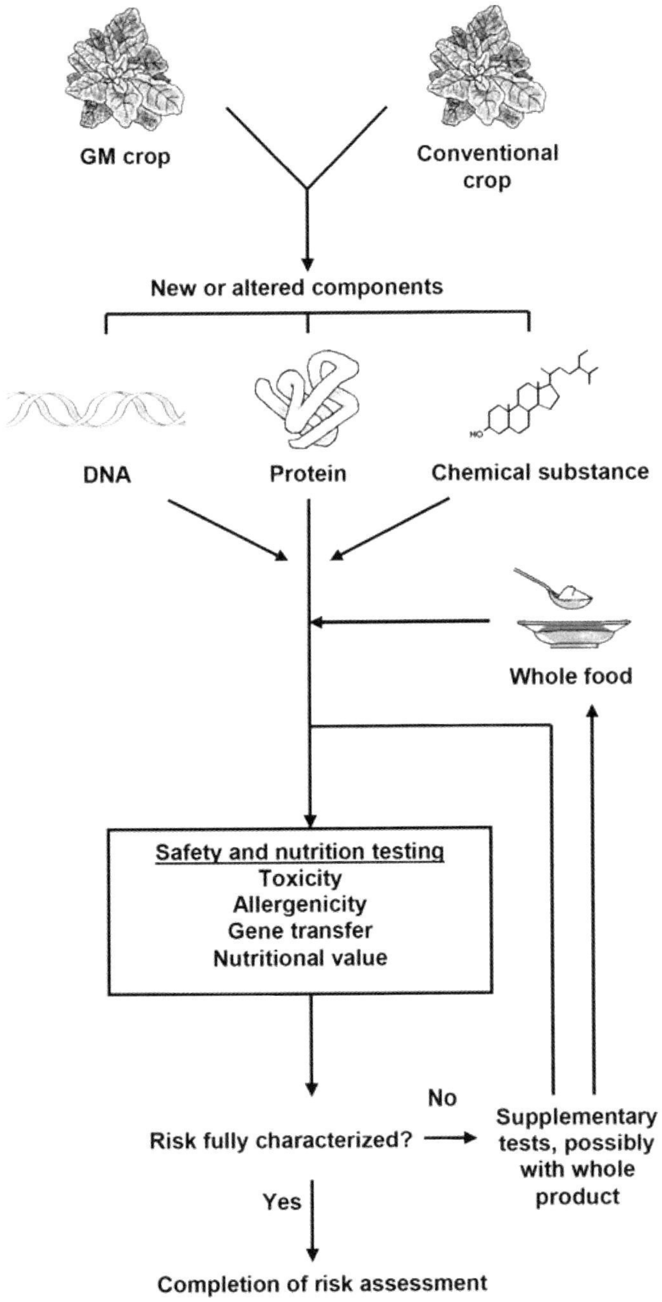

Figure 1. General outline of the comparative safety assessment of a GM crop.

38.3
General Data

The general data can provide background information on the donor organism of the "foreign" DNA and the recipient, *i.e.*, the organism that is to become genetically modified. This may include, for example, considerations of the history of domestication of the recipient crop; agricultural cultivation practices; applications for food, feed and other purposes; the extent of processing before consumption; and production and consumption figures of the pertinent crop.

38.4
Molecular Characterization of the Introduced DNA

The molecular characterization of the introduced DNA includes details on the characteristics of the DNA used for the genetic modification. These details pertain to, for example, the genes that occur in this DNA, the method by which the DNA has been introduced into the GM organism, how it has been incorporated into the genetic material of the GM organism, and the expression of genes in the GM organism.

38.5
Comparison of the GMO with a Conventional Counterpart

As stated above, the comparison of the GMO with a conventional counterpart serves as a starting point of the safety assessment. For GM crops, this usually includes the analysis of agronomic and other phenotypic traits of the GM crop, such as morphology, yield, and plant disease susceptibility. In addition, compositional analysis is usually also carried out. This may include key macronutrients, micronutrients, other bioactive compounds, antinutrients, toxins, and secondary metabolites. The OECD Task Force on the Safety of Novel Food and Feeds has issued a range of consensus documents that recommend for specific crops which key compositional items should be measured in new varieties of these crops for the comparison. The consensus documents that have thus far been published cover, for example, barley, canola, cotton, maize, potato, rice, soybean, sugar beet, and wheat [5]. In addition, ILSI has established a web-based databank with recent data from field trials on the composition of conventional varieties of crops, such as cotton, maize, and soybean (http://www.cropcomposition.org). These data have been reviewed for their quality before having been entered into the database, and the website visitors are able to select data based on crop type, geographical region, year, substance, and analytical method.

38.6
Potential Toxicity of Introduced Foreign Proteins

For the purpose of determining the potential toxicity of introduced foreign proteins, data on the toxicity of the host organism from which a gene has been derived, or of the protein itself, if already known, is useful. Usually, bioinformatics methods are used to determine if the amino acid sequence of the new protein has any similarity with sequences of proteins that are known to be toxic. In addition, the resistance to *in vitro* degradation by the stomach protein-degrading enzyme pepsin is commonly tested. Resistance to such degradation may indicate that the protein has an increased likelihood to sustain passage to thc intestinal tract after oral consumption and thus be able to exert its toxicity, if any. In addition, oral toxicity testing of the purified protein in laboratory animals may be considered, if the outcomes of the previously mentioned tests warrant this. The scientific GMO Panel of the European Food Safety Authority, for example, recommends carrying out a 28-day repeated dose toxicity study [6].

38.7
Potential Toxicity of the Whole Food

If the comparison between the GM crop and its conventional counterpart indicate substantive nonequivalence or if there are any doubts remaining, whole product testing for potential toxicity in laboratory animals may be considered. However, as stated above, foods are complex mixtures of beneficial and noxious substances, the balance of which is positive. Unlike the testing of purified chemicals, the testing of whole foods is restricted to a narrow dose range in which the food can be provided through the diet. This relates to, for example, the nutritional balance of the diet, its bulkiness, and palatability. Usually a 90-day study in rodents is carried out, after which body weight, feed consumption, organ weights, gross pathology, serum and urine chemistry, hematology, and histopathology are performed according to OECD guidelines for toxicity testing of chemicals.

38.8
Potential Allergenicity of the Introduced Foreign Proteins

Allergy is a kind of hypersensitive immune reaction, of which the symptoms include, among others, jitteriness, vomiting, skin hives, and shock. The agent that causes an allergy is designated "allergen" and all known allergens in food are proteins. Although allergens make up only a small fraction of all known proteins in foods, it is still recommended that new proteins be tested for allergenicity, i.e., the potential to act as an allergen.

For example, if the donor organism of the gene encoding the new protein is known to contain an allergen, or if the protein itself is known to be allergenic, this also provides an additional indicator of potential allergenicity. In addition, similar to toxicity testing, the new protein is submitted to a comparison with allergens by bioinformatics and to *in vitro* pepsin-using degradation assay. For the bioinformatic comparisons, the Codex Alimentarius guidelines refer to the recommendations of an FAO/WHO Expert Consultation [7]. According to these recommendations, sequences should be screened for fully identical short segments of minimally 6–8 contiguous amino acids, and for similar segments comprising minimally 35% identical amino acids in an 80-amino-acid sliding window. Various websites offer their facilities to website visitors, such as, the Allermatch™ website (http://www.allermatch.org), at which a query amino acid sequence is compared to a database of sequences of allergens ([8] and references to other websites herein). Several authors, such as Kleter and Peijnenburg [9], have proposed refinements of this approach involving bioinformatics for the prediction of potential cross-reactivity with allergens.

Depending upon the outcomes of these tests, the cross-reactivity of the new protein with known allergens can be tested by doing serum binding tests. Such tests serve to determine if sera containing immunoglobulin E (IgE) antibodies obtained from patients allergic to the known allergen bind to the new protein. IgE antibodies are associated with allergies, and their binding to allergens ultimately leads to allergic reactions.

Codex Alimentarius recognizes that none of these tests alone can be completely predictive of allergenicity; hence it recommends a "weight of evidence" approach, in which the outcomes of all these tests are taken together [3].

38.9
Potential Allergenicity of the Whole Food

Potential allergenicity of the whole food may be considered if the recipient crop already contains allergens. It may therefore be useful to check if the intrinsic allergenic properties of the crop have been changed due to the genetic modification, for example as an unintended side effect. This may entail, for example, serum screening similar as described for testing the purified protein.

38.10
Potential Horizontal Gene Transfer

Horizontal gene transfer refers to the transfer of genes between organisms belonging to different species. It is known that genetic modification by exchange of DNA between unrelated organisms can occur in nature. For example, organisms can be modified by uptake of free DNA, although this is known to be a rare event. As regards the genes introduced into GM crops, the possible transfer of antibiotic resistance genes to other organisms, particularly pathogenic microorganisms, is considered in particular, such as by the Codex Alimentarius guidelines [3]. Such transfer from GM crops to other organisms is most likely to occur through the release of free DNA from the crop. For successful transfer, the recipient organism has to take up the free DNA, incorporate it into its genetic material, and stable maintain and express the information contained by the newly introduced DNA over multiple generations.

Regulatory agencies also consider other possible health consequences of horizontal gene transfer, such as the likelihood of increased pathogenicity. A recent review provides an overview of such considerations for genes introduced in commercial GM crops [10].

38.11 Nutritional Characteristics

The compositional analysis already provides information on possible changes in the nutritional value of the GM crop associated with the contents of nutrients and anti-nutrients. Various experimental GM crops with nutritionally improved characteristics have been developed for food and feed (Table I). If the nutritional properties of the GM crop have been altered, either in the content or bio-availability of nutrients, it may be useful to test these in target domestic animals for animal feed, or in laboratory animals, or in clinical trials for human food. A popular model for testing nutritional characteristics is the rapidly growing broiler chicken. Because these animals show rapid growth, reaching their adult sizes in approximately six weeks, any changes in nutritional properties is likely to be picked up during such feeding studies. These tests usually focus on performance, such as the feed intake, body weight increase, and weights of edible body parts after termination of the study. In case of crops with altered bioavailability of nutrients, "balance" studies to test for the bio-availability can also be considered [6].

38.12 Potential Unintended Effects of the Genetic Modification

It can be envisioned that besides the intended effects of a genetic modification, such as the expression of an insecticidal protein, the modification may also have brought about unintended effects. This may relate, for example, to the insertion of the newly introduced DNA into an intrinsic gene of the recipient organism. Also interactions of introduced enzymes with pre-existing biochemical pathways, such as competition for common precursors, can in some theoretical cases be envisioned.

The extensive phenotypic, agronomic, and compositional analyses of the GM crop and its counterpart as mentioned above already provide an indication of whether unintended effects might have occurred. Whole food tests for toxicity, allergenicity, and nutritional characteristics may provide another safeguard against unintended effects, although they may in some cases not be as sensitive as chemical-analytical assays.

Table I. Examples of nutritionally improved GM crops recently reported in literature.

Category	Crop	Trait
Oil	Indian mustard [11]	Introduction of very long chain fatty acids by introduction of enzymes (fatty acid desaturases and elongases, lysophosphatidic acid acyltransferase) involved with their biosynthesis
Protein and amino acid	Alfalfa [12] (animal feed)	Increases in cysteine and methionine by overexpression of an enzyme (cystathionine γ-synthase) involved with the biosynthesis of sulfur amino acids
Fiber	Fescue grass [13] (animal feed)	Suppression of an intrinsic enzyme (caffeic acid O-methyltransferase) involved with the biosynthesis of lignin (decrease in lignin and increase of lignin digestibility)
Mineral	Maize [14]	Expression of an iron-binding protein (ferritin) and an enzyme (phytase) that degrades an iron-binding antinutrient (increase in bioavailable iron)
Vitamin	Soybean [15]	Overexpression of enzymes (chorismate mutase-prephenate dehydrogenase, homogentisate phytyltransferase, and *p*-hydroxyphenylpyruvate dioxygenase) that are involved with the biosynthesis of vitamin E precursors (increase in tocochromanols, including tocotrienol)
Vitamin, antioxidant	Tomato [16]	Suppression of photomorphogenesis-related transcription factor TDET1, involved with regulation of pigment biosynthesis (increased provitamin A, lycopene, and flavonoids)
Antioxidant	Canola [17]	Introduction of precursor-synthesizing enzyme (stilbene synthase) and silencing of alternative pathway involving sinapate glucosyltransferase (increase in resveratrol glucoside)
Antinutrient	Cassava [18]	Suppression of enzymes (CYP79D1 and CYP79D2) involved with the biosynthesis of cyanogenic glucosides (decrease in linamarin)

The GM crops currently on the market have in most cases undergone relatively minor modifications, comprising new proteins expressed at low levels and without any other effects on compositional and agronomic-phenotypic characteristics. For future crops with more complicated modifications, it may be envisioned that there is a higher likelihood of unintended effects. The potential use of advanced, holistic "profiling" techniques to detect unintended effects in GM crops is discussed in a number of recent reviews [19–21].

38.13
Pesticide Residues

Herbicide-resistant GM crops may have been modified such that they are able to metabolize the particular herbicide, rendering it innocuous for the crop plant. Alternatively, enzymes targeted by the herbicide may have been rendered insensitive to the herbicide by introduction of mutated versions or equivalents derived from other donor organisms. Both these modifications may lead to altered levels of the herbicide and its metabolites being present in the crop. The issue of the safety of the altered herbicide residue profile usually is considered as part of the pre-market assessment of the registration of the herbicide, which in many countries follow a regulatory procedure different from that for GM crops.

38.14
Research into the Safety of GM Crops

Various research projects on the safety of GM crops have been carried out or are still in progress. For example, the European ENTRANSFOOD project was recently concluded. This project brought together various existing European projects on the safety of GM foods, and provided for a platform for exchange between scientists from the various projects, as well as between scientists and other stakeholders to discuss the latest developments within various working groups focusing on the topical areas of research.

Besides the scientific outputs of the separate projects, which have been published in scientific journals and other media, the ENTRANSFOOD project has issued a series of publications. For example, a flyer and an overarching report provide its main findings and recommendations, both targeted at a broader audience. A special issue of the scientific journal Food and Chemical Toxicology features a series of scientific reviews, which highlight the topical issues and the overarching conclusions of the ENTRANSFOOD project [22–29].

For example, the paper by Koenig *et al.* describes an integrated approach towards the safety testing of GM foods [25]. In addition, the paper by Van den Eede *et al.* discusses the issue of horizontal gene transfer from GM crops to other recipients, in particular that of antibiotic resistance genes, for which a classification system is proposed [29]. This classification is based on the natural background occurrence of the antibiotic resistance, the clinical importance of the topical antibiotic, and the likelihood of horizontal transfer. Three classes of antibiotic resistance genes are thus discerned, the first of which comprises genes that can be used in commercial crops, the second for genes in crops that are only released into the field on a small scale, and the third for genes that should not be used at all.

The other reviews consider consumer perception, the use of advanced profiling techniques for the detection of unintended effects, and traceability and detection of GM foods. The quoted papers and more information on the ENTRANSFOOD project can be found on its website (http://www.entransfood.com).

Part of the ENTRANSFOOD project activities are currently extended under the umbrella of the SAFE FOODS project, the greatest part of which is funded by the European Union. For example, analytical profiling techniques are employed to test for differences between different cultivars of crops, including genetically modified, conventional, and organically grown ones. In addition, consumer perception and institutional arrangements for food risk management are investigated and recommendations made, among others for promotion of stakeholder involvement in risk management. Other topics studied include the early identification of emerging food risks, the use of advanced statistical methods for estimation of consumer exposure, and the development of a refined risk analysis model based upon an integration of the project outcomes. More information on SAFE FOODS can be found on its website (http://www.safefoods.nl).

38.15 Conclusion

Prior to their release onto the market, GM foods have to undergo a rigorous safety assessment. Although the regulations pertaining to GM foods may differ between countries, the regulatory safety assessment follows an internationally harmonized approach, as recently laid down by Codex Alimentarius in its guidelines. The currently commercialized GM crops may be mainly of agronomic importance, with modifications that have minor or no effects on crop characteristics apart from the intended effect. Future GM crops with other traits, such as consumer-oriented nutritional improvements, may be more complicated and hence increase the likelihood of unintended effects. Research is currently going on, focusing on the use of advanced supplementary methods for safety and nutritional testing that may aid in the assessment of future GM crops.

38.16 Acknowledgment

H.A.K. would like to dedicate this chapter to Prof Maurizio Brunori of the University of Rome La Sapienza on the occasion of his seventieth birthday.

Financial support from IUPAC and the Dutch Ministry of Agriculture, Nature, and Food Quality is gratefully acknowledged.

38.17 References

1 C. James, Executive Summary of Global Status of Commercialized Biotech/GM Crops in 2005, ISAAA Briefs No. 34, *International Service for the Acquisition of Agri-Biotech Applications*, Ithaca, NY, **2005**. http://www.isaaa.org/kc/bin/briefs34/es/index.htm

2 H. A. Kuiper, G. A. Kleter, H. P. J. M. Noteborn, E. J. Kok, *Plant Journal*, **2001**, *27*, 503–528.

3 Codex Alimentarius Commission, Codex Principles and Guidelines on Foods Derived from Biotechnology. Food and Agriculture Organization, Joint FAO/WHO Food Standards Programme, Codex Alimentarius Commission, Rome, **2003**. http://www.who.int/foodsafety/biotech/codex_taskforce/en/index.html

4 E. J. Kok, H. A. Kuiper, *Trends in Biotechnology*, **2003**, *21*, 439–444.

5 OECD, *Consensus Documents for the Work on the Safety of Novel Foods and Feeds*, Organization for Economic Cooperation and Development, Paris, **2006**. http://www.oecd.org/document/9/0,2340,en_2649_34385_1812041_1_1_1_1,00.html

6 EFSA, Guidance Document of the GMO Panel for the Risk Assessment of Genetically Modified Plants and Derived Food and Feed, *European Food Safety Authority*, Parma, **2004**. http://www.efsa.europa.eu/science/gmo/gmo_guidance/660_en.html

7 FAO/WHO, Joint FAO/WHO Expert Consultation on Foods Derived from Biotechnology – Allergenicity of Genetically Modified Foods – Rome, 22–25 January 2001, *Food and Agriculture Organization of the United Nations*, Rome, **2001**. http://www.fao.org/ag/agn/food/pdf/allergygm.pdf

8 M. W. E. J. Fiers, G. A. Kleter, H. Nijland, A. A. C. M. Peijnenburg, J.-P. Nap, R. C. H. J. Van Ham, *BMC Bioinformatics*, **2004**, *5*, 133. http://www.biomedcentral.com/1471-2105/5/133

9 G. A. Kleter, A. A. C. M. Peijnenburg, *BMC Structural Biology*, **2002**, *2*, 8. http://www.biomedcentral.com/1472-6807/2/8

10 G. A. Kleter, A. A. C. M. Peijnenburg, H. J. M. Aarts, *Journal of Biomedicine and Biotechnology*, **2005**, *4*, 326–352. http://www.hindawi.com/GetArticle.aspx?doi=10.1155/JBB.2005.326

11 G. Wu, M. Truksa, N. Datla, P. Vrinten, J. Bauer, T. Zank, P. Cirpus, E. Heinz, *et al.*, *Nature Biotechnology*, **2005**, *23*, 1013–1017.

12 T. Avraham, H. Badani, S. Galili, R. Amir, *Plant Biotechnology Journal*, **2005**, *3*, 71–79.

13 L. Chen, C.-K. Auh, P. Dowling, J. Bell, D. Lehmann, Z.-Y. Wang, *Functional Plant Biology*, **2004**, *31*, 235–245. http://www.publish.csiro.au/nid/102/paper/FP03254.htm

14 G. Drakakaki, S. Marcel, R. P. Glahn, E. K. Lund, S. Pariagh, R. Fischer, P. Christou, E. Stoger, *Plant Molecular Biology*, **2005**, *59*, 869–880.

15 B. Karunanandaa, Q. Qi, M. Hao, S. R. Baszis, P. K. Jensen, Y. H. H. Wong, J. Jiang, M. Venkatramesh, *et al.*, *Metabolic Engineering*, **2005**, *7*, 384–400.

16 G. R. Davuluri, A. van Tuinen, P. D. Fraser, A. Manfredonia, R. Newman, D. Burgess, D. A. Brummell, S. R. King, *et al.*, *Nature Biotechnology*, **2005**, *23*, 890–895.

17 A. Hüsken, A. Baumert, C. Milkowski, H. C. Becker, D. Strack, C. Möllers, *Theoretical and Applied Genetics*, **2005**, *111*, 1553–1562.

18 D. Siritunga, R. Sayre, *Plant Molecular Biology*, **2004**, *56*, 661–669.

19 B. Chassy, J. J. Hlywka, G. A. Kleter, E. J. Kok, H. A. Kuiper, M. McGloughlin, I. C. Munro, R. H. Phipps, *et al.*, *Comprehensive Reviews in Food Science and Food Safety*, **2004**, *3*, 35–104. http://members.ift.org/NR/rdonlyres/27BE106D-B616-4348-AE3A-091D0E536F40/0/crfsfsv3n2p00350104ms20040106.pdf

20 H. A. Kuiper, E. J. Kok, K. H. Engel, *Current Opinion in Biotechnology*, **2003**, *14*, 238–243.

21 E. J. Kok, G. A. Kleter, J. P. van Dijk, Use of the cDNA Microarray Technology in the Safety Assessment of GM Food Plants, *TemaNord*, **2003**, *558*, Nordic Council of Ministers, Copenhagen, **2003**. http://www.norden.org/pub/velfaerd/livsmedel/sk/TN2003558.asp

22 L. Breslin, *Food and Chemical Toxicology*, **2004**, *42*, 1043.

23 F. Cellini, A. Chesson, I. Colquhoun, A. Constable, H. V. Davies, K. H. Engel,

A. M. R. Gatehouse, S. Kärenlampi, *et al.*, *Food and Chemical Toxicology*, **2004**, *42*, 1089–1125.

24 L. Frewer, J. Lassen, B. Kettlitz, J. Scholderer, V. Beekman, K. G. Berdal, *Food and Chemical Toxicology*, **2004**, *42*, 1181–1193.

25 A. Koenig, A. Cockburn, R. W. R. Crevel, E. Debruyne, R. Grafstroem, U. Hammerling, I. Kimber, I. Knudsen, *et al.*, *Food and Chemical Toxicology*, **2004**, *42*, 1047–1088.

26 H. A. Kuiper, *Food and Chemical Toxicology*, **2004**, *42*, 1044–1045.

27 H. A. Kuiper, A. König, G. A. Kleter, W. P. Hammes, I. Knudsen, *Food and Chemical Toxicology*, **2004**, *42*, 1195–1202.

28 M. Miraglia, K. G. Berdal, C. Brera, P. Corbisier, A. Holst-Jensen, E. J. Kok, H. J. P. Marvin, H. Schimmel, *et al.*, *Food and Chemical Toxicology*, **2004**, *42*, 1157–1180.

29 G. van den Eede, H. Aarts, H.-J. Buhk, G. Corthier, H. J. Flint, W. Hammes, B. Jacobsen, T. Midtvedt, *et al.*, *Food and Chemical Toxicology*, **2004**, *42*, 1127–1156.

Keywords

Plant Biotechnology, Genetic Modification, Comparative Analysis, Food Safety, Crop Composition, Toxicology, Allergies, Gene Transfer, Nutrition, Analytical Profiling

39
Toxicology and Metabolism Relating to Human Occupational and Residential Chemical Exposures

Robert I. Krieger, Jeff H. Driver, John H. Ross

39.1
Introduction

Whenever commercial chemicals, including pesticides are used, accidental, unintended, or unavoidable human exposure may occur via skin contact, ingestion and/or inhalation. Knowledge of the extent and duration of exposure is essential for enlightened and responsible product stewardship, risk characterization, and management. The societal benefits of our present lifestyle that prominently features chemical technologies can be visualized by review of age-adjusted unintentional death rates that have decreased 62% in the U.S., 1910–2003 [1]. Motor vehicle accidents claimed more than 42,000 lives each year since 1999 dwarfing other prominent causes of death including falls, poisoning, ingestion of food and foreign objects, firearms, poisons (solid and liquid, gas and vapors) in descending order. Unintentional injuries are the fifth leading cause of death, exceeded by heart disease, cancer, stroke, and chronic lower respiratory disease. The reduction in death rate during this period of increased reliance on chemical technologies with population tripling represents an estimated 4,800,000 fewer people being killed due to unintentional injuries. That reality is frequently lost among current campaigners and environmental activists who regularly fail to distinguish exposure from toxicity.

In this discussion, exposure is contact with the potential for absorption (or a direct effect in the case of irritants). Contacts include ingestion, inhalation, and skin contact in addition to the parenteral route used in experimental studies. Determinants of response or descriptors of regulatory no adverse effect levels (NOAELs) include dose (μg/person), dosage (μg/kg), and time (acute, subacute, and chronic). Regardless of route and extent, the vast majority of our unknown and unnumbered chemical exposures are benign.

A scheme for classification comes from listing accidental, unintended, and unavoidable exposures that occur in our personal, occupational, community, and global environments. From this scheme, the highest degree of control of personal exposures occurs in the personal environment (food, alcohol,

Pesticide Chemistry. Crop Protection, Public Health, Environmental Safety
Edited by Hideo Ohkawa, Hisashi Miyagawa, and Philip W. Lee

ISBN: 978-3-527-31663-2

nicotine, etc.) while the least control operates with global distribution (chlorinated aromatic hydrocarbons, PCBs, and other substances refractory to environmental mineralization).

In this context, it is also possible to classify pesticides after the scheme presented in the keynote speech. Commodity chemicals were introduced in the 1950s and those in use today (or occurring in residue analysis) are described by extensive environmental and toxicological data. Generic chemicals of the 1960s and 1970s have been the subject of field studies that permit estimates of potential human exposure. The smaller class, proprietary chemicals, has robust environmental databases and sometimes can be assessed in exposed persons using biomarkers of exposure in blood and urine (as well as other biological matrices).

A useful research strategy for predicting the environmental and biological fate of new, proprietary chemicals is to gain understanding of critical factors of old chemical technologies to inform current registration, stewardship, and general pesticide science. Although risk assessment in its current form requires estimates of absorbed daily dose and dosage, the resulting estimates are often inflated due to default assumptions related to human time-activity patterns during and post-application, pesticide availability from exposure media, e.g., treated indoor surfaces, as well as toxicodynamic and toxicokinetic factors. The resulting exposure estimates often have little bearing on reality and distort risk perception by specialists and the public alike, e.g., indoor exposure estimates of organophosphorous insecticide exposure by Berteau *et al.* [2], are 2 to 3 orders of magnitude above measured levels [3].

39.2 Pesticide Handlers

Exposures of pesticide handlers first received attention in the 1950s when organophosphorous insecticides such as ethyl parathion became available for management and control of a variety of plant pests. Toxicity testing focused on the well-known “6-pack”- acute LD_{50}s by the oral, inhalation, and dermal routes of exposure, skin and eye irritation and sensitization in the guinea pig. These tests formed the experimental foundation for the signal word on U.S. EPA labels – Danger, Warning, and Caution. Human exposure studies pioneered by Durham *et al.* [4] provided worker exposure data following analysis of cotton gauze patches placed on and beneath the clothing to intercept pesticide liquids and dusts. Passive, personal dosimetry is a reproducible and standardized method of measuring pesticide handler exposure and has been adopted as a regulatory standard in the U.S. since publication of handler guidelines in 1986 by U.E. EPA [5].

Standard worker clothing includes long-sleeved shirts, long pants, socks, and shoes. These garments provide substantial exposure reduction. Personal protective equipment e.g., gloves, hats, aprons, coveralls, and layers of other clothing that is worn as protection from the elements, further diminishes human pesticide exposure potential during mixing, loading, and application of pesticide.

PHED is commonly used by registrants and government agencies to supplement and validate field exposure studies, and as an evaluation tool for analysis of exposure data. PHED contains over 1,700 records of data on measured dermal and inhalation exposures, as well as accompanying data on parameters that may affect the magnitude of exposures. Despite the numerous deficiencies of PHED as noted by Ross *et al.* [6], it still represents the best available first-tier assessment source for many different mixer/loader/applicator exposure scenarios. The resulting Pesticide Handlers Exposure Database (PHED) can be used to estimate worker exposure, but the data must be factored by clothing penetration (reliably less than 20%), dermal absorption (commonly less than 1% to about 35% of applied dose), and an estimate of pounds of active ingredient handled to obtain an estimate of Absorbed Daily Dose and Dosage. Current efforts to validate PHED exposure estimates using biological monitoring are welcomed and will substantially improve the reliability of exposure estimates for risk characterization. The "unit exposure" metrics derived from passive dosimetry are found in databases such as the PHED, the Outdoor Residential Exposure Task Force (ORETF) Database, and most currently, in the Agricultural Handlers Exposure Database (AHED©). Unit exposure values expressed as milligrams exposure per pound of active ingredient handled (mg/lb a.i.) have become the standard for first-tier, predictive dermal and inhalation exposure assessment of pesticide handlers.

Biomonitoring is generally regarded as a "gold standard" of exposure because it typically involves fewer measurements and fewer assumptions than passive dosimetry in determining absorbed dose in individuals exposed to pesticides. Comparing dose estimates from biomonitoring versus passive dosimetry requires knowledge of clothing penetration, route-specific absorption, particularly for the dermal route, metabolic pathways and terminal products, routes of elimination, and associated kinetics.

Well-studied chemicals such as chlorpyrifos, 2,4-D, and pyrethrins can be used to more fully validate exposure assessments conducted with passive dosimetry. Total absorbed daily dose values estimated using unit exposure metrics reported in PHED for common use scenarios generally exceed measurements for those same-use scenarios made using urinary biological monitoring methods. Differences in estimated (from passive dosimetry) versus measured absorbed dose values reflect, in part, actual versus estimated bioavailability from using rat rather than human dermal absorption data, and the impact of other covariates such as clothing penetration. The results of biomonitoring studies can reduce the schism between actual exposure reflected by this more direct measurement methodology and the perceived threat of adverse health affects often associated with more conservative, predictive exposure assessment methods for chemical technologies including pesticides.

39.3 Harvesters of Treated Crops

Field workers who harvest treated crops have fewer exposure mitigation options. Since pesticide decay begins at application, surface residues are the primary source of exposure once sprays have dried, dust settled, and vapors dissipated. Dislodgeable foliar pesticide residues (DFR; $\mu g/cm^2$) can be factored by an empirical transfer coefficient (cm^2/hour) and harvest time in hours to predict order of magnitude external harvester exposure following the insightful research of Nigg and Stamper [7] and Zweig *et al.* [8]. More recently, a group of transfer coefficients has been collated and actively used by government agencies to estimate post application worker exposure [9]. The relationship is ultimately expressed as follows:

$$(\mu g/cm^2)\ (cm^2/h)\ (h) = \text{Potential Dermal Dose (PDD; } \mu g/\text{person)}$$

DFR results from liquid extraction of treated leaves [Iwata *et al.* [10]. Its relationship with PDD is best studied at very short time intervals (when exposure potential is maximal). Improved short- and longer-term exposure estimates may result from physical removal of residues from leaf surfaces (Li *et al.* [11]). A detailed discussion of these factors is provided in Whitmyre *et al.* [12].

Hand harvesting results in substantial hand contact and absorption. Earlier EPA-sponsored studies registered extremely high potential hand exposure. Light cotton gloves were used as passive dosimeters in U.S. EPA and U.S. Department of Labor "Youth in Agriculture" research [13] in a variety of hand-harvested crops. The glove residue was greater than the residue on gauze patches used to assess the distribution and amount of potential exposure of harvesters. The significance of hand exposure and the mitigating properties of gloves remains an active area of research in PCEP. Biomonitoring studies have shown the importance of hands as a route of harvester exposure [14].

39.4 Residents Indoors

Two very different types of indoor exposure become regulatory concerns of major proportions. Important human exposure scenarios follow indoor use of pesticides dispensed as foggers, area sprays, perimeter (baseboard) sprays, and/or crack-and-crevice applications. Foggers represent the greatest exposure potential since they distribute and deposit $\mu g/cm^2$ surface residues throughout a treated room. The availability of surface cypermethrin residues is highest on ceramic tile and in descending order linoleum > wood > carpet. Knowledge of deposition and available surface residue levels as a function of time post-application is critical to exposure assessment. The characteristics of indoor residential environments (e.g., air exchange, surface types), product use and human activity patterns, and other modifying habits and practices (e.g., vacuuming, surface cleaning, the use of door mats) are additional determinants of potential exposure.

Determination of reliable indoor exposure estimates includes considerable uncertainty and reliance on unfounded conjecture [2]. Routine exposure estimates have ranged from toxicologically negligible levels to amounts that would produce frank systemic toxicity if they ever occurred.

In practice, if excessive resident exposures occur, they are traceable to misuse or exposure to unpleasant or obnoxious odors not attributable to the active ingredient. In spite of considerable posturing by campaigners and activists it remains that chlorpyrifos was withdrawn by the registrant rather than banned by the U.S. EPA. Safe indoor use of chlorpyrifos continues in parts of the world outside of the United States. The confusion and anxiety generated by the many disparate opinions represents the consequences of non-validated models, invalid toxicology testing procedures, ultraconservative default assumptions, and regulatory concerns dominated by worst-case assumptions.

More extensive use of situational exposure monitoring and development of biological exposure indices are means to minimize some scientific uncertainties about potential resident pesticide exposures (Keenan *et al.* [15]).

Bystander exposure that occurs as a consequence of pesticide drift is more difficult to assess in both magnitude and time. Drift occurs during application or shortly thereafter. Malodor and lachrymation, formerly considered warning properties of pesticide exposure in some cases, are regulated adverse effects. Sampling and super-sensitive analysis can also elevate otherwise benign drift to exposures of regulatory significance when label language is invoked: "Do not apply this product in a way that will contact workers or other persons, either directly or through drift." Given general public fear and anxiety concerning pesticide exposure, it is imperative to clarify the health significance of accidental, unintentional, and unavoidable pesticide drift.

Drift at its lowest confirmable levels is a consequence of the Laws of Conservation of Matter and, perhaps, careless application at higher levels of exposure. Levels of drift exposure possibly represent nanograms to micrograms per kg body weight. Inhalation and deposition on exposed skin would be the prominent routes of exposure. Drift can usually be distinguished in dose and time from those exposures that occur by overspray or result from accidental pesticide application [16]. Accidental and unavoidable exposures, assuming pesticide use, should be distinguished during the risk assessment process. The procedures used to determine acceptable food residues could productively be applied to the inevitable, unavoidable bystander exposures that result from otherwise legal pesticide use.

39.5
Estimates of Human Exposure

Measurement of human pesticide exposures at varying levels of certainty have existed for over 50 years [17]. Concern about worker exposure developed coincident with the emergence of organophosphorous insecticides in pest management. Earliest studies were based upon biological monitoring of blood cholinesterases and urine biomarkers. Estimates based upon passive dosimetry became available in research to assess workplace exposures [4]. Later exposure estimates included environmental measures of deposition and dislodgeable residues [10, 18] and permissible exposures based upon pesticides in air (with skin notation emphasizing the importance of dermal contact [19] when warranted. U.S. EPA has promulgated numerous algorithms to estimate dietary, drinking water, residential, and occupational exposures based upon concentration and time data. These algorithms are very useful, transparent means of obtaining human exposure estimates, but they often include extremely conservative default assumptions, e.g., pesticide handling without the protection of clothing, unsustainable breathing rates, and outdated average body weights for professional pesticide applicators. The National Center for Environmental Assessment has prepared an Exposure Factors Handbook [20] to address factors commonly used in exposure assessments. This handbook was first published in 1989 as regulatory guidance on how to select values for exposure assessments which have become important regulatory rudiments since 1984. The importance of aggregate exposure estimates has been pushed to prominence by the Food Quality Protection Act of 1996 in the United States and regulations of the European Commission [21].

39.6 Exposure Biomonitoring

"Biological monitoring" means measuring of any chemical marker for humans associated with dose of a chemical or physical agent and/or associated in the past, present, or future with any adverse effect (Que Hee [22]). All of the factors that can influence an adverse effect (response) can influence the biological monitoring of exposure (dose). When toxicodynamic and toxicokinetic data are available, dose can be back-calculated or reconstructed from biomarker data from blood and/or urine. This process can yield absorbed daily dosage (μg pesticide equivalents/kg body weight) for application in risk assessment. The practice is especially important for development of aggregate exposure assessments to comply with the requirements of the Food Quality Protection Act of 1996.

Since pesticide regulation is based upon No Observed Adverse Effect Levels (NOAEL, mg/kg-day), it is rarely possible to directly relate adverse health effects and exposure. The [NOAEL (mg/kg-day)/exposure (mg/kg-day)] ratio yields the Margin-of-Exposure or Margin-of-Safety (MOE, MOS). When MOE is factored by uncertainty factors representing individual variability (10x) and species-to-human uncertainty (10x) the resulting reference dosage (e.g., RfD = MOE/100) can be estimated. Accidental, catastrophic exposure and "off-label" uses are outside of a determination of a MOE. Thus margins-of-exposure (measured human dosage/ NOAEL/uncertainty factors) becomes a surrogate for "safety". Biomarkers are a surrogate for "safety" of a particular set of circumstances in which human pesticide exposure occurs. Distinct weaknesses of MOEs are the uncertainty of the NOAEL (the dosage at which no adverse effect was observed) and the loss of a factor to account for the slope of the dose-response relationship.

Toxicodynamic and toxicokinetic data may be applied to biological monitoring to evaluate serial default assumptions. Particularly important are estimates of absorbed dosage derived from passive dosimetry. Biological monitoring may also be used to establish clothing penetration and dermal absorption with lower detection limits and longer monitoring periods during which biomarker elimination represents aggregate exposure.

Biomarker levels in blood or urine (μg biomarker/person) may be a more reliable indicator of exposure than absorbed daily dosage (μg/kg-day). The utility of such a biochemical exposure index results from its simplicity when combined with a medical determination of "safety" by a physician. Biochemical Exposure Index marker concentrations (BEI, [23]) include medical judgment of "safe" levels for healthy workers' lifetime exposures that augments or replaces default driven exposure assessments derived from safety evaluation studies in laboratory animals. The present use of NOAELs divided by uncertainty factors (typically 100, but in some cases much higher) arithmetically yields a reference dose (RfD) of uncertain significance with respect to health (see definition of NOAEL). Uncertainties and extrapolations associated with the RfD are not understood (or they are ignored) by many activists/campaigners, media, some regulators, and the general public. People may be easily persuaded in all too many instances to regard the RfD as an adverse effects threshold.

When pesticide biomarkers are also food residues or other environmental products, exposure estimates may be inflated. The occurrence of dialkylphosphates and dialkylthiophosphates has been demonstrated (Krieger *et al.*, [24]). At very low levels of exposure, these preformed biomarkers may confound and inflate organophosphorous insecticide exposure estimates. The same is likely to occur in other cases where exposure biomarkers represent complex mixtures, e.g., ubiquitous traces of DDT and DDTs with very different toxicological significance, and pyrethroid degradates such as 3-phenoxybenzoic acid, a metabolite of synthetic pyrethroid pesticides such as permethrin, preformed under a variety of conditions. Preformed exposure biomarkers will be of less toxicologic importance for assessment of direct resident or worker exposure than in environmental studies dominated by barely detectable levels of no apparent toxicological significance and uncertain etiology.

It is unfortunate and misleading that the terms *Risk Assessment* and *Risk Characterization* imply an estimate of "risk," i.e., the probability or likelihood of exposure producing an adverse effect including an estimate of the severity of the illness. In fact, *hazards* only become *risks* when a susceptible population is exposed. Pesticide safety generally results from the conservative development of use patterns that minimize human exposure below experimental or epidemiological *no observed adverse effect levels factored by additional multiple uncertainty factors*. The resulting reference dose (RfD) causes some investigators, regulators, and members of the public to respond to exceedences of the RfD as though it represented a toxic clinical end point rather than very conservative health guidance.

39.7 Conclusion

When pesticide safety evaluations were initially conducted, FIFRA was intended to "prevent unreasonable adverse effects on human health or the environment." Organophosphorous insecticides were introduced to California agriculture about 1950 with accompanying determination of cholinesterase status and urine biomonitoring overseen by physicians [25]. With the passage of the Food Quality Protection Act of 1996, a still higher standard of safety is sought: "reasonable certainty of no harm." Aggregate exposure assessment using dietary food, water, and residential exposures have placed a premium on human exposure measurements. Human pesticide exposure monitoring is essential to provide exposed persons and the public the evidence of safety that they have come to demand [26]. The present system of study and ranking pesticide exposures and terming the result "risk assessment" fails to diminish public perception and may even heighten anxiety about normal pesticide exposure. Serious consideration should be given to increased participation of physicians and epidemiologists in the pesticide regulatory process to discern that toxicology per se is a small, but vital, part of assuring safe pesticide use.

39.8 References

1 National Safety Council, Injury Facts® 2004 Edition, Itasca, IL, **2004**.

2 P. Berteau, J. Knaak, D. Mengle, J. Schreider, Insecticide Absorption from Treated Surfaces. In *American Chemical Society Symposium Series 382*, R. Wang, R. Franklin Honeycutt, J. Reinert (Eds.), Washington, DC, **1989**, pp. 315–326.

3 R. I. Krieger, C. E. Bernard, T. M. Dinoff, R. L. Williams, Biomonitoring of Persons Exposed to Insecticides Used in Residences, *Ann. Occup., Hyg.*, **2001**, *45*, S143–S153.

4 W. F. Durham, H. R. Wolfe, Measurement of the Exposure of Workers to Pesticides, *Bull. Wld. Hlth. Org.*, **1962**, *26*, 75–91.

5 U.S. EPA (U.S. Environmental Protection Agency) (**1986**), Pesticide assessment guidelines, Subdivision U Applicator exposure monitoring, U.S. Environmental Protection Agency, Report #540/9-87-127, Washington, D.C.

6 J. H. Ross, J. H. Driver, C. Lunchick, C. Wible, F. Selman, Pesticide Exposure Monitoring Databases in Applied Risk Analysis, *Rev. Environ. Contam. Toxicol.*, **2006**, *186*, 107–132.

7 H. N. Nigg, J. H. Stamper, Dislodgeable Residues of Chlorobenzilate in Florida Citrus: Workers Reentry Implications, *Chemosphere*, **1984**, *13*, 1143–1156.

8 G. Zweig, J. T. Leffingwell, W. J. Popendorf, The Relationship Between Dermal Pesticide Exposure by Fruit Harvesters and Dislodgeable Foliar Residues, *J. Environ. Sci. Hlth.*, **1985**, *B30*, 27–59.

9 U.S. EPA (U.S. Environmental Protection Agency) (**2000**), Science Advisory Council for Exposure Policy Number 003.1 Regarding: Agricultural Transfer Coefficients, Revised August 7, U.S. Environmental Protection Agency, Office of Pesticide Programs, Washington, D.C.

10 Y. Iwata, J. B. Knaak, R. C. Spear, R. J. Foster, Worker Reentry into Pesticide-Treated Crops. I. Procedure for the Determination of Dislodgable

Pesticide Residues on Foliage, *Bull. Environ. Contam.Toxicol.*, **1977**, *18*, 649–655.
11 Y. Li, J. J. Keenan, H. Vega, R. I. Krieger, Human Exposure to Surface Pesticide Residues: Dislodgeable Foliar Residues and Pilot Studies to Predict Bioavailability, Abstract, American Chemical Society Meeting, San Francisco, CA., **2006**.
12 G. K. Whitmyre, J. R. Ross, M. E. Ginevan, D. Eberhart, Development of Risk-Based Restricted Entry Intervals. In *Occupational and Residential Exposure Assessment for Pesticides*, C. A. Francklin, J. P. Worgan (Eds.), John Wiley & Sons, West Sussex, England., **2005**.
13 U.S. EPA (U.S. Environmental Protection Agency) (**1980**), Youth in Agriculture, Interagency Agreement of March 17, 1980, U.S. EPA/USDOL.
14 R. I. Krieger, Pesticide Exposure Assessment, *Toxicology Letters*, **1985**, *82*, 65–72.
15 J. J. Keenan, R. S. Gold, G. Leng, X. Zhang, R. I. Krieger, Pyrethroid Exposure in the Indoor Environment Following Use of Cypermethrin Foggers, Abstract, Society of Toxicology, Annual Meeting, **2006**.
16 J. J. Van Hemmen, Pesticides and the Residential Bystander, *Annals of Occupational Hygiene*, **2006**, *50*, 651–655.
17 G. S. Batchelor, K. C. Walker, Health Hazards Involved in the Use of Parathion in Fruit Orchards of North Central Washington, *Am. Med. Assoc. Arch. Indust. Hygiene*, **1954**, *10*, 522–529.
18 D. M. Stout, M. A. Mason, The Distribution of Chlorpyrifos Following a Crack and Crevice Type Application in the US EPA Indoor Air Quality Research House, *Atmos. Environ.*, **2003**, *37*, 5539–5549.
19 American Conference of Governmental Industrial Hygienists (ACGIH®), Guide to Occupational Exposure Values. ACGIH, Kemper Meadow Dr., Cincinnati, OH, **2004**.
20 U.S. EPA (U.S. Environmental Protection Agency) (**1999**, February) Exposure Factors Handbook, EPA/600/C-99/001 (CD Rom), National Center for Environmental Assessment, Cincinnati, OH.
21 European Commission (EC) **2002**, Technical Guidance Document on Risk Assessment Office for Official Publications of the EC, Luxembourg, Internet publication <http://ecb.jrc.it/>.
22 S. S. Que Hee, Biological Monitoring, Van Nostrand Reinhold, New York, NY, **1993**.
23 V. Fiserova-Bergerova, History and Concept of Biological Exposure Indices. In *Biological Monitoring of Exposure to Industrial Chemicals*, V. Fiserova-Bergerova, M. Ogata (Eds.), ACGIH, Cincinnati, OH, **1990**, pp. 19–23.
24 R. I. Krieger, T. M. Dinoff, R. L. Williams, X. Zhang, J. H. Ross, L. S. Aston, G. Myers, Preformed Biomarkers in Produce Inflate Human Organophosphate Exposure Assessments, *Environ Health Perspec.*, **2003**, *111*, A688–689.
25 P. Washburn, Personal conversations recounting the introduction of parathion into the California citrus industry, Washburn & Sons, Highgrove, CA, **2006**.
26 R. I. Krieger, Human Test Data: Essential and Safe, *Chem. and Eng. News*, **2005**, *83*, 4–7.

Keywords

Pesticide Exposure Assessment, Biomonitoring, Pyrethroid, Situational Monitoring

40
Bioavailability of Common Conjugates and Bound Residues

Michael W. Skidmore, Jill P. Benner, Cathy Chung Chun Lam, James D. Booth, Terry Clark, Alex J. Gledhill, Karen J. Roberts

40.1
Introduction

Before any pesticide can be authorized for use, its human and environmental safety have to be considered through risk assessment; i.e., a comparison of hazard and exposure. A key element in the risk assessment process is the determination of the residue definition, i.e., the components of the residue resulting from the use of the pesticide that are considered to be relevant. Residues remaining on items for food and feed are identified in metabolism studies, in which the pesticide is radiolabeled to enable its fate and behavior to be followed. The metabolic pathways can be extremely complex but can be grouped into four distinct categories or phases [1].

Phase I metabolism mostly involves oxidation, reduction, and hydrolytic reactions introducing functional groups into the xenobiotic compound. These residues are generally extractable and can be readily characterized/identified and their relevance assessed, based on their concentration, toxicity or by structure-activity relationships.

Phase II metabolism or conjugation, occurs when parent or Phase I metabolites are covalently bound to an endogenous molecule. Conjugated residues in plants were historically perceived as terminal products generated through a detoxification mechanism and thus received little attention. This was compounded by technical difficulties in the identification of complex polar components. With current technological advances and analytical techniques, it is possible to identify conjugates intact and also to investigate their behavior. This raises the dilemma as to how these residues should be regulated. In most instances, conjugates are regulated as the exocon, on the assumption that they are deconjugated in the body.

Pesticide Chemistry. Crop Protection, Public Health, Environmental Safety
Edited by Hideo Ohkawa, Hisashi Miyagawa, and Philip W. Lee

ISBN: 978-3-527-31663-2

Phase III so-called "bound" or unextracted residues can be formed by processes which lead to either covalent binding to or physical encapsulation within a biological macromolecule. For many years, it was believed that unextractability was synonymous with a lack of bio-availability. Recent investigations have, however, shown that the bioavailability of a bound residue is dependent on the nature of the binding [2]. The critical question for Phase III metabolism is its definition and the rigor of the extraction procedure. Guidance on these questions has been given by the IUPAC Commission on Agrochemicals and the Environment [3]. Again a key question is how to regulate these residues?

Phase IV residues result from the incorporation of the radiolabel from the pesticide into naturally occurring compounds, e.g., proteins and sugars. These residues are of no toxicological concern.

The focus of the current project is on the bioavailability of specific Phase II and Phase III metabolites, derived from plants. The objective of the project is:

> To understand the behavior of glucoside conjugates and specific bound residues under conditions found in the gastrointestinal tract (GIT) of humans and livestock species. The term behavior implicitly includes stability and propensity to be systemically absorbed.

The project has been divided into 2 phases, an extensive literature review followed by an experimental phase.

40.2
Literature Search

An extensive literature review was conducted covering publications from 1970 which resulted in the following observations:

- The number of conjugate types is extensive; the most frequently reported being glycosides, sulfates, and catabolites of glutathione conjugations. The data suggest that β-D-glucosides are the most common form of conjugate reported in plants. Similarly, glucuronides are probably the most common conjugates in animals.
- Except for the recognized generalizations that glucosides are found in plants and glucuronides are found in animals, it is difficult to establish a consistent correlation of conjugate type and species.
- Glucosides appear to be readily hydrolyzed by microflora in the rumen or in the lower intestine of non ruminants.
- Different glucosidic linkages, e.g., *O*-, *N*-, and ester glucosides, demonstrate distinctly different behaviors. The nature of the glucoside conjugate can have a significant effect on its behavior as is the case with hymexazol where both *O*- and

N-glucosides are formed. On oral administration to the rat the *O*-glucoside is almost quantitatively absorbed and excreted, largely unchanged, in the urine. In the case of the *N*-glucoside, only 50% was absorbed and eliminated unchanged in the urine [4]. Examples of ester glucosides suggest that these are readily hydrolyzed in the GIT.

- There is a distinct shortage of information concerning the stability of conjugates to enzymatic and chemical conditions encountered in the GIT of humans and livestock species. The primary focus in the literature has been the characterization of the conjugate and identification of the exocon via hydrolysis. This has routinely been achieved through the use of strong mineral acids/bases at high temperatures or readily available enzymes to effect the cleavage of the conjugate. These data cannot be extrapolated to predict stability under conditions found in the GIT.
- There is a paucity of information relating to the physicochemical characteristics of conjugates. The physicochemical properties of conjugates are particularly important when considering their potential bioavailability, i.e., physical parameters such as solubility, pKa, and lipophilicity can influence absorption and lability. It is clear that further information on the physicochemical properties of conjugates would provide an additional understanding of the potential behavior of these residues in terms of stability and potential for absorption.
- Model systems to predict physicochemical characteristics and to assess likely absorption of conjugates are available but have received limited use for conjugates.
- The field of bound residues is a complex area which over many years has lacked a clear definition of what constitutes a bound residue.
- Although a significant amount of information is available on the behavior of bound residues, it has either been carried out using model compounds or ill-defined unextractable residues. The data are therefore difficult to interpret.

40.3 Experimental Phase

The experimental phase of the project has been designed to investigate the behavior of twelve glucoside conjugates and two typical bound residues, i.e., lignin and hemicellulose. For conjugates, the investigation includes stability under conditions found in the GIT, propensity for permeability, physicochemical properties, stability to typical GIT and rumen microflora and *in vivo* bioavailability of two radiolabeled conjugates. For bound residues, following thorough characterization, the investigation will include solubilization in typical GIT conditions and oral bioavailability.

40.3.1
Conjugates

The literature search demonstrated that β-D-glucosides are the most common type of conjugate derived from the metabolism of pesticides in crop plants. For this reason, β-D-glucosides were selected as the model conjugates. Selection of individual compounds was based upon a number of factors but the overall aim was to achieve diversity in terms of steric and electronic effects within the exocon moiety. Another important factor concerned compound supply and, where possible, commercially available compounds were chosen. As commercial sources did not offer sufficient diversity and could not provide glucosides with nitrogen or acyl linkages, additional compounds were prepared synthetically. The final selection of twelve compounds is shown in Figure 1.

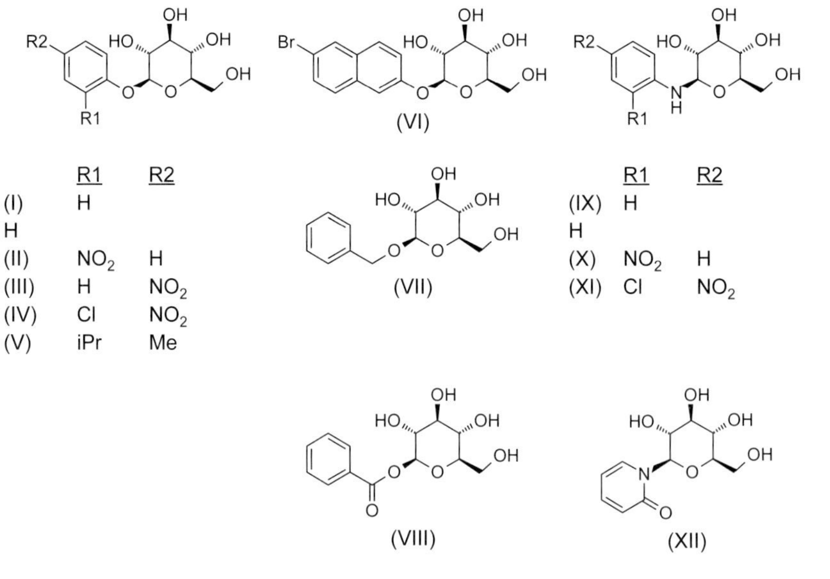

Figure 1. Structures of conjugates used for study.

40.3.2
Chemical and Enzymatic Hydrolysis

The initial series of experiments assessed the stability of the conjugates to the pH and temperature conditions (37 °C) found in the mammalian GIT; pH 1 and pH 9 were chosen to represent the normal extremes of acidity and basicity in the human. While pH 5 provided an intermediary value, it is also relevant to parts of the ruminant digestive system. The rate of degradation of the conjugates was monitored over a 24-hour period by LC-MS/MS (turbo ion spray interface). Optimization of the LC system provided an efficient method in which several compounds could be analyzed simultaneously. During the early stages of method development it became apparent that, at the low concentrations employed (5–10 μg/mL), significant proportions of the conjugates were lost from solution as a result of adsorption onto glass and other surfaces. This problem was eliminated by use of suitable plastic vessels for the incubation phase. In addition, subsamples taken at designated timepoints for LC-MS/MS analysis were diluted with acetonitrile to prevent losses within glass LC vials.

The six phenol conjugates (I)–(VI) and the benzyl alcohol conjugate (VII) were stable at all three pH values over a 24-hour period with no evidence for degradation. The aniline conjugates (IX)–(XI) behaved very differently with complete degradation of all three compounds within a few hours at pH 1, but the rate of hydrolysis was structure-dependent. Total loss of (IX) was also observed at pH 5 and pH 9. The hydrolytic stability of several other β-D-glucosides of anilines (4-chloroaniline (XIII), 3,4-dichloroaniline (XIV) and chloramben (XV), Figure 2) under acidic conditions is documented in the literature [5–6]. An assessment of the data for all six aniline conjugates shows that the pKa of the aniline correlates with the rate of hydrolysis of the corresponding glucose conjugate, indicating that as a general rule, electron withdrawal from the glycosidic nitrogen atom increases acid stability [7], Figure 2.

The benzoic acid conjugate (VIII) did not undergo significant hydrolysis but incubation of the starting material under certain conditions resulted in multiple LC peaks, all of which retained the molecular weight of the original conjugate. ^{1}H NMR studies demonstrated that the reaction products arose from migration of the benzoyl group around the glucose ring, effectively an intramolecular transesterification process. Reactions of this type are well documented for

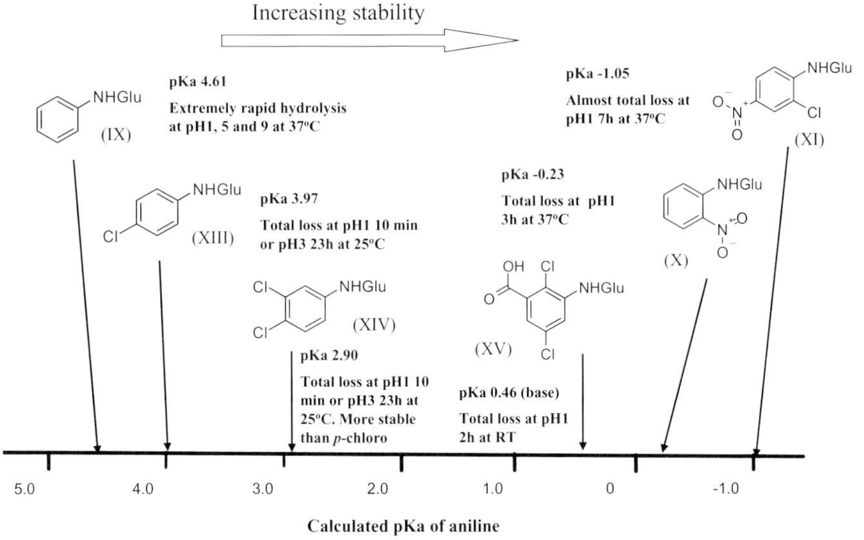

Figure 2. Relationship between the pKa of anilines and the hydrolytic stability of the corresponding β-D-glucosides.

glucuronides but the corresponding reaction of glucosides, although reported in the literature [8], is less familiar within the field of metabolism chemistry. Results for the pyridone conjugate (XII) are at present inconclusive, due to difficulties with chromatography.

A further series of experiments assessed the stability of the conjugates to the two major sets of digestive enzymes commonly present in the human digestive tract; simulated gastric fluid (SGF) and simulated intestinal fluid (SIF). For each of the incubations, the rate of degradation of the conjugate was measured over 24 hours by LC-MS/MS.

Compounds (I)–(XII) were incubated in an SGF buffer at pH 1.2 which contained 0.32% pepsin. All *O*-linked compounds and the *N*-linked pyridone, (XII), were stable to these conditions. The *N*-linked compounds (IX), (X), and (XI), however, rapidly degraded, with the rate of degradation being the same in the presence or absence of pepsin. Therefore degradation is concluded to be pH rather than enzyme mediated.

The enzymes for the SIF incubations were, in general, split into two groups: proteases and amylase plus lipase. The incubations were performed only on compounds (I)–(VIII), (XI), and (XII); not all the anilines were investigated because of their instability in the stomach. Compounds were incubated in an SIF buffer (pH 7.5–7.9) containing either trypsin, chymotrypsin, carboxypeptidase A & B, elastase or α-amylase and lipase. Compounds (I), (V)–(VII), and (XII) showed some degradation in the incubation with proteases which appeared to be enzyme mediated. Compounds (II), (III), (IV), (VIII), and XI were stable in the incubations. All the compounds investigated in the amylase/lipase incubations were stable.

40.3.3
Prediction of Permeability

A prediction of the potential for the conjugates to be absorbed from the gut was made by using the *in vitro* CACO-2 cell assay. The CACO-2 assay uses cultured human colon carcinoma cells as a surrogate for the epithelial cells found in the gut wall. CACO cells cultured in transwell plates were sourced and grown until they formed tight junctions, (approximately 21 days from first seeding) which were determined by measuring total epithelial electrical resistance. At this point, test compound was added to the culture medium on the apical side and the appearance of compound at the basolateral side of the plate was measured, after 1 hour, by quantitative LC-MS/MS. In addition to the test compounds, 6 reference compounds (antipyrin, caffeine, theophylline, hydrochlorothiazide, furosemide, and atenolol) were dosed. The reference compounds have a known permeability ranging from low to high, therefore the permeability of the test compounds was compared to that of the reference compounds allowing them to be ranked as being of low, medium or high permeability. All the conjugates tested showed low permeability in this assay.

Two software packages, ADMET predictor (Simulations plus) and ADME Boxes (Pharma Algorithms), were used to predict the potential absorption of the conjugates *in silico*. Demonstration copies of the software were obtained and the structures of the conjugates and their parent exocons were analyzed using each of the packages. For both packages, the predicted log P values of the exocons correlated extremely well with the measured values and each of the exocons was predicted to have high permeability which is commensurate with the expected behavior of these compounds. For the *O*- and *N*-glucose conjugates, the predicted log P values correlated less well with the measured values and there were some slight differences between the two packages in terms of predicted permeability. There were more marked differences between the two packages in the predictions of permeability for the conjugates, with the extremes of permeability (low and high) being predicted for some of the conjugates.

40.3.4 Bound Residues

The area of bound residues is highly complex and there is much conflicting information in the literature. There are few pesticidal compounds for which the chemical nature of binding to macromolecules is well understood and in the past there has been confusion between natural incorporation and true molecular binding of xenobiotic moieties. In order to advance knowledge of the relevance of bound residues in the mammalian diet, it was essential to prepare samples of bound residues under conditions relevant to agricultural practice. This required spray application of chemicals to whole plants and harvest of commodities at a time appropriate to commercial food or feed production. The use of radiolabeled test material was essential to allow quantification of macromolecular binding and to track release of bound residues at low levels in a complex matrix. Following discussions [9] and a further review of the literature, 3,4-dichloroaniline (XVI) and pentachlorophenol (XVII), Figure 3, were selected as the test chemicals. 3,4-Dichloroaniline is reported to bind to lignin [10], while pentachlorophenol binds principally to hemicellulose in cell cultures but is associated with a diverse range of macromolecules, including hemicellulose, protein and lignin, in whole plants [11].

The chemicals were formulated to facilitate effective application in aqueous media then sprayed onto established wheat plants grown in large containers of soil in a glasshouse. The application timing for each chemical was optimized for appropriate macromolecule biosynthesis in the crop, hence (XVI) was applied only

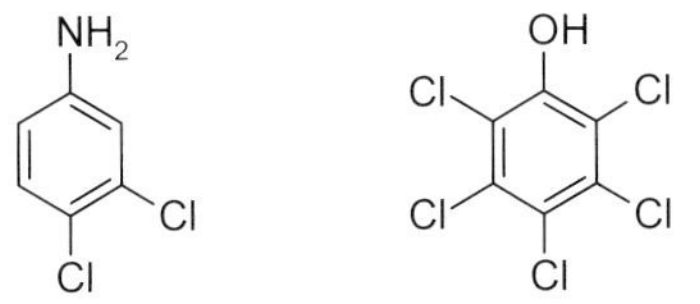

(XVI) (XVII)

Figure 3. Structures of 3,4-dichloroaniline (XVI) and pentachlorophenol (XVII).

after lignification had commenced. The extent of radiochemical binding within the crop was monitored at intervals and the wheat was harvested as it approached maturity. After removal of the grain heads, the remaining crop (hay/straw) was extracted exhaustively using a number of different acetonitrile/water mixtures and the residual debris, defined as the bound residue, was used for further experimentation. Analysis of extracts and debris indicated that 35% of the total radioactivity residue (TRR) derived from (XVII) and 58% of the TRR from (XVI) was associated with the debris.

40.3.5
Characterization of the Bound Residues

Characterization of the bound residues generated from (XVI) and (XVII) is being conducted using cell wall fractionation based on enzymic and chemical methods. Two procedures have been examined to compare and contrast practical aspects of the fractionations and to assess the degree of consistency achieved by different methods. The first procedure, which was developed originally to characterize unextractable residues of (XVII) [11], utilized enzymes (α-amylase and pronase) to release material associated with starch and protein and then employed stronger chemical methods (EGTA, dioxan/hydrochloric acid, potassium hydroxide, and sulfuric acid) to characterize residues bound to pectin, lignin, hemicellulose, and cellulose. The second procedure, which was developed by combining steps from two other literature methods, [12] and [13], employed milder conditions and included incubations with pectinase, cellulase, and hemicellulase enzymes. Initial results suggested that there was distribution of radioactivity amongst all the fractions but there were some noticeable differences between compounds and between methods. Overall the chemical method appeared to achieve a significantly greater degree of solubilization of the bound residue from (XVI). A final assessment of the results will be carried out after examining a third technique known as "Clean Fractionation" [14], which has been reported to have potential for commercial-scale purification of cellulose, hemicellulose, and lignin.

40.3.6
Chemical and Enzymatic Hydrolysis

Bound residues were incubated at pH 1, 5, and 9 at a temperature of 37 °C and the extent of hydrolysis/solubilization was determined by quantification of the radioactivity released. An additional 3–9% TRR was solubilized from the (XVI) bound residue, but at pH 9, *ca.* 18% TRR was solubilized from the (XVII) bound residue. This may be due in part to the tendency of hemicellulose to be solubilized in basic conditions. Results are displayed graphically in Figure 4.

The stability of radiolabeled bound residues to the two major groups of digestive enzymes was assessed by incubations similar to those described in 3.2, with the exception that the protease and amylase/lipase enzymes were combined into a single SIF incubation. The incubation mixtures were centrifuged and the super-

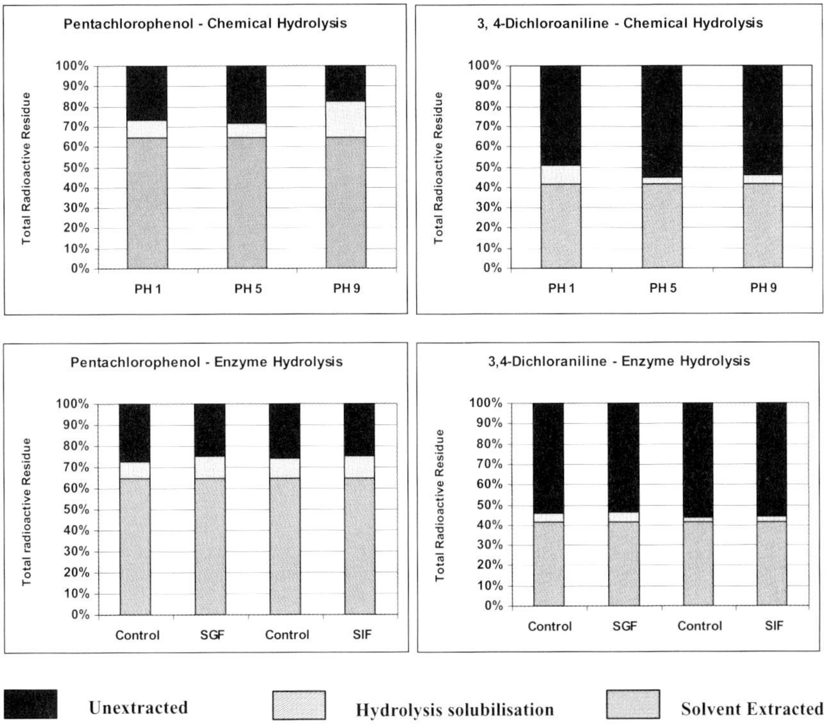

Figure 4. Extraction and hydrolysis results for wheat treated with 3,4-dichloroaniline (XVI) and pentachlorophenol (XVII).

natant analyzed for radioactivity. Control incubations were performed without enzyme. Only a small amount of radioactivity (3–5% TRR) was solubilized following incubation of the (XVI) bound residue. A greater proportion, 8–11% TRR, was recovered in the incubations using the (XVII) bound residue. Enzymes did not appear to significantly affect the total release of radioactivity over the 24-hour incubation period. Results are displayed graphically in Figure 4.

40.3.7
Bioavailability of Bound Residues

The bioavailability of the radiolabeled bound residues of (XVI) and (XVII) is being investigated following oral administration to bile duct-cannulated rats. The outcomes of these studies are still being analyzed and only initial results are presented. Experiments were conducted in accordance with the UK Home Office regulations for animal welfare.

A sample of the ^{14}C-labeled debris was formulated for dosing, as a slurry, in 0.5% aqueous carboxymethyl cellulose. Bile duct-cannulated rats (3 per sex per residue) were given two oral gavage doses, 8 hours apart, of the dose formulation to give a total dose of 2800 mg ^{14}C-residue/kg bodyweight. The dose formulation and regime was chosen to give the maximum possible amount of radiolabel without compromising the ability of the animals to feed normally. Animals were housed individually and unrestrained, in metabolism cages, with free access to food and fluids for the duration of the experiment. Urine, feces, and bile were collected at intervals up to 48 hours after dosing at which time the animals were humanely killed and the GIT and contents were excised. The radioactive content of urine, bile, feces, GIT and contents, and residual carcass was measured and the extent of absorption calculated from a summation of the radioactivity present in urine, bile, and residual carcass. In the bile duct-cannulated rat, radioactivity in the GIT and contents and feces is considered to represent unabsorbed material.

For (XVI), absorption of radioactivity was minimal with the majority (*ca.* 95%) of the dose being present in feces. Results to date indicate that the absorption of radioactivity from (XVII) bound residues is slightly higher than those from (XVI).

40.4 Conclusion

The literature review demonstrated that there is a gap in the basic understanding of the properties and behavior of conjugated metabolites. The current project is providing an opportunity to conduct a detailed investigation of twelve glucoside conjugates. Although the work is only partially complete the following observations have been made:

- Conjugation of an exocon with a glucose reduces the log P by *ca.* 2.
- Benzoic acid glucose conjugate exhibits acyl migration.
- Permeability of sugar conjugates is predicted to be low by CACO-2 cell line measurements.
- Permeability, predicted by *in silico* models, is high for certain conjugates, i.e., phenol and benzyl alcohol glucoside conjugates.
- *N*-glucose conjugates show degrees of lability to chemical hydrolysis, *O*-glucose conjugates are stable to the pH and temperature conditions found in the GIT.
- *O*-glucose conjugates are stable in gastric and pancreatic enzymes – lability in proteases is under further investigation.

The area of "bound" residues is particularly complex and requires harmonization in definition and investigation into methods of characterization.

40.5 Acknowledgment

The authors would like to express their thanks to the United Kingdom Pesticides Safety Directorate (PSD), sponsors for this project.

40.6 References

1 H. W. Dorough, *J. Environ. Pathol. Toxicol.*, **1980**, *3*, 11–19.
2 H. Sandermann, M. Arjmand, I. Gennity, R. Winkler, C. B. Struble, *J. Agric. Food Chem.*, **1990**, *38*, 1877–1880.
3 M. W. Skidmore, G. Paulson, H. A. Kuiper, B. Ohlin, S. Reynolds, *Pure and Appl. Chem.*, **1998**, *70*, 1423–1447.
4 M. Ando, M. Nakagawa, M. Ishida, *J. Pestic. Sci.*, **1988**, *13*, 473.
5 R. Winkler, H. Sandermann Jr., *Pestic. Biochem. and Physiol.*, **1989**, *33*, 239–248.
6 S. R. Colby, *Science*, **1965**, *150*, 619–620.
7 R. Winkler, H. Sandermann Jr., *J. Agric. Food. Chem.*, **1992**, *40*, 2008–2012.
8 A. Brown, T. C. Bruce, *J. Am. Chem. Soc.*, **1973**, 95, 1593–1601.
9 H. Sandermann Jr., Personal communications.
10 G. G. Still, H. M. Balba, E. R. Mansager, *J. Agric. Food Chem.*, **1981**, *29*, 739–746.
11 C. Langebartels, H. Harms, *Ecotox. and Environmental Safety*, **1985**, *10*, 268–279.
12 W. F. Feely, L. S. Crouch, *J. Agric. Food Chem.*, **1997**, *45*, 2758–2762.
13 J. B. Pilmoor, J. K. Gaunt, T. R. Roberts, *Pestic. Sci.*, **1984**, *15*, 375–381.
14 J. J. Bozell, S. K. Black, D. Cahill, D. K. Johnson, 9th *International Symposium Wood and Pulping Chemistry*, **1997**, *11*, 1–4.

Keywords

Pesticide Metabolism, Conjugate, Bound Residue, Bioavailability

41
Multiresidue Analysis of 500 Pesticide Residues in Agricultural Products Using GC/MS and LC/MS

Yumi Akiyama, Naoki Yoshioka, Tomofumi Matsuoka

41.1
Introduction

In Japan, the “Positive List” system was implemented on May 29, 2006. The Japanese Ministry of Health, Labour, and Welfare had established many provisional maximum residue limits (MRLs) in addition to present MRLs, and 586 pesticides were regulated under the Food Sanitation Law [1]. A uniform level of 0.01 ppm is established as the level having no potential to cause damage to human health [2]. A rapid and sensitive multiresidue analytical method was required to conduct efficient and effective monitoring surveys of pesticide residues to ensure food safety.

We developed a multiresidue analytical method by GC/MS in 1995 [3] and modified by the use of LC/MS in 2002. Thus, we expanded the number of pesticides tested annually and continued monitoring surveys for the last 10 years [4–6]. Our multiresidue analytical method and the monitoring data on pesticide residues in agricultural products are reported in this chapter.

41.2
Multiple Residue Analysis

Portions of ground samples (25 g) were homogenized with 60 mL of acetonitrile and filtered. The filtrates were subjected to clean up using an ODS solid-phase extraction (SPE) cartridge (1 g). The acetonitrile was separated by salting-out and 36 mL were collected. After evaporation to dryness, the residue was adjusted to a volume of 3 mL in *n*-hexane-acetone (1 : 1). A 2-mL aliquot portion was loaded onto a PSA cartridge (200 mg) and eluted 3 times with 2 mL of *n*-hexane-acetone (1 : 1) to remove fatty acids and chlorophylls. All the eluates collected were evaporated to dryness and adjusted to a volume of 2 mL with *n*-hexane-acetone (4 : 1) for GC/MS analysis. A 0.4-mL aliquot of the final solution was evaporated and redissolved in 0.2 mL of acetonitrile for LC/MS analysis. The GC/MS and

Pesticide Chemistry. Crop Protection, Public Health, Environmental Safety
Edited by Hideo Ohkawa, Hisashi Miyagawa, and Philip W. Lee

ISBN: 978-3-527-31663-2

LC/MS operating conditions are shown in Tables I and II, respectively. As data were acquired in the scan mode, the number of pesticides analyzed was unlimited. Positive analytes were confirmed by retention time and relative response ratio of main fragment ions and also by mass spectra. Rapid and reliable confirmation was available using our original macro program that enabled printing of reports containing the extracted ion chromatograms for every pesticide automatically after each injection.

Table I. GC/MS conditions.

Apparatus: Agilent 6890GC + 5973inert MS	
Column:	HP-5MS (30 min m × 0.25 mm i.d., film thickness 0.25 μm) + guard (*ca.* 50 cm × 0.25 mm i.d., non-coating)
Column temp.:	70 °C (3 min) → 30 °C/min → 160 °C (0 min) → 2.5 °C/min → 200 °C (0 min) → 8 °C/in → 300 °C (5.5 min)
Carrier gas:	He
Inlet pres.:	40 psi (1 min), 16 psi (const. press)
Flow:	5.4 mL/min (initial), 1.7 → 0.7 mL/min (during analysis)
Injection volume:	4 μL (splitless)
Purge off time:	1 min
Injector temp.:	250 °C
Interface temp.:	280 °C
Ion source temp.:	230 °C
Quadrupole temp.:	150 °C
Ionization energy:	70 eV
Scan range and scan cycle:	m/z 50–550 (2.9 cycles/sec)
(Retention Time Locking technique is adopted.)	

Table II. LC/MS conditions.

Apparatus: Agilent 1100 MSD(SL)	
Column:	Inertsil ODS3 (150 mm, 3.0 mm, 5 μm) + guard (10 mm, 3.0 mm, 5 μm)
Mobile phase:	CH_3CN – 10 mM CH_3COONH_4 [(15 : 85) → (95 : 5)] / 20 min + (95 : 5) 10 min
Flow rate:	0.5mL/min
Column temp.:	40 °C
Injector program:	Sample soln. (acetonitrile) 4 μL + Water 16 μL, mix 5 times, then inject 20 μL
Sample cooler:	15
Ionization and capillary voltage:	ESI (Positive, 4000V) (Negative, 3500V)
Nebulizer gas:	50 psi, Drying gas 10 L/min (350)
Fragmentor voltage:	100V, 200V
Scan range and scan cycle:	m/z 50–950 (0.98 cycles/sec)

Recoveries were investigated by fortifying 3 matrices (brown rice, lemon, and spinach) with 500 different kinds of compounds, all at the 0.1-ppm concentration level. Among the results calculated by the external standard in solvent, 421 compounds were recovered between 70 and 120% and 478 compounds were between 50 and 140% in average. High recoveries above 100% were due to matrix effects during GC/MS analysis. Recoveries below 50% were due to degradation during analysis. For those compounds, we tried to analyze degradation products together with their parent pesticides as possible [7].

Detection limits were defined by a signal-to-noise (S/N) ratio of 3. Among 423 compounds analyzed by GC/MS, 390 showed detection limits below 1 ppb and almost all were below 3 ppb. Even with LC/MS, among 148 compounds analyzed, 140 showed detection limits below 3 ppb. Quantitation limits defined by an S/N ratio of 10 did not exceed 0.01 ppm, for almost any pesticide using either GC/MS or LC/MS. As a result, we could set reporting limits of our monitoring survey at 0.01 ppm.

Matrix effects were investigated by fortifying 3 matrices with 20 compounds at 0.1 ppm and analyzing by both GC/MS and LC/MS. Recoveries obtained by the solvent and matrix-matched standards were calculated separately and compared. The results are shown in Figure 1. In GC/MS analysis, butroxydim, propaquizafop, and tralkoxydim showed high recoveries above 150% calculated by the solvent standard, whereas those obtained using a matrix-matched standard were nearly 100%. These compounds are easy to adsorb on the insert liner or column and showed tailing peaks without matrix. In LC/MS analysis, recoveries calculated by the solvent standard were somewhat lower than those obtained by

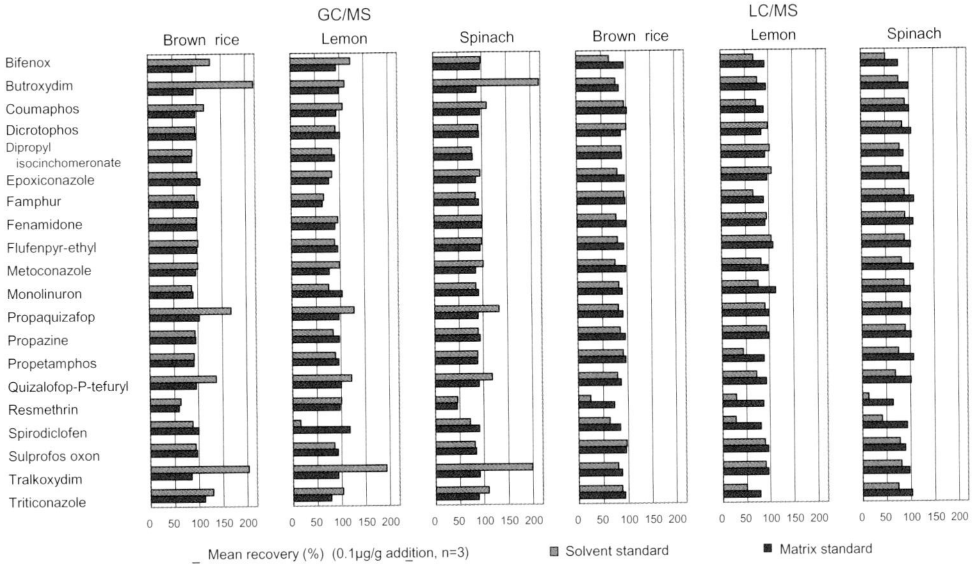

Figure 1. Comparison of matrix effects appeared in GC/MS and LC/MS analysis.

the matrix-matched standard. Resmethrin and spirodiclofen showed especially low recoveries by the solvent standard and ionization of these was supposed to be strongly suppressed by co-eluting matrix. For others, including those that showed matrix enhancement in GC/MS analysis, suppression of ionization was no more than 30%.

41.3
Monitoring Results

A monitoring survey of pesticide residues was conducted for about 200 agricultural products per year. Samples were collected in wholesale and retail markets all over the Hyogo prefecture. Using our method, 15 samples were prepared in a day and the screening data for 457 pesticides and 43 of their metabolites were obtained the next day.

Detection rates during a recent 4-year period are shown in Figure 2 for each group of food. In the left figure, trace level residues between 0.001 and 0.01 ppm were considered positive. As a whole, the single residue method yields a detection rate of about 56% whereas the multiresidue method detects positive. But in the right figure, the detection rates were decreased 10–20% by excluding the trace level residues. As *ca.* 40% of total detections were below 0.01 ppm, the detection limit was an important factor to evaluate the detection rates. Detection rates in fruits were higher than those in vegetables and cereals. Imported citrus fruits showed high detection rates, and more than 70% of samples contained multiresidues. However, for the other fruits and vegetables there were no remarkable differences between domestic and imported samples.

Pesticides frequently found in each group were as follows; acetamiprid and carbendazim were in domestic vegetables and fruits, cypermethrin was in imported frozen vegetables, and imazalil and chlorpyrifos were in imported citrus fruits.

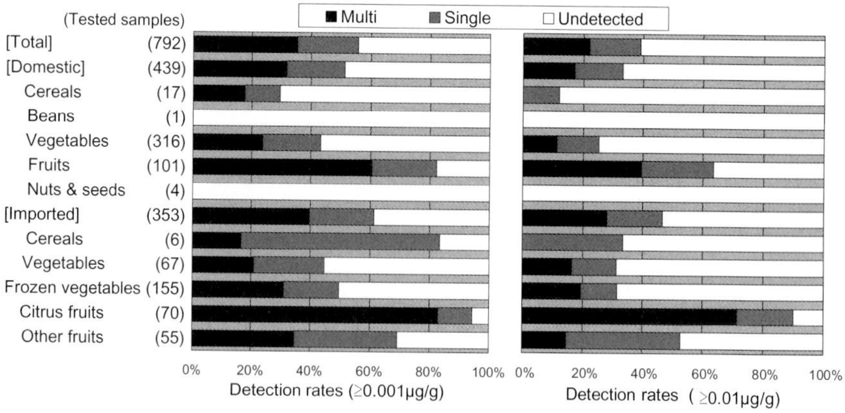

Figure 2. Detection rates of positive samples during FYs 2002–2005.

Among 792 samples, we found violations of the MRL in only 1 sample. Dieldrin was detected at 0.06 ppm in cucumber above MRL 0.02 ppm. But the violation rates were supposed to increase under the new legislation "Positive List", which were enforced on May 29, 2006. We compared the residue levels of our monitoring data with new MRLs and residues. Residue levels above MRLs were found in 4 domestic vegetables, 2 imported fruits, and 10 imported vegetables.

41.4 Conclusion

Using our multiresidue analytical method, 457 pesticides and 43 their metabolites can be extracted simultaneously and analyzed using GC/MS and LC/MS with the scan mode. Among them, 478 compounds showed recoveries between 50 and 140% and 496 compounds showed quantitation limits below 0.01 ppm.

A monitoring survey was conducted for about 200 agricultural products per year, and detection rates including trace levels were 56% in 792 samples monitored during a 4-year period. The ratio of violative samples against the Food Sanitation Law was less than 0.2%, but it is supposed to rise at about 2% under the new legislation "Positive List".

41.5 References

1 Ministry of Health, Labour and Welfare, Japan: Notification No.499 (29 November 2005).

2 Ministry of Health, Labour and Welfare, Japan: Notification No.497 (29 November 2005).

3 Y. Akiyama, M. Yano, T. Mitsuhashi, N. Takeda, M. Tsuji, *J. Food Hyg. Soc. Japan*, **1996**, *37*, 351–362.

4 Y. Akiyama, N. Yoshioka, M. Tsuji, *J. AOAC Int.*, **2002**, *85*, 692–703.

5 Y. Akiyama, N. Yoshioka, *Reviews in Food and Nutrition Toxicity*, **2003**, *1*, 400–444.

6 Y. Akiyama, N. Yoshioka, K. Ichihashi, *J. Food Hyg. Soc. Japan*, **2005**, *46*, 305–318.

7 Y. Akiyama, N. Yoshioka, M. Tsuji, *J. Food Hyg. Soc. Japan*, **1998**, *39*, 303–309.

Keywords

Pesticide Residues, Multiresidue Analysis, Monitoring, Agricultural Products, GC/MS, LC/MS, Positive List

VII
Environmental Safety

Pesticide Chemistry. Crop Protection, Public Health, Environmental Safety
Edited by Hideo Ohkawa, Hisashi Miyagawa, and Philip W. Lee

ISBN: 978-3-527-31663-2

42
Current EU Regulation in the Field of Ecotoxicology

Martin Streloke

42.1
Introduction

The whole regulatory process to reach authorizations of plant protection products within the European Union (EU) has a complicated structure and is therefore difficult to understand. The most important regulation is EU-directive 91/414/EEC [1] where all general items including the work-sharing between the EU-level and member states is described but also the data requirements and basic criteria for risk assessment and risk management. Recently a clear separation of risk assessment and management located even in different authorities was established on the EU level. As the criteria especially for standard risk assessments are well harmonized amongst OECD-member states (Organisation for Economic Cooperation and Development), there is only a need to give a short overview here. However, most interesting are those scientific issues which currently cause the biggest problems in the regulatory process and these examples will be discussed in more detail. Implementation of probabilistic risk assessments, endocrine disruption, persistent compounds in soil, refined risk assessment for birds and mammals, mesocosm studies, monitoring data or risk mitigation measures are such items.

42.2
Regulatory Process

As mentioned before, the regulatory evaluation of plant protection products in the EU is carried out by EU-authorities and member states together. The active substances are evaluated on the EU-level whereas the authorization of products falls under the responsibility of member states. At the EU-level, the European Food and Safety Authority (EFSA) organizes the risk assessment process in close connection with experts from member states and a committee of independent scientists mainly from academia. Based on the outcome of this risk assessment, the European Commission together with representatives from member states

Pesticide Chemistry. Crop Protection, Public Health, Environmental Safety
Edited by Hideo Ohkawa, Hisashi Miyagawa, and Philip W. Lee

ISBN: 978-3-527-31663-2

discuss problems of active substances in official management meetings and they are finally responsible for deciding upon the inclusion of actives into Annex I of Directive 91/414/EEC (risk management). Alongside with inclusion, special areas of concern and general risk management options are determined, which have to be recognized by member states. In principal, member states can only grant authorizations for products which contain actives of this positive list but transitional solutions are possible and, additionally, specific regulations for old compounds exist. One important issue to be dealt with on the member-state level is the setting of risk mitigation measures to protect, for example, aquatic life.

As regards the implementation of scientific progress into the risk assessment and management schemes, not only voluntary initiatives like SETAC-workshops (Society of Environmental Toxicology and Chemistry) but also the work of groups like FOCUS (Forum for the co-ordination of pesticide fate models and their use) have proven to be effective. Often experts from single member states, industry and/or academia initiated such projects. The outcome of these projects were picked up in EU-Guidance Documents (http://www.ec.europa.eu/food/plant/protection/evaluation/index_en.htm) which are not legally binding but well recognized within the regulatory community. However, these projects were slowed down over the last years due to discussions on responsibilities on the EU-level and reductions of staff in industry and regulatory authorities. Consequently, there exists usually not one official EU-opinion about a scientific problem. Therefore only trends of discussions are given here in this presentation and no official EU-statements.

42.3 Standard Risk Assessment

For most of the products currently registered within EU only conservative standard environmental risk assessments were conducted which showed that expected risk was acceptable. Basic data requirements according to Annex II (active substances) and Annex III (formulated products) for most important groups of organisms are [2]:

A complete data package for active substances must be available for nearly all intended uses and for most of the group whereas, for formulated products and metabolites, bridging studies are required. Only in cases where these substances are more toxic than the actives is a complete data package also required for these substances. For those groups where mainly tests with the formulated product are required, often no additional tests with actives are requested. Usually studies are conducted in accordance with OECD-guidelines.

For birds and mammals, estimated theoretical environmental concentrations (ETEs) are calculated for granivorous, herbivorous, and insectivorous organisms depending on their size and different groups of crops. Seed treatments are a special item. Based on a large survey of the relevant literature, food intake rates (FIR) and expected residues on feed items (RUD) for these different groups of organisms were collected and representative numbers were fixed in the relevant

Table I. Mainly standard data requirements for groups of non-target organisms according to Annex II and III of Directive 91/414/EEC (a.i. – active ingredient).

Group of organisms	Data requirement	Annotation
Birds and mammals	Acute, short-term, reproduction test	Mainly a.i. tests, data from mammalian toxicology are used
Aquatic organisms	Acute test fish (2x) and *Daphnia*, long-term/chronic test fish and *Daphnia*, algae, aquatic plant for herbicides, sediment organisms, bioaccumulation study	Mainly a.i. tests
Arthropods	Glass-plate test with *Aphidius* and *Typhlodromus*, bee tests	Mainly product tests
Soil organisms	Acute and reproduction test with earthworms, soil microflora studies	Mainly a.i. tests for earthworms but product tests for microflora
Terrestrial plants	Data from screening tests, for herbicides seedling emergence and/or vegetative vigour laboratory tests with 6 monocotyle and 6 dicotyle species	Mainly product tests

guidance document [3]. Other parameters like the fraction of diet obtained from the treated area (PT), the fraction of contaminated food type in diet (PD), or possible avoidance responses (AV) are not considered in the standard risk assessment due to its conservative nature. After having implemented this approach in 2002, it turned out that these assumptions are often too conservative and therefore improved methods for refined assessments have been developed to facilitate authorization of products.

Exposure concentrations for aquatic organisms are calculated for the Annex I listing on the EU-level in accordance with FOCUS-surface water models [4]. Whereas Step 1 is very conservative and nearly no compound fulfils the relevant requirements when using these PEC-values (Predicted Environmental Concentration), calculations with Step 2 tools lead to a more realistic exposure assessment. However, often even more realistic model calculations are needed for the 10 representative surface water scenarios of Step 3 to come to safe use within EU. Whereas in the first two steps, a 30-cm deep static waterbody is used, a few additional types are covered in Step 3. Usually only one PEC is calculated for one use and all relevant exposure routes together. In Step 4, even more specific scenarios have recently been made available which are also useful in connection with the setting of risk mitigation measures [5]. However, currently it is not clear to which extent these methods are used on the member state level or even within the EU-evaluations because the process of implementation has not been finalized so far. On the member state level, frequently different models, which had been developed much earlier than the FOCUS-proposals for calculating exposure concentrations, are used which fit better into the national schemes for setting risk mitigation measures. The exposure routes are separated; therefore, risk mitigation measures can be set specifically for single exposure routes and authorized uses.

For non-target arthropods, soil organisms and terrestrial plants, the exposure estimates in-crops are quite simple because an overspray situation is anticipated [6]. However, depending on the structure and height of the canopy, interception factors are considered to come to more realistic predictions. As regards the off-crop scenario, things are more complicated but in general the same spray-drift approach as for aquatic organisms is used. For non-target soil organisms, the amount which is predicted to reach the soil surface is considered to be equally distributed in the top layer of 5 cm.

Endpoints like mortality, growth, reproduction, etc., are regarded as important indicators to decide about possible effects on the sustainability of populations which is in general the main protection goal. Especially when using data from mammalian toxicology for wildlife assessments but, also in connection with endocrine disruption, the relevance of data for biomarkers is difficult to evaluate.

In accordance with the requirements of Annex VI of 91/414/EEC [7], where the basic principles for decision-making are laid down, Toxicity/Exposure Ratios (TER) are to be calculated. Uncertainty factors of 10 (chronic risk) and 100 (acute risk) must be applied for aquatic organisms. For terrestrial organisms, uncertainty factors of 5 and 10 are to be used, respectively. Different approaches exist for the in-crop area and non-target arthropods and bees in general. Uncertainty arises mainly from the fact that only for a few representative species toxicity data are available. If these trigger values are not breached, a listing of an active substance on Annex I or an authorization of a formulated product respectively are possible.

If predicted exposure is higher than toxicity (including the relevant uncertainty factor) for the most sensitive species and endpoint, an unacceptable risk is expected but a refinement of the assessment is possible (famous “unless clauses” of Annex VI). Another frequently used option is to set risk mitigation measures like buffer zones to protect aquatic life but also arthropods and plants.

42.4
Refined Risk Assessments

There are a lot of options for refined assessments on the exposure and effect side available. Typically, studies with a more realistic exposure regime and/or more species/endpoints and/or a longer test duration are conducted. Taking the specific fate properties of active substances or exposure conditions for species at highest risk into consideration are options on the effects side. Below important methods are briefly explained.

42.4.1
Refined Risk Assessments for Birds and Mammals

Currently refined options, especially in the case of chronic evaluations, are under discussion because the relevant trigger value has frequently been breached leading to an interruption of the process of Annex I inclusion. Options for refinement are to use more appropriate (focal) species for the special uses in order to adjust important parameters for exposure assessments. Over the last years, more information about the biology of single species has been collected and is ready for use. Recently, a database on more realistic residue values for insects as prey became available. Furthermore, industry is conducting large field trials in Europe to collect data to refine PT- and PD-values and the results are implemented continuously into the registration procedure. As regards mammals, there is room to reduce uncertainty factors because the tested species are only representative for a few species dwelling in the agricultural landscape.

42.4.2
Persistent Compounds in Soil

Especially in the case of fungal diseases, products must persist on crops for a while to become effective. Unfortunately they persist not only on crops but also in soil. The more persistent a compound, the higher is the degree of uncertainty for the risk prediction. Furthermore long-lasting residues in soil may cause problems if fields would be needed for other uses like agriculture. Recently an excellent Dutch paper was made available dealing with these topics [8]. Additionally, an EU-Guidance document was prepared in 2000 [9] and the SETAC-workshop EPFES [10] was held. If compounds are moderately persistent (DT_{90} > 100 days), additional reproduction tests with earthworms, collembola and soil mites and a litter bag test are often required. The latter one must be submitted in any case if DT_{90} > 1 year. These additional data are required to reduce uncertainty of the risk assessment.

42.4.3
Use of Probabilistic Risk Assessment Methods for Regulatory Purposes

Over the last 10 years a lot of efforts have been made to implement probabilistic methods for environmental risk assessments (ERA) into regulatory decision-making schemes. ECOFRAME in the US (www.epa.gov/oppefed1/ecorisk) was the first and largest project in this area, the EU-project EUFRAM (www.eufram.com) is a comparable initiative. Whereas on the effect side, several examples exist where species sensitivity distributions (SSDs) were used for regulatory decision-making, comparable cases on the exposure side are rare.

The Netherlands started to use probabilistic methods for regulatory purposes (HC5-method, Hazardous Concentration) [11]. As regards aquatic organisms, SSDs have been used to refine risk assessments if standard risk assessments indicated an unacceptable risk. At the HARAP-workshop [12], the use of probabilistic methods was discussed in detail. In general, at least toxicity data for eight species should be available when generating an SSD. However, as sensitivity is less variable especially for vertebrates and animal welfare more important, a lower number should be acceptable. Due to the high number of test results for different terrestrial plant species, the SSD-approach is frequently used in this area. The distribution of toxicity values should be normal distributed. Therefore it is usually not possible to put toxicity values of all groups of aquatic organisms together because especially herbicides and insecticides affect mainly plants or invertebrates. As for long-term and chronic testing, only a small number of aquatic test species and guidelines are available, the approach was mainly used to refine acute risk assessments. On the basis of comparisons with data from mesocosm studies, it was shown that when using SSDs, a reasonable protection level is kept [13]. Usual problems when using this approach for regulatory purposes are to decide about the most appropriate uncertainty factor and the number of species to be tested. Often the HC5 is used for decision-making but this could be a problem for new test methods due to considerable variability of experimental data and consequently high standard deviations around this value.

Probabilistic exposure assessments are more complicated to conduct. Some examples for the main exposure routes, spray drift, runoff and drainage, were presented in a workshop in Berlin in 2003 [14]. Whereas in the well-known atrazine case [15], partly monitoring data were used to generate a curve of exposure concentrations, later on a geographical analysis of the agricultural landscape has been used to predict exposure for single segments of waterbodies like streams, ditches, brooks etc. from nearby fields. This type of analysis was applied to pesticide uses in cotton in the US [16] but subsequently also for pyrethroids in arable crops. In Germany, large projects are under way to develop methods for conducting probabilistic exposure assessments for spray drift in different crops [17].

When using this approach, it is important that digitalized geographical data with high resolution are available. Determination of distances between crop and waterbody are simple to conduct in the case of permanent crops like orchards or grapes. However, for annual crops the situation is more challenging. Also, important data for exposure assessment like hydrological properties of waterbodies or drift reducing properties of riparian vegetation are difficult to implement because appropriate Geographic Information Systems (GIS) are usually not available. It is also difficult to decide about the area for which such an assessment should be conducted. Analysing high resolution aerial images (HR-data) for all oilseed rape fields in Germany, for example, would be much too expensive and time consuming. Therefore representative areas should be identified but criteria for doing so are difficult to find. Within such areas it must be decided whether all waterbodies should be included or only those stretches of running waters which are located directly adjacent to the relevant crop and therefore contamination is likely. Implementation of uncontaminated stretches would increase the number of zero and very low single PEC-values and finally lowering the critical percentile from the distribution curve. However, taking these exposure concentrations into consideration is a reasonable and pragmatic method to take recolonization processes indirectly into account.

With respect to decision-making, often the 90th percentile from the exposure distribution is used because there are still several conservative assumptions within the assessments. However, a reasonable decision about the percentile to be taken should consider the variability of data and uncertainties in a "weight of evidence" approach. When communicating the outcome of probabilistic assessments to non-experts, there are sometimes concerns because in theory a "slight risk" will be left in any case. This is also relevant for deterministic approaches but there this risk is hidden (e.g., 90th percentile of spray-drift residues). Therefore at least in Germany, the implementation of probabilistic approaches should be accompanied by monitoring studies to show that the legally required protection level is kept also when using this new assessment method. Probabilistic exposure assessments have been used in Germany for regulatory decision-making.

42.4.4 Microcosm/Mesocosm Testing with Aquatic Organisms

Parallel to the development of probabilistic methods, the number of submitted higher tier tests has clearly increased and they were used for regulatory purposes. Test methods were discussed at the HARAP and especially on the CLASSIC-workshop [18]. An OECD-guidance document was also prepared [19]. No clear definitions exist but microcosms could be single species laboratory tests with a more realistic exposure regime but also larger systems in glass-houses containing whole communities. Mesocosms are most often outdoor experimental ponds with a volume around 3–5 m^3. To ease a conclusive interpretation of the results, it is usually better to use this size together with more replicates instead of conducting tests in much larger systems under realistic exposure conditions. The exposure regime should not be too specific in order to facilitate the use of data for different types of uses. Due to a more realistic exposure regime, higher toxicity values – indicating lower toxicity – are often derived from these studies. In microcosm studies, this item is usually covered. Furthermore, recovery could be investigated in larger systems. If no additional species than the standard set are tested a reduction of the uncertainty factor is not possible. This increased number of species is typical for mesocosms where whole communities are tested and recovery could be investigated over a longer period. However, often clear statistical interpretations are only possible for a few species whereas for a larger number trends can clearly be identified. As uncertainty decreases if more species are tested, the safety factor should normally be reduced especially in the case of mesocosm studies.

42.4.5
Data from Monitoring Studies

Risk assessment methods have been improved considerably over the last decade but at the same time new groups of organisms were considered to be at risk and additional endpoints were regarded as important. Risk could be over- or underestimated or even overlooked. Overall this shows that quality of risk predictions must be checked continuously. From monitoring studies, rough estimates could be derived to determine the quality of risk assessment schemes. This issue was discussed on the EPIF-workshop [20] and recommendations were proposed. Within the EU, residues of pesticides and other types of chemicals in important surface waters are monitored routinely under a special water regulation. Based on toxicity data, quality standards are set to enable evaluation of measured residues. However, sampling technique and exposure regime in the relevant toxicity study used for setting a quality standard must fit with real type of exposure in waterbodies. There are two main types of monitoring which may be connected: Chemical and biological monitoring. When using only data from chemical monitoring and residues are below the quality standard, it is very likely that no effects have occurred. If measured concentrations are not clearly higher, it is difficult to decide whether aquatic organisms were really affected. On the other hand, only with data from chemical monitoring, synergism of effects for example cannot be predicted. Therefore, data from biological monitoring help to overcome all these problems. With biological data only, it is difficult to identify clear cause/effect relationships. Control sites are needed but often difficult to find. Improved statistical methods may help to solve this problem. Ideally, both methods should be used together in one study. In any case, the interpretation of data is difficult. However, there is evidence available from monitoring studies that effects on populations occurred at least when products were used directly adjacent to non-target areas.

42.4.6
Endocrine Disruption

Public interest in effects to the hormone system, as a consequence of the use of plant protection products, is high. There is no clear guidance available how to handle this problem for regulatory purposes, at least not in the EU. Like bioaccumulation and persistence, these types of effects make risk predictions more uncertain. Sublethal endpoints like reproduction become more important because long-term repercussions of populations, especially of fish, must be avoided. Chronic tests for primary producers like algae and invertebrates are usually available. For active substances of plant protection products, usually at least a 28-day fish test is available and growth as sublethal endpoint is covered. If the very early life stages are additionally tested (ELS-test), even more information as regarding effects on the sustainability of populations becomes available. However, often full-life-cycle tests with fish have been required in case findings in mammalian toxicology indicated effects to the hormone system. Rough comparisons with the aforementioned shorter tests revealed that the latter method is often not clearly more sensitive. There is always a discussion about the relevance of data for biomarkers like enzyme activities or hormone titers in the blood for decision-making if data on population-relevant endpoints are available. Occasionally, if thyroidal effects were expected, tests with the claw frog *(Xenopus laevis)* were required. An OECD-guideline for such tests is in an advanced developmental stage. A less advanced test method for a life-cycle-test with midges (*Chironomus riparius*) should be available in due course. Overall there are a considerable number of test methods available to investigate endocrine disruption and toxicity values are used for regulatory risk assessments.

42.5 Risk Mitigation Measures

Even if refined risk assessments have been conducted, there is often a need to mitigate risk to non-target life with appropriate measures to an acceptable level. Furthermore, these measures can be used instead of sometimes expensive refinement methods. Attempts have been made to collect risk mitigation measures currently used in EU member states [5] and approaches were discussed [21]. Annex V of Directive 91/414/EEC [22] contain S-phrases which might be used on labels by member states when authorising products. Unsprayed buffer zones are used as a risk mitigation measure in several member states mainly to protect aquatic life. However, it is difficult to harmonize mitigation measures because there are still clear differences in agricultural practice between member states. The availability of spray-drift reducing machinery is for example different. The same is true for the legal framework to enforce restrictions.

Over the last decade in Germany, mainly restrictions to protect aquatic life from exposure via spray drift have been set. Approaches were changed several times and no-spray buffer zone distances of up to 150 m were used in the past but currently 20 m is the maximum. Width of buffer zones depends on the drift-reducing properties of the application technique. The machinery is classified in an official list depending on the measured degree of drift reduction. A comparable approach with a maximum width of 5 m is used to protect non-target arthropods and plants in the off-crop area (hedgerows, grass strips, etc.). Regarding runoff, grassed buffer strips and conservation tillage are used as measures to mitigate risk for aquatic life. State authorities are responsible for enforcement of restrictions and offences may be punished by a fine of up to 50000 €. However, enforcement is sometimes really challenging because it is, for example, difficult to come across exactly in time when a farmer is applying a plant protection product.

42.6 Conclusions

Over the last 15 years a good deal of progress has been made to harmonize the principles of environmental risk assessment within the European Union. Basic data requirements in the area of ecotoxicology and standard risk assessment schemes were outlined in different regulations. However, there are several items like exposure assessment for birds and mammals, persistent compounds, probabilistic risk assessments, aquatic higher tier tests, monitoring data, endocrine disruption, and risk mitigation measures which are currently under discussion. Research projects are under way, workshops were organized, and other reasonable attempts have been made to come to science-based but, at the same time from a regulatory point of view, useful solutions for these highly sophisticated problems. New methods were implemented in EU-guidance documents. Other tools have partly been used on a case-by-case basis or in single member states, although official implementation on the EU-level is still pending. This will be discussed during the process of revision of Directive 91/414/EEC and its annexes, which is currently under way.

42.7 References

1 EUROPEAN COMMISSION, Council Directive 91/414/EEC of 15 July 1991 Concerning the Placing of Plant Protection Products on the Market (*Official Journal of the European Communities L 230*, 19.08.91, p. 1).

2 EUROPEAN COMMISSION, Annexes II and III, Sections 8 and 10, respectively, "Ecotoxicology" Commission Directive 96/12/EC of 8 March 1996 Amending Council Directive 91/414/EEC Concerning the Placing of Plant Products on the Market (*Official Journal of the European Communities L65*, 15.03.96, p. 20).

3 EUROPEAN COMMISSION, Directorate E – Food Safety, Guidance Document on Risk Assessment for Birds and Mammals Under Council Directive 91/414/EEC, SANCO 4145/2000, 25 September 2002.

4 FOCUS, FOCUS Surface Water Scenarios in the EU Evaluation Process Under 91/414/EEC. *Report of the FOCUS Working Group on Surface Water Scenarios*, EC Document Reference SANCO/4802/2001-rev. 2 final (**2003**). http://viso.jrc.it/focus/

5 FOCUS, Landscape and Mitigation Factors in Aquatic Ecological Risk Assessment. *Report of the FOCUS Working Group on Landscape and Mitigation Factors in Ecological Risk Assessment*, draft 18 June 2004. http://viso.jrc.it/focus/.

6 EUROPEAN COMMISSION, Council Directive, 91/414/EE – Guidance Document on Terrestrial Ecotoxicology SANCO/10329/2002 rev 2 final, 17 October 2002.

7 EUROPEAN COMMISSION, Council Directive, 97/57/EC of 22 September 1997 – Establishing Annex VI to Directive 91/414/EEC, Concerning the Placing of Plant Protection Products on the Market, *Official Journal of the European Communities*, 27 September 1997.

8 A. M. A. Van der Linden, J. J. T. I. Boesten, T. C. M. Brock, G. M. A. Van Eekelen, F. M. W. de Jong, M. Leistra, M. H. M. M. Montforts,

J. W. Pol, *Persistence of Plant Protection Products in Soil; a Proposal for Risk Assessment*, **2006**, 105 pp.

9 EUROPEAN COMMISSION, Directorate General for Agriculture, Guidance Document on Persistence in Soil 9188/VI/97 rev. 8, 12.07.2000.

10 J. Römbke, F. Heimbach, S. Hoy, C. Kula, J. Scott-Fordmand, P. Sousa, G. Stephenson, J. Week, *Effects of Plant Protection Products on Functional Endpoints in Soil (EPFES)*, SETAC, Lisbon, 24–26 April 2002.

11 T. Aldenberg, W. Slob, Confidence Limits for Hazardous Concentrations Based on Logistically Distributed NOEC Toxicity Data, *Ecotoxicol Environ Saf.*, **1993**, *25*, 48–63.

12 P. J. Campbell, D. J. S. Arnold, T. C. M. Brock, N. J. Grandy, W. Heger, F. Heimbach, S. J. Maund, M. Streloke, *Guidance Document on Higher-Tier Aquatic Risk Assessment for Pesticides (HARAP)*, France, 19–22 April 1988.

13 P. J. Van den Brink, L. Posthuma, T. C. M. Brock, The Value of the Species Sensitivity Distribution Concept for Predicting Field Effects: (Non-)Confirmation of the Concept Using Semi-Field Experiments. In *Use of Species Sensitivity Distributions in Ecotoxicology*, L. Posthuma, T. P. Traas, G. W. Suter (Eds.), Lewis Publishers, Boca Ration, Fla., USA,, **2001**, pp. 155–193.

14 A. W. Klein, F. Dechet, M. Streloke, UBA/IVA/BVL, *Probabilistic Assessment Methods for Risk Analysis in the Framework of Plant Protection Products Authorization Berlin*, 25–28 November 2003.

15 K. R. Solomon, D. B. Baker, R. P. Richards, K. R. Dixon, S. J. Klaine, T. W. La Point, R. J. Kendall, C. P. Weisskopf, *et al.*, Ecological Risk Assessment of Atrazine in North American Surface Waters, *Environ Toxicol. Chem.*, **1996**, *15*, 31–76.

16 S. J. Maund, K Z. Travis, P. Hendley, J. M. Giddings, K. R. Solomon, Probabilistic Risk Assessment of Cotton Pyrethroids: V. Combining Landscape-Level Exposures and Ecotoxicological Effects Data to Characterize Risks, *Environmental Toxicology and Chemistry*, **2001**, *20*, 687–692.

17 B. Golla, S. Enzian, V. Gutsche, *First Results of a Probabilistic Pesticide Exposure Analysis in Germany*, Poster Presentation at SETAC Europe 16th Annual Meeting, 8.–11.5.2006, The Hague.
M. Trapp, G. Tintrup, R. Kubiak, *Investigation of Methods and Geo-Databases for a Refined Probabilistic Exposure Assessment on a National Scale*, Poster Presentation at SETAC Europe 16th Annual Meeting, 8.–11.5.2006, The Hague.
T. Schad, *Introduction of Generic Landscape Characteristics in Probabilistic Aquatic Exposure and Risk Assessment – A Project of the German Crop Protection Association (IVA)*, SETAC Europe 16th Annual Meeting, 8.–11.5.2006, The Hague.

18 J. M. Giddings, T. C. M. Brock, W. Heger, F. Heimbach, S. J. Maund, S. M. Norman, H. T. Ratte, C. Schäfers, *et al.*, *SETAC, Community-Level Aquatic System Studies: Interpretation Criteria*, Proceedings from the CLASSIC Workshop, **1999**.

19 OECD Organisation for Economic Co-operation and Development, Guidance Document on Simulated Freshwater Lentic Field Tests (Outdoor Microcosms and Mesocosms), *ENV/JM/MONO*, **2006**, *17*.

20 M. Liess, C. Brown, P. Dohmen, S. Duquesne, A. Hart, F. Heimbach, J. Kreuger, L. Lagadic, *et al.*, EU & SETAC Europe Workshop, *Effects of Pesticides in the Field (EPIF)*, **2003**, 136 p., Le Croisic, France.

21 R. Forster, M., Streloke, *Workshop on Risk Assessment and Risk Migitation in the Context of the Authorization of Plant Protection Products (WORMM)*, Braunschweig, Germany, **1999**.

22 EUROPEAN COMMISSION, Commission Directive 2003/82/EC – of Commission Amending Council Directive 91/414/EEC as Regards Standard Phrases for Special Risks and Safety Precautions for Plant Protection Products, SANCO/10376/2002 rev 9.

Keywords

Environmental Risk Assessment, European Union, Ecotoxicology, Data Requirements, Directive 91/414/EEC

43
A State of the Art of Testing Methods for Endocrine Disrupting Chemicals in Fish and Daphnids

Satoshi Hagino

43.1
Introduction

In the past decade, various scientific knowledge and skills had been accumulated to detect endocrine effects on wildlife. Some authorities and organizations are making efforts to establish testing methods for endocrine disrupting chemicals (EDCs) in fish and other wildlife. In the present paper, a state of the art of testing methods for EDCs is introduced to discuss advantages and/or disadvantages of fish testing methods, from the viewpoints of endpoints, species including S-rR strain medaka for sex and thyroid hormones, as well as effects and significance of juvenile hormone (JH) mimics on daphnids, etc.

43.2
Fish Testing Methods for Sex Hormones

Regarding fish, two expert meetings on testing for endocrine disruption took place in London (EDF1, 1998) [1] and Tokyo (EDF2, 2000) [2] by the Organization for Economic Co-operation and Development (OECD). At this moment, the several tests were identified as candidates for the testing methods for EDCs in fish and categorized in three tiers to take a step-by-step approach. Tier 1 screening includes a fish prolonged toxicity test (enhanced OECD TG 204) or fish juvenile growth test (OECD TG 215), fish gonadal recrudescence assay [proposal by Environmental Protection Agency of United States (US EPA)], adult terminal reproductive assay (US EPA proposal) and sex reversal assay (our proposal), Tier 2 testing includes enhanced fish early life stage test (partial life-cycle test, PLC, based on OECD TG 210) and terminal reproductive test (partial life-cycle test), and Tier 3 testing includes fish full life-cycle test (FLC, based on US EPA, OPPTS 850.1500). Thereafter, the OECD conducted two ring tests as the Tier 1 screening, i.e., a 21-day adult toxicity test and a 21-day reproductive test to standardize and validate their protocols and the results were reported this January. Collaterally,

Pesticide Chemistry. Crop Protection, Public Health, Environmental Safety
Edited by Hideo Ohkawa, Hisashi Miyagawa, and Philip W. Lee

ISBN: 978-3-527-31663-2

Table I. Comparative chart of fish test designs for endocrine-disrupting chemicals.

Level	Test type	Exposure duration	Endpoints examined
Tier 1 Screening	Prolonged toxicity test (OECD 204)	Juvenile → 14 d (or 28 d)	VTG, GSI, gonadal histology
	Gonadal recrudescence assay	Mature → 21 d	GSI, secondary sexual characteristics, VTG
	Adult terminal reproductive assay	Mature → 21 d	Survival, growth, VTG, GSI, gonadal histology, Secondary sexual characteristics, embryo viability
	Sex reversal assay	Newly hatched → 42 dph*	Secondary sexual characteristics, gonadal histology
Tier 2 Testing	Enhanced early-life stage toxicity test (OECD 210) = Partial life-cycle test	Fertilized egg → 60 dph	Hatching success, growth, VTG, secondary sexual characteristics, gonadal histology
	Terminal reproductive test = Partial life-cycle test**	Mature → 28 d	Time to 1st spawning, spawning frequency, no. of eggs, fertility, F1 hatching success
Tier 3 Testing	Full life-cycle test	P: Fertilized egg to adult reproduction; F1: Fertilized egg to young, 180 d in total	Hatching success, growth, VTG, fecundity, fertility, secondary sexual characteristics, gonadal histology

* days after post hatch; ** enhanced OECD 210 is considered to be PLC in the latest aspect

Japan Environment Agency decided to conduct three phases of the fish tests, i.e., 1) vitellogenin (VTG) induction assay, 2) PLC and 3) FLC, and carried out these tests using 28 substances within the framework of "Strategic Programs on Environmental Endocrine Disruptors '98 (SPEED '98)" [3]. These testing methods are summarized in Table I.

43.3
S-rR Strain Medaka and Sex Reversal Test

The major fish species used on these testing methods for EDCs are Japanese medaka (*Oryzias latipes*, Adrianichthyidae), fathead minnow (*Pimephales promelas*, Cyprinidae), and zebrafish (*Danio rerio*, Cyprinidae). Although these fish species are widely used for ecotoxicology, medaka is considered to be the best fish among

Table II. Conditions of sex-reversal assay.

Item	Condition
Test fish	Medaka (*Oryzias latipes*) S-rR strain
Duration of the test	exposure: 0–28 days after hatching; recovery: 28–42 days after hatching
Test concentrations	maximum concentration: 1/10 of LC_{50}
Volume	5 L
Number of fish used	60/concentration; decreasing to 20 males and 40 females
Food	*Artemia* nauplii
Water temperature	25°C
Photoperiod	16/8 hours light/dark condition
Endpoints	Secondary sexual characteristics (see Figure 2)

these species for using assay to determine endocrine-disrupting effects due to their distinguishableness of both genotypic and functional sex [4].

In some strains of medaka, genotypic sex can be identified by their body color. About 80 years ago, the special strain of medaka of which the body color was encoded into the gene linked to sex chromosome "X" or "Y" was discovered [5] and hence, females are always white and males orange-red because the small "r" is recessive and the large "R" dominant. It is well known that the d-rR strain medaka has been established at Nagoya University in the last half of the 1940s based on the knowledge [6]. Yamamoto conducted a series of investigation for sex differentiation of medaka using the d-rR strain. The S-rR strain medaka has been also established by the Sumika Technoservice Corporation in the same manner as Yamamoto did for d-rR and applied to our proposed sex reversal assay. Their original body color remains unchanged regardless of sex-reversal and the original genotypic sex can be distinguished by the color. An outline of the method is shown in Table II and Figure 1.

We examined the effects of strong estrogen (17β-estradiol = E2, ethinyl-estradiol = EE2 and diethylstilbestrol = DES), weak estrogen (4-t-pentylphenol = 4tPP), strong androgen (methyltestosterone = MT), anti-androgen (flutamide = Flu), and suspected EDCs (4-nonylphenol = NP, bisphenol A = BPA and di-2-ethylhexyl phthalate = DEHP) by the secondary sexual characteristics of medaka. The results showed that endpoints we selected were effective to a great extent to assess endocrine-disrupting actions by sex reversal because the endpoints measured were clearly different between male and female [7–9]. These experiments gave that the no-observed-effect concentrations (NOECs) of E2, EE2, DES, 4tPP, MT, Flu, NP, BPA, and DEHP resulted in 0.01, 0.01, 0.01, 1, < 0.01, < 1000, 10, 100, and 10000 μg/L, and the lowest-observed-effect concentrations (LOECs) of the chemicals were 0.032, 0.032, 0.032, 10, 0.01, > 1000, 100, 1000, and > 10000 μg/L, respectively (Figure 2). One of the most remarkable points would

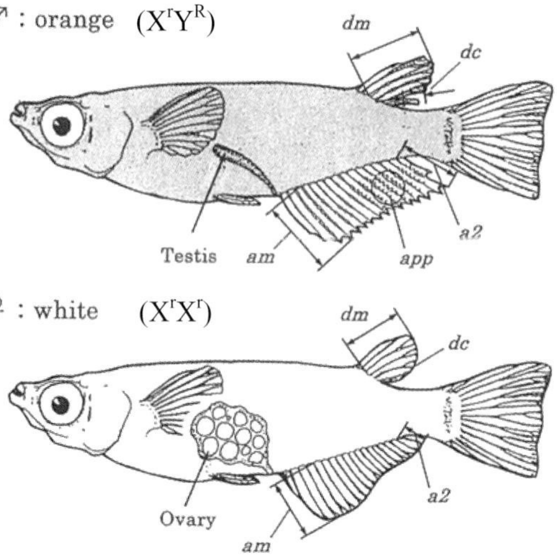

Figure 1. Endpoints of sex reversal assay (Hagino *et al.* [7]).

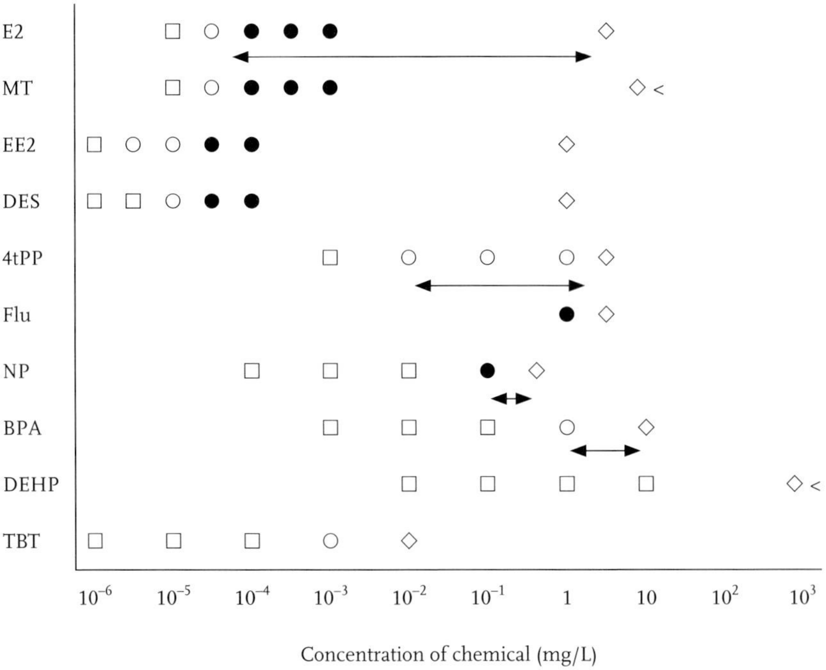

Figure 2. Summary diagram of effect concentrations of the chemicals selected on sex reversal.
□: normal, ○: partial sex-reversal, ●: complete sex-reversal, ◇: 96hr-LC_{50}
↔ range of LC_{50}/LOEC.

be the difference in these values on sex-reversal/acute toxicity between strong estrogen (or androgen) and weak estrogen (or suspected EDCs). 4tPP, NP, and BPA were recognized to have weak estrogenic effects to fish because the LOEC values were much closer to their LC_{50} values, being the ratio of only 260, 6.8, and 10, respectively, while the ratio of more than 43000 for strong estrogen (or androgen). This means that the suspected endocrine-disrupting effects are not so severe ("hormonal") as thought since the ratios of most chemicals between general long-term toxicity and acute toxicity are experientially 10 to 100. In the case of anti-androgen, Flu, the effects were measurable in combination with androgen. Although MT induced entire masculinization of female medaka at 0.1 μg/L, the combination of 0.1 μg MT/L and 1 mg Flt/L did not induce any changes in females. Based on these results, the assay is useful in determining not only estrogenic and androgenic effects, but also anti-androgenic and anti-estrogenic effects of certain chemicals.

43.4
Effects of Pesticides Listed in SPEED '98

Within the framework of SPEED '98, Japan Environment Agency conducted PLC and/or FLC with 28 of 67 chemicals listed [3]. The results showed that only 3 chemicals, i.e., 4-nonylphenol, 4-t-octylphenol, and bisphenol A were suggested as EDCs. We also conducted sex reversal assay or PLC test (56-day exposure) to evaluate the effects of some pesticides (malathion, benomyl, cypermethrin, permethrin, esfenvalerate, fenvalerate) listed in SPEED '98. The results showed that these pesticides gave no endocrine-disrupting effects even at one-seventh to one-twentieth of the acute toxicity values (Table III). The effect of benomyl on hatchability is likely to be caused by inhibition of cell division. Based on these findings, it is unlikely that these chemicals listed affected as EDCs.

43.5
Advantages and Disadvantages of the Endpoints Selected

The OECD selected (1) gross morphology (including determination of the gonado-somatic index (GSI) and appearance and disappearance of secondary sexual characters), (2) vitellogenin (VTG), and (3) gonad histology as the core endpoints for endocrine disruption. Based on the results from many authors, GSI is not so an effective endpoint due to the low sensitivity. Since VTG is induced only in mature females in principle, this endpoint is only applicable to estrogenic action. Also, there are many problems in VTG, i.e., different measured values between the determination method or tissues selected, existence of normal induction in male, etc. Although VTG has an ability to screen the suspected endocrine disrupting chemicals even now, it is necessary to re-consider much more biological significance. The current histological method has the disadvantage, i.e., oversight

Table III. Evaluation on the endocrine-disrupting effects of malathion, benomyl, cypermethrin, permethrin, esfenvalerate, and fenvalerate.

Pesticide	**Acute toxicity**	**Sex reversal assay, PLC* or VTG induction assay****		
	96hr-LC_{50} (μg/L)	**Concentration tested (μg/L)**	**Endpoints and effects (+: positive, –: negative)**	**LC_{50}/LOEC ratio**
Malathion	4200	20	*dm/tl* (–), *dc/dm* (–), *am/tl* (–), *a2/tl* (–), *app* (–)	< 210
		50	*am/tl* (–)*, *app* (–)*, testis-ova (–)*, fecundity (–)*, fertility (–)*	< 84
		200	*dm/tl* (–), *dc/dm* (–), *am/tl* (–)*, *a2/tl* (–), *app* (–)*, testis-ova (–)*, fecundity (–)*, fertility (–)*	< 21
Benomyl	2300	10	Hatchability (–)*, VTG (–)**	< 230
		20	Hatchability (–)*, *am/tl* (–)*, *app* (–)*, testis-ova (–)*, fecundity (–)*, fertility (–)*	< 84
		30	Hatchability (+)*, *am/tl* (–)*, *app* (–)*, testis-ova (–)*, fecundity (–)*, fertility (–)*	< 84
		100	VTG (–)**	< 23
Cypermethrin	24	0.2	*dm/tl* (–), *dc/dm* (–), *am/tl* (–), *a2/tl* (–), *app* (–)	< 120
		2	*dm/tl* (–), *dc/dm* (–), *am/tl* (–), *a2/tl* (–), *app* (–)	< 12
Permethrin	75	0.5	dm/tl (–), dc/dm (–), am/tl (–), a2/tl (–), app (–)	< 150
		1	*am/tl* (–)*, *app* (–)*, VTG (–)*, testis-ova (–)*, fecundity (–)*, fertility (–)*	< 75
		5	*dm/tl* (–), *dc/dm* (–), *am/tl* (–), *a2/tl* (–), *app* (–)	< 15
		10	*am/tl* (–)*, *app* (–)*, VTG (–)*, testis-ova (–)*, fecundity (–)*, fertility (–)*	< 7.5
Esfenvalerate	4.2	0.02	*dm/tl* (–), *dc/dm* (–), *am/tl* (–), *a2/tl* (–), *app* (–)	< 210
		0.2	*dm/tl* (–), *dc/dm* (–), *am/tl* (–), *a2/tl* (–), *app* (–)	< 21
Fenvalerate	11	0.1	*dm/tl* (–), *dc/dm* (–), *am/tl* (–), *a2/tl* (–), *app* (–)	< 110
		1	*dm/tl* (–), *dc/dm* (–), *am/tl* (–), *a2/tl* (–), *app* (–)	< 11

in the testis-ova due to the limited observation of the testis region. A quantitative evaluation technique, called the fragmented method, we proposed would be helpful in solving this problem [10]. In conclusion, many secondary sexual characters are supposed to be the most useful endpoint to determine endocrine disrupting effects.

43.6
Consistency of the Results Obtained Between Sex Reversal Assay, PLC, and FLC

Prolonged toxicity tests, gonadal recrudescence assays, adult terminal reproductive assays and terminal reproductive assays are inadequate since three core endpoints are not subject to these testing methods or it is impossible to induce these effects during the set-up periods. On the other hand, all authorities and organizations positioned FLC as the definitive test for EDCs in sex hormones although the test needs a very long period (180 days). We compared remaining testing methods, i.e., sex reversal assay (or PLC) and FLC from the viewpoints of effectiveness and manpower. Data from both testing methods showed that the sensitivities of fish were overlapped and no strengthening was observed with test duration and/or fish generation (Table IV) [4, 7, 9, 11–12]. The former test is considered to be able to select true endpoints concerning endocrine disruption although much more endpoints can be evaluated in the latter test. These results suggest that sex reversal test is very useful to evaluate the effects of EDCs without waste of money, skills and time.

Table IV. Summary of LOEC values determined for various endpoints in sex reversal assay and FLC test with 17β-estradiol, ethinylestradiol, methyltestosterone, 4-nonylphenol, and bisphenol A.

Chemical	LOEC (μg/L) and the endpoint (in parenthesis)		
	Sex reversal assay	Full life cycle	
17β-Estradiol	0.032 [7] (sex differentiation)	> 0.0094 [4] (sex differentiation in F1)	0.016 [12] (sex differentiation and testis-ova in F1–F3)
Ethinylestradiol	0.032 [7] (sex differentiation)	> 0.0101 [4] (sex differentiation in F1)	
Methyltestosterone	0.01 [7] (sex differentiation)	0.00998 [11] (sex differentiation in F1)	
4-Nonylphenol	10 [9] (sex differentiation)	8.2 [4] (sex differentiation in F1)	16 [12] (testis-ova in F1–F3)

43.7
Development of Test Method for Thyroid Hormone

Much research has been conducted on sex hormones, namely estrogenic chemicals in fish. In other hormones including thyroid hormones (THs), however, there is little knowledge about fish under the present circumstances. THs, however, are associated with sex hormones and growth hormones through the intermediary of feedback mechanisms. There is strong evidence that THs accelerate the metamorphosis of flounder and other fishes [13–14]. Furthermore, the latest knowledge shows that THs regulate not only metamorphosis but also early development [15]. We have established the standard testing method of THs using early life stages of medaka, and have determined the most appropriate morphological endpoints and duration of exposure. For example, abdominal fin folds in TH groups disappeared earlier than those of the control and the differences between both groups maximized on day 7–15 (Figure 3). Scales in TH groups began to differentiate on day 9, i.e., approximately a 4-day acceleration over the control group, and the differences in diameter, the number of ridges and the number of rows were apparent up to 15 days after hatching [16]. These findings suggested that thyroid hormones accelerated metamorphosis of medaka, and the observation of these parameters were effective to detect TH actions.

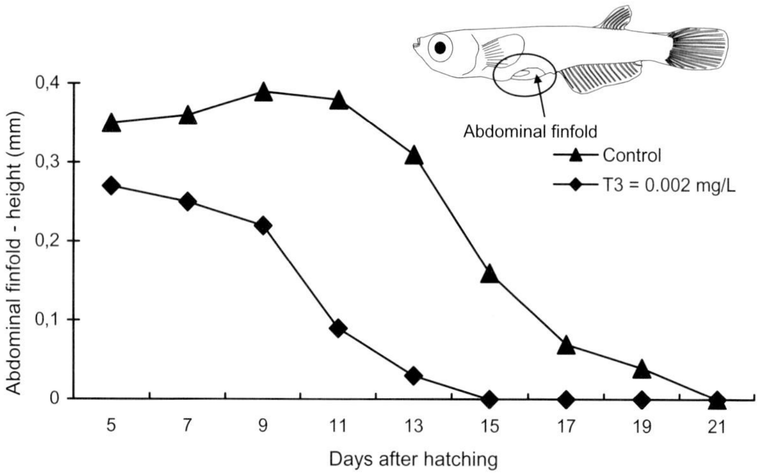

Figure 3. Effects of triioidothyronin (T3) on the development of abdominal finfold in medaka.

43.8
Endocrine Disrupting Effect of JH Mimics to Daphnids

In contrast to vertebrates, knowledge of the hormonal systems in arthropods is quite limited and only two hormones are investigated. Ecdysone acts as a

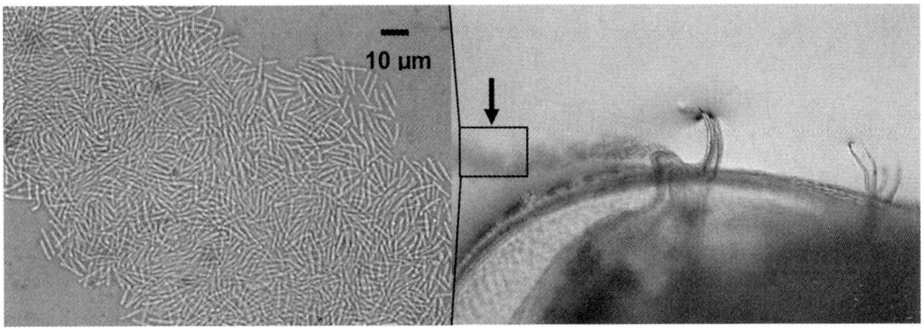

Figure 4. Photographs of male *D. magna* releasing sperms.

molting hormone and JH (including some types) acts as a "rejuvenating drug" in insects. Methyl farnesoate, a precursor of JH is supposed to be a sex hormone in crustaceans. Some researchers discovered that daphnids exposed to JHs and their mimics produce male neonates even though their reproduction phase is parthenogenesis and they produce only females under laboratory conditions [17–18]. Based on these findings, an enhanced daphnid reproduction test (revised OECD Guideline 211) was proposed by the National Institute for Environmental Studies, Japan [19]. We traced the test with pyriproxyfen and the results were comparable. After transfer to chemical-free water, however, the female daphnids reproduced female neonates within a few days. Additionally, males from a mother exposed developed and matured normally, and active spermatogenesis was observed (Figure 4). Hence, it is questionable that JH mimics act as an endocrine disruptor because male daphnids are produced under field conditions to maintain its species due to food deficiency, low temperatures, short durations of sunshine, etc. The production of male daphnids would be an adaptation to environmental variations and contributed to avoiding adverse conditions, and to deleting genetic disorders. Of course, it would not be ready to generalize that JH mimics are sex hormones of crustaceans based on the reproductive particularity of daphnids among them.

43.9 Conclusion

In the case of fish, although major testing methods have been proposed in relation to sex hormones, there would be plenty of scope for the selection of biologically relevant endpoints. And the research for effects of the other hormones would be the next step. Medaka methods have been comprehensively examined and found to be one of the most excellent protocols. In the case of crustaceans, since it is unclear whether the effects detected are endocrine disrupting or not on daphnids or other aquatic arthropods species, the clarification of fundamental hormonal systems would be imperative.

43.10 References

1 OECD Test Guidelines Programme, *Report from the OECD Expert Consultation on Testing in Fish*, London, 28–29 October 1998, **1999**.
2 OECD Test Guidelines Programme, *Record from the 2nd OECD Expert Consultation on Testing in Fish (EDF2)*, Tokyo, 15–16 March 2000, **2000**.
3 Ministry of Environment, Japan, SPEED '98/*JEA Strategic Programs on Environmental Endocrine Disruptors '98*, **1998**.
4 T. H. Hutchinson, H. Yokota, S. Hagino K. Ozato, *Pure Appl. Chem.*, **2003**, *75*, 2343–2353.
5 T. Aida, *Genetics*, **1921**, *6*, 554–573.
6 T. Yamamoto, *J. Exp. Zool.*, **1953**, *123*, 571–594.
7 S. Hagino, M. Kagoshima, S. Ashida, *Environ. Sci.*, **2001**, *8*, 75–87.
8 S. Hagino, M. Kagoshima, S. Ashida, S. Hosokawa, *Environ. Sci.*, **2002**, *9*, 475–482.
9 S. Hagino, M. Kagoshima, S. Ashida, SETAC *20th Annual Meeting Abstract Book*, **1999**, 59.
10 B.-L. Lin, S. Hagino, M. Kagoshima, S. Ashida, T. Iwamatsu, A. Tokai, K. Yoshida, Y. Yonezawa, *et al.*, *J. Japan Soc. Water Environ.*, **2003**, *26*, 725–730.
11 M. Seki, H. Yokota, H. Matsumoto, M. Maeda, H. Tadokoro, K. Kobayashi, *Environ. Tox. Chem.*, **2004**, *23*, 774–781.
12 B.-L. Lin, S. Hagino, M. Kagoshima, S. Ashida, T. Hara, T. Iwamatsu, A. Tokai, Y. Yonezawa, *et al.*, *J. Japan Soc. Water Environ.*, **2004**, *27*, 727–734.
13 Y. Inui, S. Miwa, *Gen. Comp. Endocrinol.*, **1985**, *60*, 1000–1001.
14 S. Miwa, Y. Inui, *Gen. Comp. Endocrinol.*, **1987**, *67*, 356–363.
15 N. Okada, T. Morita, M. Tanaka, M. Tagawa, *Fisheries Sciences*, **2005**, *71*, 107–114.
16 S. Hagino, M. Kagoshima, S. Ashida, Y. Takimoto, International Symposium on Standardization of Medaka Bioresources, *Abstract Book*, **2005**, 63.
17 A. W. Olmstead, G. A. LeBlanc, *Environ. Tox. Chem.*, **2000**, *19*, 2107–2113.
18 N. Tatarazako, S. Oda, H. Watanabe, M. Morita, T. Iguchi, *Chemosphere*, **2003**, *53*, 827–833.
19 N. Tatarazako, National Institute for Environmental Studies, Japan, October, **2005**, 1–42.

Keywords

Endocrine-Disrupting Chemicals, Fish, Medaka, Secondary Sexual Characters, Thyroid Hormones, Daphnids

44
Pesticide Risk Evaluation for Birds and Mammals – Combining Data from Effect and Exposure Studies

Christian Wolf, Michael Riffel, Jens Schabacker

44.1
Introduction

Birds and mammals may be exposed to toxic effects of active substances following the field use of plant protection products. In current ecotoxicological risk assessments for pesticide registration endpoints, of toxicity tests are compared with estimations of the expected exposure of wildlife species in the field. From the data on toxicity and exposure, a risk quotient (e.g., TER: Toxicity Exposure Ratio) is calculated and compared to safety factors (e.g., 10 for acute risk). If the quotient is larger than the safety factor, the risk is considered to be acceptable. On the other hand, if the quotient is below the safety factor, a possible risk is indicated and further refinement of the input parameters is necessary to show that no risk for wildlife species will exist when the substance is applied under practical field conditions.

Input parameters for the risk quotient calculation (i.e., TER) are derived from animal toxicity tests at exposure levels which differ in time and applied dose. For the assessment of an acute exposure, the LD_{50} value of an acute oral test for birds or mammals is taken. Regarding only birds, for short-term time scales, a five-day dietary test [LC_{50}] is conducted according to OECD Method No. 205. For the long-term time scale, the NOEL of a reproduction test (birds, OECD Method No. 206), a multi-generation study or teratology studies (mammals) are used. Tests from the human toxicological package may have several insufficiencies when being used for ecotoxicological risk assessments of wild mammals.

The dietary exposure is estimated as outlined in the EU Guidance Document SANCO 4145/2000 [1]. The estimated exposure is computed by multiplying the food intake of a focal species and the concentration of a particular compound in the diet. Several factors (PT, PD, see below) can be included into the calculation to model the theoretical exposure in a more realistic way. Exposure will be expressed as daily dose for all time scales based on the following equation:

$$ETE = (FIR / bw) \cdot C \cdot PT \cdot PD \qquad \text{[Equation 1]}$$

Pesticide Chemistry. Crop Protection, Public Health, Environmental Safety
Edited by Hideo Ohkawa, Hisashi Miyagawa, and Philip W. Lee

ISBN: 978-3-527-31663-2

in which:

ETE	Estimated theoretical exposure
FIR	Food intake rate of indicator species [g ww/day]
bw	Body weight [g]
C	Concentration of compound in diet [mg/kg ww] C = Application rate [kg/ha] · RUD (residues per unit dose, e.g., 1 kg a.s./ha)
PT	Portion of diet obtained in treated area
PD	Portion of food type in diet

44.2 Principles of the Risk Assessment within the EU

The risk assessment is conducted in two tiers. In the first tier, fixed conservative default input parameters given in the EU Guidance Document SANCO 4145/2000 are used for the ETE-calculation [1]. The first tier is meant to be a 'screening' process, selecting 'uncritical' products and uses that will pass the process fast and easily. However, experience shows that actually about 75% of the first tier risk assessments do not pass one of the TER-trigger values. When a product does not pass the Tier 1, it enters the higher tier or refined risk assessment. In the Tier 2, a full justification is needed for all assumptions which differ from the Tier 1 default values.

44.3 Refined Risk Assessment

Several data sources for a refinement of risk assessments for birds and mammals can be used: (1) information from other parts of the data package, for example, from the residue section of the dossier, (2) data from scientific literature or official research projects [e.g., CSL (DEFRA-UK) projects on wildlife in agricultural landscape], (3) results from (generic) field studies, which can obtain very focused data sets for refinements (to refine focal species (FIR/bw), PT, PD or RUD (e.g., arthropods)). In all cases, it is necessary to quantify parameters so that they can be used in the ETE-calculation.

44.4 Higher-Tiered Studies

A refined risk assessment and specific higher-tiered studies related to this assessment (e.g., field studies) should not only demonstrate that no unacceptable effects on birds/mammals occur under practical conditions of use, but also demonstrate why those effects will not occur. One possibility to describe the exposure situation of birds and wild mammals in a particular cropping scenario is the determination of the general ecological parameters of a focal species, especially its feeding ecology. By determining the spatial pattern of feeding sites and the food items ingested in different habitats, a much more precise exposure pattern can be estimated. This information can be of a generic nature, independent from a specific pesticide used. Thus, generic field studies, describing the feeding ecology of potentially exposed wildlife species are an efficient tool to obtain realistic and quantifiable input parameters for the ETE-calculation. In generic field studies with birds and mammals, the resource requirements (manpower, costs) are similar to other higher-tiered studies in the ecotoxicological data package. However, they have a higher return on investment, because one study can be used for several compounds.

There are several circumstances under which a generic wildlife study might be necessary: Higher-tiered studies for birds and mammals will be necessary, if a lack of data in specific crop scenarios is obvious: e.g., when the species that can be found in the target crop are unknown, when information on the ecology (exposure) for relevant species in a certain crop is missing, or when regulatory authorities request to justify a refined risk assessment. Field studies will generate a more robust data set to maintain registration since they can be designed to answer specific questions arising during the risk assessment which are not covered by general ecological studies.

Prior to starting a wildlife field study, a survey of existing scientific literature will lead to a definition of clear study objectives (e.g., for ECPA members using the AGROBIRD and AGROMAM database from RIFCON GmbH). If possible, it is advisable to involve regulatory authorities to discuss the concept and protocol of study. Moreover, visiting field study sites to enhance the acceptability of the study is recommended. Only a few institutes and CRO's in Europe are able to perform wildlife field studies according to acceptable quality standards. It is advisable to use well-experienced contract research organizations to perform wildlife field studies.

44.5 Case Study for Combining Effects and Exposure Studies

As an example for the use of field data (in this case, a combination of a generic study, a field-effect study and a field residue study), the following case study is presented: the agricultural scenario was a spray application of an organothiophosphate insecticide in arable crops. Due to the inherent bird toxicity (e.g., LD_{50}: 10 mg a.s./kg bw) estimated according to the EU Guidance Document SANCO/4145/2000, the acute TER is < 0.5 within a Tier 1 risk assessment [1]. Since the trigger value within the EU (≥ 10) is not met, a refined risk assessment is necessary. As a consequence, a field study was conducted to derive refined exposure parameters for small insectivorous birds in an arable crop. Furthermore, the effects of the product on the population were recorded. The field work of the higher-tiered studies resulted in a set of data which was used as modified input parameters for exposure calculation.

Generic information (not compound specific) of the study include the identification of focal species, the time individual birds spent foraging in treated areas and the diet composition of the focal species. First, field monitoring (transect counts) identified the focal species (*yellow wagtail, Motacilla flava*). Together with literature data, information on the focal species was used for the refinement of the FIR/bw quotient. Radio tracking of individuals of the focal species before and after application of the product estimated the time individuals spent foraging in treated areas (PT). Diet composition of the focal species was analyzed by stomach flushing and feces analysis (PD).

Monitored compound-specific data were the survival rates of adults and nestlings within the treated area (effects under field conditions). Furthermore, residues in/on food items (arthropod prey) within treated fields (term 'C' within the ETE-calculation) were analyzed in arthropod samples collected with suitable methods (e.g., pitfall traps, inventory spray) from treated fields.

Based on data for a relevant indicator species, refined values for body weight, fraction of diet obtained in treated area (PT) and fraction of food type in diet (PD), a refined TER was calculated. As a result of the intensive field work, a set of realistic dietary exposure levels could be calculated for the birds which were observed within 1–10 days after the application of the respective pesticide, resulting in acute TER-values between 12 and 697. These acute TER values, which substantially exceed the value obtained in the Tier 1 assessment (acute TER < 0.5), are all higher than the trigger of 10 set by Annex VI of Directive 91/414 EEC for a refined risk assessment. Furthermore, pesticide exposure did not adversely affect avian survival (neither adults nor young birds). In conclusion, an unacceptable risk for birds following the use of the compound under practical field conditions is not to be expected.

44.6 Conclusion

Field studies are highly recommended to provide ecological (generic) data on potentially exposed wildlife species for refined risk assessments. As can be seen from the case study, the required input data (generic and/or compound specific) on relevant species and crop scenarios can be generated by tailor-made field studies. These quantitative data are very suitable for ETE-calculations. Therefore, specific data obtained from field studies may contribute to more scientific and realistic risk evaluations.

However, it should be noted that field studies have limitations, e.g., the sample size (individuals of focal species) is limited by resources and time, the exposure pattern in laboratory studies are often not comparable with the exposure pattern in a real field situation (e.g., NOED from reproduction study with 22 weeks of continuous exposure will be compared with a exposure time of 16 days according to food residue data obtained from the field). Some studies are raising further questions indicating a need for further research. Some work still has to be done to assure mutual agreement of the refinement approaches between notifiers and regulators.

44.7 Reference

1 Anonymous, Guidance Document on Risk Assessment for Birds and Mammals under Council Directive 91/414/EEC – SANCO 4145/2000, European Commission, Health and Consumer Protection Directorate General, **2002**.

Keywords

Pesticide Risk Assessment, Birds and Wild Mammals, Generic Field Studies, EU Guidance Documents

45
Bioassay for Persistent Organic Pollutants in Transgenic Plants with Ah Receptor and GUS Reporter Genes

Hideyuki Inui, Keiko Gion, Yasushi Utani, Hideo Ohkawa

45.1
Introduction

Contamination of the environment and agricultural products with persistent organic pollutants (POPs), including dioxins and the pesticides used in the past, is a serious global problem. Aldrin, dieldrin, endrin, DDT, chlordane, heptachlor, mirex, toxaphene, polychlorinated biphenyls (PCBs), polychlorinated dibenzo-*p*-dioxins (PCDDs), polychlorinated dibenzofurans (PCDFs), and hexachlorobenzene are examples of POPs. Recently, pentabromodiphenyl ether, hexachlorocyclohexanes (including lindane), chlordecone, and hexabromobiphenyl have also been discussed for listing in known POPs. Although POPs are present at nano level concentrations, they accumulate at the tops of food chains, particularly in aquatic ecosystems, owing to their high liposolubility and persistence. Some POPs are highly toxic to mammals, including humans. Therefore, monitoring of POP contamination at nano level concentrations is necessary for evaluating the risks of these compounds to the environment and human health.

Analytical instruments such as high-resolution gas chromatography/mass spectrometry (HRGC/MS) have been used for quantification and qualification of POPs in the environment. However, HRGC/MS analysis is quite expensive (approx. 1000 USD/sample), and results take a long time to obtain. In contrast, bioassays are inexpensive, simple, and quick, and the risk of secondary contamination with chemicals is low. Various bioassays have been developed for direct detection of the biological effects of POPs. In this study, we used transgenic plants to develop an on-site bioassay for dioxins in soil. We introduced aryl hydrocarbon receptor (AhR) and β-glucuronidase (GUS) reporter genes into plants to construct a mammalian dioxin-dependent inducible expression system.

Pesticide Chemistry. Crop Protection, Public Health, Environmental Safety
Edited by Hideo Ohkawa, Hisashi Miyagawa, and Philip W. Lee

ISBN: 978-3-527-31663-2

45.2 Dioxins

Among POPs, coplanar PCBs, PCDDs, and PCDFs (dioxins) are of particular concern because they are toxic to mammals: these compounds bind to and activate the AhR.

There are many dioxin congeners, and the toxicity of the congeners varies depending on the number and positions of the chlorine substituents. 2,3,7,8-Tetrachlorodibenzo-*p*-dioxin (2,3,7,8-TCDD) shows the highest toxicity to mammals. Values for dioxin toxicity were reassessed by the World Health Organization in 2005 [1]. The toxic equivalency factor (TEF) of dioxin congener represents its toxicity relative to that of 2,3,7,8-TCDD, which is defined as having a TEF value of 1. Another parameter is the toxic equivalent quantity (TEQ), which is the total toxicity of a mixture of compounds represented as the sum of the concentrations of each compound multiplied by its TEF. Dioxin contamination is usually represented in terms of TEQ values.

In some locations in Japan, dioxin concentrations exceed the environmental standard set for Japan (< 1000 pg-TEQ/g soil) (Table I). These areas are the sites of chemical plants and industrial-waste incinerators. Usually, the soils from these sites were excavated and sequestered.

Table I. Japanese locations where dioxin contamination in soil exceeds 100,000 pg-TEQ/g.

Place	Year	Concentration (pg-TEQ/g)
Osaka	1998	52,000,000
Tokyo	2001	570,000
Wakayama	2002	100,000
Tokyo	2006	600,000
Tokyo	2006	590,000

45.3 Dioxin Bioassays

In 2005, the Ministry of the Environment of Japan evaluated four novel simple bioassays for monitoring dioxin contamination in effluent gas, fly ash, and cinders. In the CALUX assay, recombinant mouse hepatoma cells (H1L6.1c2) with four dioxin-responsive elements upstream of the luciferase gene are treated with extracts from contaminated samples, and the luciferase activity is then measured [2]. Luciferase activity induced in the recombinant human hepatoma cell line 101L and the recombinant mouse hepatoma cell line HeB5 is utilized in the P450 Human Reporter Gene System (HRGS) and Ah luciferase assays, respectively [3].

These three bioassays are suitable for direct detection of dioxin toxicity through the AhR in recombinant mammalian cells. The fourth novel assay, DioQuicker, is an enzyme-linked immunosorbent assay (ELISA) kit for dioxin analysis, and it utilizes competitive reaction of a sample and coating antigens with an anti-dioxin antibody and subsequent reaction with a secondary antibody [4]. The AhR assays and the ELISA are highly sensitive, inexpensive, simple, and quick compared with HRGC/MS analysis. However, both HRGC/MS and the bioassays require pretreatment of samples by column chromatography on sulfuric acid-impregnated silica gel, multilayered silica gel, or active carbon.

45.4
The AhR

The AhR is a well-researched receptor in a mammalian dioxin-dependent inducible expression system (Figure 1) [5]. Dioxins that are transferred into mammalian cells specifically bind to the AhR complex, which then is transported into the nucleus, where it forms a heterodimer with an AhR nuclear translocator (Arnt). The heterodimer binds to a xenobiotic responsive element (XRE) upstream of the gene for a drug-metabolizing enzyme (CYP1A1). This binding induces the production of CYP1A1 mRNA and thus the production of the CYP1A1 enzyme, which metabolizes some dioxins.

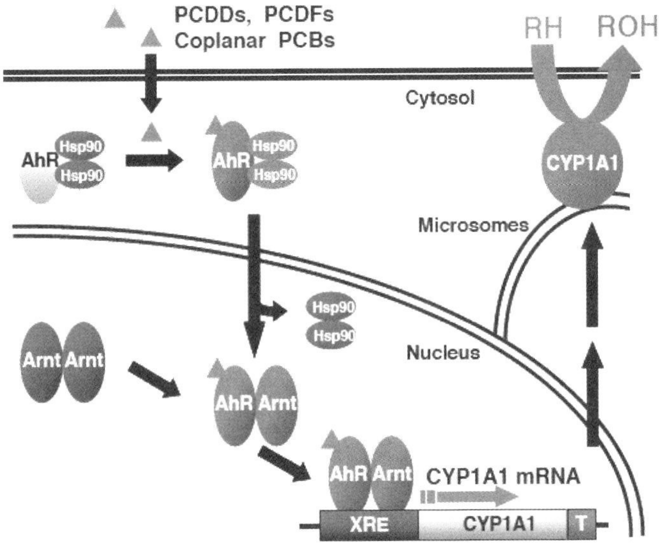

Figure 1. Dioxin-dependent AhR-mediated expression of the *CYP1A1* gene in mammals.
Arnt = AhR nuclear translocator: XRE = xenobiotic responsive element: Hsp90 = heat shock protein 90: T = terminator: R = substrate.

Dioxin toxicity varies with species (Table II). Guinea pigs are the animals most sensitive to 2,3,7,8-TCDD and PCB. In contrast, rabbits, mice, dogs, and hamsters are not highly sensitive to 2,3,7,8-TCDD. Interestingly, the mouse AhR shows the highest binding affinity (Kd) for 2,3,7,8-TCDD (Table III), even though the dioxin sensitivity of this species is relatively low. The guinea pig AhR shows a relatively high binding affinity for 2,3,7,8-TCDD.

The mouse and guinea pig AhRs are 805- and 846-amino-acid proteins, respectively, with transactivation domains in the C-terminal region, DNA-binding and dimerization domains in the N-terminal region, and ligand-binding domains in the central region [22–23] (Figure 2A).

Table II. LD_{50} values for 2,3,7,8-TCDD and PCB in various animal species.

Animal	**LD_{50} (mg/kg body weight)**	
	2,3,7,8-TCDD	**PCB**
Guinea pig	0.0006 [6] – 0.002 [7]	0.5 [8] – 10 [8]
Monkey	0.002 [9] – 0.070 [10]	
Rat	0.020 [9] – 0.022 [6]	1000 [11] – 19000 [12]
Rabbit	0.12 [6]	
Mouse	0.28 [7]	800 [13]
Dog	1.0 [9]	
Hamster	1.2 [14] – 5.1 [15]	

Table III. Kd values of various animal AhRs for 2,3,7,8-TCDD.

Animal AhR	**Kd (nM)**
Mouse	0.034 [16] – 1.7 [17]
Human	1.6 [17] – 18.6 [18]
Guinea pig	2.5 [19]
Rat	3.3 [20] – 4.7 [20]
Monkey	16.5 [21]
Dog	17.1 [21]
Pig	17.5 [18]

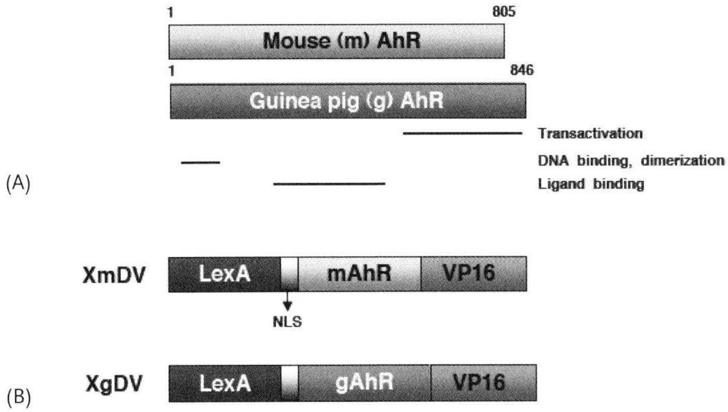

Figure 2. (A) Mouse and guinea pig AhRs and (B) the recombinant AhRs constructed to increase the sensitivity of the bioassay. NLS = nuclear localization signal.

45.5
POP Bioassay Using Transgenic Plants

Certain plant species that are creeping and have deep roots absorb and accumulate nano-level concentrations of POPs from a wide area through their highly developed root systems. Therefore, an *in situ* bioassay using such plants does not require the extraction of POPs from environmental samples, because the POPs are absorbed into the plants through the roots.

We introduced the novel AhR-mediated GUS reporter gene expression system into *Arabidopsis* and tobacco plants for bioassay of POPs (Figure 3). The mouse AhR was used in this expression system because of its high binding affinity for 2,3,7,8-TCDD, and the guinea pig AhR was used because of this species' high sensitivity to this compound (Tables II, III). To develop a highly sensitive bioassay using these transgenic plants, we replaced the DNA-binding and transactivation domains in the native AhRs with the bacterial LexA DNA-binding domain and the virus VP16 transactivation domain, resulting in the recombinant AhRs XmDV and XgDV, containing mouse and guinea pig AhRs, respectively (Figure 2B). We expected that these recombinant AhRs would have higher binding affinity for the DNA sequence upstream of the GUS reporter gene and higher transactivation activity than the corresponding native AhRs. In a previous study, we introduced recombinant AhR containing the native AhR into tobacco plants [24]. When treated with 3-methylcholanthrene, a typical AhR agonist, these transgenic tobacco plants showed a dose-dependent increase in GUS activity. We expected that the transgenic plants with XmDV and XgDV would show higher sensitivity in the POP bioassay than plants with AhRV, containing the AhR and virus VP16.

The two genes for the recombinant AhRs were inserted into a plant expression plasmid having the LexA target sequence combined with the GUS gene. The

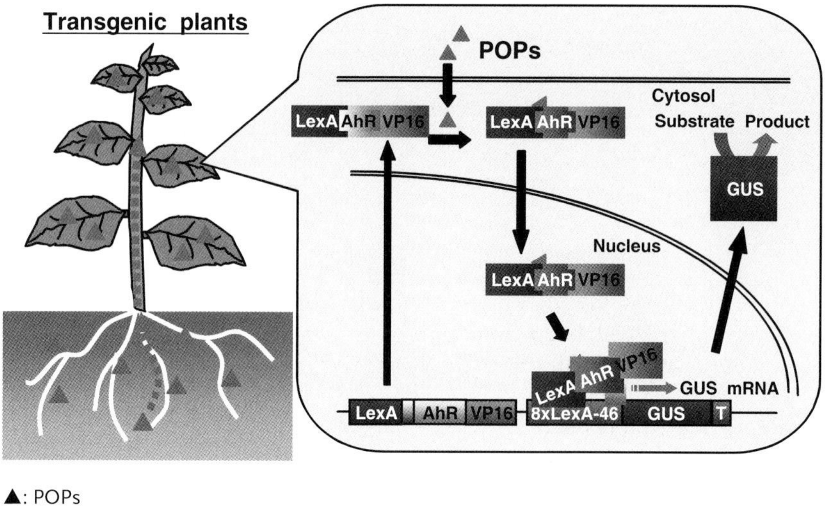

Figure 3. Bioassay of POPs using the transgenic plants.

resulting plasmids were each introduced into *Agrobacterium tumefaciens* to transform *Arabidopsis* and tobacco plants. The transgenic plants were selected several times with kanamycin and subjected to further experiments. Dose- and time-dependent increases in GUS activity were observed upon treatment of the plants with 3-methylcholanthrene and PCB126. In contrast, the transgenic plants did not show a dose-dependent increase in GUS activity upon treatment with PCB180. These results strongly suggest that the transgenic plants are suitable for bioassay of dioxin toxicity, because the TEF values of PCB126 and PCB180 are 0.1 and 0, respectively. Furthermore, we found that these transgenic plants could detect *p,p'*-DDT and dieldrin, although *p,p'*-DDT is an AhR antagonist. The transgenic tobacco plants clearly detected dioxin concentrations below 1000 pg-TEQ/g, which is the environmental standard for soil in Japan [submitted]. On the basis of these results, the transgenic plants carrying the recombinant AhR appear to be useful for *in situ* bioassay without prior extraction of POPs from environmental samples, thus reducing the cost of the assay as well as the risks of secondary contamination. The method should also permit continuous bioassay of POPs at contamination sites near incinerators. This bioassay is environmentally benign because it is driven by photosynthesis.

However, the use of these transgenic plants does have several drawbacks. First, the rate-limiting step for this bioassay is the absorption and translocation of the POPs into plants. POPs with high liposolubility are not easily absorbed and translocated into plants. In particular, absorption of POPs from weathered soil and soil with a high organic content is difficult. *Cucurbitaceae* species, which are known to actively absorb some POPs, may be useful for this bioassay [25]. Second, climate-independent assay is required for consistent results. Of course, assay results will have to be fully validated by comparison with the results of the standard analytical method. Third, and most important, the public must accept release of these transgenic plants into the environment. Extensive research on the safety of these plants should be conducted before this bioassay is released to the market.

45.6 Prospects

Bioassays involving the measurement of GUS activity in transgenic plants are not always convenient, because such assays must be conducted in the laboratory. Therefore, we proposed that using a flower color gene as a reporter gene instead of the GUS gene might be a simple method for confirming contamination. There have been many studies of the biosynthesis of flower color and identification of related genes. Flower color change will be brought about by inducible overexpression, or reduction in expression, of transcripts of these genes in transgenic ornamental plants under the control of recombinant AhRs when dioxin contamination is present.

45.7 Acknowledgments

This project was supported in part by a Grant-in-Aid for Scientific Research (A) and by the Bio-oriented Technology Research Advancement Institution (BRAIN).

45.8 References

1 http://www.who.int/ipcs/assessment/tef_values.pdf.
2 I. Windal, M. S. Denison, L. S. Birnbaum, N. Van Wouwe, W. Baeyens, L. Goeyens, *Environ. Sci. Technol.*, **2005**, *39*, 7357–7364.
3 J. W. Anderson, S. I. Hartwell, M. J. Hameedi, *Environ. Sci. Technol.*, **2005**, *39*, 17–23.
4 http://www.k-soltech.co.jp/dioquiker_english.htm.
5 J. P. Whitlock, Jr., *Annu. Rev. Pharmacol. Toxicol.*, **1999**, *39*, 103–125.
6 B. A. Schwetz, J. M. Norris, G. L. Sparschu, U. K. Rowe, P. J. Gehring, J. L. Emerson, C. G. Gerbig, *Environ. Health Perspect.*, **1973**, *5*, 87–99.
7 E. E. McConnell, J. A. Moore, J. K. Haseman, M. W. Harris, *Toxicol. Appl. Pharmacol.*, **1978**, *44*, 335–356.
8 J. D. McKinney, K. Chae, E. E. McConnell, L. S. Birnbaum, *Environ. Health Perspect.*, **1985**, *60*, 57–68.
9 M. H. Bickel, *Experientia*, **1982**, *38*, 879–882.
10 E. E. McConnell, J. A. Moore, D. W. Dalgard, *Toxicol. Appl. Pharmacol.*, **1978**, *43*, 175–187.
11 L. H. Garthoff, F. E. Cerra, E. M. Marks, *Toxicol. Appl. Pharmacol.*, **1981**, *60*, 33–44.
12 U. Seidel, E. Schweizer, F. Schweinsberg, R. Wodarz, A. W. Rettenmeier, *Environ. Health Perspect.*, **1996**, *104*, 1172–1179.

13 R. Hasegawa, Y. Nakaji, Y. Kurokawa, M. Tobe, *Sci. Rep. Res. Inst. Tohoku Univ. [Med.].*, **1989**, *36*, 10–16.
14 J. R. Olson, M. A. Holscher, R. A. Neal, *Toxicol. Appl. Pharmacol.*, **1980**, *55*, 67–78.
15 J. M. Henck, M. A. New, R. J. Kociba, K. S. Rao, *Toxicol. Appl. Pharmacol.*, **1981**, *59*, 405–407.
16 A. Poland, D. Palen, E. Glover, *Mol. Pharmacol.*, **1994**, *46*, 915–921.
17 M. Ema, N. Ohe, M. Suzuki, J. Mimura, K. Sogawa, S. Ikawa, Y. Fujii-Kuriyama, *J. Biol. Chem.*, **1994**, *269*, 27337–27343.
18 P. Lesca, R. Witkamp, P. Maurel, P. Galtier, *Biochem. Biophys. Res. Commun.*, **1994**, *200*, 475–481.
19 P. A. Bank, E. F. Yao, H. I. Swanson, K. Tullis, M. S. Denison, *Arch. Biochem. Biophys.*, **1995**, *317*, 439–448.
20 R. Pohjanvirta, M. Viluksela, J. T. Tuomisto, M. Unkila, J. Karasinska, M. A. Franc, M. Holowenko, J. V. Giannone, *et al.*, *Toxicol. Appl. Pharmacol.*, **1999**, *155*, 82–95.
21 C. Sandoz, P. Lesca, J. F. Narbonne, *Toxicol. Lett.*, **1999**, *109*, 115–121.
22 M. Ema, K. Sogawa, N. Watanabe, Y. Chujoh, N. Matsushita, O. Gotoh, Y. Funae, Y. Fujii-Kuriyama, *Biochem. Biophys. Res. Commun.*, **1992**, *184*, 246–253.
23 M. Korkalainen, J. Tuomisto, R. Pohjanvirta, *Biochem. Biophys. Res. Commun.*, **2001**, *285*, 1121–1129.
24 H. Inui, H. Sasaki, S. Kodama, N.-H. Chua, H. Ohkawa, American Chemical Society: Washington D.C., **2005**, *Vol. ACS Symposium Series 892*, 40–47.
25 A. Hulster, J. F. Muller, H. Marschner, *Environ. Sci. Technol.*, **1994**, *28*, 1110–1115.

Keywords

Persistent Organic Pollutants (POPs), Aryl Hydrocarbon (Ah) Receptor, Transgenic Plants, Reporter Gene, Bioassay, Dioxins

46
Recent Developments in QuEChERS Methodology for Pesticide Multiresidue Analysis

Michelangelo Anastassiades, Ellen Scherbaum, Bünyamin Taşdelen, Darinka Štajnbaher

46.1
Introduction

The French author and pilot Antoine de Saint-Exupéry once described development as "the path from the primitive via the complicated to the simple". This quotation also very much reflects the evolution of sample preparation methodologies for pesticide multiresidue analysis in the past 50 years: Early methods involved simple liquid-liquid partitioning to cover a narrow spectrum of exclusively non-polar compounds. However, following the introduction of highly polar pesticides in the late 1960s, complex methodologies involving numerous troublesome partitioning and cleanup steps had to be introduced to enable adequate determinative analysis of the target pesticides using the "primitive" instrumentation available at this time. Variations of these types of methods are still widely in use today but are gradually being replaced by novel approaches that focus on simplification, miniaturization, and automation and take advantage of the enhanced possibilities offered by modern analytical instrumentation especially in terms of detection selectivity and sensitivity. Indeed, the boom noticed in the development of simplified sample preparation methodologies in the past decade has been closely related to the dramatic pace of innovation in instrumental analysis techniques, with the most significant impact in this respect being attributed to the LC/MS(/MS) technology, which opened the door for an easy and reliable analysis of numerous traditionally "difficult" pesticides. The growing need to lower costs and turnaround times and to increase sample throughput in laboratories of course further accelerated the acceptance and implementation of those novel sample preparation methodologies.

QuEChERS, which stands for Quick, Easy, Cheap, Effective, Rugged, and Safe, is one of those new-generation sample preparation methods for pesticide multiresidue analysis [1]. Although very recently introduced (development between 2000–2002, publication in 2003), the method has been widely embraced by the international pesticide residue analysts community and is already being used in numerous laboratories worldwide [2–6]. Aiming to deliver an economical and

Pesticide Chemistry. Crop Protection, Public Health, Environmental Safety
Edited by Hideo Ohkawa, Hisashi Miyagawa, and Philip W. Lee

ISBN: 978-3-527-31663-2

robust methodology that is fit for purpose, the development of QuEChERS focused on streamlining the procedure wherever possible by simplifying or omitting impractical, laborious, and time-consuming steps. In principle, QuEChERS constitutes a simplification of traditional sample preparation methods and briefly involves an initial extraction with acetonitrile, liquid-liquid-partitioning after addition of a mixture of $MgSO_4$ and NaCl, followed by a simple "dispersive-SPE" clean-up step, in which a portion of the raw extract is mixed with bulk SPE sorbent which is subsequently separated by centrifugation. The advantages of the QuEChERS method include: (a) rapidity (sample preparation of six previously homogenized samples in *ca.* 30 min), (b) simplicity and robustness (few, simple steps), (c) low solvent consumption (only 10-mL acetonitrile), (d) low costs, (e) practically no glassware needs, (f) amenability of acetonitrile extracts to GC- and LC-applications as well as to (dispersive) SPE cleanup, and (g) coverage of a very broad pesticide spectrum (including basic, acidic, and very polar pesticides).

With the implementation of the original QuEChERS-method in our pesticide residue analysis laboratory in 2002, and the associated validation experiments for numerous pesticides in different representative commodities, it soon became clear that some amendments to the original procedure had to be introduced to improve the recoveries of certain pH-dependent pesticides and to expand the spectrum of commodities amenable to the method. The routine use of the method furthermore raised the need to further improve its selectivity in order to enhance the robustness of determinative analysis. The modifications introduced are subject of this paper.

46.2
Reagents

Water, acetonitrile, and methanol of HPLC quality; dry ice; ammonium formate; magnesium sulfate anhydrous grit (for example, Fluka No. 63135); sodium chloride; disodium hydrogencitrate sesquihydrate (for example, Aldrich No. 359084); trisodium citrate dihydrate (for example, Sigma No. S4641); magnesium sulfate anhydrous fine powder (for example, Merck No. 106067); sodium hydroxide solution in water (5N); 5% formic acid in acetonitrile (v/v); amino-sorbent (for example, Bondesil-PSA 40-µm Varian No. 12213023); graphitized carbon black sorbent (GCB) (for example, Supelco Supelclean Envi-Carb SPE Bulk Packing, No. 57210U); ODS-(C18)-Sorbent (for example, Macherey and Nagel, CHROMABOND, article No. 730602, particle size 45 µm).

Pesticide standards: Prepare stock solutions thereof in acetonitrile or acetone (e.g., 1 mg/mL); working standard solutions of individual pesticides or mixtures thereof are prepared by appropriately diluting the stock solutions with acetonitrile.

Internal Standards (ISTDs): for GC/MS: triphenylphosphate (TPP), PCB 18, PCB 8, triphenylmethane (TPM); for LC/MS ESI(+): TPP, tris(1,3-dichloro-isopropyl)phosphate; for LC/MS ESI(–): nicarbazine. Prepare solutions containing one or more of the compounds proposed with concentrations of 10 to 50 µg/mL for the solution to be added during sample preparation. An appropriate dilution (e.g., factor of 10) is prepared to be used for the preparation of calibration solutions.

Quality Control (QC) standards: PCB 138, anthracene or d10-anthracene

Buffer-Salt-Mixture: To induce phase separation, the following mixture of salts is required per sample test portion (containing approx. 10 g of water): 4 g magnesium sulfate anhydrous grit, 1 g of sodium chloride, 1 g of disodium hydrogen citrate sesquihydrate, and 0.5 g of trisodium citrate dihydrate. It is advisable to prepare in advance a sufficient number of portions of this mixture, the preparation of which is immensely facilitated if a sample divider (see apparatus) is used.

Dispersive SPE Mixtures:
PSA/$MgSO_4$-mixture: For dispersive SPE, most samples require a mixture of 25 mg PSA and 150 mg $MgSO_4$ anhydrous grit per mL sample extract (e.g., for 6-mL extract, this corresponds to 150 mg PSA and 900 mg $MgSO_4$). Also here, the preparation of multiple mixtures in advance (e.g., by means of a sample divider) is indicated.

PSA/$MgSO_4$/ODS-mixture: For extracts of samples with high lipid content a mixture of 25 mg PSA, 25 mg ODS and 150 mg $MgSO_4$ is required per mL extract.

PSA/$MgSO_4$/GCB-mixtures (GCB-Mixtures): For extracts of samples containing high amounts of chlorophylls or carotinoids, dispersive SPE is performed with mixtures containing 25-mg PSA, and 150 mg of GCB-Mixture 1 or 2 (see procedure), with **GCB-Mixture 1** containing 1 part of GCB sorbent and 59 parts of $MgSO_4$ powder and **GCB-Mixture 2** containing 1 part of GCB sorbent and 19 parts of $MgSO_4$ powder.

46.3
Apparatus

Usual laboratory apparatus and, in particular, the following: Sample processing equipment (for example, Stephan UM 5 universal); high-speed dispersing device (for example, Ultra-Turrax, the diameter of the dispersing elements should fit the openings of the centrifuge tubes used); automatic pipettes (suitable for handling volumes of 10 to 100 µL, 200 to 1000 µL and 1 to 10 mL); 50-mL centrifuge tubes with screw caps (for example, 50-mL Teflon® centrifuge tubes with screw caps e.g., Nalgene/Rochester, USA; Oak-ridge, article no. 3114-0050 or disposable 50-mL centrifuge tubes, e.g., Sarstedt/Nümbrecht, Germany, 114 × 28-mm, PP, article no. 62.548.004); 10-mL PP-single use centrifuge tubes with screw caps (for example, Greiner Bio One/Kremsmünster, Austria; 100 × 16-mm, article no. 163270 or Simport/Canada, 17 × 84-mm, article no. T550-10AT, when using the Bürkle sample divider, see below); 10-mL solvent-dispenser for acetonitrile; centrifuges suitable for the centrifuge tubes employed in the procedure and capable of achieving at least 4000 rcf; powder funnel to fit to the openings of the centrifuge tubes; 1.5 mL GC/LC autosampler vials (if necessary with micro-inserts); 20-mL screw cup vials for extract storage (for example, EPA-vials G24, Ziemer GmbH/Langerwehe, Germany, article no. 1.300160); plastic cups (stackable) (for example, flame photometer cups 25-mL article no. 10-00172 from JURO-LABS/ Henfenfeld, Germany, these are used for the storage of the buffer-salt mixture portions); sample divider, to automatically portion salts and sorbents (for example, from Retsch/Haan, PT 100 or Fritsch/Idar-Oberstein, Laborette 27 or Bürkle/ Lörrach, Repro high-precision sample divider); vibration device (for example, Vortex, to distribute the fortified pesticides in for recovery studies); HPLC-MS or HPLC-MS/MS-System, equipped with electrospray ionization (ESI) interface; gas chromatographic system equipped with appropriate detectors, e.g., MS, MS/MS, TOF, ECD, NPD, FPD, and with PTV-injector with solvent vent mode.

46.4 Procedure

Preparation of a representative sample portion: The reduction of the test sample shall be carried out in such a way that representative portions are obtained. In the case of fruits and vegetables, cryogenic milling (e.g., using dry ice) is highly recommended to reduce particle size and thus enhance residue accessibility and extractability as well as sample homogeneity, leading to reduced sub-sampling variability. Cutting the samples coarsely (e.g., 3 × 3-cm) with a knife and putting them into the freezer (e.g., –18 °C overnight) prior to cryogenic milling reduces the amount of dry ice required and facilitates processing. In the case of dried fruits, 500 g of material are taken and mixed with 850 g of cold water, intensively with a powerful mixer.

Scaling: The described extraction and cleanup steps are scalable as desired as long as the amounts of reagents used remain in the same proportion. It should be kept in mind, however, that the smaller the amount of the employed test portion is, the higher the sub-sampling variability will be. Thus, during validation, each laboratory should investigate if the sub-sampling variability achieved for representative samples containing incurred residues is acceptable.

Weighing of test portion and water addition: In the case of samples containing more than 80% of water and less than 4% of lipids (most fruits, vegetables, juices etc.), weigh 10 g ± 0.1 g test portion of the comminuted homogenous and frozen sample into a 50-mL centrifuge tube. For samples not belonging to this group, the addition of water and/or the reduction of the sample size may be necessary as shown in Table I.

Table I. Grouping of plant products for QuEChERS sample preparation.

Group	Content of: Water	Content of: Lipid	Size of Sample Portion	Addition of: Water?	Addition of: Partitioning salts?	Examples	Remarks
Commodities with low lipid and high or intermediate water content							
A	> 80%	< 4%	10 g	no	yes	Most fruits and vegetables, juices	
B	30–80%	< 4%	10 g	To reach approx. 10-g water in total	yes	Bananas, potatoes, fresh bread	
Commodities with low lipid and low water content							
C	15–30%	< 8%	5 g	8.5 g (may be also added during processing)	yes	Raisins and other dried fruits, dates	Employ a blender if required to assist extraction
D	< 15%		5 g	10 g	yes	Cereals, dry pulses, dry mushrooms, honey	
Extract-rich commodities with low water content							
E	< 15%	depends on sample size if 2 g < 20%	1–3 g	10 g	yes	Spices, fermented products (tea, coffee)	
Commodities with high fat content							
F	> 8%	> 8%	5 g/2 g	To achieve approx. 10-g water in total	yes	Avocado, olives, margarine (2-g sample)	Add ISTD to extract aliquot after cleanup. Employ a blender if required to assist extraction (not for oils)
G	< 8%	> 8%	3 g/5 g	optionally 10 mL	yes	Oil-seeds, nuts, peanut butter soja flour (5-g sample)	
H	0	100%	2 g	No	no	Oils	

First extraction step: 10-mL acetonitrile followed by the ISTD solution (e.g., 100 μL) are added, the tube is closed and shaken vigorously by hand for 1 min.

Notes:
(a) Should the test portion contain more than a certain amount of lipids, the ISTD should be added to an aliquot of the separated acetonitrile phase after cleanup, assuming that the volume of the acetonitrile phase is identical to the volume of acetonitrile employed for the initial extraction (i.e., 10 mL). The lipid tolerance depends on the ISTD's lipophilicity. For TPP, the limit is *ca.* 1 g, for TPM *ca.* 0.1 g, and for the PCBs 8 and 18 *ca.* 0.05 g. The latter three are to be added at the first extraction step only in case of low lipid content matrices such as fruits and vegetables.
(b) If the sample's degree of comminution is insufficient, the extraction can be assisted by a high-speed disperser (e.g., Ultra-Turrax) to obtain better accessibility of the residues. The dispersing element is immersed into the sample/acetonitrile mixture and comminution is performed for about 2 min. at high speed. If the ISTD solution has already been added, no rinsing of the dispersing element is necessary. Nevertheless, it still has to be cleaned thoroughly before being used for the next sample to avoid cross-contamination.

Second extraction step and partitioning: Add one portion of the buffer-salt mixture to the suspension derived from the first extraction, close the tube, immediately shake vigorously for 1 min. and centrifuge for 5 min. at 4000 rcf. The upper phase can be directly employed for LC-MS(/MS) measurement.

Cleanup: There are various options for cleanup depending on the type of sample:

(a) *Freezing out (for removal of lipids, waxes, sugars, and other matrix co-extractives with low solubility in acetonitrile)*: An aliquot of the acetonitrile phase is transferred into a centrifuge tube and stored overnight in a freezer (for fat 2 h are normally sufficient), wherewith the major part of fat and waxes precipitate. Should the precipitates not separate by decantation, they may be separated either by a quick centrifugation or by filtering the still cold extract through a piece of cotton wool. The extract can be used for further cleanup by dispersive SPE according to (b) or (d).
Note: When only lipids are to be removed, freezing out may be replaced by a dispersive SPE (D-SPE)-cleanup, where C18 (ODS) sorbent is used as described in (c).
(b) *D-SPE with a PSA/$MgSO_4$-mixture (for most samples)*: An aliquot of the acetonitrile phase is transferred into a PP-single use centrifugation tube already containing 25-mg PSA and 150-mg magnesium sulfate per mL extract. The tube is closed, shaken vigorously for 30 sec. and centrifuged (for 5 min. at 4000 rcf).

Note: It is helpful to load the centrifuge tubes with the dispersive SPE sorbents before beginning the extraction procedure needed for one batch of samples. Instead of PSA, other amino-type sorbents may also be employed for cleanup.

(c) *D-SPE with a PSA/$MgSO_4$/ODS-mixture (for removal of lipids)*: Proceed as described in (b) but additionally use 25-mg ODS sorbent per mL extract. This type of cleanup is recommended for extracts of test samples containing more than 50 mg of lipids (see also 5.3). This type of cleanup is superfluous if freeze-out (a) was performed.

(d) *D-SPE with PSA/$MgSO_4$/GCB-mixtures (removal of chlorophyll and carotinoids)*: Proceed as described in (b) employing 25-mg PSA and 150 mg of GCB-Mixture 1 or 2 depending on the pigment content. GCB-Mixture 1 is used for carrots and *Lactuca* varieties (except iceberg lettuce and lettuce hearts), while GCB-Mixture 2, which contains a higher GCB content, is used for crops with very high pigment content such as red sweet pepper, spinach, lamb's lettuce, rucola, and vine leaves.

Extract stabilization: Following cleanup with PSA, extracts have to be re-acidified to protect pesticides that are sensitive to degradation at high pH values. For this purpose, an aliquot of the cleaned-up extract is transferred into a screw cap storage vial, taking care to avoid sorbent particles being carried over, and slightly acidified by adding 10 µL of a 5% formic acid solution in acetonitrile per mL extract. The pH-adjusted extract is filled into autosampler vials to be used for GC- and LC-based determinative analysis. The residual extract may be stored in a refrigerator to be used later on if needed.

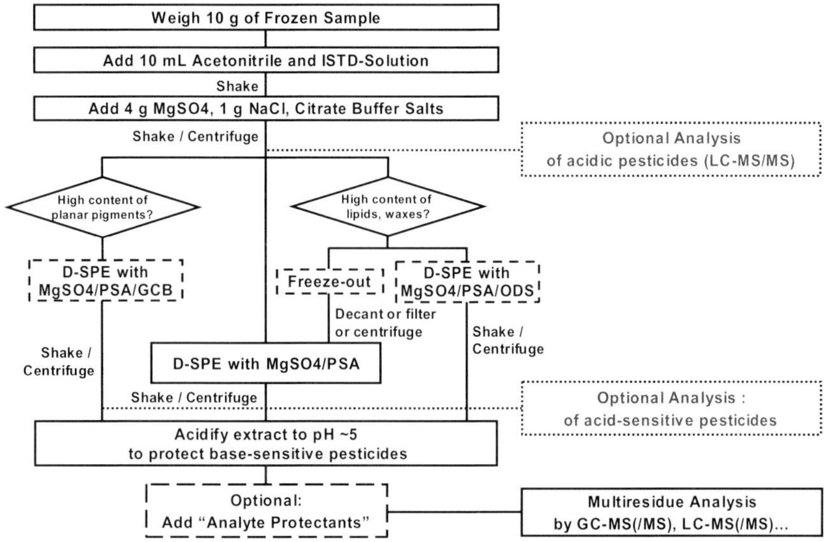

Figure 1. Flowchart of the QuEChERS method.

46.5 Discussion

As regards the strategy to be followed for the modification of the original QuEChERS-method, it was decided not to seek a single all-embracing procedure that would cover "all analytes in all commodities", but rather to develop several simple variations of the method, each one optimized for a specific commodity group. At first glance, one would assume that the implementation of several of such commodity-specific methods, instead of only one universal method in a laboratory, would significantly increase the workload for validation and on-going sample analysis, but this is not necessarily the case. As regards validation, the effort is not expected to noticeably increase, since according to most quality assurance protocols, including the DG-SANCO quality control procedures, laboratories have to validate at least one representative commodity per group, irrespective if the method employed is claimed to be generic or not. Of course, having just one universal procedure for all situations makes life easier, but given the great diversity in the composition of commodities and the physicochemical properties of pesticides, such a procedure would likely contain just too many compromises regarding selectivity and scope: Not considering the particularities of the various types of commodities in the extraction and cleanup strategy compromises the selectivity, robustness, and accuracy of the method. Furthermore, the narrower the scope of analytes covered by the multiresidue method employed in a laboratory, the greater the need to perform additional procedures, in order to cover the entire spectrum of target pesticides.

46.5.1 Improving the Recoveries of Certain Pesticides

pH-dependent pesticides: Although most pesticides are not noticeably affected by any pH-extremes that may occur during QuEChERS sample preparation, there are some, that need special attention in this respect. While certain pesticides are prone to significant degradation at high or low pH values, others tend to get ionized by protonation (basic pesticides, at low pH) or deprotonation (acidic pesticides, at high pH). Since the ionic form has higher affinity towards aqueous rather than organic surroundings, this effect may negatively affect the analyte transfer into the organic layer during liquid-liquid partitioning leading to lower recoveries. As numerous pH-sensitive pesticides are residue-relevant in real samples, any broad-scope multiresidue method should address pH-issues to ensure acceptable recoveries. To be considered in this respect is of course the natural pH of the commodities in question (typically ranging between 2.5–6.5), but also the substantial pH-elevation which occurs following the contact of sample extracts with amino-sorbents in dispersive SPE cleanup, that negatively affects the stability of certain pesticides during extract storage. In the development of the original QuEChERS method, the abovementioned effects of pH on pesticides as well as the effect of pH on the selectivity of extraction (see below) were contemplated.

However, it was decided to keep the method simple and not to introduce any pH-adjustment in the extraction/partitioning step, since the base-sensitive pesticides tested gave acceptable recoveries when extraction and injection were performed fast, and since the recoveries of the basic pesticides were shown to be completely unaffected by pH. As to the protection of base-sensitive pesticides following PSA cleanup, the addition of acetic acid to the final extracts was suggested in the original procedure.

(a) *pH-Adjustment during extraction/partitioning*: Validation studies with the original QuEChERS-method, showed fluctuating recoveries of base-sensitive pesticides such as captan and folpet, especially in the case of commodities with high pH. Initially we addressed this issue by skipping PSA cleanup and/or by acidifying samples, having natural pH > 5, with acetic acid (that was added to the acetonitrile) or sulfuric acid. These measures had a clearly positive effect on the recoveries of base-sensitive pesticides. A similar approach was also introduced by Schenk *et al.* [7], whereas Lehotay *et al.* [8] introduced buffering with sodium acetate and acetic acid to adjust the pH of all samples between 5 and 6 achieving good recoveries for both, base- and acid-sensitive pesticides regardless of the initial pH of the commodity in question. This buffering procedure had the advantage of requiring the addition of only one solid component (sodium acetate) while the acetic acid was added in liquid form together with the acetonitrile, thus keeping the procedure simple. However, parts of the acetate buffer obviously partition into the organic phase, where they exhibit a strong buffering activity. As a result, the measured pH value of the acetonitrile extract remains virtually constant even when using double the amount of PSA per extract volume compared to the original procedure. This may be an advantage regarding the stability of base-sensitive pesticides, but it also is a disadvantage as regards the cleanup performance of PSA, which is dramatically impaired by the strong buffer activity of acetate, resulting in visibly worse cleanup results compared to the original QuEChERS method. This observation was recently confirmed by Herzegova [3] and Hajslova [5] in the latter case also as regards to LC-MS/MS suppression phenomena.
While developing the present modified QuEChERS protocol, the aim was to find a buffer exhibiting less or no negative impact on PSA cleanup performance. As shown by Lehotay *et al.* [8] and results from our laboratory, a pH higher than 6 is required for sufficient protection of the acid-sensitive compound pymetrozine. Same applies to dioxacarb and ethoxyquin. Further experiments [9] suggested a pH of less than 5.5 to achieve good recoveries for the most acidic pesticides such as imazapyr, picloram, clopyralid, 2-CPA and dicamba, which showed a dramatic drop in their recoveries at pH 5.5 and above (also in the case of the abovementioned acetate-buffered version). Furthermore, base-sensitive compounds such as captan, folpet, chlorothalonil, prefer pH lower than 6, and dithianon even lower than 5.5. Finally, a pH range of 5 to 5.5 was considered as the best compromise, at which both quantitative extraction of sour herbicides and protection of alkali labile (e.g., captan, folpet) and acid

labile (e.g., pymetrozine, dioxacarb, ethoxyquin) compounds is satisfactorily achieved.

Our initial experiments focused on acetate buffering, however, at lower concentrations than suggested in [8], since we have observed that this measure drastically reduces the negative impact on PSA cleanup performance. The intention was to keep the magnesium sulfate/sodium chloride composition of the original method and treat commodities differently depending on their pH. Commodities with high pH would be acidified with acetic acid solution as briefly described above, while acidic samples would be buffered by addition of a highly concentrated potassium-acetate solution. Potassium-acetate was preferred over sodium-acetate due to its better solubility in water (253/492 g instead of 119/170 g per 100-mL cold/hot water), which enabled its addition in liquid state, thus avoiding the troublesome weighing step. Only the most acidic samples (e.g., lemons and currants) required additional sodium hydroxide to reach the target pH area. This procedure gave much better cleanup results than the version described in [8] but still worse results compared to the original procedure.

Aiming to find a buffer with practically no impact on the cleanup performance of PSA and with a high buffering capacity at pH 5–5.5, various additional buffers were tested including phosphate buffer ($pK_a = 2.2/7.2/12.4$), and several organic acid buffers with multiple carboxy or hydroxy groups including malate ($pK_a = 3.4/5.0$), citrate ($pK_a = 3.1/4.8/6.4$) and succinate ($pK_a = 4.3/5.6$). A mixture of disodium and trisodium citrate was finally chosen as the best generic option to adjust the pH of various samples to the desired range. Only the most acidic commodities with pH < 3 (i.e., lemons/limes, currants, and raspberries), constitute an exception requiring the addition of some sodium hydroxide (see procedure). Among the advantages of citrate buffering is that the salts are readily available at moderate prices and that there is no negative impact on the subsequent PSA-cleanup step. On the contrary, buffering enhances the selectivity of the partitioning step, thus avoiding over-saturation of PSA (see below). A negative aspect, however, is that two additional solid components have to be employed, which complicates the preparation of the buffer-salt portions. However, the use of rotary sample dividers or of commercially available ready-to-use buffer-salt mixtures circumvents this problem.

(b) *pH-Adjustment of final extracts*: Another important pH-related aspect with even more influence on pesticide quantification than sample pH itself is the degradation of pesticides in the final sample extracts. Following PSA contact, the measured pH of the extracts reaches values typically between 8 and 9, which compromises the stability of alkaline-sensitive pesticides such as captan, folpet, dichlofluanid, tolylfluanid, pyridate, methiocarb sulfone, and chlorothalonil. The pH values in the extracts were measured with a standard pH-meter calibrated using an aqueous buffer solution. In laboratory practice, time intervals of one week or longer between the preparation of sample extracts or calibration solutions and their injection in the chromatographic instruments are not uncommon. Proper quality control requires one to ensure that the pesticide losses during extract storage and the potentially associated quantification errors remain minimal. In order to assess the best pH-compromise for the final QuEChERS extracts, several representative compounds including alkaline and acid-sensitive pesticides were tested as to their stability during storage in QuEChERS extracts previously adjusted at various pH-values ranging from 4 to 9. While some pesticides were more unstable at acidic conditions (e.g., sulfonylureas, carbosulfan, ethoxyquin) and others at high pH (e.g., captan, folpet, dichlofluanid), most compounds, including the ISTDs, were sufficiently stable throughout the tested pH range over a period of 14 days. Few compounds were unstable at both high and low pH (e.g., amitraz), suggesting the need for immediate measurement.

Based on these results, adjustment of the QuEChERS extracts to a pH around 5 was deemed to be a good compromise to slow down the degradation rate of most susceptible pesticides, so that extract storage over several days at room temperature becomes possible. Following PSA cleanup, both the withdrawal of extract aliquots and the subsequent acidification should be performed quickly to minimize degradation of highly alkaline-sensitive compounds, and especially chlorothalonil which is known to react with amino-groups. Acid-labile pesticides such as pymetrozine, dioxacarb, and thiodicarb were also sufficiently stable over several days in pH 5-adjusted extracts. However, some very acid-sensitive compounds such as most sulfonylurea herbicides, carbosulfan, benfuracarb and ethoxyquin, are not sufficiently protected at pH 5. If the measurement can be performed quickly, the acidified extract can be employed; otherwise analysis should be performed from a non-acidified aliquot of the PSA-cleaned-up extract (pH > 8). At these conditions, the compounds were shown to be stable over several days. It should be noted in this context that some of the most acidic sulfonylurea pesticides may experience losses during PSA cleanup. Carbosulfan and benfuracarb (both having individual MRLs) are degraded to carbofuran not only in the extracts at pH 5 but also previously in the samples. Thus, merely if carbofuran is present in the acidified extract, an additional run of the alkaline aliquot is needed to check for the presence of the precursor compounds.

(c) *Improving recoveries of special pesticides*: While most pesticides give very good recoveries using the generic procedure described, some very polar ones (with log $K_{ow} < -2$) such as chlormequat, mepiquat or glyphosate give low recoveries and require separate procedures. However, there are also certain "difficult" pesticides that require slight modifications of the QuEChERS partitioning, cleanup or extract storage conditions to give sufficient recoveries (> 70%). In a routine laboratory work, the use of these modified procedures would be indicated if the normal procedure signifies critical levels.

Pesticides with acidic groups (e.g., phenoxyalcanoic acids) interact with amino-sorbents such as PSA. Thus, if such pesticides are within the scope of analysis, their determinative analysis should be performed directly from the raw extract after centrifugation but prior to cleanup. For this, an aliquot of the raw extract is directly filled into a vial and analyzed preferably by LC-MS/MS in the ESI (–) mode. Covalently bound acidic pesticide residues can be easily released by alkaline hydrolysis prior to extraction. The sample is adjusted to pH 12 with NaOH (in case of dry samples after water addition) and left with occasional stirring for 30 minutes at room temperature. After neutralization of the previously added NaOH with H_2SO_4-solution, the QuEChERS procedure is normally performed as described above.

The recoveries of the acid-labile pesticides ethoxyquin and pymetrozine can be raised if 1.5 g of trisodium citrate is used, instead of using 1 g of the di- and 0.5 g of the trisodium citrate, while keeping low temperatures is additionally helpful especially for ethoxyquin. Furthermore, as ethoxyquin degrades at pH 5-adjusted extracts, measurement should be performed immediately or alternatively directly from the non-acidified extract. There are also compounds normally giving recoveries slightly above 70% with the normal procedure, but much better recoveries if slightly modified. For example, chlorothalonil and dithianon give best recoveries if PSA cleanup is not performed. Very polar compounds, such as acephate and methamidophos give higher recoveries if the amount of NaCl is reduced or totally skipped.

46.5.2
Improving Selectivity

The dramatic improvements in the field of instrumental analysis in terms of detection sensitivity, chromatographic and mass spectrometric resolution, and computational signal deconvolution, have definitely helped to substantially simplify sample preparation procedures by reducing the required degree of selective pesticide enrichment. Nevertheless, experience has shown that a certain degree of selectivity in sample preparation is still indispensable as it helps to slow down instrument deterioration and to reduce interferences in determinative analysis including the so-called matrix-induced signal suppression and enhancement effects as well as the matrix-induced signal diminishment effect caused by increasing contamination of the GC-system surfaces. In this respect, it has to be kept in mind, however, that the broader the scope of pesticides to be covered by a multiresidue method is, the more restricted the freedom to remove interfering matrix co-extractives becomes. The selectivity of any broad-scope pesticide multiresidue procedure will, thus, always be a matter of compromise. The modifications introduced to improve the selectivity of the QuEChERS method concerned both partitioning and cleanup.

Selectivity of partitioning: The addition of citrus buffering salts to elevate the pH of sour fruits has also dramatically reduced the amount of co-extractives in the raw extracts. This observation was also made during the development of the original QuEChERS method where the peaks of fatty acids as well as of maleic and fumaric acid in full-scan GC/MS became smaller as the pH of acidic samples

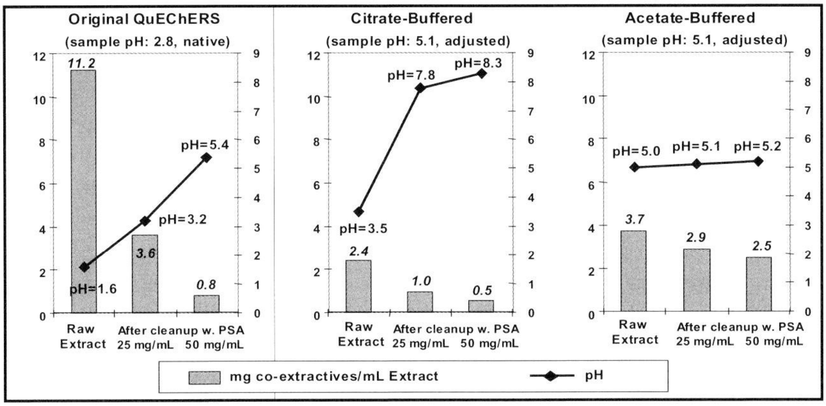

Figure 2. Influence of buffering on the selectivity of the QuEChERS method.

was raised, and was reconfirmed by Lehotay *et al.* when employing acetate buffer to raise the pH between 5.1 and 6.0 [8].

Apart of the pH, the type and amount of salts employed during partitioning also have a great influence on the selectivity of partitioning. As demonstrated in the original QuEChERS publication, the polarity-scope of the method becomes narrower towards the polar end of the spectrum the more NaCl is added at the partitioning step, since NaCl forces water out of the acetonitrile phase making it less receptive for polar compounds. A ratio of 4 g $MgSO_4$ and 1 g NaCl was finally chosen in the original procedure as a compromise between lowering the amount of co-extracted sugars in the raw extracts and keeping the recovery of the most polar pesticide methamidophos above 80%. In the present modified procedure, the question arose whether or not to keep using NaCl together with the citrate buffer salts, which also exhibit a salting out effect, thus further narrowing the polarity scope towards the polar end of the spectrum. Although the use of NaCl caused a recovery-drop of methamidophos and acephate from 80–85% to 70–75%, it was kept in the procedure for the sake of overall selectivity.

Selectivity of cleanup: The PSA/$MgSO_4$ composition used for dispersive SPE cleanup was kept the same as in the original QuEChERS method. The cleanup results were, however, much better due to the use of citrate buffer, which significantly reduced the amount of PSA-removable co-extractives in the raw extracts as described above. Especially in the case of commodities with high acidity or high load of PSA-removable co-extractives such as phenolic anthocyanidines (e.g., strawberries) the PSA amount employed in the original procedure was not sufficient, so that the measured extract pH following cleanup remained comparably low. This effect can be nicely seen in Figure 2, where using the original QuEChERS-procedure, the pH of the red currant extract remained clearly below 7, even when employing double the amount of PSA than prescribed (50 mg/mL). With the use of the citrate buffer, this PSA oversaturation problem does not appear any more. Searching for alternatives for PSA, a number of amino-sorbents from various companies were tested. In principle, all amino-type sorbents remove the same type of co-extractives (acids including fatty acids, anthocyan-pigments, sugars) and elevate extract pH. In general, cleanup efficiency was found to increase in the row mono- < di- < tri-amino-sorbents. Bifunctional sorbents containing both amino and reversed-phase functionalities (C8 or C18), were also evaluated. However, both amino and reversed-phase (see also below) cleanup efficiency was less efficient than when using the individual sorbents, so that more than 75 mg sorbent/mL extract were needed to achieve a similar cleanup effect.

In addition to the amino-sorbents, other sorbent types have also been tested to exploit possibilities to remove co-extractives not or marginally removed by PSA, such as lipids, waxes, chlorophyll. Acidic, neutral, and basic alumina sorbents were found to remove, although to a slightly smaller extent, similar co-extractives as amino-sorbents, however, with less influence on pH. These sorbents, being much cheaper than PSA, will thus be more intensively studied in future.

Various carbon-based sorbents have also been tested including active carbon powder and carbon pellets. Graphitized carbon black (GCB) was found to be the most appropriate in terms of overall performance and handling. However handling still remains delicate with sorbent amount, cleanup-time, and amount of co-extractives having carbon-affinity being important parameters to consider in order to achieve good cleanup results but still avoid unacceptable losses of planar pesticides also exhibiting strong affinity towards carbon (e.g., thiabendazole, chlorothalonil, HCB, quintozene, coumaphos, cyprodinil). The PCBs 8 and 18 show also a very strong affinity towards carbon and should thus not be used as ISTDs when GCB cleanup is performed. Triphenylphosphate and triphenylmethane have a moderate affinity towards GCB, and can be used. Nevertheless, recovery studies showed that no noteworthy losses of pesticides occur if the extract still maintains some visible amount of chlorophyll or carotinoids following GCB-cleanup. This can be explained by the much stronger affinity of those planar pigments towards GCB than any of the pesticides. The same applies to anthracene (or d10-anthracene) which may be used as QC standard, since it was shown, that if more than 70% of anthracene is recovered, this will also be the case even for the pesticides with the highest affinity towards carbon (e.g., HCB and chlorothalonil). The use of toluene to moderate the interaction of carbon with pesticides would also be an option, but it was not considered since this would affect the amenability of the extracts to LC-applications.

For the removal of lipids, various silica-based (C8 and C18) and polymeric (PS-DVB) sorbents were tested. Recently Lehotay *et al.* [10] also showed the effectiveness of C18 for the removal of lipids from QuEChERS extracts. Although PS-DVB type sorbents are regarded as more lipophilic than C18 (ODS) sorbents, the PS-DVB sorbents tested were found to be less efficient in removing fatty lipids than ODS-type sorbents, with the exception of a PS-DVB with 1500 m^2 surface from Interchim®, which showed a similar effect to ODS. The better efficiency of ODS in this respect is obviously related to the fact that triglycerides are structurally very closely related to ODS, both containing fatty acid chains. Nevertheless,

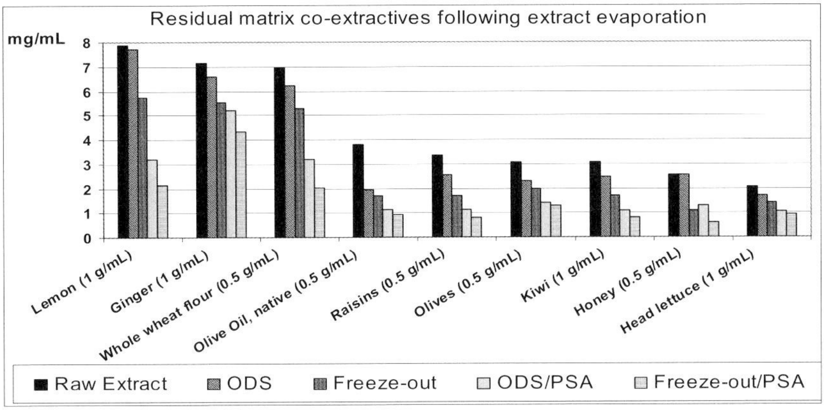

Figure 3. Comparison of cleanup efficiency achieved by freeze-out and D-SPE using ODS.

PS-DVB-sorbents also showed some influence on the recoveries of non-polar compounds (e.g., HCB, DDE, ethofenprox, halfenprox). C18 or C8 did not affect the recovery of any pesticide. Whereas PSA cleanup efficiency improves the less water is contained in the extracts (see below), lipid removal using reversed-phase sorbents worsens, and better results are achieved when skipping MgSO4 in D-SPE. Nevertheless, for practicality reasons, PSA and ODS cleanup is combined.

In addition to the dispersive SPE, freezing-out was also tested as a simple cleanup approach. Freezing out also helps to partly remove various sample co-extractives with limited solubility in acetonitrile such as lipids, waxes, and sugars by simply shifting the saturation equilibrium. In terms of lipid removal, freezing-out was shown to give similar results to C18-cleanup, the removal of additional co-extractives during freeze-out, however, results in much better overall cleanup results as shown in Figure 3. In the case of olive oil, the co-extracted matrix (mostly triglycerides and fatty acids) dropped from 4.6 mg/mL to 0.9 mg/mL following C18/PSA or freezing-out/PSA cleanup, which corresponds to ~99.55% matrix removal referred to the initial 2-g oil matrix. In comparison, following gel permeation chromatography cleanup (GPC) of 2×0.5 g of oil dissolved in a 1 : 1 cyclohexane:ethyl acetate mixture, the residue remainder following evaporation was 11 mg, which corresponds to only 98.89% matrix removal efficiency. A positive aspect concerning both C18 and freeze-out cleanup is that neither pesticides nor the proposed internal and QC-standards are affected. In GPC, however, losses of early eluting high molecular size pesticides (e.g., pyrethroids) as well as adsorptive losses of basic pesticides are frequently observed.

46.5.3
Expanding the Commodity Spectrum Covered by QuEChERS

The original QuEChERS procedure only focused on high water and low fat containing commodities such as fruits, vegetables, and juices (group A in Table I).

Other types of commodities, however, require some modifications to account for a low water content, a high lipid content or a high load of co-extractives. In Table I, the various commodities are grouped based on their lipid and water content and the method modifications to be applied in each case are shown. The table may also serve as guidance on how to handle commodities not explicitly named here such as commodities of animal origin. Some of the proposed modification will be discussed in the following:

Commodities containing less than 80% of water generally require the addition of water so that its total mass in the extraction batch reaches approximately 10 g. An exception is dry samples with very high fat content such as oils. An addition of water in this case would negatively influence the recoveries of highly non-polar compounds as shown in Figure 4. For commodities having water contents lower than 30%, the weight of the test portions is typically reduced (groups C to H in Table I).

Especially challenging are **commodities with intermediate or high fat content.** According to our experiments, the solubility of vegetable oil in raw QuEChERS extracts is ~4 mg/mL (~40 mg in 10 mL). In the presence of water, the solubility is slightly lower (~2 mg/mL). Due to this low solubility, excess lipids form an additional layer into which highly lipophilic pesticides tend to partition. The partitioning rate depends on the pesticide's lipophilicity, the lipid/acetonitrile ratio as well as on the presence or absence of water, which obviously modifies the polarity of the acetonitrile phase, making it less receptive for lipophilic compounds.

The partitioning behavior of lipophilic pesticides, both in presence and absence of water, was studied in order to define the limits of lipid tolerance. Figure 4 shows the recoveries achieved for various representative lipophilic compounds in the presence of various amounts of oil, both in the presence and absence of water. The most affected of the compounds shown in Figure 4 seems to be the PCB 138 followed by hexachlorobenzene (HCB, log $K_{ow} = 5.66$), p,p'-DDE (6.51), α-endosulfane (4.74) and lambda-cyhalothrin (7.0). Interestingly, this order does not correlate either with the order of log K_{ow} or with the water solubility values with highly chlorinated compound showing a higher affinity towards the oil-phase that their octanol:water partitioning co-efficient would suggest. Considering the general validation requirement of at least 70% mean recovery, the amount of lipid tolerated to be present within one test portion will mainly depend on which pesticides are included in the target spectrum. For example, should the environmental contaminants HCB and p,p'-DDE be included, the approximate lipid tolerance will be 0.1 g in presence and 0.5 g in absence of water. However, if endosulfane, which is still used in agriculture, is set as border, the lipid tolerance enhances to approximately 0.5 and 2.0 g, respectively.

The sample weights suggested in Table I constitute a compromise between the recoveries of the lipophilic pesticides, the limits of detection, and the sub-sampling variability. For the commodity groups A–E, the sample weights given in the table ensure HCB-recoveries > 70%. For the commodity groups F–H,

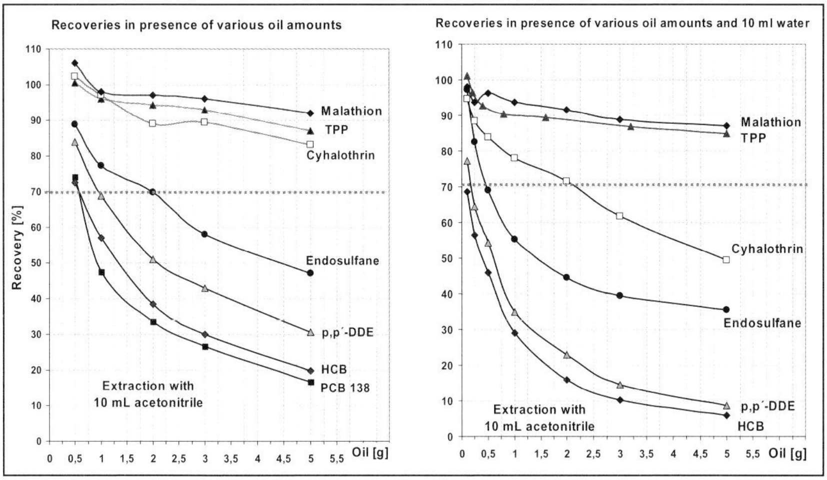

Figure 4. Recoveries of lipophilic pesticides at various oil amounts in presence or absence of water.

however, the situation asks for compromises. For example, in the case of avocado flesh with approximately 20% fat content, 5 g sample will contain approx. 1 g of lipids which theoretically would mean that the 70% recovery requirement would be met by endosulfane but not by p,p'-DDE and HCB. In the case of vegetable oils (group H), where the test portion is 2 g and water is not added, p,p'-DDE and HCB recoveries are clearly below 70%. However, since the partitioning system is very constant, the recoveries are highly reproducible so that recovery correction is justified. Group G is a special case since water content is very low (< 8%) and fat content high. If extraction is performed without the addition of water, lipophilic pesticides give higher recoveries. On the other hand, water is often indispensable when extracting polar pesticides from dry commodities and furthermore it visibly improves the extraction of samples with peanut-butter- or tahina-like consistency. To clarify this, experiments with such type of samples containing incurred residues of polar pesticides should by performed.

In the presence of fat, special attention should be paid on the choice of the ISTD as well as on the stage of the analytical procedure in which the ISTD is added to the sample. Any ISTD added at the beginning of the procedure should not experience any significant losses during partitioning or cleanup so that the recovery correction factor introduced remains small. The recovery tolerance for the ISTD will not be 70%, as for the pesticides, but rather 95%. Thus, when dealing with commodities with a high fat content, it is better to add the ISTD to an aliquot of the end-extract assuming its volume is identical to the acetonitrile volume added at the beginning of the procedure. Since PCB-138 has higher lipid-affinity than any other pesticide, it can be used as a QC-standard (see also Figure 4). A PCB-138 recovery exceeding 70% will indicate that there were no unacceptable losses, even of the most lipophilic pesticides, due to partitioning into the lipid phase.

Apart from the dry and high lipid content commodities, commodities with a high amount of co-extractives are also difficult to handle. Commodities with a high chlorophyll content require cleanup with GCB as described above. Commodities with very high sugar content do not pose many problems as sugar is removed both during partitioning as well as during cleanup with $MgSO_4$/PSA. Using both enzymatic and gravimetric analysis it was shown, that the more $MgSO_4$ and/or PSA is used during D-SPE, the more sugar is removed. The lower the water content in the acetonitrile extracts, the higher the activity of PSA, and the lower is the solubility of various polar compounds such as sugars. Fermented commodities such as tea contain a multitude of polyphenols and other fermentation products. A reduction of the sample size is thus indicated but the use of more PSA for cleanup is also very helpful. Even more effective is the use of $CaCl_2$ instead of $MgSO_4$ as drying salt due to its higher water affinity. A disadvantage of $CaCl_2$ is that it becomes liquid in the hydratized form and that it induces losses of polar compounds such as methamidophos. If such polar compounds are not part of the target spectrum, $CaCl_2$ is a serious option for cleanup. 25-mg $CaCl_2$ together with 25-mg PSA per mL extract are usually enough. Commodities with a high content in essential oils, such as spices, remain a problem since typical essential oils components, such as terpenoids, cannot be separated from pesticides neither via polarity differences nor by size (using GPC). Chromatographic and mass-spectrometric separation play a key role here.

It should be mentioned, that the commodity grouping shown in Table I is not the same as the commodity grouping typically suggested for method validation purposes where, in addition to the water and lipid content, also the sugar content, the acidity, and sometimes also the chlorophyll content are considered. For example, group A in Table I covers commodities of "high water content" (e.g., cucumber), "high acidity" (e.g., citrus fruits) and "high chlorophyll content" (e.g., spinach). "High sugar content" commodities would be covered by the groups C (e.g., raisins) and D (e.g., honey), "dry" commodities (e.g., wheat flour) by group D and "high fat commodities" by group F (e.g., avocado), G (e.g., peanuts) or H (e.g., oil). The so called "difficult commodities" would correspond to group E.

46.6 Measurement

As the QuEChERS-extracts are solved in acetonitrile, they are directly amenable to GC- and LC-applications. However, since acetonitrile is rather difficult to handle by GC using split/splitless inlets, the use of a PTV with solvent vent possibility is highly recommended. Should a PTV not be available and the desired pesticide detection limits cannot be achieved using the split/splitless technique, extract concentration followed by a solvent exchange, if necessary, may be considered. If GC-MSD is employed, a simple evaporative concentration of the extracts by a factor of four should be sufficient. To achieve this, e.g., a 4-mL extract (acidified to pH 5) is transferred into a test tube and reduced to *ca.* 1 mL at 40 °C using a slight nitrogen flow. Solvent exchange is an option if GC performance using acetonitrile is not satisfactory or if GC-NPD is employed without PTV-inlet. For this, an extract aliquot is evaporated to almost dryness at 40 °C using a slight nitrogen flow and redissolved in 1 mL of an appropriate solvent (some droplets of a keeper e.g., dodecane, can help to reduce losses of the most volatile compounds). The blank extract (needed for the preparation of calibration solutions) should be treated the same way. In any case, the use of analyte protectants has been shown to significantly improve chromatography and reduce matrix-induced effect related errors in GC analysis [11–12].

As regards LC-MS/MS, most compounds can be analyzed in the ESI (+) mode. There are several acidic compounds, however, that are more sensitive in the negative ESI (–) mode. To avoid precipitation of non-polar pesticides due to solubility shift, any dilution of extract with the aqueous component of the LC-eluent is better to be performed automatically in the instrument injector itself.

46.7 Validation

At the CVUA Stuttgart, the present procedure has been validated for more than 500 pesticides and metabolites. The method was also validated in four interlaboratory studies performed within the aim of the Pesticide Working Group of the German Chemical Society (GDCh), the Pesticide Working Group of the German Official Laboratories (BLAPS) and the EU-Community Reference Laboratory for Pesticide Analysis using Single Residue Methods. These studies included recovery experiments of different pesticide mixtures fortified on various representative commodities (acidic, dry, high-water content and high-sugar content) at different levels. One study involved GC-MS and LC-MS/MS analysis at levels 0.25 and 0.025 mg/kg using lettuce, cucumber, and orange matrices and the other three only LC-MS/MS analysis at levels 0.01 and 0.1 mg/kg using lemons, cucumber, wheat flour, and raisins. In total, 134 pesticides representing various classes were validated so far by 3–8 laboratories and more than 23,000 individual recovery values were collected within this frame. Average recoveries were in most cases higher than 95% and variations in most cases lower than 6%, thus showing the method's ability to deliver accurate and precise results. More details can be found in the web-site www.quechers.com as well as in a CEN-procedure, which is currently in preparation.

46.8 Conclusion

The modifications introduced to the QuEChERS method in order to prevent degradation of pH sensitive pesticides, improve selectivity of partitioning and cleanup, and to expand the spectrum of commodities covered, clearly improved the applicability of the method. A compromise pH range of 5–5.5 during extraction is adjusted by a citrate buffer and contributes in improving the selectivity of partitioning as well as the stability of alkaline- and acid-labile compounds. The degradation of alkali-labile compounds during extract storage following dispersive SPE-cleanup with PSA is avoided by the addition of formic acid. Various options for cleanup of special commodity co-extractives are presented such as GCB for chlorophyll, ODS for lipids, and PSA/$CaCl_2$ for fermented products. Freezing-out was demonstrated as a highly efficient way of removing various types of co-extractives such as oils, waxes, and sugars. The presented method has been successfully validated in various inter-laboratory studies and will be soon adapted as a CEN standard method and in Germany as an official national method.

46.9 Acknowledgment

We would like to thank Dr. Erhard Schulte from the Institute of Food Chemistry of the University of Münster/Germany for fruitful discussions.

46.10 References

1 M. Anastassiades, S. J. Lehotay, D. Stajnbaher, F. J. Schenck, *J. AOAC Int.*, **2003**, *86*, 412.
2 M. Okihashi, Y. Kitagawa, K. Akutsu, H. Obana, Y. Tanaka, *J. Pest. Sci.*, **2005**, *30*, 368.
3 A. Hercegova, M. Domotorova, D. Kruzlicova, E. Matisova, *J. Sep. Sci.*, **2006**, *29*, 1102.
4 S. J. Lehotay, A. de Kok, M. Hiemstra, P. Van Bodegraven, *J. AOAC Int.*, **2005**, *88*, 595.
5 J. Hajslova, T. Cajka, O. Lacina, J. Ticha, *European Pesticide Residues Workshop*, **2006**, Korfu, Book of Abstracts, C. Lentza-Rizos (Ed.).
6 C. Diez, W. A. Traag, P. Zommer, P. Marinero, J. Atienza, *J. Chromatogr. A*, **2006**, *1131*, 11.
7 F. J. Schenck, J. E. Hobbs, *Bull. Environ. Contam. Toxicol.*, **2004**, *73*, 24.
8 S. J. Lehotay, K. Mastovska, A. R. Lightfield, *J. AOAC Int.*, **2005**, *88*, 615.
9 M. Anastassiades, *MGPR 2003 Aix en Provence*, Book of Abstracts, M. Montury (Ed.).
10 S. J. Lehotay, K. Mastovska, S. J. Yun, *J. AOAC Int.*, **2005**, *88*, 630.
11 M. Anastassiades, K. Maštovska, S. J. Lehotay, *J. Chromatogr. A*, **2003**, *1015*, 163.
12 K. Maštovska, S. J. Lehotay, M. Anastassiades, *Anal. Chem.*, **2005**, *77*, 8129.

Keywords

QuEChERS, Dispersive SPE, D-SPE, Pesticides, Multiresidue Analysis, LC-MS/MS, PSA, GCB, Freeze-Out

47
Summary of Scientific Programs in 11th IUPAC International Congress of Pesticide Chemistry

Hisashi Miyagawa, Isao Ueyama

47.1
Introduction

Under the theme of "Evolution for Crop Protection, Public Health and Environmental Safety", the 11th IUPAC-International Congress of Pesticide Chemistry was held from August 6th to 11th, 2006, in Kobe, Japan. The scientific program opened with the Keynote Address entitled "Challenges and Opportunities in Crop Production Over the Next Decades" presented by J. C. Collins* (DuPont Crop Protection, USA). The scientific program of this Congress included 4 plenary lectures, a total of 114 lectures in 20 technical sessions and poster sessions composed of more than 575 papers. Two special workshops were also conducted to focus on the newly introduced residue management system in Japan and on the topic of mosquito vector control. Furthermore, a total of 28 luncheon/evening seminars were held to address interdisciplinary issues around current crop protection and production. This overview highlights the scientific programs of this Congress.

47.2
Plenary Lectures

First, K Mori* (Tokyo Univ., Japan), the chairperson of the Executive Committee of this Congress, summarized the history of pesticide use in Japan and contributions from Japanese chemists to new pesticides discovery. He reviewed his 50 years of synthetic natural product chemistry research in his quest for a new type of environmentally benign pesticides. S. Pandey* (FAO) presented a talk "Hunger and malnutrition amidst plenty: what must be done?" After he briefly surveyed the present status of hunger and poverty in the world, the activities and priorities of FAO were introduced along with what we can do to eradicate poverty and hunger in the world. Y.-Z. Yang* (ICAMA, China) reviewed the current status of pesticide management of China, which attracted great attention with a detailed

Pesticide Chemistry. Crop Protection, Public Health, Environmental Safety
Edited by Hideo Ohkawa, Hisashi Miyagawa, and Philip W. Lee

ISBN: 978-3-527-31663-2

background on China's dramatic agriculture system progress over the last 25 years. K. D. Racke* (Dow AgroSciences and IUPAC) presented on "Food safety assessment and international trade implications of pesticide residue in food" to highlight the necessity of more globally harmonized MRLs.

47.3 Session Lectures and Special Workshops

Session 1 Drug design based on agrogenomics

Organizers: U. Schirmer (Germany), M. Akamatsu (Japan)

First, U. Schirmer (Consultant, Germany) reviewed the recent advances in modern technologies such as functional genomics, transcriptomics, proteomics, metabolomics and bioinformatics that have a striking impact on drug design research. A. Klausener* (Bayer CropScience, Germany) stated how modern techniques made the mode-of-action studies of biologically active compounds more efficient, and demonstrated how the better knowledge of mode of action facilitated the development of flubendiamide. An increasing availability of genome information of insects and other animals also provides more chances to find new targets of drugs. H. Noda (NIAS, Japan) overviewed the way of applying genome information to discover new insecticides. The utility of a unique model organism, *Caenorhabditis elegans*, for identifying new potential targets was exposited by R. C. Ackerson (Devgen, Belgium). The similar approaches in the field of herbicide and fungicide research were surveyed by T. Ehrhardt (BASF, Germany) and by S. J. Dunbar* (Syngenta, UK), respectively. In spite of the technical advances and much hope therewith, successful examples of target-based screening are yet to materialize. All the speakers pointed out the necessity of an integrated approach of target-based and classical screening using smaller libraries with higher quality.

Session 2 Biopesticides and transgenic crops

Organizers: T. Yamamoto (USA), H. Ohkawa (Japan), J. E. Dripps (USA)

This session presents biological approaches for crop protection/production. A rapid growth in cultivating transgenic crops containing insect-resistant and herbicide-tolerant genes has caused certain changes in the practice of agriculture. After T. Yamamoto* (Pioneer HiBred International, USA) presented an overview on the subject matter, three lectures were given relating to the application of *Bacillus thuringiensis* (Bt). P Warrior* (Valent BioSciences, USA) reviewed sprayable biopesticides and M. J. Adang (Univ. of Georgia, USA) talked about his recent findings on the mode of action. A rapid and efficient way of discovering bacterial genes that are useful in transgenic crops was offered by M. Koziel (Athenix, USA). Concerning the development of herbicide tolerant crops, B. K. Singh (BASF Plant Science, USA) discussed non transgenic and transgenic approaches to endow crops with tolerance to imidazolinone herbicide. A unique approach of utilizing a transgenic plant was introduced by H. Inui* (Kobe Univ., Japan), in which an animal gene of arylhydrocarbon receptor was applied for sensitive detection of environmental contaminants.

Session 3 New chemistry

Organizers: E. Kuwano (Japan), G. D. Crouse (USA), U. Mueller (Switzerland)

Undoubtedly, this Kobe Congress will be best remembered for the chemistry and biochemistry of new ryanodine receptor-acting insecticides. Monumental lectures were given by G. P. Lahm* (DuPont Crop Protection, USA) and A. Seo* (Nihon Noyaku, Japan), dealing with chlorantraniliprole (Rynaxypyr™) and flubendiamide, respectively. These products are effective for controlling a variety of lepidopteran species, with a lack of cross-resistance with existing insecticides. The session also highlighted innovative new herbicides and fungicides. M. Muehlbach* (Syngenta, Switzerland) presented a talk about the chemistry of pinoxaden and its combination with a safener and an adjuvant to create a highly effective cereal herbicide with control of grass weeds. T. C. Johnson* (Dow AgroSciences, USA) described how reversing the direction of the sulfonamide group relative to previous broadleaf sulfonamide herbicides results in improved grass weed activity, leading to penoxsulam, a herbicide that controls grasses, sedges, and broadleaf weeds in rice. As to the new fungicides, synthesis of boscalid was described by M. M. Keil (BASF, Germany). Sophisticated application of the Suzuki coupling reaction was a key to obtain an important intermediate, which led to the development of a multipurpose fungicide used for specialty crops. T. Wegmann (Bayer CropScience, Germany) presented the chemistry and biology of fluopicolide, a highly effective oomycete fungicide. Fluopicolide represents a new chemical family with a novel mode of action showing no cross-resistance with other oomycete fungicides. It provides consistent high level performance often able to set a new standard in oomycete control in a wide range of crops.

Session 4 Natural products

Organizers: M. Iwata (Japan), J. R. Coats (USA), P. Lewer (USA)

A wide range of approaches in attempts to develop new products from natural sources was demonstrated. S. O. Duke (USDA-ARS, USA) introduced their ongoing research to increase the amount of allelopathy substances in crops by genetic manipulation, in order to make the crops resistant to weeds. M. J. Everett (MerLion Pharmaceuticals, Singapore) made a presentation on their huge sample collection and how they are proceeding with the screening in a highly efficient manner to find novel bioactive compounds. N. Orr (Dow AgroSciences, USA) reported the findings on the mode of action of spinosad, one of the most successful insecticides of natural product origin in recent years, whereby spinosad was demonstrated to affect the nicotinic acetylcholine receptors in insects' nervous systems in a different manner from neonicotinoids. N. B. Perry (Univ. of Otago, New Zealand) delivered a talk about the potential of bioactive molecules found in the unique indigenous plant species of New Zealand. M. B. Isman* (Univ. of British Columbia, Canada) reviewed the biological activity of essential oil from the aspect of pesticidal use and focused on the effectiveness of rosemary oil for controlling two spotted spider mites on vegetable crops without a negative impact on their natural enemies. A. Fukuzawa* (Hokkaido Tokai Univ., Japan) described the chemical interaction between potato cyst nematode and its host plant. With the increasing knowledge of chemical factors that play a critical role in establishing the parasitism, a possibility of eco-chemical control was mentioned.

Session 5 Bioregulator for crop protection
Organizers: K. Yoneyama (Japan), R. A. Menendez (USA)

The session was one of those designed in this congress for the consideration of a new facet of utilization of chemicals in agriculture. Regulation of plant growth and development by chemicals is an important strategy to enhance and improve the quality of plant production. T. Asami* (RIKEN, Japan) talked of chemical biology in the field of plant science. He demonstrated how the specific inhibitors of plant hormone biosynthesis contributed to clarify the hidden details in the hormone action and perception. E. A. Curry (USDA-ARS, USA) showed successful use of plant growth regulators, in particular, 1-methylcyclopropene (1-MCP), in fruit quality improvement. P. Hedden (Rothamsted Research Station, UK) described the effectiveness of overexpression of gibberellin (GA) 20-oxidase for elevating the GA content and enhancing the growth, while the overexpression GA 2-oxidase, GA-deactivating enzyme, was the most efficient for depletion and useful in improving grain quality in wheat. P. D. Petracek (Valent BioSciences, USA) reported the modulation of herbicide efficacy by salicylate. Salicylate reduced efficacy of paraquat by inhibiting transport of paraquat to its site of action. On the contrary, it potentiated the actions of photosystem II inhibitors, protoporphyrinogen oxidase inhibitors and glyphosate, by the inhibition of stress tolerance system in plants. T. Tanaka (Cosmo Oil, Japan) discussed plant growth promotion by 5-aminolevulinic acid (ALA), a precursor of chlorophylls and heme. These talks definitely demonstrated the potential of synthetic and natural plant growth regulators to contribute to plant production.

Session 6 Control agents for vectors and communicable diseases
Organizers: J. M. Clark (USA), N. Matsuo (Japan)

This session convened for the first time in the history of the IUPAC/ICPC meetings to cover public health issues and the role of chemical control in dealing with them. Y. Shono* (Sumitomo Chemical, Japan) presented information on metofluthrin, a novel pyrethroid and innovative mosquito control agent with high vapor pressure and potency, which is effective in mosquito coils, fan vaporizers, etc. Y. Eshita* (Oita Univ., Japan) reviewed the vector competency of common species of Japanese mosquito to transmit dengue and West Nile virus and found that only *Aedes* transmitted dengue virus due to midgut barriers in the other genus of mosquitoes. *Culex* mosquitoes and *Aedes albopictus* allowed the replication and transmission of West Nile virus. E. D. Walker (Michigan State Univ., USA) presented data supporting the notion that commercial long-lasting insecticide-treated nets protect human hosts by attracting mosquitoes and killing them rather than repelling mosquitoes. Additionally, the operational dose on the nets appears to control pyrethroid resistant mosquitoes. J. G. Vontas (Agricultural Univ. of Athens, Greece) presented a functional genomics approach in identifying non target site (metabolic) resistance mechanisms in *Anopheles* malaria vectors. Microarray analysis using the "Detox Chip" identified a number of metabolic resistance mechanisms. W. S. Leal (Univ. of California-Davis, USA) used a "reverse chemical ecology' approach to screen for novel mosquito ovipositional attractants by studying the binding to odorant-binding proteins and activation of odorant receptors. J. M. Clark (Univ. of Massachusetts, USA) has developed an *in vitro* rearing system for human head lice and established that permethrin resistance occurs worldwide, is due to *kdr* nerve insensitivity, and can be efficiently and affordably monitored using SISAR diagnostic technology.

Special workshop for "Mosquito Control"

This session was linked with the special workshop featuring mosquito control, in which 6 speakers presented the topics on the current epidemiological situation, insecticide resistance and technologies for the control. P. Paeporn (Ministry of Public Health, Thailand) explained about deltamethrin and permethrin resistant strains of *Aedes aegypti* in Thailand. The resistance development was analyzed both behaviorally (attractancy and repellency) and physiologically (penetration into cuticle, target site, and metabolic detoxification). S. Kasai (National Institute of Infectious Diseases, Japan) discussed the insecticide susceptibilities of the West Nile virus vector mosquitoes collected from Japan. The results of resistance development of *Culex* colonies to etofenprox were rather high, and cross-resistance to other pyrethroids was also high. T. Tomita (National Institute of Infectious Diseases, Japan) presented on the pyrethroid resistance of *Culex pipens molestus* and *C. p. pallens* colonies collected around Tokyo, and analyzed on degrees of target sensitivity (sodium channel) and increased detoxifying metabolism (cyctochrome P450). H. Kawada (Nagasaki Univ., Japan) talked on field evaluation of spatial repellency of metofluthrin-impregnated plastic strips against vector mosquitoes. The device disrupted orientation but was positively affected by average room temperature and negatively affected by opening areas of the room. J. Nash, (Bayer CropScience, Singapore) introduced film forming aqueous spray technology, or FFAST as a major technique for control of the mosquitoes. Finally, T. Itoh (Sumitomo, Japan) presented the effectiveness of their product Olyset®, which is used as a major tool in the RBM (Roll Back Malaria) campaign.

These two sessions dealing with vector control issues were further supplemented by a luncheon seminar entitled "Prospects for use of insecticide treated materials in personal protection and chemical control of vector borne diseases" by P. F. Guillet (WHO, Switzerland). The importance of technology using insecticide-incorporated or -coated polymer fibers was pointed out. Comments were also given that WHO was in favor of the indoor use of DDT for controlling malaria vectors.

Session 7 Mode of action and resistance mechanism – insect control
Organizers: Y. Ozoe (Japan), J. G. Scott (USA), P. Maienfisch (Switzerland)

S. W. Chouinard (Cambria Bioscience, USA) presented his study of elucidating the site-of-action of insecticides, based on the genetics of *Drosophila*. Topics on the targets of picrotoxin and spinocyn A were dealt with. S. R. Palli (Univ. of Kentucky, USA) described the recent findings on the action mechanism of juvenile hormone analogs. The effect of methoprene on the transformation of midgus tissues were demonstrated in the context of programmed cell death caused by molting hormone. Concerning neonicotinoids, three talks were given: H. Kayser (Syngenta, Switzerland) suggested the presence of multiple action sites of neonicotinoids, based on the analysis of interaction between thiamethoxam and membrane fraction from aphids. K. Matsuda* (Kinki Univ., Japan) examined the molecular interaction between neonicotinoids and acetylcholine receptor in detail to elucidate the amino acid residues in the receptor that play determinant role in the expression of selective toxicity. R. Nauen (Bayer CropScience, Germany) overviewed the occurrence of neonicotinoid resistance, and demonstrated that a mutation in the alpha-subunit of acetylcholine receptor was one of the factors that gave rise to the resistance. Y. Kono (Tsukuba Univ., Japan) presented a topic on organophosphate/carbamate resistance, showing that mutation in *Ace*-paralogous acetylcholinesterase relates to the reduced sensitivity. J. G. Scott (Cornell Univ., USA) described the molecular mechanism of pyrethroid resistance, in which the involvement of the mutations in the sodium channel gene and a cytochrome P450 gene (*CYP6D1*) in the resistant houseflies in the US was discussed.

Session 8 Mode of action and resistance mechanism – weed control
Organizers: H. Matsumoto (Japan), S. B. Powles (Australia), G. Donn (Germany)

G. Donn (Bayer CropScience, Germany) described gene expression profiling as a diagnostic tool for studying herbicide mode-of-action. A commercially available chip loading all the genes of Arabidopsis was effectively used for grouping the herbicide actions through fingerprinting analysis of the expression profile on the microarray after the treatment of the plant with herbicides. K. Grossmann (BASF, Germany) extended the topic of mode-of-action diagnosis by presenting the utility of physiological profiling, or physionomics. The response patterns of herbicides to a defined set of 10 conventionally used bioassays were successfully integrated to build up a database, which provided a typical fingerprint per mode-of-action and thus was useful in categorizing the new compounds. S. B. Powles (Univ. Western Australia, Australia) reviewed major target site resistance mechanisms/ mutations concerning PS II inhibitors, acetolactate synthase inhibitors, acetyl CoA carboxylase inhibitors and glyphosate. K. Kreuz (BASF, Germany) discussed non-target site resistance to herbicides. Metabolism-based resistances associated with cytochrome P450's, glutathion-S-transferases, and aryl acylamidases were exemplified, along with a specific resistant mechanism against bipyridyl herbicides. P. C. C. Feng (Monsanto, USA) demonstrated that the mechanism conferring glyphosate resistance in Conyza populations was not associated with the mutation of the target site but with the decreased transport to the target organs and/or tissues.

Session 9 Mode of action and resistance mechanism – plant disease control
Organizers: H. Ishii (Japan), K. Kuck (Germany), B. A. Fraaije (U.K.)

S. Mitani (Ishihara Sangyo, Japan) described that the target site of newly developed cyazofamide was located in complex III in mitochondria. The site was referred to as Qi center, different from that of strobilurin fungicides, Qo, offering the basis for the absence of cross-resistance. Y. Yanase* (Mitsui Chemicals, Japan) addressed an evolution of complex II inhibitor and demonstrated the advantages of penthiopyrad over the old type inhibitors used for long time in terms of broader spectrum of control. R. Beffa (Bayer CropScience, France) reviewed the action of fluopicolide. The compound is active on oomycetes, but has a different mode of action from those of existing fungicides. Implication of cytoskeletal spectrin-like proteins was suggested. B. T. Navé (BASF, Germany) reported that a new compound, metrafenone, developed for the control of powdery mildew induced swelling and collapse of the hyphal tips as a result of weakening of the cell wall by disruption of the F-actin cap. H. Ishii (NIAES, Japan) reviewed the disease-resistance inducers including probenazole, tiadinil, isotianil, and acibenzolar-S-methyl. Recent findings on the mechanism and the effective application of acibenzolar-*S*-methyl were focused. M. Fujimura (Toyo Univ., Japan) showed that iprodione and fludioxonil interfere with osmotic signal transduction pathway in fungi, and mutations in the relevant proteins give rise to the resistance to these fungicides. A new PCR-based method was also presented to detect the field isolates of *Botrytis cynerea* that are resistant to dicarboximides and benzimidazoles. The current resistance status was reviewed for the two most important fungicide classes. B. A. Fraaije (Rothamsted Research Station, UK) surveyed the studies on the resistance mechanisms of sterol demethylase inhibitors in plant pathogenic fungi, in comparison with the cases of human pathogens. In both cases, resistance is associated not only with the mutation in the target molecule, but also with the enhanced elimination by ABC transporters. K. H. Kuck* (Bayer CropScience, Germany) summed up the resistance mechanisms to QoI fungicides including strobilurins. Among the reported mutations in the cytochrome b gene, the change from glycine to alanine at position 143 is most frequently observed and of practical importance.

Session 10 Advances in formulation and application technology
Organizers: J. A. Zabkiewicz (New Zealand), L. D. Gaultney (USA), X.-K. He (China), K. Isono (Japan)

The session was started with a review by P. J. Mulqueen (Syngenta, UK) focusing on the current smart formulations involving a.i. combinations, stable emulsions with graft copolymer technology, and controlled release microencapsulation controlled by pH effects. W. L. Geigle* (DuPont Crop Protection, USA) gave an exposition of homogeneous granule blend technology which has simplified operational practices, cut down on regulatory lead in times and is providing tailored product compositions for different end-users. M. Bratz (BASF, Germany) talked of the potential of nanotechnology. The huge change in physical, electrical, optical, absorptiveness, and bioavailability properties of 100-nanometer particulates is mind-boggling, and left an impression of a future megashift in formulation potential. T. Watanabe (Agro Kanesho, Japan) presented a 'logistic-kinetic' dynamic model of pesticide foliar uptake. The effect of different adjuvants can influence critical uptake parameters. H.-Z.Yuan (Chinese Academy of Agriculture Science, China) noted microemulsion formulation registrations had gone from 33 in 2000 to 353 in 2005. An example of the benefits of such formulations was given using a chlorpyrifos formulation *vs.* a standard EC product. T. M. Wolfe (Agriculture & Agri-Food Canada, Canada) calculated that air-inclusion nozzle technology could lead to 612,000 kg less each year of pesticide pollution in Canada. However, an understanding of the influences of carrier volume, product type, and target plant to get best results is required.

Session 11 Metabolism and toxicology

Organizers: K. Tanaka (Japan), Y. Kim (Republic of Korea), J. H. Krauss (Switzerland), R. I. Krieger (USA)

M. W. Skidmore* (Syngenta, UK) presented the results from an ongoing research project on the bioavailability of conjugated metabolites and bound residues. Y. Yamazoe (Tohoku Univ., Japan) gave an overview over cytochrome P450 enzymes in the human metabolism. Prediction of cyctochrome P450 enzyme induction by pharmaceuticals or agrochemicals was also discussed. Presentation by R. I. Krieger* (Univ. of California- Riverside, USA) dealt with the issues of chemical exposure to the human in relation to the occupation and the residence near the site of pesticide use. The importance of biomonitoring was advocated to reduce the dissociation between the actual exposure and the perceived risk to the public. D. Myers (Huntingdon Life Sciences, UK) gave a detailed and informative insight into the study design of milk transfer studies and discussed its relevance for developmental neurotoxicity. An originally programmed presentation by R. L. Rose (North Carolina St. Univ., USA) entitled "Pesticide metabolism and metabolic interactions in humans" was not performed due to the unexpected sudden death of the speaker by a car accident just before the congress. We take this opportunity to offer our sincerest condolences to his family.

Session 12 Resistance management and IPM

Organizers: R. ffrench-Constant (U.K.), H. Nemoto (Japan)

P. C. Jepson (Oregon State Univ., USA) discussed the effects of pesticides on natural enemies that play a critical role in IPM. The factors governing the key processes underlying pesticide impact were analyzed to explore ways of mitigating the side effects on beneficial organisms. R. ffrench-Constant* (Univ. of Bath, UK) described that neonicotinoid resistance was conferred by DDT-R gene in *Drosophila,* and therefore required no fitness cost. Y. Suzuki (NARO, Japan) reviewed the history and status of rice IPM in monsoonal Asia including Japan. Problems in keeping good implementation of IPM currently arise from economic difficulties of farmers as well as the shrinking government services, and the measures for overcoming such problems were discussed. M. Osakabe (Kyoto Univ., Japan) demonstrated the changes in species composition of spider mites in fruit orchard in association with pest management. The manner of divergence of spider mites depends on species, which affects the manner of acaricide resistance dispersion. R. Nauen (Bayer CropScience, Germany) introduced the activity of the Insecticide Resistance Action Committee of CropLife International to promote the development of resistance management strategies in crop protection and non-agricultural insecticide use. Finally, S. M. Williamson (Rothamsted Research, UK) kindly presented on a mechanism-based approach to monitoring for insecticide resistance in the peach-potato aphid, *Myzus persicae*, as an urgent substitute to the originally programmed talk by I. Denholm.

Session 13 Environmental chemistry/Residue analysis

Organizers: R. A. Yokley (USA), F.-M. Liu (China), S. Jackson (USA), I. Saito (Japan)

J. D. Vargo (Univ. of Iowa, USA) described the modifications and improvements to US EPA Method 535 (analysis of chloroacetanilides, chloroacetamides, and their chloroethanesulfonic and oxanillic acids degradates in water), resulting in the increase in sensitivity as well as the significant decrease in the sample analysis time. J. S. Corley (Rutgers, State Univ. of New Jersey, USA) presented a discussion of the need for standardized definitions of limits of detection, limit of quantification, and the error associated with each measurement of analytical methods, from a viewpoint of global harmonization of maximum residue levels (MRLs) and uniform risk assessments. H. Tamura (Meijo Univ., Japan) presented his work on the analysis of alkylphenols and their polyethoxylates (endocrine disruptors in fish) in soil using matrix-assisted laser desorption ionization-mass spectrometry (MALDI). Due to the difficulties associated with analyzing these compounds using traditional GC and LC techniques, the proposed technique is supposed to make a significant contribution to clarify their fate in soil, which has been less studied compared to the fate and behavior in water. M. Anastassiades* (CVUA Stuttgart, Germany) discussed the QuEChERs (quick, easy, cheap, effective, rugged, and safe) method and how its applicability had been expanded to include additional analytes and sample matrices. Details of the procedure were described and how the final fractions could be analyzed using either GC/MS or LC/MS/MS. A new aspect of residue analysis was shown by G.-M. Shan (Dow AgroSciences, USA) in the presentation on the monitoring of Bt proteins in soil, in connection with the risk assessment of GM crops. A novel biomimetic extraction procedure using an artificial work gut fluid was developed, which enabled an efficient quantification by coupling with immunoassay.

Session 14 Environmental risk assessment, regulatory aspects and risk communication

Organizers: H. Yamamoto (Japan), M. Streloke (Germany)

This session convened in association with Session 15, and specific issues concerning the environmental fate of chemicals and new risk assessment methods were covered. Presentations were given on regulatory requirements and risk assessment schemes from the US, Europe, and Japan. The status in the US was described by M. A. Corbin (US EPA, USA) that there is specific attention given to endangered species, and the legislation specifically mentions risk-benefit analysis. M. Streloke* (Federal Office of Consumer Protection & Food Safety, Germany) reported the system of environmental risk assessment in the EU. The inter-relationships and scientific differences in opinion between EU-level and member states was explained, with reference to the currently ongoing revision regarding the data requirements and principles of risk assessment to harmonize the regulation within the EU. J. Koide (Ministry of Environment, Japan) reviewed the regulation in Japan, focusing on risk assessment on paddy rice uses and aquatic organisms. Overall it was shown that data requirements and the basics of risk assessment were similar in the EU and US. Japan is moving in the same direction. K. R. Solomon (Univ. of Guelph, Canada) explained the technique of probabilistic ecological risk assessment. This allows the assessment to be more realistic by taking into consideration the variability and uncertainty in exposure values. Methods for communicating the risks to decision makers as well as to the public were also suggested, which becomes more important as the risk assessment becomes more complicated.

Session 15 Environmental fate and ecological effect

Organizers: A. Barefoot (USA), T. Katagi (Japan), C. Romijn (Germany)

This session preceded Session 14 to review the recent advances in the technology serving for a higher level of environmental risk assessments. S. Hagino* (Sumika Technoservice, Japan) covered the detecting methods for endocrine disrupting effects on aquatic organisms. Efficacy and validity of the assays using medaka (killifish) and daphnids were shown. C. Wolf* (RIFCON, Germany) reviewed the risk evaluation methods toward birds and wild mammals, and discussed the results and the problems thereof. S. Ishihara (NIAES, Japan) demonstrated a new method to evaluate the effects on diatom, the major algal species in river systems in Japan. The results on the effects of herbicides were presented. Current European testing and risk assessment methodologies on honey bees were overviewed by C. Maus (Bayer CropScience, Germany). It was stated that the current set of testing methods almost sufficiently work for pertinent assessment, although further research is needed to define the ecologically significant adverse effects of chemicals. W. C. Koskinen (USDA-ARS, USA) reported an ingenious method to evaluate the sorption of pesticides to soil. Using this method, the effects of aging on the sorption of herbicides including triazines and sulfonylureas were examined, revealing that the soil sorption of pesticides and their bioavailability by the soil microorganisms were governed by extremely complex factors. M. A. Corbin (US EPA. USA) explained the NAFTA (North American Free Trade Agreement)-harmonized guidance for the studies of pesticide field dissipation, developed by the collaboration between the agencies of the USA and Canada. Concepts as well as the practical instructions for conducting tests were presented.

Session 16 Monitoring and remediation of POPs
Organizers: A. Katayama (Japan), A. S. Felsot (USA), L. L. McConnell (USA)

Global and regional trends in the sources and distribution of persistent organic pollutants (POPs) were highlighted. S. Tanabe (Ehime Univ., Japan) reported the spatial distribution of POPs in the Asia-Pacific region on the basis of their monitoring studies for various sea animals. The contamination in waters near the source areas was shown to be decreased compared to that in the 1970s, but that in remote oceans it continues. A. Chowdhury (Univ. Calcutta, India) reviewed the sources of POPs in tropical Asia including India, Pakistan, and Bangladesh. DDT, which is manufactured and applied for vector control against malaria disease, represents the prime source in these countries. N. Seike (NIAES, Japan) reported the status of contamination in the agricultural fields in Japan from the 1960s to present, based on the analyses of historical soil collections. Concentrations of dioxins in paddy soils increased during the 1960s and 1970s, due to the impurities in herbicides such as PCP and CNP, but decreased thereafter, while the major source of dioxins in soil was changed to the combustion process after the 1980s. Following these, 2 lectures were given to focus on the decontamination process. P. Fogg (ADAS, UK) presented the results of studies on fate and behavior of pesticides in Biobeds. Effectiveness was demonstrated for the degradation of most of the pesticides, so the biobed technology definitely provides a useful measure of treating pesticide waste in farms, thus reducing the risk to aquatic environment. Relative to wastewater treatment, A. T. Lemley (Cornel Univ., USA) described the effectiveness of chemical oxidation using anodic Fenton treatment for the degradation of pesticides. The study results were presented on the reaction mechanisms, the effects on improvement of biodegradability, and the construction of flow-through reaction system that can be adopted for future field use.

Session 17 Emerging technologies in crop protection and production
Organizers: K.-J. Schleifer (Germany), H. Miyagawa (Japan), H. M. Brown (USA)

This session was divided into two completely separate parts. The first part covered recent virtual screening techniques and their growing impact in the classical research process. After K.-J. Schleifer* (BASF, Germany) overviewed the principles of virtual screening techniques and some successful examples, M. Asogawa (NEC, Japan) demonstrated the effectiveness of their originally developed drug-screening system using active learning methods. Combination with the descriptor sampling strategy, which allows more flexible learning, successfully resulted in the selection of more diverse lead compounds from the library in search for the new ligands of G-protein coupled receptors. R. Viner (Syngenta, UK) presented examples of a fragment-based design approach to find inhibitors of acyl-carrier protein enoly reductase, a potential target for new herbicides. The approach contrasts with the conventional lead-optimization strategy. The estimation of reactivity of a compound toward cytochrome P450-catalyzed oxidation was dealt with by M. E. Beck* (Bayer CropScience, Germany). Application of Fukui function to predict the site of reaction in a series of agrochemicals was demonstrated along with the discussion of its validity. The second part of the session focused on natural product chemistry concerning the germination control of plant seeds. Y. Sugimoto (Kobe Univ., Japan) described the biosynthesis and biological activity of strigolactones, the germination stimulants toward weeds *Striga* and *Orobanche* which parasitize the roots of many food crops. It was suggested that germination stimulation occurs through an activation of ethylene biosynthesis in the weed seeds by strigolactone, providing a clue of future protection strategy against this weed. K. W. Dixon (Kings Park and Botanic Garden/Univ. Western Australia, Australia) presented the isolation and identification of butenolide, a unique phytoreactive compound from cellulose-derived smoke. This compound enhances seed germination and deemed as the active principle in smoke. Future application possibilities were also discussed.

Session 18 Genomics, proteomics, and metabolomics
Organizers: D. Ohta (Japan), R. Feyereisen (France), E. Ward (USA), T. A. Walsh (USA)

D. C. Boyes (Monsanto, USA) described their (formerly Icoria) program for the discovery of genes with utility in crop improvement, based on high-throughput precise morphometric and phenotypic analysis of *Arabidopsis* mutants and transformants. Its efficiency was exemplified by the functional analysis of trimeric G-protein in *Arabidopsis*. Ongoing application of this technology to soybean growth and yield was also mentioned. D. Shibata (Kazusa DNA Research Inst., Japan) detailed the integration of transcriptomic and metabolomic data involved in plant secondary metabolism using *Arabidopsis* and Lotus model systems. An extensive network of mapped interactions was defined, requiring the development of a novel strategy to achieve multigene transformations to further probe this system. T. A. Walsh* (Dow AgroSciences, USA) introduced the concept of chemical genetics and described a successful example of identifying several novel herbicide targets using this strategy.

Session 19 Global food quality and human health protection issues
Organizers: M. W. Skidmore (U.K.), K. Watanabe (Japan), B. J. Petersen (USA)

Sessions 19 and 20 were held in a series, the former representing the second part and focusing on issues of risk assessment methods, use of the results for risk management decisions and to communicate with the public. E. D. Caldas (Univ. of Brasilia, Brazil) presented the new JMPR methods for estimating consumer exposures and examples of the impact of better data. Y.-B. He* (Ministry of Agriculture, China) presented methods used by China to estimate exposure and to establish MRLs. N. Tayaputch* (Lab. Center for Food & Agric. Products, Thailand) summarized the extensive work of the ASEAN countries that has resulted in many harmonized MRLs. All of the speakers highlighted the reliance on CODEX and the need for rapid decisions by CODEX. A. S. Takei (ICaRus Japan) explained the new "Positive List" system used by the Japanese government for pesticides, feed additives, and veterinary medicines. G. A. Kleter* (Wageningen Univ., Netherlands) provided a comprehensive look at the rigorous approaches used to confirm the safety of new GM foods as well as identifying some of the unique challenges in safety testing. The session concluded with a panel discussion that highlighted the need for globally harmonized MRLs, better data, as well as better risk assessment methods.

Session 20 Global food safety and trade issues
Organizers: P. W. Lee (USA), E. Carazo (Costa Rica), Y.-B. He (China), N. K. Umetsu (Japan)

J. J. Baron* (USDA, USA) presented an operation model showing how industry, government agencies (e.g., US EPA) and IR-4 can partner together in the generation of GLP residue data for registration of specialty crops. J. N. Seiber* (USDA-ARC, USA) provided an overview of food safety in terms of microbial toxins in food items. P. S. Villanueva (US EPA, USA) described the history of the project to develop a statistically based method for establishing MAFTA-harmonized tolerance levels, principles behind the methods, and the way of actual implementation. Y. Yamada (MAFF, Japan) reviewed the policies and processes of establishing Codex MRLs in food by the Codex Committee on Pesticide Residues, or CCPR. C. P. Gaston* (Exponent, USA) presented the examples of efforts by developing countries to protect their export crops from rejection in more industrialized countries due to pesticide residues. The needs of international harmonization of MRL calculation were highlighted.

Special workshop
"Japan positive MRL system: Regulatory and trade considerations"
S. Miyagawa (MHLW, Japan) explained the background and implementation experience for the new system of food safety regulation, which came into force on May 29, 2006, for some 745 pesticides on both domestic and imported foods. He reported that the first several months of implementation of the positive list resulted in an increase in violations, which came from a trade standard issue. Then, the lecturers from China, USA, and Australia discussed implementation issues and future needs related to agricultural practices and compliance with the new positive MRL system of Japan on the part of farmers and exporters. W. L. Chen (China Agric. Univ., China) stressed the importance of education and stewardship on the part of farmers in China and other Asian nations in achieving a high level of compliance for foods exported to Japan. H. W. Ewart (California Citrus Quality Council, USA) emphasized the difficulties faced by growers and exporters in complying with disharmonized MRL standards of various trading

partners, and stressed the importance of the Codex system of world food MRLs as the best hope for global harmonization. K. Bodnaruk (AKC Consulting, Australia) discussed steps that grower organizations were taking to implement best export practices for meeting export needs for Japan and other markets. In the following panel discussion, needs of compliance with the new positive list system within food exporting countries through effective communication and grower/exporter education were suggested, together with the needs of creative approaches to manage the disparities between the Japan positive MRL list and the GAP of trading partners.

47.4
Poster Session

A total of 577 posters were submitted, and they were categorized into 3 groups. Subcategories in each category are as follows, along with the numbers of posters in the parentheses:

Category I
1. Chemistry Including Natural Products: Control of Insect (65), Weed (20), Disease (33), Plant Growth (21), Vector (3)
2. New Technologies for Lead Generation and Drug Design (20)
3. Biopesticides and Transgenic Crops (13)
4. New Technologies for Pest Control (5)

Category II
1. Mode of Action, Resistance Mechanism and New Targets: Control of Insect (45), Weed (12), Disease (33), Plant Growth (3), Vector (3)
2. Resistance Management and IPM (16)
3. Metabolism and Toxicology (20)
4. Formulation and Application (19)

Category III
1. Residue Analysis (71)
2. Human Exposure (15)
3. Environmental Fate and Ecological Effect (59)
4. Risk Assessment and Regulation (26)
5. Monitoring and Remediation of POPs (17)

Late submission (submitted after April 18, 2006)
Category I (11)
Category II (15)
Category III (32)

- A total of 89 posters were selected to organize 15 workshops for oral presentation (see Appendix) by the following poster session organizers: H. Miyoshi (Japan, Chair)
- Category I: T. Haga (Japan), I. Kumita (Japan), T. Ando, (Japan), J. Lyga (USA), U. Schirmer (Germany), T. Yamamoto (USA), P. Maienfisch (Switzerland), T. Seitz (Germany)
- Category II: K. Matsuda (Japan), Y. Yogo (Japan), K. Motoba (Japan), R. Nauen (Germany), K.-H. Kuck (Germany), L. Gaultney (USA), M. Skidmore (U.K.)
- Category III: Y. Matoba (Japan), K. Sato (Japan), T. Nagayama (Japan), L. McConnell (USA), M. Schocken (USA), M. Anastassiades (Germany), E. Capri (Italy)

Poster Award

The Poster Award Committee [J. B. Unsworth (IUPAC, UK, Chairperson), E. Carazo (Costa Rica), J. M. Clark (USA), D. Hamilton (Australia), B. Hock (Germany), B-W. Krueger (Germany), Z.-M. Li (China), L. McConnell (USA), H. Miyagawa (Japan), H. Miyoshi (Japan), D. R. Wauchope (USA), and B. Rubin (Israel)] selected Gold-, Silver-, and Bronze-prize posters, along with a special prize, from each category. Gold-prize posters were also awarded by IUPAC. Award-winning posters are as follows:

Category I

Gold & IUPAC prize: "Δlac-Acetogenins are a new class of inhibitors of mitochondrial complex" by N. Ichimaru *et al.*, Kyoto Univ., Japan.

Silver prize: "Managing grass weeds in cereals: chemistry and biology of the novel herbicide pinoxaden" by F. Cederbaum *et al.*, Syngenta, Switzerland.

Bronze prize: "5-(2,6-Difluorobenzyl)oxymethyl-5-methyl-3-(3-methylthiophen-2-yl)-1,2-isoxazoline as a useful rice herbicide" by I. T. Hwang *et al.*, KRICT, Korea.

Special prize: "Structure based molecular design of AHAS inhibitors" by J.-G. Wang *et al.*, Nankai Univ., China.

Category II

Gold & IUPAC prize: "A nicotinic acetylcholine receptor point mutation (Y151S) conferring insecticide resistance causes reduced agonist potency to a range of neonicotinoids" by Z-W. Liu *et al.*, Univ. College London and two other organizations, UK and China.

Silver prize (shared by two posters): "Flubendiamide stimulates Ca^{2+} pump activity coupled to RyR-mediated calcium release in lepidopterous insects" by T. Masaki *et al.*, Nihon Nohyaku and Bayer CropScience, Japan and Germany, and "Elucidation of the mode of action of Rynaxypyr^{TM}, a selective ryanodine receptor activator" by D. Cordova *et al.*, DuPont Crop Protection, USA.

Bronze prize: "A new evaluation method for plant defense activators based on potentiation of elicitor-responsive photon emissions (ERPE) in rice cells" by H. Iyozumi *et al.*, Shizuoka Agric. Exp. Station, Japan.

Special prize: "Study of acetohydroxyacid synthase through mutation and QSAR" by Z. Xi *et al.*, Nankai Univ., China.

Category III

Gold & IUPAC prize: "Multiresidue analysis of 500 pesticide residues in agricultural products using GC/MS and LC/MS" by Y. Akiyama *et al.*, Hyogo Pref. Inst. of Public Health and Environ. Sci., Japan.

Silver prize: "Aerobic mineralization of hexachlorobenzene by *Nocardioides* sp. PD653" by K. Takagi *et al.*, NIAES and two other organizations, Japan.

Bronze prize: "Development of acetylcholine esterase detection kit for the determination of organophosphorus and carbamate residues in agricultural samples" by B. M. Kim *et al.*, Chonnam National Univ. and Cairo Univ., Korea, and Egypt.

Special prize: "Modelling environmental behaviour of photosensitive pesticides by revealing features of degradation and possible ways of interactions" by Attila Kiss, Esterházy Károly Univ., Hungary.

Papers of the Gold-prize winners are included in this Proceeding.

47.5
Luncheon and Evening Seminars

A total of 18-luncheon and 10-evening seminars were held, and updated information on new technologies, environmental issues, food safety, and corresponding regulations was offered. They were all sponsored, which represented the first case in the Congress history, and definitely provided strong financial support of this Congress. Followings is the list of topics noteworthy from the aspect of public interest.

For better public relations and communication, N. K. Umetsu, President of Pesticide Science Society of Japan, demonstrated the efforts by the Society to convince, both scientifically and emotionally, the public about issues concerning the safety of pesticide and pesticide residue on food at the seminar entitled "Risk assessment of pesticides – Risk of eating". N. Motoyama (Chiba Univ., Japan) stated the importance of dialogue with consumers for a successful campaign at the seminar co-sponsored by Japan Crop Protection Association and Pesticide Science Society of Japan, based on the outcomes of campaigns in which he has taken part as a crop protection science expert. He reported that the perception on pesticides among the participating consumers greatly depends on presentation methods, and "Face to Face" and "Animated" presentations were especially effective. Also, the IUPAC/FAO/IAEA project aiming at giving easier access to information on agrochemicals via an IUPAC web page was introduced. J. Unsworth (IUPAC) gave an outline of the information which is available on the internet and the care that must be taken when using it. Then, I. Ferris (IAEA/FAO) demonstrated the INFOCRIS database which can provide a comprehensive profile of the properties of agrochemicals.

As to the regulatory aspect, J. J. Baron (USDA, USA) introduced the IR-4 project that provided a practical solution to agrochemical use for specialty crops in the US, and thereby, attracted attention of many specialty crop growers and regulatory agencies from different countries. Issues of crop grouping and representative crops in residue studies were discussed, referring to standardizing of MRLs. L. Rossi (US EPA, USA) presented a comprehensive overview of the history and current use of probabilistic approaches to dietary risk assessment in the USA, and its advantage was demonstrated as a more realistic assessment than the established deterministic methods. A seminar organized by S. Impithuksa (Ministry of Agric. & Coop., Thailand), F.-M. Liu (China Agric. Univ., China) and I. Yamamoto (Tokyo Univ. of Agric., Japan) addressed the issues on the control of pesticide residues in exporting crops from China and Thailand, and their programs to enforce food safety were introduced.

The current situation of the production, regulation, and perception of GM crops around the world was also overviewed, in which the situations in Northern America, Europe, and Japan were described by G. R. Stephenson (Univ. of Guelph), J. Unsworth (UK) and K. Tanaka (Sankyo Agro, Japan), respectively. In addition, some results of the IUPAC-funded project on the environmental consequences of the changed usage of pesticides on GM crops were demonstrated by Stephenson and G. A. Kleter (Wageningen Univ., Netherlands) that pesticide usage generally appears to be decreased on GM crops in terms of quantities applied to these crops, so likely resulting in a less environmental impact. In this regard, concerns about other environmental hazards besides changes in pesticide usage were addressed in the following discussion.

Concerning technological advances, H. Ohkawa (Fukuyama Univ., Japan), the chairperson of the Organizing Committee of this Congress, organized two seminars to focus on monitoring techniques for POPs. One of the seminars featured biomonitoring using transgenic plants, in which H. Inui (Kobe Univ., Japan) presented a lecture entitled "Monitoring of estrogenic compounds in transgenic plants carrying ER and reporter genes", followed by "Monitoring of dioxins in transgenic flowering plants carrying AhR and reporter genes" by Y. Tanaka (Suntory, Japan). The other covered immunochemical technique. Y. Goda (Japan EnviroChemicals, Japan) demonstrated "The use of enzyme-linked immunosorbent assays (ELISA) for the detection of pollutants in environmental and industrial wastes". B. Hock (Technische Univ. München, Germany) reviewed "Immuno- and receptor-assays for environmental compounds".

Seminars also worked well to deepen the international exchanges. Two joint seminars were held under co-sponsorship by the Pesticide Chemistry Societies of China and Japan. At the first one dealing with herbicides, Z.-M. Li (Nankai

Univ., China) presented his recent research results entitled "Structure-activity relationship of novel sulfonylurea inhibitors on AHAS". H. Matsumoto (Univ. of Tsukuba, Japan) then introduced his study "Mode of action of several classes of herbicides causing photooxidative injury in plants". The second one dealt with insecticides, where X.-H. Qian (East China Univ. of Science and Technology, China) and K. Matsuda (Kinki Univ., Japan) presented their recent findings on chemistry and biochemistry of neonicotinoids. Interchange between Pesticide Science Societies of Korea and Japan was also maintained at a seminar entitled "Current and future R&D activities in agrochemical area in Korea and Japan". D. S. Kim (LG Life Sciences, Korea) and R. Ichinose (Sankyo, Japan), respectively, highlighted the priority subjects in the field of fungicide research in their own countries. P. Zaprong (T.J.C. Chemical, Thailand) and Q. T. Dong (HAI Agrochem Joint Stock Company, Vietnam) presented agriculture and crop protection in their countries. In both countries, most of pesticides are imported from China, India, and other countries, with the total imported pesticides by value being increased year after year. Safer pesticides for the environment and humans are required, while both governments are promoting to reduce the agrochemical usage, and to expand the organic agriculture and IPM.

A special seminar was held in honor of achievements by two great Japanese pesticide chemists. M. Eto (Kyushu Univ.) and T. Fujita (Kyoto Univ.) were invited to give lectures entitled "Significant feature of organophosphorus agrochemicals in structure and activity" and "Proposal of SAR-omics as a paradigm for the lead evolution in drug design", respectively.

Other technical seminars sponsored by commercial organizations included:

- Registration and commercialization of agrochemicals in Europe by M. Weidenauer (Battelle, Switzerland) and the importance of formulation development in agchem product life cycle management by M. Bell (Battelle, UK)
- Specialty of Nisso chemical analysis services – qualitative and quantitative analyses of trace amounts of ingredients in complex matrices by T. Gomyo, K. Miya and S. Kobayashi (Nippon Soda, Japan)
- Physical, chemical, and environmental properties of selected chemical alternatives for the pre-plant use of methyl bromide as soil fumigant by L. O. Ruzo (PTRL West, USA)
- Introduction of sample cleanup method related to the positive list system in Japan by T. Imanaka (GL Sciences, Japan)
- How to refine and evaluate a TER – aquatic risk assessment for pesticides in Europe by C. Schaefers and W. Koerdel (Fraunhofer IME, Germany)
- New R&D cooperation models between CRO's and industry – where do we stand and where do we go? by A. Wais (RCC, Switzerland)
- Advance in analytical methods for metabolism by D. J. Lankester (Covance, UK)
- Addressing ecotoxicology issues to meet 21st century needs by C. Hutchinson (Wildlife International, USA)

- The Agilent 6000 series LC/MS solutions for pesticides and veterinary medicines in food and the environment by J. Zweigenbaum (Agilent, USA)
- Design considerations for radiolabelled plant metabolism and related studies by S. Chapleo (Charles River Lab., UK)
- Finding solutions in global agrochemical development: case studies from our experience by D. Kirkpatrick (Huntingdon Life Sciences, UK)
- Eurofins agrochemicals group presents an update on EU regulatory requirements by F. Bodzian (Eurofins Agrochemicals., Germany)

47.6 Other Scientific Programs

As the first attempt at an IUPAC Congress, the Research Directors of the major agri-businesses assembled and gave individual presentations outlining company strategy and their research and development successes at "Research Director Forum". Contributors to the session were: P. Eckes (BASF), A. Klausener (Bayer CropScience), D. Kittle (Dow AgroSciences), P. Confalone (DuPont Crop Protection), T. Haga (Ishihara), T. Umemura (Sumitomo Chemical), and G. Ramos (Syngenta). Common themes in the presentations included an emphasis on the need to develop an innovative culture and the importance of recruiting and retaining highly motivated people. Although it was clear that each company had access to similar tools and information, they were being applied differently resulting in a diversity of approach. An active panel discussion followed under the moderation by N. K. Umetsu and K. D. Racke, where a range of questions and comments were raised from the audience. They included clarification of individual research strategies and a positive message for careers in the industry.

Taking an opportunity of this Congress, an open seminar for the public also convened as a satellite event to achieve better recognition among targeted consumers. Under moderation by M. Kitano (Meiji Univ., Japan), N. Motoyama (Chiba Univ., Japan) and K. Maita (Inst. of Environ. Tox., Japan) explained in plain words the role of pesticides in crop protection and the management concerning pesticide use to secure food safety to an audience of approximately 370. F. Fukunaga, a farmer of Hyogo Prefecture, Japan, represented the necessity of proper means to protect crops from the viewpoint of a farm producer.

47.7 Acknowledgments

In preparing this manuscript, the articles in "Kobe Gazette" published on site were fully referred. Some of the summaries are taken from the articles with modification. We appreciate the authors of these meeting summaries and articles for their contributions.

We gratefully acknowledge the financial contributions from the following organizations for this Conference:

Agrochemical Division of the American Chemical Society Crop Life International
- BASF
- Bayer CropScience
- Dow AgroSciences
- DuPont Crop Protection
- FMC
- Sumitomo Chemical
- Syngenta

Hoh Noh Kai
The International Union of Pure and Applied Chemistry (IUPAC)
Japan Crop Protection Association
The Japan Food Chemical Research Foundation
Japan Plant Protection Association
Kato Memorial Bioscience Foundation
Kobe City
National Institute for Agro-Environmental Sciences
Nakauchi Foundation
Pesticide Science Society of Japan
Sankyo Foundation of Life Science

Appendix

Title List of Presentations in Selected Poster Workshops

Category I

A) Insect Control

- Discovery, synthesis and acaricidal activity of cyflumetofen and its derivatives (N. Takahashi, Otsuka Chemical, Japan)
- Substituted arylpyrazole anthranilic diamides: novel insecticidal activators of the ryanodine receptor (T. Selby, DuPont Crop Protection, USA)
- Synthesis and insecticidal activity of new hydrazone derivatives (Y. Masuzawa, Nissan Chemical, Japan)
- Essential structural factors of mitochondrial complex I inhibitor acetogenins (M. Abe, Kyoto Univ., Japan)
- Identification and characterization of a novel peptide (pp 3158) from the starved larval hemolymph of the silkworm, *Bombyx mori* (S. Nagata, Univ. of Tokyo, Japan)

B) Weed and Plant Growth Control

- Managing grass weeds in cereals: chemistry and biology of the novel herbicide pinoxaden (M. Muehlebach, Syngenta, Switzerland)
- Synthesis and herbicidal activity of novel phytoene desaturase inhibitors: 3-(substituted oxy)pyrazole-4-carboxamide derivatives (R. Ohno, Sagami Chemical Res. Center, Japan)
- One-pot tandem aldol-alkylation routes to (dehydro)dioxopiperazines: novel analogues of the thaxtomin family of phytotoxins (A. Plant, Syngenta, UK)
- Inhibitors of IAA-amino acid conjugate synthetases and hydrolases as chemical probes to study IAA homeostasis (L. H. Tai, Kyoto Univ., Japan)
- Synthesis and molecular design of novel agrochemicals containing selenium (Z. Li, East China Univ. of Sci. & Tech., China)
- Synthesis and germination stimulating activity of some imino-analogs of strigolactones (Y. Kondo, Kobe Univ., Japan)

Pesticide Chemistry. Crop Protection, Public Health, Environmental Safety
Edited by Hideo Ohkawa, Hisashi Miyagawa, and Philip W. Lee

ISBN: 978-3-527-31663-2

C) Disease Control

- Synthetic studies of biologically active natural products of agricultural interest (H. Kiyota, Tohoku Univ., Japan)
- The antifungal 3-aryl-5-methyl-2,5-dihydrofuran-ones related to incrustoporin (P. A. Worthington, Syngenta, UK)
- Structure-activity relationship study of flagellin-derived elicitor peptides whose conformations were constrained (M. Miyashita, Kyoto Univ., Japan)
- Orysastrobin – an effective fungicide in rice (J. Rheinheimer, BASF, Germany)
- Isolation and structural properties of aerial mycelium differentiation inhibitory substance of *Streptomyces scabiei* causing potato common scab (M. Natsume, Tokyo Univ. of Agriculture and Technology, Japan)
- Synthesis and fungicidal activities of 12-alkoxy(benzyloxy)imino-1,15-cyclopentadecanlactams (D.-Q. Wang, China Agricultural Univ., China)

D) New Technologies for Lead Generation and Drug Design

- Rational design of novel herbicides based on ligand-receptor interaction studies (G.-F. Yang, Central, China Normal Univ., China)
- Metabolic phenotyping and biomarker identification through FT-ICR MS-based metabolomics studies (D. Ohta, Osaka Prefecture Univ., Japan)
- Virtual target-based screening in combinatorial library design: enhancing *in-vitro* and *in-vivo* hit rates (S. Lindell, Bayer CropScience, Germany)
- Novel genes, *FMI1, FMI2,* and *MGH61A,* strongly expressed at early infection stage of the rice blast fungus – have the gene products the potentiality to be new target sites of fungicides? (T. Teraoka, Tokyo Univ. of Agric. & Tech. Japan)
- Structure based molecular design of AHAS inhibitors (J.-G. Wang, Nankai Univ., China)
- Structure-activity relationship for the activity of non-steroidal ecdysone agonists and the prediction of the ligand binding to the *Bombyx mori* ecdysone receptors (Y. Nakagawa, Kyoto Univ., Japan)

E) Miscellaneous Technologies for Pest Control Including Transgenic Crops

- Cloning and characterization of NRPS genes in fungal herbicide *Exserohilum monoceras* (A. Morita, Kyoto Univ., Japan)
- Dimethyldisulfide: a new soil fumigant (R. M. Bennett, Cerexagri, USA)
- An oxidoreductase is highly expressed in a virulence – loss – mutant of the cabbage yellows fungus having biocontrol activity (A. Okabe, Tokyo Univ. of Agric. & Tech., Japan)
- Lepidopteran insect-killing *Bacillus sphaericus* with no mosquitocidal activity and a novel insecticidal factor (H. Nishiwaki, Kinki Univ., Japan)
- Biolistic transformation and detection of jellyfish green fluorescent and chitinase proteins in Indian Basmati rice (J. Tarafdar, Bidhan Chandra Krishi Viswavidyalaya, India)
- Development of defense gene expression monitoring system by the bioluminescence reporter genes in higher plants (T. Tanaka, Yokohama National Univ., Japan)

Category II

A) Discovery of Targets and Emergence of Unexpected Resistance

- Significance of the sulfonylurea receptor (SUR) as the target of diflubenzuron in chitin synthesis inhibition in *Drosophila melanogaster* and *Blatella germanica* (F. Matsumura, Univ. of California, Davis, USA)
- Characterization of cytotoxicity of pyridalyl, an insecticidal agent, to cultured insect and mammalian cells (T. Utsumi, Yamaguchi Univ., Japan)
- A nicotinic acetylcholine receptor point mutation (Y151S) conferring insecticide resistance causes reduced agonist potency to a range of neonicotinoids (N. S. Millar, Univ. College London, UK)
- Flubendiamide stimulates Ca^{2+} pump activity coupled to RyR-mediated calcium release in lepidopterous insects (T. Masaki, Nihon Nohyaku, Japan)
- Elucidation of the mode of action of Rynaxypyr^{TM}, a selective ryanodine receptor activator (D. Cordova, DuPont Crop Protection, USA)
- Phthalic acid diamides: Mode of action and target selectivity (U. Ebbinghaus-Kintscher, Bayer CropScience, Germany)
- Flubendiamide, a novel insecticide, selectively activates lepidopterous ryanodine receptor (S. Kiyonaka, Kyoto Univ., Japan)
- Endocrine disruption, resistance and induction of cytochrome P450 genes by xenobiotics in *Drosophila* (R. Feyereisen, INRA/Universite de Nice Sophia Antipolis, France)

B) Herbicidal Action and Plant Growth Regulation

- Interactions between the parasitic weed *Striga hermonthica* and its host *Sorghum bicolor* at a molecular level (Y. Hiraoka, Kobe Univ., Japan)
- Dichloromethyl ketal structure affects the expression of glutathione *S*-transferase isoforms in maize (I. Jablonkai, Inst. of Biomolecular Chemistry, Hungary)
- Reduced sensitivity of [*Amaranthus palmeri* S. Wats] to glyphosate results in lack of control in genetically modified crops (T. C. Mueller, Univ. of Tennessee, USA)
- Screening of *Arabidopsis* mutant by phytohormone brassinosteroid biosynthesis inhibitor (Brz) (T. Komatsu, Tokyo Univ. of Agric. & Tech., Japan)
- ζ-Carotene desaturase inhibition by 4-(3-fluorophenyl)-3-(4-trifluoromethylbenzylthio)-4*H*-1,2,4-triazole (Y. Watanabe, Kyushu Univ., Japan).

C) New and Established Target and Resistance of Disease Control

- Reliable^{TM}: labor saving disease control agent for *Phytophthora infestans* on potatoes in Japan (H. Hadano, Bayer CropScience, Japan)
- Antifungal activity and mode of action of tolnifanide (C. Tanaka, Kyoto Univ., Japan)
- Development of QoI resistant alleles in populations of strawberry powdery mildew (J. M. Fountaine, NIAES, Japan)
- Fungicide resistance in the barley pathogen *Rhynchosporium secalis* in Northern Ireland (C. McCartney, Queen's Univ. of Belfast, UK)

- Histidine biosynthesis in *Magnaporthe grisea*: a potential target for plant protectants? (E. Thines, Inst. for Biotechnology and Drug Research, Germany)
- Transcriptome profiling of the response of *Mycosphaerella graminicola* isolates to triazole fungicides using cDNA microarrays (B. Fraaije, Rothamsted Research, UK)

D) Alternative in Metabolism & Toxicology Study

- Oxidative metabolic profiling of pesticides using transgenic tobacco cell suspension cultures, which express human cytochrome P450 isozymes (B. Schmidt, RWTH Aachen Univ., Germany)
- Iron-porphyrin catalyzed oxidation model for plant metabolism of pesticides (M. Fukushima, Sumitomo Chemical, Japan)
- Comparison of oral absorption and systemic availability (A. Langford-Pollard, Huntingdon Life Sciences, UK)
- Species difference in excretion and PK of procymidone and its metabolites (Y. Tomigahara, Sumitomo Chemical, Japan)
- Sex-dependent difference of methoxychlor *O*-demethylation by rat liver microsomes (K. Ohyama, Inst. of Environ. Tox., Japan)
- Evaluation of estrogen receptor binding activity of metabolites of DDT analogs (M. Akamatsu, Kyoto Univ., Japan)

E) Topics in Formulations, Application, Resistance, and Integrated Pest Management

- Detection and management of the newly established Q biotype of *Bemisia tabaci* (Gennadius) in the USA (T. J. Dennehy, Univ. of Arizona, USA)
- Screening of rhizo-functional bacteria against *Fusarium oxysporum* f. sp. *lycopersici* and their antifungal and hyphal branching-inducing principle(s) (A. Asante, Hokkaido Univ., Japan)
- Encapsulation technologies for crop protection (W. L. Geigle, DuPont Crop Protection, USA)
- Factors influencing the association between a.i. and adjuvant in the leaf deposit for adjuvanted SEs (M. Faers, Bayer CropScience, Germany)
- Measurement of collection efficiency on rotating rods, strings, and horizontal surfaces (W. C. Hoffmann, USDA-ARS, USA).

Category III

A) Fast Procedures for Pesticide Multi-residue Analysis

- Multiresidue analysis of 500 pesticide residues in agricultural products using GC/MS and LC/MS (Y. Akiyama, Hyogo Pref. Inst. of Public Health and Environ. Sci., Japan)
- Multiresidue method for the determination of pesticide residues in food by GC/MS, GC/FPD and LC/MS/MS (M. Okihashi, Osaka Pref. Inst. of Public Health, Japan)

- Simultaneous determination of pesticides in crops by LC/MS/MS using a cleanup step with ultrafiltration (H. Kajita, Res. Inst. for Environ. Sci. and Public Health of Iwate Pref., Japan)
- Multi-residue pesticide analysis in fruits and vegetables by LC/MS using triple stage quadrupole mass spectrometry and time of flight mass spectrometry (M. Takino, Yokogawa Analytical Systems, Japan)
- Rapid and simple method for the determination of postharvest fungicides in citrus fruits by LC/MS/MS (Y. Okamoto, Osaka Pref. Inst. of Public Health, Japan)
- A model study of residue analysis for 26 pesticides applied to peach, Japanese pear and apple (T. Takino, Bayer CropScience, Japan)

B) Improving Analysis, and Assessing Human Exposure Considering Sampling and Processing

- Evaluation of large volume injection in GC-MS analysis of pesticide multi-residues using a PTV injector with automated liner exchange (N. Ochiai, GERSTEL, Japan)
- Elution patterns of multiclass pesticides from three types of anion exchange cartridges (K. Iijima, Inst. of Environ. Tox., Japan)
- Daily intake of pesticides based on the market basket method in Hyogo prefecture, Japan (N. Yoshioka, Hyogo Pref. Inst. of Public Health and Environmental Sciences, Japan)
- Fungicide and insecticide residues in wine grape by-products following field applications (G. Rose, Primary Industries Research Victoria, Australia)
- Estimation of sampling uncertainty for determination of pesticide residues in plant commodities (A. Ambrus, Hungarian Food Safety Office, Hungary)
- Monitoring of pesticide residues in fresh vegetables and fruits in the Philippines 2000–2005 (N. C. Chen, National Pesticide Analytical Lab., Philippines)

C) Fate and Bioremediation of Selected Pesticides

- The biotransformation of ethaboxam in plants, animals and the environment; Ethaboxam: fate and effects in the aquatic environment (D. Kirkpatrick, Huntingdon Life Sciences, UK)
- Pyridaben: fate in the environment (J. O'Connor, Huntingdon Life Sciences, UK)
- Amisulbrom: metabolic fate in soil (Y. Ijima, Nissan Chemical, Japan); Mitigation of the point source contamination by pesticides in Mediterranean conditions (E. Capri, Catholic Univ. of the Sacred Heart, Italy)
- Persistence and dissipation behavior of novel insecticide A9908 in Liuyang river water, P R China (X.-M. Ou, Hunam Research Inst. of Chemical Industry, China)
- Anaerobic biodegradation of 4-alkylphenols by a nitrate-reducing enrichment culture (A. Shibata, Nagoya Univ., Japan)

D) Risk Assessment and Mitigation of Pesticide and POPs Contamination

- Predictive value of aquatic toxicity testing with additional invertebrate species (K. Barrett, Huntingdon Life Sciences, UK)
- Remediation of simazine and 4-nonylphenol with transgenic plants carrying drug-metabolizing cytochrome P450 genes (H. Inui, Kobe Univ., Japan)
- Biodegradation of chlorinated aliphatic hydrocarbons and benzene in an anaerobic river sediment (F.-M. Liu, China Agricultural. Univ., China)
- Phototransformation of oryzalin in aqueous isopropanol and acetonitrile (S. K. Pramanik, Bidhan Chandra Krishi Viswavidya, India)
- Bioavailability of synthetic pyrethroids in surface aquatic systems (J. Gan, Univ. of California, USA)

E) Higher Tier Exposure Assessment

- Steps 1234 – a new model for the estimation of PEC-surface water (M. Klein, Fraunhofer IME, Germany)
- Probabilistic risk assessment through pesticide fate modeling for evaluating management practices to prevent pesticide runoff from paddy fields (S. H. Vu, Tokyo Univ. of Agric. & Tech., Japan)
- Investigation of new methods and geo-databases for a refined GIS-based probabilistic exposure assessment (R. Kubiak, RLP AgroScience, Germany)
- Is the resulting residue proportional to pesticide application rate? (P. S. Villanueva, US EPA, USA)
- Engineering of the transgenic plants carrying the receptor-mediated reporter gene expression systems for bioassay of POPs (K. Gion, Kobe Univ., Japan)
- Establishing the relevance of FOCUS surface water scenarios for pesticide risk assessment in the UK landscape (O. R. Price, Cambridge Environmental Assessments, UK)

Author Index

a
Akiyama, Y 395
Ambrus, A 349
Anastassiades, M 439
Asami, T 177

b
Bainard, LD 201
Baron, JJ 323
Beck, ME 227
Bellin, CA 111
Benner, EA 121
Benner, JP 383
Booth, JD 383
Boutsalis, P 101
Brunner, H-G 101

c
Caspar, T 121
Cederbaum, F 101
Chen, H 323
Chung Chun Lam, C 383
Clark, CE 111
Clark, T 383
Collins, JC Jr. 3
Cordova, D 111, 121
Cornes, D 101
Corran, AJ 65

d
Denholm, I 271
Devisetty, BN 249
Driver, JH 373
Dubas, CM 111
Dunbar, SJ 65

e
Ebbinghaus-Kintscher, U 137
Eshita, Y 217

f
ffrench-Constant, R 305
Flexner, L 111, 121
Freudenberger, JH 111
Friedmann, AA 101
Fujioka, S 127
Fukuzawa, A 211
Furuya, T 127

g
Gaston, CP 349
Geigle, WL 241
Gion, K 431
Gledhill, AJ 383
Glock, J 101
González, RH 349
Gutteridge, S 121

h
Hagino, S 415
Han, Z 271
He, Y 331
Hirooka, T 127
Hofer, U 101
Hole, S 101
Hollingshaus, JG 111
Holm, RE 323
Huang, Q 159
Hughes, KA 141

i
Ichimaru, N 171
Inui, H 431
Isman, MB 201
Iwasaki, T 149

j
Johnson, TC 89

Pesticide Chemistry. Crop Protection, Public Health, Environmental Safety
Edited by Hideo Ohkawa, Hisashi Miyagawa, and Philip W. Lee

ISBN: 978-3-527-31663-2

k
Katsuta, H 295
Kiser, J 189
Kishi, J 295
Kitahata, N 177
Klausener, A 55
Kleter, GA 361
Kodama, Hiroki 127
Kodama, Hiroshi 127
Komalamisra, N 217
Krieger, RI 373
Kuck, K-H 275
Kuiper, HA 361
Kunkel, DL 323
Kurane, I 217

l
Lahm, GP 111, 121, 141
Lansdell, SJ 271
Li, Z 159
Liu, Z 271
Luemmen, P 137

m
Machial, CM 201
Maetzke, T 101
Mann, RK 89
Martin, TP 89
Masaki, T 137
Matsuda, K 261
Matsunaga, T 149
Matsuo, N 149
Matsuoka, T 395
McCart, C 305
Millar, NS 271
Miresmailli, S 201
Miyoshi, H 171
Mori, K 13
Mori, T 149
Muehlebach, M 101
Murai, M 171
Mutti, R 101

n
Nakano, T 127, 177
Niderman, T 101
Nishimatsu, T 127

o
Ohkawa, H 431

p
Pandey, S 43
Pingali, P 43
Pugh, LM 241

q
Qian, X 159
Quadranti, M 101

r
Racke, KD 29
Ragghianti, JJ 121
Raming, K 55
Rauh, JJ 121
Rhoades, DF 121
Riffel, M 425
Roberts, KJ 383
Ross, JH 373

s
Sacher, MD 121
Schabacker, J 425
Scherbaum, E 439
Schindler, M 227
Schleifer, K-J 77
Schnyder, A 101
Seiber, JN 315
Selby, TP 111, 121, 141
Seo, A 127
Shao, X 159
Shen, J 159
Shono, Y 149
Skidmore, MW 383
Smith, BK 111
Smith, RM 121
Song, W 331
Sopa, JS 121
Štajnbaher, D 439
Stenzel, K 55
Stevenson, TM 111, 121
Stoller, A 101
Streloke, M 403
Sugano, M 149

t
Takasaki, T 217
Takashima, I 217
Tao, Y 121
Taşdelen, B 439
Tayaputch, N 341
Tian, Z 159
Tohnishi, M 127
Tsubata, K 127

u

Ujihara, K 149
Ushijima, H 217
Utani, Y 431

w

Walsh, TA 285
Wang, Y 159
Warrior, P 249
Wendeborn, S 101
Wenger, J 101
Williamson, MS 271
Wolf, C 425
Wu, L 121

y

Yamamoto, T 189
Yanase, Y 295
Yang, YZ 23
Yasokawa, N 137
Yoshikawa, Y 295
Yoshioka, N 395

Subject Index

a
abamineSG 185
abscisic acid (ABA) 177 ff.
– biosynthesis inhibitor 184 f.
acceptable daily intake (ADI) 31, 334
Accord PCR (polymerase chain reaction) diagnostic 307
acetamiprid 166
acetogenin 171 ff.
– Δlac-acetogenin 171
– mode of action 172
– structure-activity relationship 173
– synthesis 171
acetolactate synthase (ALS) 89 ff.
acetyl-coenzyme A carboxylase 101 ff.
acetylcholine
– binding protein (AChBP) 264, 273
– quaternary ammonium 261
acrylamide 319
acute reference dose (ARfD) 31
ADI, *see* acceptable daily intake
adigor® 107
adjuvant effect 107
ADME-Tox (absorption, distribution, metabolism, excretion, and toxicological) 87
Aedes
– *aegypti* 157, 217
– *albopictus* 155 ff., 217
aflatoxin 318
agricultural handlers exposure database AHED© 375
agricultural research 55
– service (ARS) 316
agricultural technology 48
agrochemical 10
agrophore model 82
Ah, *see* aryl hydrocarbon
AhR, *see* aryl hydrocarbon receptor
allergenicity 365
– genetically modified food 366
– introduced protein 365
d-allethrin 152
ALS, *see* acetolactate synthase
Alternaria solani 281
alternative oxidase (AOX) 276
ambrosia beetle 18
Ames test 134
2-amino-5-methoxytriazolo[1,5-c]pyrimidine 90
2-aminotriazolo[1,5-c]pyrimidine 89
analytical method 316
androgen 417
Anopheles
– *balabaciensis* 157
– *sundaicus* 157
anthranilic diamide 111 ff., 121 ff., 141 ff.
– insecticidal activity 144
– synthesis 142
anti-androgen 417
Anticarsia gemmatalis 117
antimicrobial activity 207
antiseptics 25
aquatic organism 409
Arabidopsis 67, 292, 436
– *thaliana* 182
ARfD, *see* acute reference dose
aryl hydrocarbon (Ah) 433
– luciferase assay 432
aryl hydrocarbon receptor (AhR) 431 ff.
2-aryl-1,3-dione 101
– biocidal 101
aryl-dione (AD or den) 105
4-aryl-pyrazolidine-3,5-dione 102
aryl-pyrazolidine-dione 103
arylpyrazole 141 ff.
– anthranilic diamide 147
arylpyrimidine 141 ff.
– anthranilic diamide 147

Pesticide Chemistry. Crop Protection, Public Health, Environmental Safety
Edited by Hideo Ohkawa, Hisashi Miyagawa, and Philip W. Lee

ISBN: 978-3-527-31663-2

aryloxy-phenoxy-propionate (AOPP or fop) 105
ASEAN expert working group (EWG) 343 ff.
Asian tiger mosquito 155
Aspergillus 318
assay design 68
ATA-7 290
ATP synthesis 171
AUR1 67
auxin response F-box protein (AFB) 293
axial® 108
azoxystrobin 69, 276

b
Bacillus
– *israelensis* (Bti) 190
– *kurstaki* (Btk) 190 ff.
– *sphaericus* 189
– *subtilis* 249 ff.
– *tenebrionis* (Btt) 190
Bacillus thuringiensis (Bt) 189 f., 249 ff., 361
– crop 195
– mode of action 192
– resistance management 197
– transgenic Bt-crop 193 ff.
– transgenic cotton 194
barley 106
barnyard grass 96
BASF Protox inhibitor (BPI) 85
Basidiomycetes 296
BC723 295
Beauveria bassiana 189
beetle 18
Bemisia tabaci 166
benzamide 295
1,2-benzenedicarboxamide 127
benzoxazinone 143
bialaphos 15
bil mutant (Brz-insensitive-long hypocotyl) 183
binding site pocket 84
bioassay 431 ff.
– dioxin 432
– POP 435
bioavailability 383 ff.
– bound residue 383 ff.
– common conjugate 383 ff.
biochemical exposure index (BEI) 379
biofuel 8
biological monitoring 378
biological pesticide
– biological activity 253
– quality control 253
– stabilization 253
biomonitoring 378
biopesticide 189, 249 ff.
– formulation 254 f.
– sprayable 249 ff.
biotic degradation 231
bisphenol A 417
Blumeria graminis 278
Botryotinia fuckeliana 68
Botrytis cinerea 295 ff.
bound residue, *see* residue
Brassica napus 17
brassinolide (BL) 17, 178 ff.
brassinosteroid (BR) 17, 178 ff.
brassinosteroid biosynthesis inhibitor 178 f.
– target site 180
brevicomin 18
broadleaf weed activity (BW) 93
brown planthopper 166
Brz220 181

c
CACO-2 cell assay 388
caffeine 61, 125
calcium 59 ff., 121 ff., 127, 137 ff.
– channel 132, 141
– homeostasis mechanism 121
– mobilization assay 118
– mobilization threshold (CMT) 116
– pump 137 f.
– release 137 f., 141
Campylobacter 315
carboxanilide fungicide family 302
carboxin 302
carotenoid cleavage dioxygenase (CCD) 184
carryover stock 5 ff.
Catharanthus roseus 180
chemical exposure 373 ff.
chemical genetics 177, 285 ff.
– target identification 288
chemical library 287
chemistry
– green 21
Chilo suppressalis 13
Chlorella vulgaris 182
4-chloro-6-methyl-anthranilic acid 116
chloropyridyl 273
2-chlorosulfonyl benzoate 92
chlorothiazolyl 273
chlorpyrifos 231
chrysanthemum dicarboxylate 150
cis nitro configuration 165
Clostridium bifermentan 189

cluster sampling method 79
Codex Committee on Pesticide Residues (CCPR) 327
Codex MRL 31 ff., 324, 341 ff., 349 ff.
– dried chilli peppers 352
– spice 352
Coleoptera 189
conjugate 386
consumer safety 35
corn insect pest 195
cotton 323
counterfeit product 9
CPD protein 180 f.
crop
– Bt-crop 195
– harvester 376
– genetically modified 361 ff.
– grouping 326
– production 3
– tolerance 106
– transgenic 189
crop protection 119
– market 6
– product 323
– research 77 ff.
Cry protein 191
cryptone 18
Culex pipiens 155
Culex quinquefasciatus 149 ff.
cyclo-hexanedione (CHD or dim) 105
cyhalofop-butyl 98
cyst nematode 17
cytochrome 69
– P450 179, 227, 432
– compound I (cpdI) 227
– CYP1A1 433
– *Cyp6g1* 306

d
Danio rerio 416
daphnid 415
DAS534 291
DEHP (di-2-ethylhexyl phthalate) 417
Dendroctonus frontalis 18
Dengue virus (DEV) 217 ff.
Denguri paper strip 156
density functional theory
– conceptual 228 ff.
dermestid beetle 18
descriptor 78
– 1-D and 2-D 78 ff.
– 3-D 81 ff.
– molecular 79
– selection 87
– space 80
deterrent effect 203
deuteromycetes 296
developing country
– trade 349 ff.
development assistance
– official (ODA) 49
DeViner 252
Diabrotica 195
Diadegma insulare 189
1,3-dialkoxybenzene 91
diamondback moth, *see Plutella xylostella*
3,4-dichloroaniline 389
dichlorobenzenesulfonyl chloride 94
dioxin 432 f.
– bioassay 432
Dipole_Mopac value 167
Diptera 189
direct sampling method 79
Directive 91/414 EEC 428
dispersive SPE 440
DiTera 256
DNA
– molecular characterization 364
docking 84 ff.
Drosophila
– Dα2 subunit 265
– *melanogaster* 309
drug discovery 55 ff.
– integration of technology 58
– tool 56
drugability 66 f.
DTT (dichlorodiphenyltrichloroethane) 305 ff.
– resistance 305 ff.

e
early life stages (ELS)-test 410
ecotoxicology 403 ff.
ELISA
– dioxin 433
emodepsid 233
endocrine disrupting chemicals (EDC) 415 ff.
– daphnid 415
– fish 415
endocrine disruption 410
energy supply 48
enrichment factor 86
environmental risk assessment (ERA) 407
9-*cis*-epoxycarotenoid dioxygenase (NCED) 184
Escherichia coli O157:H7 315 ff.
essential oil 201

– antimicrobial activity 207
– deterrent effect 203
– herbicidal activity 203
– insecticidal effect 203
estimated theoretical environmental concentration (ETE) 404, 425 ff.
estrogen 417
etofenprox 15
European ENTRANSFOOD project 369
European Union (EU) 34 f., 403 ff.
– MRL 34
– regulation 403 ff.
exchange rate 6 f.
exposure
– biomonitoring 378
– study 425

f

fall armyworm, *see Spodoptera frugiperda*
fan emanator 155
FAO (food and agriculture organization) 350, 362
fenarimol 181
fenitrothion 15
fenvalerate 15
field trial 333
fish 415
Flavivirus 217
FLC 416
florasulam 89 ff.
fluazinam 15
flubendiamide 58 ff., 127 ff., 137 ff.
– insecticidal activity 128 ff.
– mode of action 132
– structure-activity relationship 128
– toxicological property 134
flumorph 23
food 29 ff.
– chain 30
– chain compromise 29
– pesticide 29 ff.
food action plan 27
– pollution free 27
food quality 341
– research 315 ff.
food safety 315 ff., 341 ff., 361 ff.
frontalin 18
Fukui function 227 ff.
– biotic degradation 231
– site of metabolism 230
fungal esterase 277
fungicidal inhibitor 71
fungicide 56, 65
– high-risk 282
– QoI 275 ff.
– resistant strain 300

g

GAP, *see* good agricultural practice
gastrointestinal tract (GIT) 384 ff.
GC/MS 395 ff.
GCB (graphitized carbon black) 452
GDP, *see* gross domestic product
gene transfer
– horizontal 366
genetic modification
– unintended effect 367
genetics
– chemical 177
gibberellin (GA) 16, 179
– biosynthesis inhibitor 179
β-D-glucoside 386
β-glucuronidase (GUS) reporter gene 431 ff.
glutamine phosphoribosylpyrophosphate amidotransferase (GPRAT) 291
Glycine max. 323
glycinoeclepin 17
glyphosate 11
GM (genetically modified) 361 ff.
GMO (genetically modified organism) 362
Gnathotrichus sulcatus 18
good agricultural practice (GAP) 29 ff., 334
Gossypium 323
– *hirsutum* 89
grain stock 5
granule blend homogeneity 242 ff.
granule formulation
– dry flowable (DF) 255
– wettable (WG) 255
grass herbicide 98
grass weed
– activity (GW) 93
– spectrum 106
grid based sampling 79
Grobodera
– *pallida* 211
– *rostochiensis* 211
gross domestic product (GDP) 4
growth reduction (GR) 93
guide of 2 78

h

3-halo-5-pyrazolecarboxylic acid 115
harvester
– treated crop 376
hash coding 78
hatching
– stimulant 211 ff.

– synergist 214
healthy food constituent 320
Helicoverpa
– *armigera* 196
– *zea* 194
Heliothis 59 f.
– *virescens* (Hv) 112, 144, 190 ff.
herbicidal activity 96, 203
herbicidal inhibitor 178
herbicide 15, 25, 56, 105, 207
– cereal 101
– clove oil-based 207
– rice 89 ff.
Heterodera glycines 212
hormone
– juvenile 17
HSAB (hard and soft acids and bases) principle 227 f.
human exposure 377
human reporter gene system (HRGS) 432
hunger 43 ff.
hydrolysis
– chemical and enzymatic 386 ff.
hydroxy-aryl-oxo-pyrazoline 105

i

IAEA (International Atomic Energy Agency) 350
identification 316
imidacloprid 15, 166
infrastructure 47
– irrigation 47
– rural 47
inositol phosphorylceramide synthase 67
insect resistance to Bt 196
insecticide 15, 24 f., 56, 111, 127, 141 ff., 203
– spray-on formulation 190
interaction energy 163
internal standard (ISTD) 440
IPM, *see* pest
IR-4 model 325
irrigation 47

j

Japan
– MRL 33 f.
Japanese mosquito 217 ff.
– susceptibility 221
– transmission 223
juvenile hormone (JH) 415

k

Kodiak® 255
kresoximmethyl 277

l

LC/MS 395 ff.
lead optimization 128
– strategy 77
Lepidoptera 116, 137, 189
lepidopterous pest 58
Leptinotarsa
– *decemlineata* 117
– *disseminate* 194
liquid vaporizer 155
Listeria monocytogenes 207, 315

m

Magnaporthe grisea 276
malaria 157
MALDI time-of-flight mass spectrometry 316
malnutrition 43 ff.
Manduca sexta 121
margin-of-exposure, *see* MOE
maximun residue limit (MRL) 31 ff., 324
– ASEAN 341
– database 38
– EU 34 f.
– harmonization 344 f.
– Japan 33 f.
– positive list system 34
– regionalization 37
– setting 331 ff.
– U.S. 33
mechanism–based approach 77
mesocosm 409
metabolism 70 ff., 373 ff., 383
– *in vitro, in vivo* 231
– site 230
metabolization
– fungal esterase 277
methamidofos 24
methyl parathion 24
metofluthrin 15, 149 ff.
– biological activity 153
– intrinsic insecticidal activity 152
– liquid vaporizer 155
– pyrethroid insecticide 149
microcosm 409
minor crop 343
mode of action (MoA) 71, 75, 105
MOE (margin-of-exposure) 378
molecular biology 319
monitoring 35, 398
– biological 410
monocrotophos 24
Monte-Carlo/simulated annealing algorithm
– in-house 230

MOS (margin-of-safety) 378
mosquito
– coil 153
– control agent 149 ff.
MRL, *see* maximun residue limit
MTF-753 295 ff.
multiresidue analysis 395 ff., 439 ff.
– pesticide 439
mutation upstream complex I 276
Mycena galapoda 277
mycoherbicide 252
Mycosphaerella graminicola 280
N-myristoyltransferase (NMT) 71
Myrothecium 251
Myzus persicae 166

n

NADH-Ubiquinone Oxidoreductase (Complex-I)
– Mitochondrial 171 ff.
NAFTA (North America free trade agreement) 326
natural resource base
– protection 50
NCED (nine-*cis*-epoxycarotenoid dioxygenase) 184
– inhibitor 185
NDGA (nordihydroguaiaretic acid) 185
nematicide 256
nematode 252
neonicotinoid
– *ab initio* quantum chemical calculation 161
– binding model 159 ff.
– bioinformatic analysis 159
– biological activity 166
– chemical modification for *cis* nitro configuration 165
– insecticide 271 ff.
– quantitative structure-activity relationships (QSAR) 167
– selectivity 261 ff.
– selectivity mechanism 159
– structure-activity relationship 159 ff.
– synthesis 165 f.
– target-site resistance 271 ff.
nicotinic acetylcholine receptor (nAChR)
– homology model 265
– insect 159, 263, 271
– loop 263
– mutation 272 ff.
– vertebrate nAChR agonist 263
Nilaparvata lugens 166, 271 ff.
nithiazine 261
nitromethylene heterocycle 261
NMT, *see* *N*-myristoyltransferase
NOAEL (no observed adverse effect level) 378
4-nonylphenol 417
norchrysanthemic acid 150
nordihydroguaiaretic acid (NDGA) 185
Nosema locustae 189
NP-1 289

o

octopamine 205
ODA, *see* development assistance
orobanchol 17
Oryza sativa 17, 94, 323
oryzalexin 17
Oryzias latipes 416
oxa-diazepane 102
[1,4,5]oxadiazepane 104
oxadiazinane 102
oxazoline 92

p

paclobutrazol 180
Paenibacillus popilliae 189
parathion 13, 24, 231
Pasteuria penetrans 252
penoxsulam 89 ff.
pentachlorophenol 389
penthiopyrad 295 ff.
– biological attribute 298
– mode of action 299
– resistance 301
– target site 299
percent in growth reduction (GR) 93
Periplaneta americana 118, 122 ff.
– neuron 122
permeability 388
persistent compound
– soil 407
persistent organic pollutant (POP) 305, 315, 431 ff.
– bioassay 435
pest
– control 13, 261
– integrated pest management (IPM) 29, 132, 250
– lepidopterous 58
pesticide 13 ff., 129, 419
– agenda 357
– biological, *see* biological pesticide
– chemical genetics 285 ff.
– China 23 ff.
– discovery 14 f.

– essential oil-based 201 ff.
– food 29 ff.
– granule 241 ff.
– guideline 26
– handler 374
– handler exposure database (PHED) 375
– hygiene 25
– lipophile 454
– management 23 ff.
– MRL 31 ff.
– multiresidue analysis 439 ff.
– pH-dependent 446
– production 24
– registration 332
– regulation 24 ff.
– research&development 23
– safety 27, 29
– science 16 ff.
– standard 440
– usage 24
pesticide residue 29 ff., 323 ff., 369
– assessment 331 ff.
– China 331
– developing country 342
– management 331
pesticide risk 425 ff.
– birds 425
– mammals 425
pharmaceutical 10
pharmacophore model 82
phenazino-L-carboxylic acid 23
phenyl pyrimidine anthranilic diamide 145
pheromone 17 f.
– enantioselective synthesis 18
PHI, *see* pre-harvest interval
phosphamidon 24
photomorphogenesis 182
Photorhabdus luminescens 190
phthalic diamide 111
phytoalexin 17
phytocassane 17
phytohormone 16
Phytophthora
– *infestans* 207
– *palmivora* 252
phytosulfokine (PSK) 177
phytotoxicity 104, 132
picolinate auxin 291
α-pinene 18
pinoxaden 101 ff.
– biological performance 106
Pisum sativum 17
pivaloate 103
plant biology 184
plant chemical biology 177 ff.
Plasmopara viticola 69
Platypus quercivorus 18
PLC 416
Plodia interpunctella 196
Plutella xylostella (Px) 112 ff., 129, 144, 189 ff., 205
pollution free food action plan 27
polyaromatic hydrocarbon (PAH) 315
POP, *see* persistent organic pollutant
positive list 395
postemergence 101 ff.
potato cyst nematode (PCN) 211 ff.
– eco-chemical control 211 ff.
potato root diffusate (PRD) 212
prallethrin 155
pre-harvest interval (PHI) 334, 356
predicted environmental concentration (PEC) 405
principal algorithmic approach 84
probenazole 15
product commercialization 10
propiconazole 181
Pseudomonas fluorescens 255
pyrazolidinedione 105
pyrethroid 15
– insecticide 149 ff.
– photo-stable 15
pyridyl pyrazole 118
N-pyridylpyrazole diamide 141
pyridyl pyrimidine anthranilic diamide 145

q
quality control (QC) standard 441
quantitative structure-activity relationships (3-D QSAR) 82
quantitative trait loci (QTL) 183
QuEChERS (quick, easy, cheap, effective, rugged and safe) 439 ff.
– validation 457
Quercus crispula 18
QoI (quinine outside inhibitor) fungicide 275 ff.
– resistance mechanism 275 ff.

r
Raffaelea sp. 18
rational design 66
refined risk assessment 406 f., 426
– birds 407
– mammals 407
regioselective introduction of an iodine atom 131
rejuvenating drug 423

residue
– analysis method 333
– bound 389 ff.
– data requirement 333
resistance management 197, 280, 309
– DDT 305
resistance mechanism 275 ff.
resistance risk assessment 275
respiratory chain 171
RfD (reference dose) 378 f.
Rhizoctonia 255
ribulose-1,5-bisphosphate carboxylase-oxygenase 182
rice 323
– herbicide 89 ff.
risk assessment
– EU 426
risk mitigation measure 411
rodenticide 25
rule of 3 78
rule of five 78
ryanodine 125
ryanodine receptor (RyR) 111 ff., 121 ff., 132
– activator 121
– cloning and expression 124
– insect 124
rynaxypyr™ 111 ff., 121 ff., 141, 330
– mechanism of action 118, 121 ff.
– toxicology 117

s

S-rR strain mekada 416
Saccharomyces cerevisiae 66 ff.
safety
– GM crop 369
SAR 101
scoring 84
screening
– classical 286
– genomic 286
– high throughput (HTS) 66 ff.
– *in vitro* 66
– ligand-based 81
– phenotype 287
– structure-based 83
– virtual 77 ff.
segregation 242
selectivity 450
Septoria tritici 280
Serratia entomophila 190
sex reversal test 416 ff.
soil
– fertility 47
– persistent compound 407
Solanaceae plant 211
source tracking 316
Southern house mosquito, *see Culex quinquefasciatus*
soybean 323
– cyst nematode (SCN) 212
sphingolipid 67
spice 351
spironolactone 182
Spodoptera frugiperda (Sf) 59, 112, 144
– SF9 cell 124
Spodoptera
– *exigua* 190
– *litura* 129
standard
– private 38
– risk assessment 404
Steinernema carpocapsae 189
strigol 17
strigolactone 17
strobilurin 69, 276
structure based design (SBD) 71 ff.
structure-activity relationship 105
sulfonylalkylamine 129
Synechocystis 289
systems biology 72 f.

t

Tanimoto coefficient 80
target protein 83
target site identification 288
target-based research 65
testing method 425 ff.
2,3,5,6-tetrafluorobenzyl ester 151
tetrahydro-pyridine nitromethylene derivative 167
Thermus thermophilus 171
thiamethoxam 166
Three High product (high toxic, high pollution, and high energy cost) 23
thylakoid membrane 182
thyroid hormone 422
TMDI (theoretical maximum daily intake) 334
tobacco budworm, *see Heliothis virescens*
tolerance 33 ff.
tomato root diffusate (TRD) 212
toxic protein (Tc) 190
toxicity
– exposure ratio (TER) 406, 425
– GM food 365
– introduced protein 365
toxicology 373 ff.
toxophore 262, 277

trade of food 323
transfluthrin 150 ff.
transgenic plant 431 ff.
translation
– *in vitro* 70
– *in vivo* 70
triadimefon 181
triazolo[1,5-a]pyrimidine 89
triazolo[1,5-c]pyrimidine sulfonamide 89 ff.
3-trifluoromethyl-5-pyrazolecarboxylic acid 115
Trifolium pratense 17
1,2,3-trisubstituted benzene 93
Triticum aestivum 94
two-state reactivity (TSR) 227
tyramine receptor cascade 205

u

ubiquinone 172
uniconazole 179 f.
United States (U.S.)
– MRL 33
USMILES (Unique Simplified Molecular Input Line Entry System) 78

r

vaporization device 156
– liquid vaporizer 155
Venturia inaequalis 277
VIP (vegetatively produced insecticidal protein) 194
vitellogenin (VTG) induction assay 416

w

West Nile virus (WNV) 217 ff.
wheat 106
World Food Code 31

x

xenobiotic responsive element (XRE) 433
xylem 182

y

Yersinia pestis 190

z

Zea mays 323

Related Titles

Wiley-VCH (Ed.)

Ullmann's Agrochemicals

2007

ISBN 978-3-527-31604-5

Franklin, C., Worgan, J.

Occupational and Residential Exposure Assessment for Pesticides

2005

ISBN 978-0-471-48989-4

Milne, GWA

Pesticides – An International Guide to 1,800 Pest Control Chemicals

2004

ISBN 978-0-566-08542-0

Wilson, M. F. (Ed.)

Optimising Pesticide Use

2003

ISBN 978-0-471-49075-3

Lee, P. W., Aizawa, H., Barefoot, A. C., Murphy, J. J. (Eds.)

Handbook of Residue Analytical Methods for Agrochemicals

2 Volume Set

2003

ISBN 978-0-471-49194-1

Plimmer, J. R. (Ed.)

Encyclopedia of Agrochemicals

3 Volume Set

2003

ISBN 978-0-471-19363-0

Voss, G., Ramos, G. (Eds.)

Chemistry of Crop Protection

Progress and Prospects in Science and Regulation

2003

ISBN 978-3-527-30540-7

Müller, F. (Ed.)

Agrochemicals

Composition, Production, Toxicology, Applications

2000

ISBN 978-3-527-29852-5